Human Genomics: Principles and Clinical Practice

Human Genomics: Principles and Clinical Practice

Edited by Alyssa Boyle

AMERICAN
MEDICAL PUBLISHERS
www.americanmedicalpublishers.com

American Medical Publishers,
41 Flatbush Avenue,
1st Floor, New York,
NY 11217, USA

Visit us on the World Wide Web at:
www.americanmedicalpublishers.com

ISBN: 978-1-63927-253-2

Cataloging-in-Publication Data

Human genomics : principles and clinical practice / edited by Alyssa Boyle.
p. cm.
Includes bibliographical references and index.
ISBN 978-1-63927-253-2
1. Human genome. 2. Genomics. 3. Medical genetics. 4. Human genetics.
5. Human chromosomes. I. Boyle, Alyssa.
RB155.5 .H87 2022
616.042--dc23

Table of Contents

Preface

The genetic material of an organism, including the DNA, is known as a genome. The interdisciplinary field of biology which focuses on the structure, function and evolution of genomes is known as genomics. It deals with the sequencing and analysis of genomes as well as the intragenomic phenomena that occur within the genomes such as epistasis, pleiotropy and heterosis. The study that particularly focuses on the applications of genomic analysis to all aspects of human health and disease is referred to as human genomics. The human genome is a complete set of nucleic acid sequence of humans. It is encoded as DNA in 23 pairs of chromosomes. These pairs are present in cell nuclei and inside a small DNA molecule within individual mitochondria. Human genomes include two forms of DNA genes, namely, protein-coding and non-coding DNA. This book provides comprehensive insights into the field of human genomics. It will also provide interesting topics for research which interested readers can take up. This book is meant for students who are looking for an elaborate reference text on human genomics.

This book unites the global concepts and researches in an organized manner for a comprehensive understanding of the subject. It is a ripe text for all researchers, students, scientists or anyone else who is interested in acquiring a better knowledge of this dynamic field.

I extend my sincere thanks to the contributors for such eloquent research chapters. Finally, I thank my family for being a source of support and help.

Editor

1

Extrapolating the effect of deleterious nsSNPs in the binding adaptability of flavopiridol with CDK7 protein: a molecular dynamics approach

C George Priya Doss[1*], N Nagasundaram[1], Chiranjib Chakraborty[2,3], Luonan Chen[4,5] and Hailong Zhu[2*]

Abstract

Background: Recent reports suggest the role of nonsynonymous single nucleotide polymorphisms (nsSNPs) in cyclin-dependent kinase 7 (*CDK7*) gene associated with defect in the DNA repair mechanism that may contribute to cancer risk. Among the various inhibitors developed so far, flavopiridol proved to be a potential antitumor drug in the phase-III clinical trial for chronic lymphocytic leukemia. Here, we described a theoretical assessment for the discovery of new drugs or drug targets in CDK7 protein owing to the changes caused by deleterious nsSNPs.

Methods: Three nsSNPs (I63R, H135R, and T285M) were predicted to have functional impact on protein function by SIFT, PolyPhen2, I-Mutant3, PANTHER, SNPs&GO, PhD-SNP, and screening for non-acceptable polymorphisms (SNAP). Furthermore, we analyzed the native and proposed mutant models in atomic level 10 ns simulation using the molecular dynamics (MD) approach. Finally, with the aid of Autodock 4.0 and PatchDock, we analyzed the binding efficacy of flavopiridol with CDK7 protein with respect to the deleterious mutations.

Results: By comparing the results of all seven prediction tools, three nsSNPs (I63R, H135R, and T285M) were predicted to have functional impact on the protein function. The results of protein stability analysis inferred that I63R and H135R exhibited less deviation in root mean square deviation in comparison with the native and T285M protein. The flexibility of all the three mutant models of CDK7 protein is diverse in comparison with the native protein. Following to that, docking study revealed the change in the active site residues and decrease in the binding affinity of flavopiridol with mutant proteins.

Conclusion: This theoretical approach is entirely based on computational methods, which has the ability to identify the disease-related SNPs in complex disorders by contrasting their costs and capabilities with those of the experimental methods. The identification of disease related SNPs by computational methods has the potential to create personalized tools for the diagnosis, prognosis, and treatment of diseases.

(Continued on next page)

* Correspondence: georgecp77@yahoo.co.in; hlzhu@comp.hkbu.edu.hk
[1]Medical Biotechnology Division, Centre for Nanobiotechnology, School of Biosciences and Technology, VIT University, Vellore 632014, Tamil Nadu 632014, India
[2] Department of Computer Sciences, Hong Kong Baptist University, Kowloon Tong, Hong Kong
Full list of author information is available at the end of the article

(Continued from previous page)

Lay abstract: Cell cycle regulatory protein, CDK7, is linked with DNA repair mechanism which can contribute to cancer risk. The main aim of this study is to extrapolate the relationship between the nsSNPs and their effects in drug-binding capability. In this work, we propose a new methodology which (1) efficiently identified the deleterious nsSNPs that tend to have functional effect on protein function upon mutation by computational tools, (2) analyze d the native protein and proposed mutant models in atomic level using MD approach, and (3) investigated the protein-ligand interactions to analyze the binding ability by docking analysis. This theoretical approach is entirely based on computational methods, which has the ability to identify the disease-related SNPs in complex disorders by contrasting their costs and capabilities with those of the experimental methods. Overall, this approach has the potential to create personalized tools for the diagnosis, prognosis, and treatment of diseases.

Keywords: nsSNPs, CDK7, Flavopiridol, Molecular dynamics, Docking

Introduction

Cyclin-dependent kinase 7 (Cdk7), a regulatory enzyme for the initiation of cell cycle progression, was initially identified from a search for cDNA encoding protein kinase(s) related to Cdk1 [1]. For activation of Cdk1, Cdk2, Cdk4, and Cdk6, the catalytic subunit of the Cdk-activating kinase requires the association of Cdk7 with a regulatory subunit, cyclin H and the phosphorylation of a conserved threonine residue at position 170 within its own T loop [2,3]. Subsequently, both CDK7 and the partner cyclin H were found to be associated with the general transcription factor TFIIH, suggesting additional roles of CDK7 in transcription. Given that CDK7 activates the main CDKs at different cell cycle transitions, it is possible to assume that the over expression of CDK7 also contributes to breast cancer cell proliferation [4]. In addition, CDK7 plays a vital role in human DNA repair mechanism (in NER pathway). Evidences support the hypothesis that mutations are early events in carcinogenesis, so the defects in DNA repair probably represent a high risk factor for many types of cancer [5,6]. Consistent with these actions, CDK7 was treated as a potent therapeutic target to inhibit the activity of cell cycle in cancerous cells. Currently, in phase-III trials for chronic lymphocytic leukemia, flavopiridol a potential antitumor drug has shown better inhibitory effect towards CDK7 [7]. It is known that flavopiridol decreases transcription by inhibiting CDK7 [8], which is responsible for the phosphorylation of the C-terminal domain of the largest subunit of RNA polymerase II, an activity essential for both transcriptional initiation and elongation [9,10]. Analyzing the human genetic variation promises to have a significant impact on the ability to understand the basis of individual variation in response to therapeutics. As we are entering the age of "personalized genomics", it is expected that the knowledge of human genetic variations could provide a basis for understanding the differences in susceptibility to diseases and designing individualized therapeutic treatments [11,12]. It was estimated that 90% of human genetic variations were caused by single nucleotide polymorphisms (SNPs) [12]. For example, changes in amino acids of proteins, such as the nonsynonymous single nucleotide polymorphisms (nsSNPs) in the gene coding regions could account for nearly half of the known genetic variations linked to human inherited diseases [13]. The nsSNP might change the physicochemical property of a wild-type amino acid that affects the protein stability and dynamics and disrupts the interacting interface, protein-small molecule, and protein-protein interaction [14-17]. Taken together, single mutation may affect binding ability of the inhibitory molecule. Recent progress in high throughput human genome research has provided a wealth of information detailing tens of millions of human genetic variations between individuals, including SNPs [11,18]. Numerous efforts have been carried out to illustrate how nsSNPs produce deleterious effects on the stability and function of a protein [19-23]. Given the large number of SNPs, a detailed experimental study on the effect of mutation in biological function is a daunting task. An effective alternative is the use of *in silico* methods. These approaches were based on the biochemical severity of the amino acid substitution, as well as the protein sequence and/or structural information, which can provide a more feasible method for phenotype prediction.

Recently, more sophisticated *in silico* algorithms were developed to predict the impact of amino-acid substitutions on protein structure and function. Some of the variation tolerance methods follow a similar procedure, in which a missense variant is first labeled with properties, related to the damage it may cause to the protein structure or function [24]. However, in other methods, predictions are based on the difference in the free energy of unfolding (DDG) between a native-type and mutant protein. The methods that use energy functions can be subdivided into physical, statistical, and the empirical potential approaches [25]. The ultimate goal of all these approaches is to determine the deleterious nsSNPs from the neutral ones. In general, *in silico* methods can provide a feasible and the high-throughput way to determine the impact of large numbers of nsSNPs on protein function. To understand the atomistic level changes and the dynamic behavior of the molecule with respect to the potential mutations, we conducted molecular dynamics (MD) simulations analysis. MD simulations can help us understand the effects of mutation on protein structure, which allow exploring how one amino acid substitution can create a ripple effect throughout the protein structure. Offman et al. found a strong correlation between MD analysis and the experimental work on the molecular basis of the most common protein upon *N370S* mutation in causing Gaucher's disease [26,27]. Thus, we assume that MD simulation analysis might provide more reliable structural information upon CDK7 mutations.

Although deleterious nsSNPs of *CDK7* gene have received considerable attention from experimental biologists, the functional consequence of most of the nsSNPs in CDK7 at the structural level is still unknown. The main goal of this *in silico* analysis is to determine the most deleterious variants in *CDK7* gene. In this context, publicly available *in silico* tools such as Sorting Intolerant From Tolerant (SIFT; J. Craig Venter Institute, Rockville, USA) [28], Polymorphism Phenotyping (PolyPhen) version 2 [29], PANTHER [30], I-Mutant3 [31], SNPs&GO (Bologna Biocomputing Group, Bologna, Spain) [32], predictor of human deleterious single nucleotide polymorphisms (PhD-SNP; Bologna Biocomputing Group) [33], and screening for non-acceptable polymorphisms (SNAP; Bologna Biocomputing Group) [34] were used to analyze the nsSNPs in *CDK7* gene. As a next step, we subjected MD simulation study in the native and mutant models of CDK7 proteins using GROMACS 4.5.3 package [35,36]. MD simulations will reveal the level of structural conformations changes with respect to the incorporation of deleterious mutations in CDK7 protein. Finally, the binding capability of CDK7 inhibitor, flavopiridol, was analyzed with respect to the structural mutations. Docking study was carried out with the help of AutoDock4 (The Scripps Research Institute, La Jolla, USA) and PatchDock [37-39]. The proposed protocol is represented schematically in Figure 1.

Results

Dataset

Dataset for the evaluation of potential nsSNPs in *CDK7* gene was retrieved from dbSNP [40] and SwissProt [41] database. We selected 14 nsSNPs for further consideration, and their associated biomedical informations were retrieved from OMIM (Johns Hopkins University, Baltimore, USA) [42], PubMed, and Swiss-Prot database. Related experimental data about the CDK7 protein and Protein Data Bank (PDB) structural information with PDB ID 1UA2 [43] were obtained from Swiss-Prot database and PDB, [44] respectively. The ligand molecule, flavopiridol, was obtained from Drug Bank database [45].

Prediction of deleterious nsSNPs by *in silico* tools

Identifying the deleterious nsSNPs has become possible with the aid of improved *in silico* algorithms. Here, we analyzed 14 nsSNPs of *CDK7* gene with seven different *in*

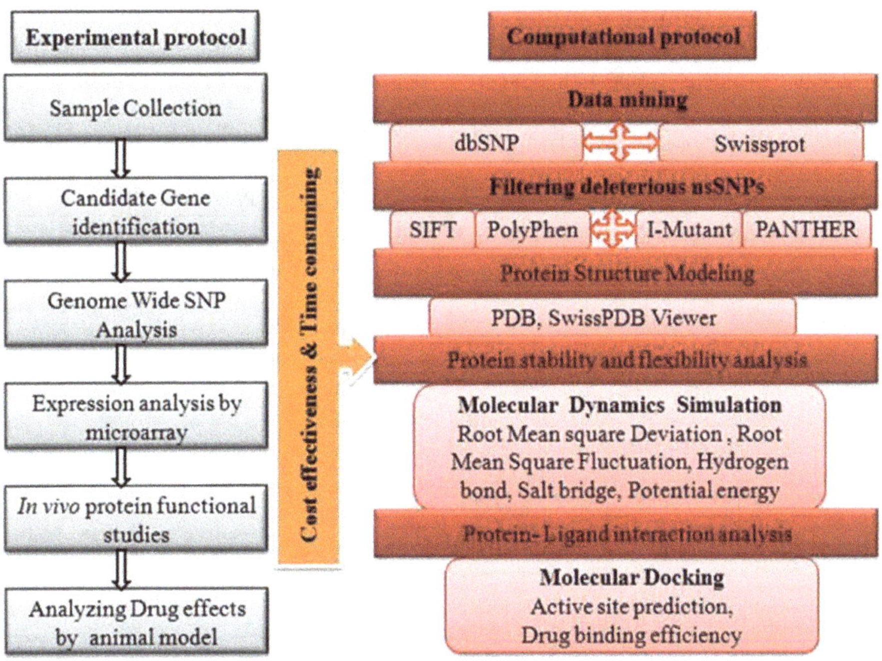

Figure 1 Outline of proposed protocol for nsSNPs analysis. This protocol explains the different steps followed in nsSNP analysis via experimental (grey color) and computational methods. Box displayed in orange color indicates the effectiveness of computational over experimental methods.

silico tools specifically SIFT, PolyPhen2, I-Mutant3, PANTHER, SNPs&GO, PhD-SNP, and SNAP to determine the protein structural and functional significance. Table 1 displays the distribution of the deleterious and neutral variations of *CDK7* gene with the corresponding amino acid substitution. SIFT makes inferences from sequence similarity using mathematical operations. SIFT constructs a multiple sequence alignment (MSA) and considers the position of the missense variants. Based on the amino acids appearing at each position in the MSA, SIFT calculates the probability and classifies a missense variant 'tolerated'. SIFT can be applied not only to naturally occurring nsSNPs but also to identify artificial missense mutations. Among the 14 nsSNSPs analyzed by SIFT, six were identified as deleterious which obtained a SIFT score ≤0.05. PolyPhen2 utilizes a combination of sequence and structure-based attributes for the description of an amino acid substitution, and the effect of mutation is predicted by a native Bayesian classifier. The sequence-based features include position specific independent count (PSIC) scores, MSA properties, and position of mutation in relation to domain boundaries as defined by Pfam [46]. The structure-derived features are solvent accessibility, changes in solvent accessibility for buried residues, and crystallographic B-factor. By PolyPhen2, nine nsSNPs were predicted as probably and possibly damaging, having the effect on protein structure and function of CDK7 protein; the remaining five were classified as benign which obtained score less than 0.15. In order to verify the prediction accuracy of SIFT scores, we used hidden Markov model (HMM)-based evolutionary approach PANTHER to verify the effect on protein function upon a single point mutation. Out of 14 nsSNPs, 7 were designated as deleterious with a score of ≤−3. In order to improve overall prediction accuracy, we used I-Mutant3, a support vector machine-based stability prediction tool. A score less than '0' means the mutation decreases the stability. The smaller the score, the more certain is the prediction. Conversely, a score more than '0' means mutation increases the protein stability. Among the 14 nsSNPs of the *CDK7* gene, 13 nsSNPs showed negative DDG values, were considered to be less stable and deleterious. The remaining one nsSNP showed a positive DDG value and classified as non-deleterious. SNPs&GO is an support vector machines (SVM) classifier based on variation type and sequence environment information, sequence profiles taken from MSAs, predictions from the program PANTHER, and a function-based log-odd score describing information about protein function defined by Gene Ontology (GO) terms. SNPs&GO predicted four nsSNPs of *CDK7* gene, which are related to a disease condition. SNAP is a neural network-based method that uses *in silico* derived protein information (e.g., secondary structure, conservation, solvent accessibility, etc.) in order to make predictions regarding functionality of the mutated proteins. The network takes protein sequences and lists of mutants as input, returning a score for each substitution. These scores can then be translated into binary predictions of effect (present/absent) and reliability indices. SNAP screened four nsSNPs of *CDK7* gene as non-neutral, and it may cause phenotypic changes. PhD-SNP is a prediction method based on single sequence profile-based SVM, trained on Swiss-Prot variants. The single sequence SVM classifies the missense variant to be pathogenic or neutral, based on the nature of substitution and properties of the neighboring sequence environment. PhD-SNP classified four nsSNPs as deleterious. Comparing the results of all seven prediction

Table 1 List of nsSNPs showing deleterious/non-deleterious scores by SIFT, PolyPhen2, I-Mutant3, PANTHER SNP&GO, SNAP and PhD-SNP

Rs IDs and variants information	Amino acid position	SIFT	Polyphen2	I-Mutant3	PANTHER	SNPs&GO	SNAP	PhD-SNP
rs193107048	V5M	0.84	0.039	−0.7	−4.76277	Neutral	Neutral	Neutral
rs145665301	I43L	0.01	0.765	−0.58	−1.92299	Neutral	Neutral	Neutral
rs137960738	*I63R*	*0.00*	*1.000*	*−1.49*	*−8.80751*	*Disease*	*Non- neutral*	*Disease*
rs17849960	Q130R	0.5	0.000	0.09	−4.57361	Neutral	Neutral	Neutral
VAR_023118	G163A	0.12	0.849	−0.94	−4.76811	Neutral	Neutral	Neutral
rs201535403	N166S	0.21	0.014	−0.95	−2.03457	Neutral	Neutral	Neutral
rs180962343	V174A	0.07	0.417	−2.26	−2.27717	Neutral	Neutral	Neutral
rs142560750	*H135R*	*0.00*	*1.000*	*−0.02*	*−11.54699*	*Disease*	*Non-neutral*	*Disease*
rs202024894	L229W	0.00	0.98	−0.04	−4.40368	Disease	Non- neutral	Disease
rs201088666	P238L	0.00	0.997	−0.04	−2.34273	Neutral	Neutral	Neutral
rs201381439	M240I	0.12	0.11	−0.03	−1.56205	Neutral	Neutral	Neutral
rs34584424	*T285M*	*0.00*	*1.000*	*−0.22*	*−8.58685*	*Disease*	*Non-neutral*	*Disease*
rs200939840	R298Q	0.32	0.006	−0.63	−0.67728	Neutral	Neutral	Neutral
rs200143477	T332A	0.80	0.000	−0.23	−0.20296	Neutral	Neutral	Neutral

Highly deleterious by SIFT, PolyPhen2, I-Mutant3, PANTHER, SNPs&GO, SNAP, and PhD-SNP are italicized.

tools, three nsSNPs at corresponding amino acid position I63R, H135R, and T285M with a highest SIFT tolerance index of 0.00 and PSIC score difference 1.0 were selected for structural analysis.

Analysis of secondary structure and surrounding amino acid changes

Structural information could play a vital role in unveiling the molecular mechanisms leading to a disease. Based on this, we proposed modeled structures for all the three mutants (I63R, H135R, and T285M) of CDK7 protein using PyMOL (Schrödinger, Bangalore, India) [47]. Substitution of an amino acid may produce changes at the structural level. Changes in the secondary structure with respect to the substituted amino acid were analyzed in PDBsum (Cambridge, UK). Additional file 1: Figure S1 displays the secondary structural elements of the native and mutant models. The number of secondary structure elements such as beta sheets, beta hairpins, beta bulges, strands, helices, helix-helix interactions, beta turns, and gamma turns was calculated for both the native and mutant models (Table 2). It has to be noted that the observed numbers of secondary structural elements are equal in both native and mutant models except the turns. There was a slight increase in the number of beta turns in all the three mutant models. The native protein exhibited only 30, while mutants obtained 31 beta turns. Substitution of arginine in the modeled H135R protein leads to decrease in the number of gamma turns as six, whereas the native and remaining mutants I63R and T285M obtained seven gamma turns. Further, the surrounding amino acid residue changes were visualized from the point of mutational position. A residue change within 4A° surroundings was observed through PyMOL (Figure 2A–C). In addition, the number of cation-pi interacting residues for both the native and three mutants was calculated using Protein Interactions Calculator server [48]. A cation-pi interaction plays a vital role in maintaining the protein structural stability and is recognized as an important non-covalent binding interaction in structural biology [49,50]. Change in the secondary structural elements may bring about some changes in the cation-pi interacting residues in the mutant models. It has to be noted that the number of intramolecular cation-pi interactions in the native protein is seven. The substitution of deleterious amino acid increased the number of cation-pi interactions in the mutants I63R, H135R, and T285M as eight, nine, and eight, respectively (Additional file 2: Table S1). Cation-pi interacting residue distances and angles varied in mutant model showed the deleterious effects of substituted amino acid. Overall, the structural analysis results inferred that the three deleterious mutations had brought a drastic change in the CDK7 protein, and it could affect the protein function.

Docking analysis

In vitro studies of flavopiridol showed inhibitory activity towards CDK7 protein and lead to programmed cell death in cancerous cells [51]. Substitution of deleterious amino acid in CDK7 protein may affect the binding ability of CDK7 with flavopiridol. This has to be analyzed to improve the potentiality of the drug to inhibit CDK7 protein. Hence, we analyzed the binding ability of flavopiridol with native and mutant models of CDK7 protein using *in silico* docking tool, Autodock4 and PatchDock [39]. Before entering into docking analysis, we evaluated the binding sites of native CDK7 protein. Flavopiridol binds at the ATP binding site of the native CDK7 protein and made contact with 12 amino acid residues. Twelve residues specifically GLY21, GLN22, PHE23, ALA24, VAL26, LYS41, PHE91, ASP97, ASN141, LEU144, ALA154, and SER161 were involved in protein-ligand interaction. This information was in concordance with the study conducted by Carlson et al. [51] and Worland et al. [52]. In their analysis, it was observed that flavopiridol directly inhibits CDK7 by competing for to the ATP binding site. In addition, these 12 residues were also involved in protein-ATP interaction observed by Lolli et al. [2] in their crystallography analysis. In the mutant models (I63R, H135R, and T285M), we observed the number of contact residues as six, seven, and six, respectively (Additional file 3: Table S2). Decrease in the number of residue contacts will definitely affect the complementarities between mutant protein and flavopiridol compound as shown in Figure 3A–D. Further, we observed the interaction of flavopiridol with native and mutant proteins by LIGPLOT (Additional file 4: Figure S2). Shape complementarity and non-covalent interactions were believed to drive protein-ligand interaction. Non-covalent bonds such as hydrogen bonds, van der Waals contacts and electrostatic forces are the dynamic forces involved in protein-ligand interactions. Calculating the interaction energies of non-covalent bonds is a key point in understanding the binding

Table 2 The number of secondary structure element in the native and mutant structures of CDK7 protein

Native and mutant models	Sheets	Beta hairpins	Beta bulges	Strands	Helices	Helix-helix interacts	Beta turns	Gamma turns
Native	2	5	4	8	12	15	30	7
I63R	2	5	4	8	12	15	**31**	7
H135R	2	5	4	8	12	15	**31**	**6**
T285M	2	5	4	8	12	15	**31**	7

Changes in the secondary elements are italicized.

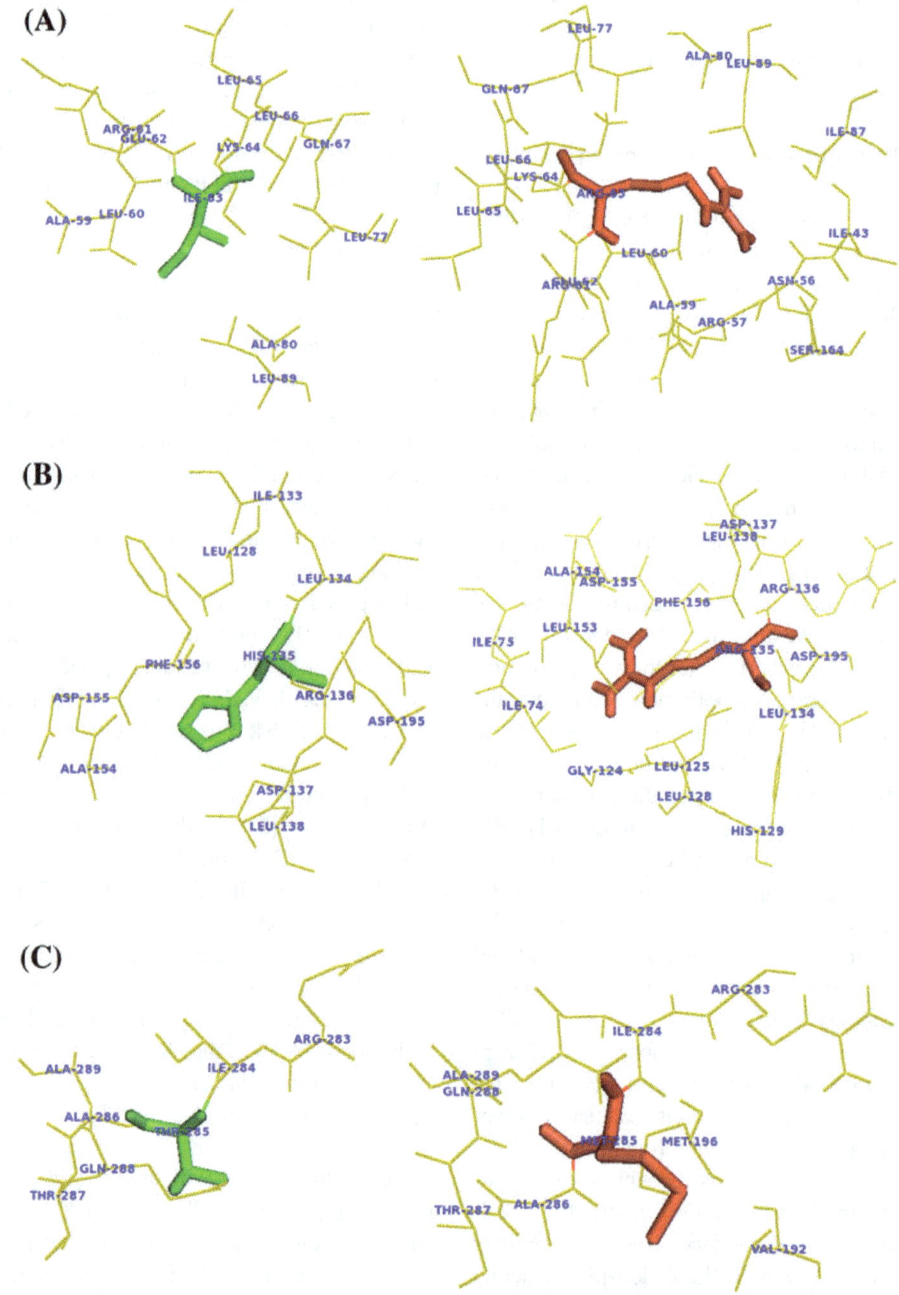

Figure 2 Change in the surrounding amino acid residues in CDK7 protein by the substitution of deleterious amino acid. (A) The native type isoleucine residue (green) at position 63 and the surrounding residues. Substitution of I63 residue with arginine (red) brings more surrounding residues in contact at position 63. **(B)** The native type histidine residue (green) at position 135 and its surrounding amino acid residues. Substitution of arginine (red) at position 135 brings more amino acids in the surrounding region. **(C)** Native type residue threonine (green) at position 285 and its surrounding amino acid residues. Substitution of methionine (red) at position 285 brings two more residues val192 and met196 within the 4 A^0 surrounding.

efficiency of a ligand molecule. The number of hydrogen bonds formed between protein and ligand, and van der Waals interacting energies and electrostatic interacting energies was computed using Autodock4. The binding energy and the non-covalent bond interaction energy between CDK7 protein (native and mutant) and flavopiridol molecule were calculated and shown in Table 3. In the native complex, the significant contribution of van der Waals and electrostatic energy was observed as −9.18 and −9.07 kcal/mol, respectively. On the contrary, mutant models I63R,

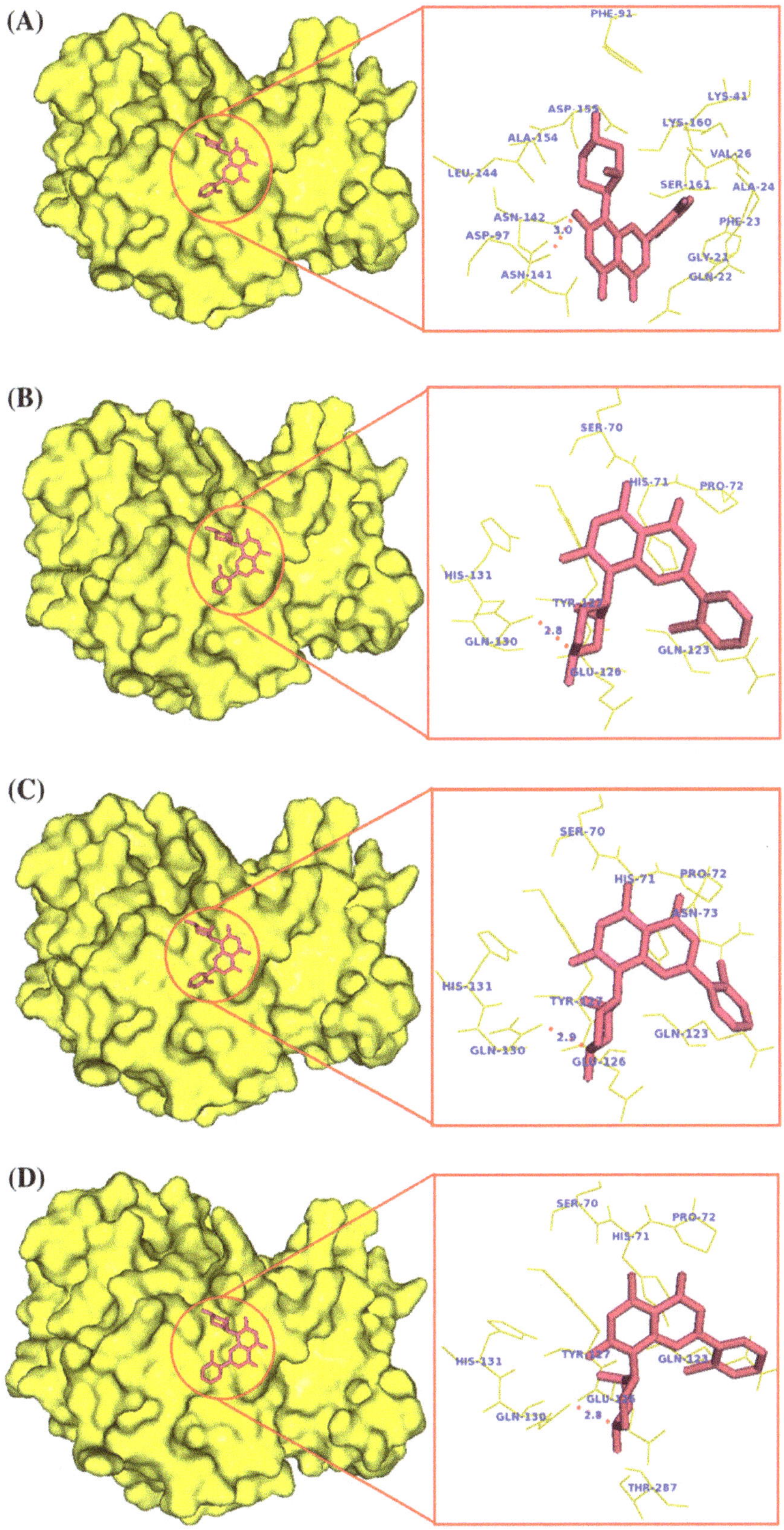

Figure 3 (See legend on next page.)

(See figure on previous page.)

Figure 3 Interaction of flavopiridol with native and mutant models of CDK7 protein. (A) Flavopiridol binds deeply with native CDK7 protein and makes contact with 12 amino acid residues. **(B)** Substitution of I63 with arginine reduced the binding affinity of flavopiridol in mutant model I63R. **(C)** Substitution of H135 with arginine, results in weak interaction of ligand flavopiridol with H135R model. **(D)** Flavopiridol binds shallowly on the surface of mutant model T285M and the number of amino acid contact become reduced.

H135R, T285M interacting with flavopiridol showed an increase in van der Waals and electrostatics energies as −5.52, −5.53, and −5.57, and −0.65, −0.6, and −0.68 kcal/mol, respectively. The total ligand receptor binding energy exhibited by the native, I63R, H135R, and T285M complexes were −8.27, −5.57, −5.53, and −5.65 kcal/mol, respectively. Lower binding energy of native complex indicates better interaction and good compatibility with the flavopiridol compound.

In order to determine the consistency in docking analysis, in addition to Autodock4 analysis, we performed docking in PatchDock [39]. Docking was performed between the drug flavopiridol with both the native type and mutant modeled structures of CDK7 protein to find out the binding efficiency in the form of PatchDock score and atomic contact energy (ACE) values (Table 4). In this analysis, we found that native type CDK7 protein obtained high PatchDock score and ACE as 5,956 and −0.25, respectively. But, all the three mutant structures (I63R, H135R, and T285M) obtained less PatchDock scores (4,148, 4,136, and 4,978) and high ACEs (8.98, 19.92, and 21.05). Notably, high PatchDock score and less ACE value were obtained in the native complex which is considered as a good docked complex than those of other three mutant complexes. This implies the concordances of computational algorithms in docking analysis and gives a 'theoretical quantitative' assessment on the binding efficiency of CDK7 mutant protein with flavopiridol.

Molecular dynamics, structural stability, and flexibility analysis

Molecular dynamic simulations studies were carried out to unravel the atomic level changes in the CDK7 protein with respect to the time scale. The overall protein stability changes upon mutation were evaluated by root mean square deviation (RMSD) values. We calculated the backbone RMSD for all the atoms from the initial structure, and this is considered to be a primary criterion to measure the convergence of the protein system concerned. The backbone RMSD was calculated for both the native and mutant models from the appropriate trajectory files (Figure 4). We observed a significant structural deviation in the mutant proteins I63R, H135R, and T285M when compared to native CDK7 protein structure. All the four structures attained a significant deviation at last 5 ns. The native and T285M mutant structure obtained a mean RMSD of approximately 0.35 nm in the last 5 ns, and mutant models I63R and H135R exhibited a deviation range from about 0.25 to 0.3 nm. This difference in the deviation range in mutant model explains the stability change and reflects the impact of substituted amino acid in the protein structure. In order to determine the structural flexibility of both the native and mutant models of CDK7 protein, we calculated the root mean square fluctuation (RMSF) values from the 10 ns simulation trajectory data. The RMSF values of native and mutant models are shown in Figure 5. In the entire 10 ns simulation period, native residues from approximately 25 to 150 nm showed a high fluctuation in comparison with I63R, H135R, and T285M mutant models. In the remaining residue range from around 150 to 260 nm, the mutant model I63R exhibited high fluctuation. Overall, RMSFs of all the mutant models were significantly deviated from the native structure in the entire simulation period. A change in the RMSFs specify the mode of flexibility changes in the mutant models and reflects the impact of deleterious amino acid substitution in CDK7 protein.

Table 3 Binding and non bonded interaction energies of native and mutant proteins of CDK7 with flavopiridol

Proteins	Binding energy (Kcal/mol)	van der Waals energy (Kcal/mol)	Electrostatic energy (Kcal/mol)	H-bond between protein and ligand	Inhibition constant	Inter molecular energy (Kcal/mol)	Internal energy (Kcal/mol)	Torsional energy (Kcal/mol)	Ref RMS
Native	−8.27	−9.18	−9.07	ASP97: OD1	869.56 nM	−8.86	−0.85	0.6	38.92
I63R	−5.57	−5.52	−0.65	GLN130: OE1 SER70: O	82.18 μM	−6.17	−0.7	0.6	8.55
H135R	−5.53	−5.53	−0.6	GLN130: OE1	87.9 μM	−6.13	−0.68	0.6	8.34
T285M	−5.65	−5.57	−0.68	GLN130: OE1	71.61 μM	−6.25	−0.56	0.6	8.58

Table 4 PatchDock scores and ACE values of CDK7 wild type and mutant type complexes

Native and mutant proteins	PatchDock score	Area	ACE Kcal/mol
Native	5965	667.90	−0.25
I63R	4148	514.50	8.98
H135R	4136	594.30	19.92
T285M	4978	575.30	21.05

Effects of deleterious mutations in hydrogen bonding, salt bridges and electrostatic potential of CDK7 protein

Hydrogen bonds and salt bridges are the key parameters in determining the stability of protein [53,54]. Non-synonymous SNPs can affect wild type protein function by affecting hydrogen bond formation [55-57]. Figure 6 depicts the number of hydrogen bonds formed in native and mutant structures of CDK7 protein. Native structure of CDK7 protein exhibits an average number of approximately 160 to 200 hydrogen bonds throughout the 10 ns simulation period. Mutant models I63R and H135R obtained more close number of hydrogen bonds, about 160 to 195 in comparison with native structure. Interestingly, T285M mutant model showed less number of hydrogen bonds around 160 to 190, when compared with the native and remaining two mutant models. Overall, it has to be noted that all the three mutant models obtained less number of hydrogen bonds in comparison with the native protein. The reduction in the number of hydrogen bonds in mutant proteins might be due to the incorporation of deleterious amino acid, and it may destroy the ability of hydrogen bond formations in CDK7 protein.

Salt bridge distances of CDK7 protein in both native and mutant proteins were calculated from the 10 ns trajectory data and shown in Additional file 5: Figure S3. In a period of about 1,000, 5,000, and 7,000 ps, native protein obtained a low salt bridge distance of approximately 0.2 nm and maintained an average range distance around 0.45 nm throughout the 10-ns simulation period. Two mutant models (I63R and H135R) maintained a similar distance like the native protein. The mutant model T285M exhibited high range of salt bridge distance in the maximum simulation period when compared to the native, I63R, and H135R structures. From this analysis, we conclude that salt bridges are more stable in I63R and H135R mutant models when compared to the mutant model T285M.

Discussion

The central objective in molecular biology and population genetics is to identify and characterize the nsSNPs that are functionally related from those that are not. This understanding not only provides insight into cancer biology but also highlights the anticancer therapeutic targets and diagnostic markers. NsSNPs in coding region can lead to amino acid change. This can lead to alterations in protein function and account for susceptibility to disease and altered drug response. Identification of deleterious nsSNPs from tolerant nsSNPs is ideal for analyzing individual susceptibility to disease, understanding the pathogenesis of

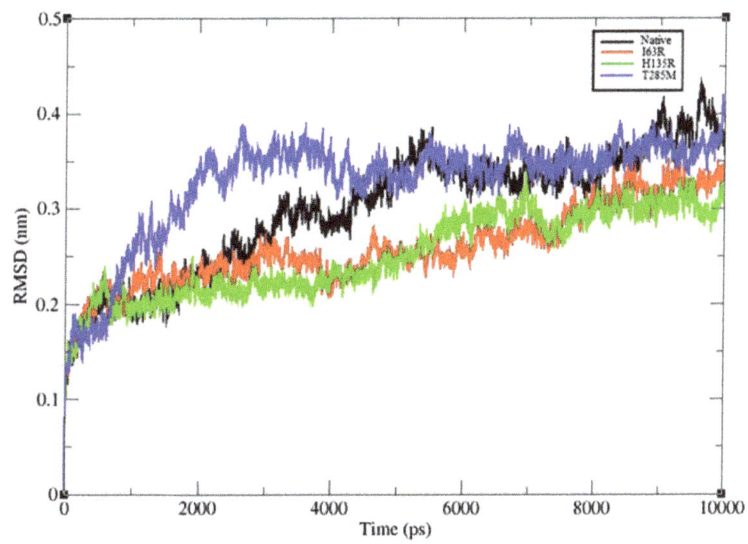

Figure 4 Backbone RMSD of wild type and mutant structure of CDK7 protein. The ordinate is RMSD (nm), and the abscissa is time (ps). Black, red, green, and blue lines indicate native, I63R, H135R, and T285M mutant structures, respectively.

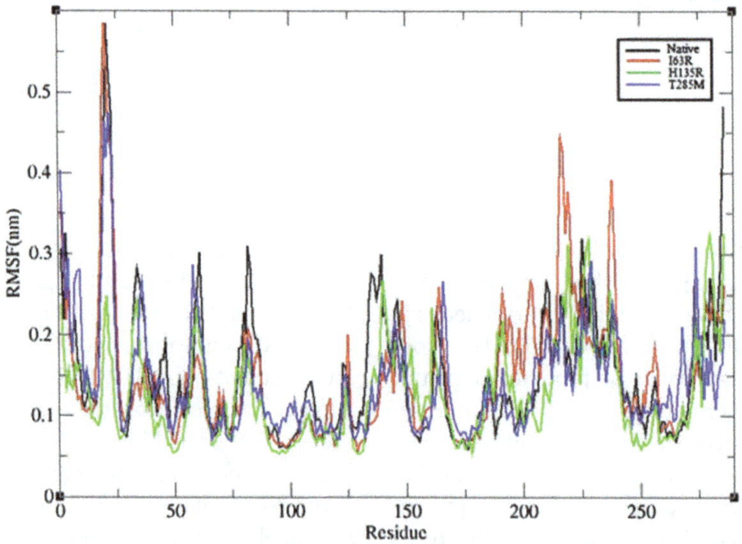

Figure 5 Carbon alpha RMSF of wild type and mutant structure of CDK7 protein. The ordinate is RMSF (nm), and the abscissa represents the residues. Black, red, green, and blue lines indicate the native, I63R, H135R, and T285M mutant structures, respectively.

disease, identifying molecular targets for drug treatment, and conducting individualized pharmacotherapy. Several experimental studies were carried out to analyze the relationship between nsSNPs and drug response in cancer treatments. Chambers et al. [58] reported the involvement of nsSNPs in the modulation of protein structure and function. Another finding by Giovannetti et al. [59] demonstrated the role of nsSNPs in the DNA-repair protein to be the potential biomarkers of primary resistance to gemcitabine/cisplatin-based polychemotherapeutic agent in the treatment of pancreatic cancer. In another analysis, Wang and Moult [60] reported the role of nsSNPs in individuals by inducing or influencing the disease by affecting protein-protein interactions, protein expression, alternative splicing, stability, folding, and ligand binding or catalysis. These mounting studies on nsSNPs assert their role in better understanding the resultant phenotypic variations among individuals with an endeavor towards new drug design and development. The exponential increase in the number of SNPs makes the determination of biological significance of each nsSNP by wet laboratory experiments impossible. Alternatively, *in silico* programs and statistical methods may be used to predict the effects caused by mutations and elucidate the underlying biological mechanisms. However, *in silico* tools can be used to examine the potentially deleterious nsSNPs that might affect important drug targets before further investigation by wet laboratory techniques. Previously, our group also identified and analyzed the effects of deleterious nsSNPs in several proteins at structural and functional level and drug binding

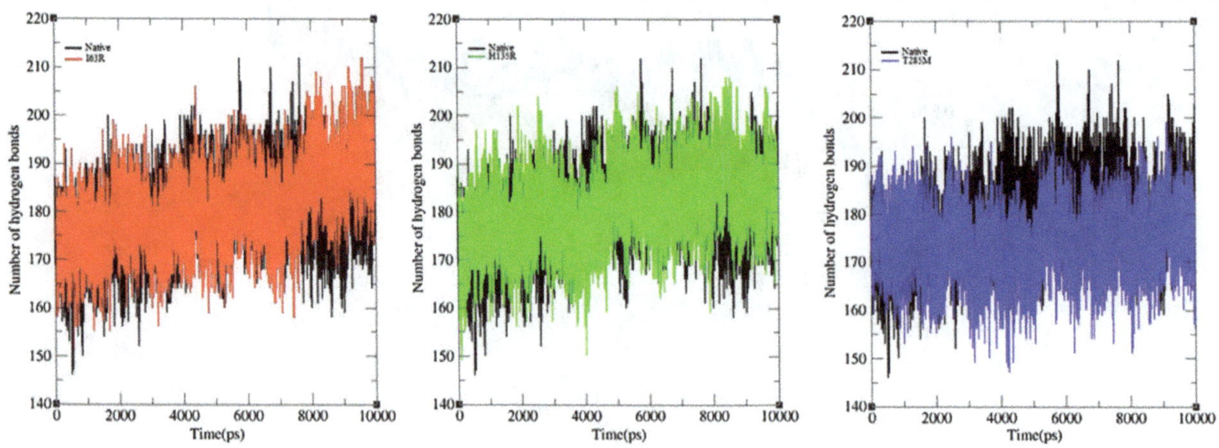

Figure 6 Number of hydrogen bond formed in wild type and mutant structure of CDK7 protein. The ordinate is the number of hydrogen bond and the abscissa is time (ps). Black, red, green, and blue lines indicate the native, I63R, H135R, and T285M mutant structures, respectively.

capability using various *in silico* tools [61-63]. In this paper, we performed a systematic *in silico* analysis to determine the potential deleterious and functional nsSNPs in CDK7 protein along with molecular dynamics and docking study. To determine the possible effects of nsSNPs in *CDK7* gene, we employed seven widely used *in silico* tools specifically SIFT, PloyPhen2, I-Mutant3, PANTHER, SNPs&GO, SNAP, and PhD-SNP. SIFT predicted 6 (42.85%) nsSNPs as deleterious, PolyPhen2 identified 9 (64.28%) substitutions which affect protein structure and function, I-Mutant3 identified 13 (92.85%) substitutions which affected the stability of protein, PANTHER predicted 7 (50%) nsSNPs to be deleterious, SNPs&GO, SNAP, and PhD-SNP identified 4 (28.57%) nsSNPs related to the disease condition. The basis for the predicting impact of nsSNPs in these seven algorithms was different, and we would expect the outcomes to occur in some ways, dissimilar. However, the positive predictions that overlap all these seven *in silico* tools would provide high reliability to behave similarly. The difference in their predictions might be due to the difference in features utilized by the methods or the training dataset. Comparing the prediction of all the seven methods, three nsSNPs (I63R, H135R, and T285M) were identified as highly deleterious and selected for further structure and functional investigations. To gain insight knowledge on the protein structure and what kind of harmful modulation these mutations give rise, the CDK7 protein was analyzed by MD approach. In the 10 ns simulation trajectory, different parameters were applied to analyze the level of structural changes. Molecular stability and flexibility changes were observed by RMSD and RMSF analyses. Stability is a fundamental property affecting the bimolecular function, activity, and regulation. Protein stability analysis results inferred that the stabilities of I63R and H135R RMSD are less deviated than those of the native and T285M protein. High or less deviation implies increase or decrease in the stability of protein. Hence, we believe that reduction in the stability of I63R and H135R models could affect the CDK7 protein structure. From the fluctuation analysis, we observed a decrease in the flexibility for all the three mutant models in first half of the residues (approximately 25 to 150) and increase in flexibility for the mutant model I63R in the rest of the residues (about 150 to 260). Increase in the flexibility could make the protein more flexible, and decrease in the flexibility makes protein more rigid. Conformational changes are required for many protein functions [64-66], but the conformational flexibility and rigidity must be well balanced [67]. The flexibility of all the three mutant models of CDK7 protein is heterogeneous in comparison with the native protein. Thus, from the RMSD and RMSF analysis, it is confirmed that substitution of amino acid adversely affected the stability and flexibility of CDK7 proteins. Beside the different electrostatic interactions, the hydrogen bonds and the salt bridges across the binding interfaces and in the protein interiors serve as main contributors in maintaining the protein structural conformation. Furthermore, incorporation of deleterious nsSNP might change the original electrostatic formations and distances that could affect the protein native structure. Consequently, CDK7 native protein obtained maximum of around 200 hydrogen bonds in the 10 ns simulation period. The mutant models I63R, H135R, and T285M obtained less hydrogen bonds approximately 195 and 190, respectively. The decrease in the number of hydrogen bonds may affect the protein structure. In salt bridge analysis, both the native and mutant models of CDK7 protein maintained the different patterns of salt bridge distances. Changes in the salt bridge distances reflect the displacement of cationic or anionic side chain residues in mutant models. In conclusion, we observed change in bonding distance by hydrogen bonding and salt bridge analysis. Change in residue distance might lead to the loss of thermodynamic stability. The main aim of this study is to extrapolate the relationship between the nsSNPs and their effects in drug-binding capability. In docking analysis, several factors involved between protein-ligand interactions were analyzed, and the analysis revealed the less binding ability of mutant models. In particular, electrostatic potential showed substantial agreement with MD analysis. In conclusion, the given *in silico* tools can indicate possible deleterious nsSNPs in CDK7 protein. Then, MD studies support the structural and conformational changes for the CDK7 deleterious nsSNP incorporated model. Finally, the binding ability of mutant model with the drug was validated to facilitate the study of new drug-targets and discovery of new drugs for CDK7 protein. *In silico* approaches reviewed here generated not only a considerable amount of valuable data but also the need for further validation by experimental methods such as *in vitro* binding/activity assays.

Materials and methods

Computational methods for finding deleterious variants

The ability to distinguish pathogenic and benign variants from a pool of data is a daunting task. Recently, many computational algorithms have been developed for the feasible prediction of disease-associated variants. Some of the methods classify deleterious variants according to the predicted pathogenicity, and other methods predict the deleterious variants based on protein-stability changes upon mutation. We used both these approaches to identify deleterious variants in the *CDK7* gene. Sequence evolutionary information-based methods (SIFT, PANTHER, and PhD-SNP) and the combination of protein structural and functional parameter-based methods (PolyPhen2, I-Mutant3, SNAP, and SNPs&GO) are some of the most reliable tools used for deleterious nsSNP prediction. SIFT, PANTHER, PhD-SNP, SNAP, SNPs&GO, and I-Mutant3 give results in two prediction categories, either tolerated or deleterious, while PolyPhen2 gives results in three categories: benign (probably lacking any phenotypic effect), possibly damaging, and probably damaging (should affect protein function). Sequence-based prediction includes all types of effect at the protein sequence level and can be applied to any human protein with known relatives. Structure-based approach is feasible to implement for proteins with 3D structures. Analyzing deleterious nsSNPs by both sequence and structure level has the added advantage of being able to assess the reliability of the generated prediction results by cross-referencing the results from both approaches. SIFT predicts whether an amino acid substitution affects protein function based on sequence homology and the physical properties of amino acids. A SIFT score ≤0.05 indicates that the amino acid substitution is intolerant or deleterious, whereas a score ≥0.05 is predicted as tolerant [68,69]. PANTHER estimates the likelihood of a particular nsSNP causing a functional impact on the protein. PANTHER uses HMM-based statistical modeling methods and multiple sequence alignments to perform evolutionary analysis of coding nsSNPs. PANTHER subPSEC scores vary from 0 (neutral) to about −10 (more likely to be deleterious). Protein sequences having subPSEC values ≤−3 is said to be deleterious. PolyPhen2.0 uses sequence, phylogenetic, and structural information in characterizing the deleterious substitution. A mutation is classified as 'probably damaging' if the probabilistic score is above 0.85 to 1; mutation is classified as 'possibly damaging' if the probabilistic score is above 0.15 to 0.84; the remaining mutations are classified as benign. I-Mutant3 is an SVM-based method for the automatic prediction of protein stability changes upon a single point mutation. The output file shows the predicted free energy change (DDG) which is calculated from the unfolding Gibbs free energy change of the mutated protein minus the unfolding free energy value of the native protein (Kcal/mol). DDG >0 means that the mutated protein has high stability and vice verse. PhD-SNP is a single sequence SVM method (SVM sequence) that discriminates disease-related mutations based on the local sequence environment of the mutation at hand and a sequence-profile-based SVM. The tool aims to predict whether an nsSNP causing a single point protein mutation would be a neutral polymorphism or one that is deleterious. SNPs&GO is a method based on SVMs, which predicts disease-associated mutations from protein sequence, evolutionary information, and functions as encoded in the gene ontology terms. The use of functional GO terms is the main aspect of novelty of this tool over other existing bioinformatics tools. From the output of the programs, we only took the binary prediction (pathogenic/neutral) into consideration without taking into account any confidence values provided by some of the programs. SNAP is based on neural network and advanced machine-learning approach to predict the functional effects of nsSNPs in proteins. It uses sequence, functional and structural (secondary structure, solvent accessibility) annotations, and biophysical and evolutionary (residue conservation within sequence families) characteristics to predict a gain or loss in protein function. SNAP predicts whether the mutation is neutral or non-neutral with expected accuracy.

Protein-ligand docking analysis

Protein-ligand interaction study was performed between the native and mutant models of CDK7 protein with the inhibiting compound, flavopiridol. In order to carry out the docking analysis, we used the AutoDock4 suite as a molecular-docking tool. AutoDock4 is a suite of programs making it possible to predict how ligands bind to large macromolecules. In this docking simulation, we used semi-flexible docking protocols. Throughout the docking simulation, the target protein is kept rigid. The ligand being docked is usually flexible and, therefore, explores an arbitrary number of torsional degrees of freedom in addition to the six spatial degrees of freedom spanned by the translational and rotational parameters. AutoDock4 provides different optimization algorithms to search the space of possible protein-ligand combinations, such as simulated annealing, genetic algorithm (GA), and hybrid evolutionary algorithms EA termed Lamarckian GA (LGA) combining the GA with a local search strategy [70]. The Lamarckian Genetic Algorithm (LGA) was chosen to search for the best conformers. The best docking solution (minimum docked free energy) is reported by AutoDock for each GA run. The total number of clusters and the rank of each docking mode (cluster rank) are also reported in the cluster analysis performed by AutoDock. Docking modes were selected on the basis of two criteria: extent of ligands associations with the key residues of the receptor and the thermodynamic stability of the docked complex so obtained. The lowest energy docking mode that would conform to the above said two parameters was selected from over 10 GA runs and hence 10 total docking mode times. The grid boxes were centered on the root of macromolecule with spacing of 0.375 Å. The estimated binding free energies were calculated using the following equation: $E_{binding} = E_{intermolecular} + E_{internal} + E_{torsional} - E_{unbound}$. The unbound structure of ligand is the same as the bound state (crystal structure), so the $E_{internal}$ is equal to $E_{unbound}$, and they do not contribute to the total energy. On the other hand, the $E_{torsional}$ is calculated based only on the number of torsional bond in ligand, so this term remains the same in each complex. It is clear that there are significant differences between nine charge methods in the estimated binding free energies, so the difference should come from the $E_{intermolecular}$, including energies of dispersion/

repulsion, hydrogen bonding (hbond), desolvation potential, and electrostatic interactions. The energy functions used in docking simulations attempt to account for the intermolecular energies between the protein and the ligand, as well as the intramolecular energies arising from the ligand conformation itself. AutoDock4 uses a grid-based approach to approximate the energy calculations used by the energy function. During the evaluation of a candidate conformation, the grids were used as lookup tables which store the values used in the calculation, thus making the overall docking simulation exceptionally fast. The Graphical User Interface program 'Auto Dock Tools' was used to prepare, run, and analyze the docking simulations. Kollman united atom charges, solvation parameters, and polar hydrogens were added into the receptor PDB information for the preparation of protein in docking simulation. Gasteiger charges were added in the ligand PDB file.

In addition to the Autodock4 study, we used PatchDock for docking native and mutant CDK7 proteins with the drug flavopiridol. PatchDock performs docking based on molecular shape representation and surface patch matching plus filtering and scoring. PatchDock is more reliable because of its fast transformational search, which is driven by local feature matching rather than brute force searching for the six-dimensional transformation space. It further speeds up the computational processing time by utilizing advanced data structures and spatial pattern detection techniques, such as geometric hashing and pose clustering. Protein and the ligand molecule were given as input in performing the docking experiments with default root-mean-square deviation (RMSD) value (4.00 Å). It generated several complex structures based on docking scores. The complex structure file, with the best docking score was selected for further analysis. The geometry of both wild type and mutant type CDK7 structures were optimized through Steepest Descent method with 1000 steps each of GROMACS 4.5.3 package. Each minimization was carried out with GROMOS-96 [71] 43a1 force field.

Molecular dynamics simulation protocol

Molecular dynamics simulations for the native and mutant models were done with MD simulation package GROMACS 4.5.3 that adopts GROMOS96 43a1 force field parameter for energy minimizations. Energy minimized structures of the native CDK7 and three mutant models were used as a starting point for MD simulations. All the proteins were solvated in a cubic box with wall extending at least 0.9 nm from all atoms and filled with SPC [72] water molecules. A periodic boundary condition was applied that the number of particles, pressure and the temperature were kept constant in the system. In order to obtain electrically neutralized system, we utilized GENION procedure from the GROMACS package to replace random water molecule with Na^+ or Cl^- ions. The temperature was kept constant by using a Berendsen algorithm [73] with a coupling time of 0.2. The minimized system was equilibrated for 10,000ps each at 300 K by position restrained molecular dynamics simulation in order to soak the water molecules into the macromolecules. The equilibrated systems were then subjected to molecular dynamics simulations for 10 ns each at 300 K. In all simulations, the temperature was kept constant at 300 K. The particle mesh Ewald method [74] was used to treat long-range Coulombic interactions and the simulations performed using the SANDER module [75]. The SHAKE algorithm was used to constrain bond lengths involving hydrogen's permitting a time step of 2fs. The coordinates were saved at regular time intervals of 1ps. The van der Waals force was maintained at 1.4 nm, and Coulomb interactions were truncated at 0.9 nm.

Analysis of molecular dynamics trajectories

Structural properties of the native and mutant models of CDK7 protein were calculated from the trajectory files with the built-in functions of GROMACS 4.5.3 The trajectory files were analyzed through the use of g_rmsd and g_rmsf GROMACS utilities in order to obtain the RMSD and RMSF values. The number of distinct hydrogen bonds formed in the protein during the simulation was calculated using g_hbond utility. The number of hydrogen bond was determined on the basis of donor-acceptor distance less than 3.9 nm and of donor-hydrogen-acceptor angle larger than 90° [76]. Salt bridge formed in CDK7 protein was analyzed using g_salt GROMACS. If the distance is ≤4.0 nm, the pair is counted as a salt bridge [77]. In order to generate the three-dimensional backbone RMSD, RMSF of carbon alpha-carbon, hydrogen bond and salt bridge analysis, and motion projection of the protein in phase space of the system were plotted for all four simulations using Graphing, Advanced Computation and Exploration program.

Additional files

Additional file 1: Figure S1. Secondary structural elements changes in mutant models of CDK7 protein. **(A)** Secondary structural elements of native CDK7 protein. **(B)** Secondary structural element changes in the mutant model I63R due to the substituted of arginine. **(C)** Substitution of histidine at position 135 showing the secondary structural changes in H135R mutant model. **(D)** Substitution of methionine at position 285 showing the secondary structural changes in T285M mutant model. All the substituted amino acids are indicated by green boxes.

Additional file 2: Table S1. Involvement of cation-Pi interaction in wild type and mutant structure of CDK7 protein.

Additional file 3: Table S2. Flavopiridol interacting residues with CDK7 wild type and mutant type structures.

Additional file 4: Figure S2. LIGPLOT analysis of CDK7-flavopiridol complex in both native and mutant states. **(A)** Native complex showing high number of residues interacting with ligand. **(B)** LIGPLOT showing interaction between mutant model I64R and flavopiridol. **(C)** LIGPLOT showing interaction between mutant model H135R and flavopiridol. **(D)** LIGPLOT showing interaction between mutant model T285M and flavopiridol.

Additional file 5: Figure S3. Salt bridge forming distances of wild type and mutant structures of CDK7 protein. The ordinate is distance (nm) and the abscissa is time (ps). Black, red, green, and blue lines indicate native, I63R, H135R, and T285M structures, respectively.

Abbreviations

ACE: Atomic contact energy; CDK7: Cyclin-dependent kinase 7; SNPs: Single Nucleotide Polymorphisms; nsSNPs: Nonsynonymous Single Nucleotide Polymorphisms; GRACE: Graphing Advanced Computation and Exploration; MSA: Multiple Sequence Alignment; RMSD: Root mean square deviation; RMSF: Root mean square fluctuation.

Competing interests

The authors declare that they have no competing interests.

Authors' contributions

NN, CGPD, CC, HZ, and LC were involved in the design of the study, drafting the manuscript, and the acquisition, analysis, and interpretation of the data. HZ, CC, and LC supervised the entire study. CGPD, CC, and HZ were involved in the final drafting of the manuscript. All authors read and approved the final manuscript.

Acknowledgments

This work was supported by the Research Grants Council of Hong Kong (212111) and Faculty Research Grants of Hong Kong Baptist University (3011299), and partially supported by National Natural Science Foundation of China (91029301, 61134013 and 61072149), the Chief Scientist Program of Shanghai Institutes for Biological Sciences, Chinese Academy of Sciences (2009CSP002), Shanghai Pujiang Program, and the FIRST program from JSPS initiated by CSTP. The authors take this opportunity to thank the management of VIT University for providing the facilities and for encouragement to carry out this work.

Author details

[1]Medical Biotechnology Division, Centre for Nanobiotechnology, School of Biosciences and Technology, VIT University, Vellore 632014, Tamil Nadu 632014, India. [2]Department of Computer Sciences, Hong Kong Baptist University, Kowloon Tong, Hong Kong. [3]Department of Bio-informatics, School of Computer and Information Sciences, Galgotias University, Uttar Pradesh, India. [4]Key Laboratory of Systems Biology, SIBS-Novo Nordisk Translational Research Centre for PreDiabetes, Shanghai Institutes of Biological Sciences, Chinese Academy of Sciences, Shanghai, China. [5]Collaborative Research Center for Innovative Mathematical Modelling, Institute of Industrial Science, The University of Tokyo, Tokyo 153-8505, Japan.

References

1. Shuttleworth J: **The regulation and functions of Cdk7.** *Prog Cell Cycle Res* 1995, **1**:229–240.
2. Lolli G, Johnson LN: **CAK-cyclin-dependent activating kinase: a key kinase in cell cycle control and a target for drugs?** *Cell Cycle* 2005, **4**:572–577.
3. Garrett S, Barton WA, Knights R, Jin P, Morgan DO: **Reciprocal activation by cyclin-dependent kinases 2 and 7 is directed by substrate specificity determinants outside the T loop.** *Mol Cell Biol* 2001, **21**:88–99.
4. Bartkova J, Zemanova M, Bartek J: **Expression of CDK7/ CAK in normal and tumour cells of diverse histogenesis, cell cycle position and differentiation.** *Int J Cancer* 1996, **66**(6):732–737.
5. Yao X, Buermeyer A, Narayanan L, Tran D, Baker S, Prolla T, Glazer P, Liskay R, Arnheim N: **Different mutator phenotypes in MLH1 - versus PMS2-deficient mice.** *Proc Natl Acad Sc* 1999, **96**:6850–6855.
6. Rajewsky M, Engelbergs J, Thomale J, Schweer T: **Relevance of DNA repair to carcinogenesis and cancer therapy.** *Recent Results Cancer Res* 1998, **154**:127–146.
7. Chen R, Keating MJ, Gandhi V, Plunkett W: **Transcription inhibition by flavopiridol: mechanism of chronic lymphocytic leukemia cell death.** *Blood* 2005, **106**(7):2513–9.
8. Carson BA, Pearlstein RA, Nail RG, Sedlacek HH, Sausville EA: **Inhibition of CDK2, CDK4 and CDK7 by flavopiridol and structural analogs.** *Proc Amer Assoc Cancer Res* 1996, **37**:424.
9. Palancade B, Bensaude O: **Investigating RNA polymerase II carboxyl-terminal domain (CTD) phosphorylation.** *Eur J Biochem* 2003, **270**:3859–3870.
10. Sims RJ III, Belotserkovskaya R, Reinberg D: **Elongation by RNA polymerase II: the short and long of it.** *Genes Dev* 2004, **18**:2437–2468.
11. Cargill M, Altshuler D, Ireland J, Sklar P, Ardlie K: **Characterization of single-nucleotide polymorphisms in coding regions of human genes.** *Nat Genet* 1999, **22**:231–238.
12. Collins FS, Brooks LD, Chakravarti A: **A DNA polymorphism discovery resource for research on human genetic variation.** *Genome Res* 1998, **8**:1229–1231.
13. Stenson PD, Ball EV, Mort M, Phillips AD, Shiel JA: **Human Gene Mutation Database (HGMD): 2003 update.** *Hum Mutat* 2003, **21**:577–581.
14. Yue P, Moult J: **Identification and analysis of deleterious human SNPs.** *J Mol Biol* 2006, **356**:1263–1274.
15. Stitziel NO, Binkowski TA, Tseng YY, Kasif S, Liang J: **topoSNP: a topographic database of non-synonymous single nucleotide polymorphisms with and without known disease association.** *Nucleic Acids Res* 2004, **32**:D520–D522.
16. Uzun A, Leslin CM, Abyzov A, Ilyin V: **Structure SNP (StSNP): a web server for mapping and modeling nsSNPs on protein structures with linkage to metabolic pathways.** *Nucleic Acids Res* 2007, **35**:W384–W392.
17. Kono H, Yuasa T, Nishiue S, Yura K: **coliSNP database server mapping nsSNPs on protein structures.** *Nucleic Acids Res* 2008, **36**:D409–D413.
18. Hinds DA, Stuve LL, Nilsen GB, Halperin E, Eskin E: **Whole-genome patterns of common DNA variation in three human populations.** *Science* 2005, **307**:1072–1079.
19. Reumers J, Maurer-Stroh S, Schymkowitz J, Rousseau F: **SNPeffect v2.0: a new step in investigating the molecular phenotypic effects of human non-synonymous SNPs.** *Bioinformatics* 2006, **22**:2183–2185.
20. Reumers J, Schymkowitz J, Ferkinghoff-Borg J, Stricher F, Serrano L: **SNPeffect: a database mapping molecular phenotypic effects of human non synonymous coding SNPs.** *Nucleic Acids Res* 2005, **33**:D527–D532.
21. Packer BR, Yeager M, Burdett L, Welch R, Beerman M: **SNP500Cancer: a public resource for sequence validation, assay development, and frequency analysis for genetic variation in candidate genes.** *Nucleic Acids Res* 2006, **34**:D617–D621.
22. Jegga AG, Gowrisankar S, Chen J, Aronow BJ: **PolyDoms: a whole genome database for the identification of non-synonymous coding SNPs with the potential to impact disease.** *Nucleic Acids Res* 2007, **35**:D700–D706.
23. Yang JO, Hwang S, Oh J, Bhak J, Sohn TK: **An integrated database-pipeline system for studying single nucleotide polymorphisms and diseases.** *BMC Bioinformatics* 2008, **12**:S19.
24. Thusberg J, Olatubosun A, Vihinen M: **Performance of mutation pathogenicity prediction methods on missense variants.** *Hum Mutat* 2011, **32**:358–68.
25. Capriotti E, Fariselli P, Casadio R: **A neural-network-based method for predicting protein stability changes upon single point mutations.** *Bioinformatics* 2004, **20**:63–68.
26. Offman MN, Krol M, Silman I, Sussman JL, Futerman AH: **Molecular basis of reduced glucosylceramidase activity in the most common Gaucher disease mutant, N370S.** *J Biol Chem* 2010, **285**:42105–42114.
27. Offman MN, Krol M, Rost B, Silman I, Sussman JL: **Comparison of a molecular dynamics model with the X-ray structure of the N370S acid-b -glucosidase mutant that causes Gaucher disease.** *Protein Engineering Design & Selection* 2011, **24**(10):773–5.
28. Kumar P, Henikoff S, Ng PC: **SIFT: predicting the effects of coding non-synonymous variants on protein function using the SIFT algorithm.** *Nat Protoc* 2009, **4**:1073–81.
29. Ramensky V, Bork P, Sunyaev S: **Human non-synonymous SNPs: server and survey.** *Nucleic Acids Res* 2002, **30**:3894–3900.
30. Mi H, Guo N, Kejariwal A, Thomas PD: **PANTHER version 6: protein sequence and function evolution data with expanded representation of biological pathways.** *Nucleic Acids Res* 2007, **35**:247–52.
31. Capriotti E, Fariselli P, Rossi I, Casadio R: **A three-state prediction of single point mutations on protein stability changes.** *BMC Bioinformatics* 2008, **9** (Suppl 2):S6.

32. Calabrese R, Capriotti E, Fariselli P, Martelli PL, Casadio R: **Functional annotations improve the predictive score of human disease-related mutations in proteins.** *Human Mutation* 2009, **30**:1237–1244.
33. Capriotti E, Calabrese R, Casadio R: **Predicting the insurgence of human genetic diseases associated to single point protein mutations with support vector machines and evolutionary information.** *Bioinformatics* 2006, **22**:2729–2734.
34. Bromberg Y, Rost B: **SNAP: predict effect of non-synonymous polymorphisms on function.** *Nucleic Acids Res* 2007, **35**:3823–3835.
35. Hess B, Kutzner C, van der Spoel D, Lindahl E: **GROMACS 4: algorithms for highly efficient, load-balanced, and scalable molecular simulation.** *J Chem Theory Comput* 2008, **4**:435–447.
36. Kutzner C, van der Spoel D, Fechner M, Lindahl E, Schmittn UW: **Speeding up parallel GROMACS on high-latency networks.** *J Comp Chem* 2007, **28**:2075–2084.
37. Morris GM, Goodsell DS, Halliday RS, Huey R, Hart WE: **Automated docking using a Lamarckian Genetic Algorithm and empirical binding free energy function.** *J Computational Chemistry* 1998, **19**:1639–1662.
38. Morris GM, Huey R, Lindstrom W, Sanner MF, Belew RK: **AutoDock4 and AutoDockTools4: automated docking with selective receptor flexibility.** *J. Comput Chem* 2009, **30**:2785–2791.
39. Schneidman D, Inbar Y, Nussinov R, Wolfson HJ: **PatchDock and SymmDock: servers for rigid and symmetric docking.** *Nucleic Acids Res* 2005, **33**:363–367.
40. Sherry ST, Ward MH, Kholodov M, Baker J, Phan L, Smigielski EM, Sirotkin K: **dbSNP: the NCBI database of genetic *variation*.** *Nucleic Acids Res* 2001, **29**:308–311.
41. Amos B, Rolf A: **The SWISS-PROT protein sequence data bank and its new supplement TREMBL.** *Nucleic Acids Res* 1996, **24**:21–25.
42. Amberger J, Bocchini CA, Scott AF, Hamosh A: **McKusick's Online Mendelian Inheritance in Man (OMIM).** *Nucleic Acids Res* 2009, **37**:793–796.
43. Lolli G, Lowe ED, Brown NR, Johnson LN: **The crystal structure of human CDK7 and its protein recognition properties.** *Structure* 2004, 12:2067–79.
44. Kouranov A, Xie L, de la Cruz J, Chen L, Westbrook J: **The RCSB PDB information portal for structural genomics.** *Nucleic Acids Res* 2006, **34**: D302–305.
45. Knox C, Law V, Jewison T, Liu P, Ly S: **DrugBank 3.0: a comprehensive resource for 'Omics' research on drugs.** *Nucleic Acids Res* 2011, **39** (Database issue):D1035–41.
46. Finn RD, Mistry J, Tate J, Coggill P, Heger A: **The Pfam protein families database.** *Nucleic Acids Res* 2010, **3**:D211–D222.
47. DeLano WL: *The PyMOL Molecular Graphics System, Version 0.99.* San Carlos, CA: DeLano Scientific; 2002.
48. Tina KG, Bhadra R, Srinivasan N: **PIC: Protein Interactions Calculator.** *Nucleic Acids Research* 2007, **35**:473–476.
49. Dougherty DA: **Cation–p interactions in chemistry and biology: a new view of benzene, Phe, Tyr, and Trp.** *Science* 1996, **271**:163–168.
50. Gallivan JP, Dougherty DA: **Cation–p interactions in structural biology.** *Proc Natl Acad Sci* 1999, **96**:9459–9464.
51. Carlson B, Pearlstein R, Naik R, Sedlacek H, Sausville E: **Inhibition of CDK2, CDK4 and CDK7 by flavopiridol and structural analogs.** *Proc Am Assoc Cancer Res* 1996, **37**:424.
52. Worland PJ, Kaur G, Stetler-Stevenson M, Sebers S, Sartor O: **Alteration of the phosphorylation state of p34cdc2 kinase by the flavone L86-8275 in breast carcinoma cells. Correlation with decreased HI kinase activity.** *Biochem Pharmacol* 1993, **46**:1831.
53. Fersht AR: **Basis of biological specificity.** *Trends Biochem Sci* 1984, **9**:145–147.
54. Shirley BA, Stanssens P, Hahn U, Pace CN: **Contribution of hydrogen bonding to the conformational stability of ribonuclease T1.** *Biochemistry* 1992, **31**:725–32.
55. Zhang Z, Teng S, Wang L, Schwartz CE, Alexov E: **Computational analysis of missense mutations causing Snyder-Robinson syndrome.** *Hum Mutat* 2010, **31**:1043–9.
56. Hunt DM, Saldanha JW, Brennan JF, Benjamin P, Strom M: **Single nucleotide polymorphisms that cause structural changes in the cyclic AMP receptor protein transcriptional regulator of the tuberculosis vaccine strain Mycobacterium bovis BCG alter global gene expression without attenuating growth.** *Infect Immun* 2008, **76**:2227–34.
57. Chen H, Jawahar S, Qian Y, Duong Q, Chan G: **Missense polymorphism in the human carboxypeptidase E gene alters enzymatic activity.** *Hum Mutat* 2001, **18**:120–131.
58. Chambers JC, Zhang WLY, Sehmi J, Wass MN, Zabaneh D: **Genome-wide association study identifies variants in TMPRSS6 associated with hemoglobin levels.** *Nat Genet* 2009, **41**:1170–2.
59. Giovannetti E, Pacetti P, Reni M, Leon LG, Mambrini A: **Association between DNA-repair polymorphisms and survival in pancreatic cancer patients treated with combination chemotherapy.** *Pharmacogenomics* 2011, **12**:1641–52.
60. Wang Z, Moult J: **SNPs, protein structure, and disease.** *Hum Mutat* 2001, **17**:263–270.
61. Rajasekaran R, George Priya Doss C, Sudandiradoss C, Ramanathan K, Purohit R, Sethumadhavan R: **Effect of deleterious nsSNP on the HER2 receptor based on stability and binding affinity with Herceptin: a computational approach.** *C R Biol* 2008, **331**(6):409–417.
62. Rajasekaran R, George Priya Doss C, Arun Prasad G, Sethumadhavan R: **In silico identification and analysis of drug resistant mutants of ABL tyrosine kinase based on detrimental missense mutations.** *Curr Signal Transd T* 2011, **6**:396–404.
63. George Priya Doss C, Sudandiradoss C, Rajasekaran R, Rituraj P, Ramanathan K, Rao S: **Identification and structural comparison of deleterious mutations in nsSNPs of ABL1 gene in chronic myeloid leukemia a bio-informatics study.** *J Biomed Inform* 2008, **41**(4):607–612.
64. Hsu YH, Johnson DA, Traugh JA: **Analysis of conformational changes during activation of protein kinase Pak2 by amide hydrogen/deuterium exchange.** *J Biol Chem* 2008, **283**:36397–36405.
65. Mohamed AJ, Yu L, Backesjo CM, Vargas L, Faryal R: **Bruton's tyrosine kinase (Btk): function, regulation, and transformation with special emphasis on the PH domain.** *Immunol Rev* 2009, **228**:58–73.
66. Muller CW, Schlauderer GJ, Reinstein J, Schulz GE: **Adenylate kinase motions during catalysis: an energetic counterweight balancing substrate binding.** *Structure* 1996, **4**:147–156.
67. Vihinen M: **Relationship of protein flexibility to thermostability.** *Protein Eng* 1987, **1**:477–480.
68. Ng PC, Henikoff S: **SIFT: predicting amino acid changes that affect protein function.** *Nucleic Acids Res* 2003, **13**:3812–3814.
69. Ng PC, Henikoff S: **SIFT: predicting deleterious amino acid changes that affect protein function.** *Genome Res* 2001, **11**:863–874.
70. Rosin CD, Scott Halliday R, Hart WE, Belew RK: **A comparison of global and local search methods in drug docking.** In *Proceedings of the 7th International Conference on Genetic Algorithms (ICGA-Y7).* Edited by Bäck T.; 1997:221–228.
71. Van Gunsteren WF, Billeter SR, Eising AA, Hunenberger PH, Kruger P: *Biomolecular Simulation: The GROMOS96 Manual and User Guide.* Zurich, Groningen: vdf Hochschulverlag AG an der ETH Zurich and BIOMOS b.v; 1996.
72. Jorgensen WL, Chandrasekhar J, Madura JD, Impey RW, Klein ML: **Comparison of simple potential functions for simulating liquid water.** *J Chem Phys* 1983, **79**:926.
73. Berendsen HJC, Postma JPM, van Gunsteren WF, DiNola A, Haak JR: **Molecular dynamics with coupling to an external bath.** *J Chem Phys* 1984, **81**:3684–3690.
74. Essmann U, Perera L, Berkowitz ML, Darden T, Lee H: **A smooth particle mesh Ewald method.** *J Chem Phys* 1995, **103**:8577–8593.
75. Case DA, Pearlman DA, Caldwell JW, Wang J, Ross WS: *AMBER Simulation Software Package.* San Francisco, CA: University of California; 2002.
76. Baker EN, Hubbard RE: **Hydrogen bonding in globular proteins.** *Prog Biophys Mol Biol* 1984, **44**:97–179.
77. Barlow DJ, Thornton JM: **Ion-pairs in proteins.** *J Mol Biol* 1983, **168**:867–885.

2

Whole exome sequencing identifies novel candidate genes that modify chronic obstructive pulmonary disease susceptibility

Shannon Bruse[1†], Michael Moreau[2†], Yana Bromberg[3†], Jun-Ho Jang[1,4], Nan Wang[2], Hongseok Ha[2], Maria Picchi[1], Yong Lin[1], Raymond J. Langley[1], Clifford Qualls[5], Julia Klensney-Tait[6], Joseph Zabner[6], Shuguang Leng[1], Jenny Mao[4], Steven A. Belinsky[1], Jinchuan Xing[2*] and Toru Nyunoya[1,4*]

Abstract

Background: Chronic obstructive pulmonary disease (COPD) is characterized by an irreversible airflow limitation in response to inhalation of noxious stimuli, such as cigarette smoke. However, only 15–20 % smokers manifest COPD, suggesting a role for genetic predisposition. Although genome-wide association studies have identified common genetic variants that are associated with susceptibility to COPD, effect sizes of the identified variants are modest, as is the total heritability accounted for by these variants. In this study, an extreme phenotype exome sequencing study was combined with in vitro modeling to identify COPD candidate genes.

Results: We performed whole exome sequencing of 62 highly susceptible smokers and 30 exceptionally resistant smokers to identify rare variants that may contribute to disease risk or resistance to COPD. This was a cross-sectional case-control study without therapeutic intervention or longitudinal follow-up information. We identified candidate genes based on rare variant analyses and evaluated exonic variants to pinpoint individual genes whose function was computationally established to be significantly different between susceptible and resistant smokers. Top scoring candidate genes from these analyses were further filtered by requiring that each gene be expressed in human bronchial epithelial cells (HBECs). A total of 81 candidate genes were thus selected for in vitro functional testing in cigarette smoke extract (CSE)-exposed HBECs. Using small interfering RNA (siRNA)-mediated gene silencing experiments, we showed that silencing of several candidate genes augmented CSE-induced cytotoxicity in vitro.

Conclusions: Our integrative analysis through both genetic and functional approaches identified two candidate genes (*TACC2* and *MYO1E*) that augment cigarette smoke (CS)-induced cytotoxicity and, potentially, COPD susceptibility.

Keywords: Whole exome sequencing, Cigarette smoke, COPD, Susceptible smokers, Resistant smokers

* Correspondence: xing@biology.rutgers.edu; nyunoya.toru@gmail.com
†Equal contributors
[2]Department of Genetics, Rutgers, The State University of New Jersey, 145 Bevier Road, Piscataway, NJ 08854, USA
[1]Lovelace Respiratory Research Institute, 2425 Ridgecrest Drive SE, Albuquerque, NM 87108, USA
Full list of author information is available at the end of the article

Background

Chronic obstructive pulmonary disease (COPD), characterized by a permanent airflow limitation, is a growing public health threat and a leading cause of disability and mortality worldwide [1, 2]. Although cigarette smoking is a major risk factor [3], evidence suggests that genetic factors modulate smoking-induced COPD [4]. For example, alpha1 antitrypsin deficiency (A1ATD), caused by a homozygous missense mutation of *SERPINA1* (PiZ; Glu342Lys) or compound heterozygosity of the PiZ and PiS variants, contributes to lung function decline among smokers [5, 6]. However, A1ATD accounts for only 2–3 % of COPD cases [7]. Genome-wide association study (GWAS) and fine-mapping studies have revealed common (>5 % minor allele frequency) single nucleotide polymorphisms (SNPs) that are associated with COPD in a number of candidate genes, including *CHRNA3/5*, *FAM13A*, and *HHIP* [8–11]. However, the effect sizes for these associated common variants are small relative to those of *SERPINA1*-mediated A1ATD and COPD. Sequencing studies of COPD hold the promise of identifying rare variants with large effects on disease susceptibility.

Recent developments in DNA sequencing technologies have dramatically reduced the cost of acquiring genetic data. In particular, whole exome sequencing (WES) has become a valuable tool for studying both rare and common diseases [12]. WES allows interrogation of the entire coding portion of the human genome at a fraction of the cost of whole-genome sequencing. In a recent study, WES led to the identification of a non-synonymous SNP in the gene *CCDC38* that was associated with resistance to cigarette smoke (CS)-related airflow obstruction, assessed by sequencing heavy smokers with normal lung function [13]. Therefore, to identify potential causal variants for both COPD and CS resistance, we conducted WES of 62 smokers with very advanced COPD and 30 resistant smokers. We sought to sample from the extremes of the phenotypic distribution, under the assumption that this would enrich sampling of rare causal variants with large effect sizes [14]. The COPD group was thus selected to contain the youngest, lightest smoking, and most severely diseased individuals from available cohorts; the resistant group was selected to contain the oldest, heaviest smoking, and healthiest individuals (i.e., no comorbidities) with normal lung function. The primary analytical approach was aimed at identifying rare variants (and the associated genes) contributing to these two phenotypes. Additionally, we employed approaches focusing on the collective impact of multiple weakly deleterious variants (both common and rare). Candidate genes were further filtered using gene expression analysis to retain a total of 81 genes, which were functionally tested in CS-exposed, immortalized human bronchial epithelial cells (HBECs) using a gene knockdown approach (Fig. 1). This systematic multi-layered approach may help remove false-positives and prioritize true COPD (or CS-induced damage resistance) genes. HBECs were chosen as an in vitro screening model because airway epithelial cells are the primary target of CS exposure. CS exposure induces inflammation, DNA damage, and autophagy that causes lung epithelial cells to undergo various cell fates, including cell death, cellular senescence, and/or transformation [15]. Although there is no sine qua non cellular phenotype of lung epithelial cells in COPD, the lungs of patients with COPD exhibit a significant

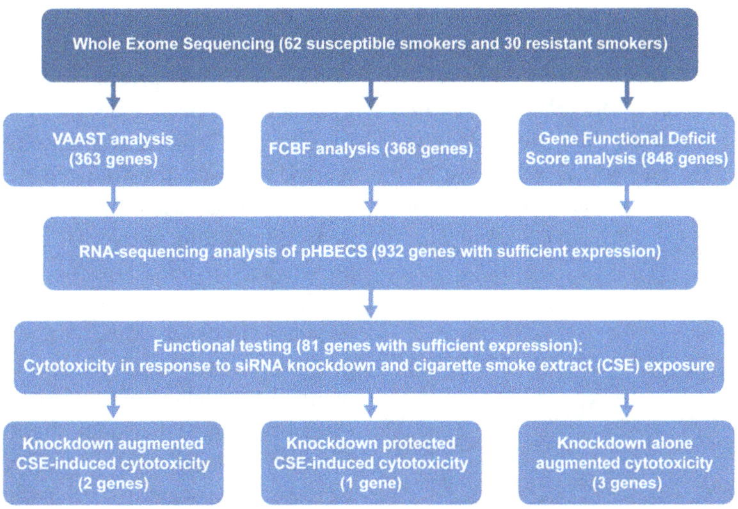

Fig. 1 Identifying candidate COPD genes through genomic and functional approaches. WES in 62 highly susceptible smokers and 30 exceptionally resistant smokers were conducted to identify exonic variants that may contribute to disease risk or resistance to CS. Top scoring candidate genes from the rare variant and gene set analyses were further filtered by requiring that the gene be expressed in primary HBECs, and 81 candidate genes were selected for in vitro functional testing in CSE-exposed HBECs. Using siRNA-mediated gene silencing experiments, we identified candidate genes whose knockdown augmented CSE-induced cytotoxicity, protected CSE-induced cytotoxicity, or alone-reduced cell viability

increase in apoptotic cells [16, 17]. We thus chose in vitro cell viability as an endpoint.

Results

VAAST analysis

We performed WES on 62 susceptible smokers with COPD and 30 resistant smokers with normal spirometries in the absence of significant comorbidities (Table 1). All individuals were of self-reported non-Hispanic white ethnicity. We used the Variant Annotation, Analysis, and Search Tool (VAAST) program to prioritize candidate genes (Additional file 1) [18, 19]. Given allele frequency and amino acid substitution severity, VAAST prioritizes genes based on their disease-causing likelihood. We performed VAAST analysis under different settings, including different modes of inheritance, penetrance models, and individual sets. From the VAAST output, several hundred genes had a p value <0.05 under each analysis set-up (Table 2). We did not use a more stringent cutoff or a false discovery ratio (FDR) correction for the p value because these candidate genes would be further prioritized in the downstream analyses and validated by an in vitro assay.

Some of the top-ranking candidate genes from the VAAST analysis contain rare deleterious mutations in multiple susceptible individuals. For example, the gene *TACC2* was ranked 6th with complete penetrance and 22nd with 5 % prevalence in our VAAST analysis of the susceptible group under a recessive inheritance model. One of the susceptible individuals carries a nonsense stop-gain mutation (chr10: 123842508) in *TACC2* (a glutamine mutated to a stop codon). Six additional susceptible individuals carry novel mutations at different positions of this gene, making *TACC2* a promising candidate. Using Sanger sequencing, we validated all nonsense and non-synonymous *TACC2* variants where the DNA samples were available, for a total of seven variants in eight COPD samples, including four novel variants (Additional file 7: Table S1). The 100 % validation rate for the seven variants suggests that the variant calls are of high quality.

Table 1 Demographic data of resistant smokers and COPD subjects

	Resistant smokers	COPD	*p* value
Gender (M/F)	7/23	34/28	<0.01*
Age	65.7 ± 5.3	57.7 ± 7.0	<0.0001#
Pack year	62.2 ± 20.6	45.7 ± 19.0	0.0003#
FEV1/FVC%	76.9 ± 5.9	29.5 ± 9.5	<0.0001#
% predicted FEV	103.9 ± 11.4	25.8 ± 11.4	<0.0001#
% predicted DLCO	N/A	32.8 ± 16.5	N/A

*Fisher exact test
#*t* test

Gene functional deficit score-based TTest and FCBF analysis

To determine the collective impact of all variants, including common and weakly deleterious variants, a gene functional deficit score was calculated for all genes. Of our set of 15,594 variant-affected genes in at least one individual in either the susceptible smokers or the resistant smokers, 848 genes attained p values of <0.05 (*TTest set*); 194 genes attained a p value <0.01 for the t test comparing resistant individuals to susceptible individuals (TTest tab in Additional file 2). However, none of the p values were significant after Bonferroni correction for multiple comparisons. To account for lack of variability and small numbers of samples in the score distributions, we further reduced the reference gene set by keeping only those affected in at least 10 individuals in both resistant and susceptible cohorts (8877 genes affected in at least 20 individuals). This reduction did not produce any genes with significant corrected p values either. This result suggests that either more individuals are necessary for improved resolution of the study or that the resistant and susceptible phenotypes are the result of additive or epistatic interaction of altered function of multiple genes.

Next, we performed fast correlation-based filter solution (FCBF) attribute analysis and the FCBF selection resulted in a set of 368 relevant genes (*FCBF set*) (of these, 59 attained an [average merit – standard error] > 0; FCBF tab in Additional file 2). Of the entire set, 62 % (229) and 38 % (139) of the genes had higher average scores in susceptible and resistant individuals respectively (FCBF tab in Additional file 2). The large proportion of genes showing higher scores in resistant individuals suggests that COPD resistance may be a separate variation-conferred phenotype, as opposed to solely reflecting the absence of COPD-predisposition variants.

To determine if the *FCBF* genes are often found together in biological interaction and regulation networks, we used the induced network modules analysis in ConsensusPathDB [20], an integrative database for gene interaction networks. The vast majority of *FCBF* genes (336 of 368, 91.3 %) were mappable in ConsensusPathDB (FCBF tab in Additional file 2). The induced network modules analysis connects *FCBF* genes/proteins (seed network nodes) into modules using knowledge of different types of interactions (network edges) between these seeds and other intermediate network nodes, such as other genes, chemical ligands, and proteins. A single network is composed of one module or several disconnected modules. In our analysis, we considered binary interactions only, including protein-protein interactions, genetic interactions, biochemical reactions, and gene regulatory interactions. The first step in the analysis induced

Table 2 The number of candidate genes under different VAAST analysis categories

VAAST analysis category			Analysis	Candidate genes
Inheritance	Parameters	Samples		
Dominant	5 % background allele frequency	Resistant vs. susceptible background	1	534
		Susceptible vs. resistant background	2	757
	Complete penetrance	Resistant vs. susceptible background	3	343
		Susceptible vs. resistant background	4	366
Recessive	5 % background allele frequency	Resistant vs. susceptible background	5	532
		Susceptible vs. resistant background	6	759
	Complete penetrance	Resistant vs. susceptible background	7	195
		Susceptible vs. resistant background	8	269

three modules containing 160 (44 % of our set of 368) interacting seeds (Z-score of intermediate nodes ≥ 20; network1 FCBF tab in Additional file 2). The largest module contained 96 % (154) of the seeds and 50 intermediate nodes; the remaining two modules contained a total of six seeds and two intermediates. Removing low confidence interactions (IntScore <0.5) [21] maintained 138 (38 %) seeds (network2 FCBF tab in Additional file 2; one module of 37 intermediate nodes and 126 seeds, plus five smaller networks encompassing the remaining 12 seeds and 2 intermediates). Building a network of solely high-confidence (IntScore >0.95) protein-protein and biochemical interactions forms a module of 111 (30 %) connected seeds (network 3 FCBF tab in Additional file 2; 24 intermediate nodes, 96 seeds in a single network, plus six smaller modules with 15 seeds and 3 intermediates). Finally, leaving only high-confidence protein-protein interactions retains 43 (12 %) interacting seeds (network 4 FCBF tab in Additional file 2; 12 intermediates and 20 seeds in a single module, plus nine smaller modules of 23 seeds and 7 intermediates, total). Eliminating all intermediates resulted in a high-confidence module of 8-seed proteins (network 5 FCBF tab in Additional file 2). Mutagenesis of interacting proteins without a link to a common phenotype is an unlikely event. Thus, these results suggest that, in the absence of otherwise unifying features of tested individuals, a disruption of an as yet undescribed molecular pathway may be involved in generating the COPD phenotype.

Candidate gene filtering using RNA-Seq in vitro analysis

Both bronchial and alveolar epithelial cells are the primary target of CS exposure. To determine the abundance of gene expression in primary human bronchial epithelial cells (pHBECs), we conducted RNA-Seq using pHBEC cultures isolated from five nonsmoking donors. We found that 96.2 % of all genes are expressed within a tenfold difference among the five donors. Therefore, we prioritized the candidate genes expressing at an average fragments per kilobase of transcript per million mapped reads (FPKM) of 1.0 or more (50.4 % of 32,457 transcripts) in pHBECs, assuming that transcripts with an FPKM value below 1.0 have low expression [22] (Additional file 3).

Three sets of candidate genes were screened for their expression levels: genes from the (1) VAAST analysis (Additional file 1) and the (2) gene functional deficit score analysis using the *t* test (TTest tab in Additional file 2) and (3) FCBF (FCBF tab in Additional file 2). For the VAAST analysis, we first filtered candidate genes using the following criteria: (1) present in both dominant and recessive analyses under the 5 % background allele frequency setting (i.e., present in analyses 1 and 5 or in analyses 2 and 6 in Table 2) and (2) having a gene ranking <300 in at least one of the analyses. These selection criteria resulted in 363 candidate genes from VAAST analysis. For the TTest and FCBF analyses, we included all candidate genes. Among the three sets of candidate genes, 363, 498, and 199 genes have an average FPKM value larger than 1.0, including 932 unique genes. To generate a set of candidate genes for in vitro functional analysis, we further examined the biological relevance of these candidate genes based on current knowledge of COPD. In the end, we selected a total of 81 of the highest scoring candidate genes for siRNA screening from (1) VAAST analysis ($n = 42$) and (2) FCBF and/or *t* test analysis ($n = 45$). Six genes overlapped between the VAAST and FCBF or *t* test analyses (Additional file 4). All variants with annotations within the 81 candidate genes are listed in Additional file 5.

To confirm these genes are relevant to COPD, we performed Gene Ontology (GO) term and pathway enrichment analysis on the 81 candidate genes. Among the 81 genes, a number of GO terms showed significant enrichment (Additional file 6). The top genes are associated with cell death, which supports our hypothesis that the COPD susceptible genes are associated with cell viability and response to cigarette smoke extract (CSE) toxicity. In addition, using the induced network modules analysis in ConsensusPathDB, 48 out of the 81 genes are connected in a network when allowing intermediate nodes, supporting their involvement in a common pathway (data not shown). Note that since the candidate genes were selected from the computationally identified gene set using expert knowledge, this enrichment in relevant GO terms does not indicate discovery but rather confirms our findings.

siRNA screening in vitro analysis

To determine whether the selected candidate genes (n = 81) are important for cell survival against cigarette smoke exposure, we examined the effects of 2 % CSE on cell viability in cultured HBECs (HBEC2) in which the individual gene expression had been suppressed using siRNA (Additional file 4). According to the gene silencing-CSE interaction analysis, we identified two candidate genes whose gene silencing augmented CSE-induced cytotoxicity (*TACC2* and *MYO1E*) and one candidate gene whose gene silencing protected against CSE-induced cytotoxicity (*SLC7A1*) (Tables 3 and 4). Among the candidate genes, suppression of *TACC2* expression induced the most pronounced effect of CSE on cell viability (Table 3 and Additional file 4). We confirmed the effects of siRNA transfection on CSE-induced cytotoxicity (Fig. 2a) and gene silencing by RT-PCR (Fig. 2b). We also performed flow cytometric analysis with Annexin V and propidum iodide staining to characterize the CSE-induced cytotoxicity in HBEC2 cells transfected with TACC2 siRNA. The percentage of early and late apoptotic cells in response to CSE was significantly increased by *TACC2* knockdown (52.2 ± 11.6 %) as compared to scrambled controls (8.1 ± 1.7 %) (Fig. 2c). Furthermore, *TACC2* knockdown activated caspase-3 in CSE-exposed HBEC2 (Fig. 2d). These data suggest that the effects of CSE on cytotoxicity of *TACC2* knockdown cells may largely be due to apoptosis.

We also identified three candidate genes whose suppression alone (in the absence of CSE) significantly reduced cell viability (*PLCH2*, *KIAA1919*, and *MRPS34*) (Table 5). These results demonstrate that WES can identify candidate genes whose gene silencing significantly alters spontaneous or CSE-induced cytotoxicity.

Discussion

We conducted a WES study using an extreme phenotype design (susceptible vs. resistant smokers) to prioritize candidate genes that potentially modify COPD susceptibility. Among the prioritized candidate genes expressed in pHBECs, we identified two genes (*TACC2* and *MYO1E*) that exhibited a strong interaction of gene (siRNA transfection) with environment (CS exposure) on cytotoxicity in vitro.

Table 3 Top five candidate genes whose gene silencing augmented CSE-induced cytotoxicity

	Gene ID	Interaction (AD/CB)	Z-score	*p* value	BH FDR
1	*TACC2**	0.196	−6.090	0.0000	0.0010
2	*MYO1E**	0.614	−2.924	0.0017	0.0020
3	*NPLOC4*	0.727	−2.070	0.0192	0.0030
4	*FUT2*	0.763	−1.796	0.0363	0.0040
5	*PEX26*	0.797	−1.537	0.0622	0.0050

*p value < BH FDR

Table 4 Top five candidate genes whose gene silencing protected against CSE-induced cytotoxicity

	Gene ID	Interaction (AD/CB)	Z-score	*p* value	BH FDR
1	*SLC7A1**	1.334	2.959	0.0015	0.0016
2	*INO80*	1.213	1.882	0.0300	0.0032
3	*AXIN1*	1.190	1.878	0.0302	0.0048
4	*EREG*	1.189	1.673	0.0471	0.0065
5	*MAP3K10*	1.177	1.519	0.0644	0.0081

*p value < BH FDR

Missing heritability is a well-recognized phenomenon in GWAS of complex phenotypes [23, 24]. Potential explanations for missing heritability include epistatic interactions (GxG) and gene-environment interactions (GxE), as well as experimental limitations, e.g. discarding informative variants due to stringent GWAS significance thresholds and insufficient linkage disequilibrium between typed and "causal" variants. Rare disease-causing variants may also be responsible for a substantial proportion of the missing heritability [25–27].

To identify candidate genes from the WES dataset, we used two complementary approaches that focused on either rare variants with presumably large effect sizes (VAAST analyses) or on multiple weakly deleterious variants causing molecular function changes (gene functional deficit score-based TTest and FCBF analyses). We used these diverse approaches because the genetic model for our phenotypes of interest is largely unknown. For example, if COPD is caused by the accumulation of multiple mutations that weakly alter the function of several genes in a certain molecular pathway, the VAAST approach is unlikely to identify the underlying causes and the gene deficit analysis is more appropriate. Unfortunately, the hypothesis of multiple interacting mutations giving rise to a phenotype is very challenging to test with both in vitro and in vivo models, as simultaneous evaluation of multiple mutations is necessary. Nevertheless, we expect that mapping these candidate genes will identify crucial molecular pathways responsible for the development of the resistance and COPD phenotypes.

The current state of the art, as described in the scientific literature, confirms some genes identified by our analysis: of the complete set of *FCBF* genes, 32 were found by a computational parsing of articles identified by the MeSH term "pulmonary disease, chronic obstructive"—the explicit term referring to COPD; another 20 genes were

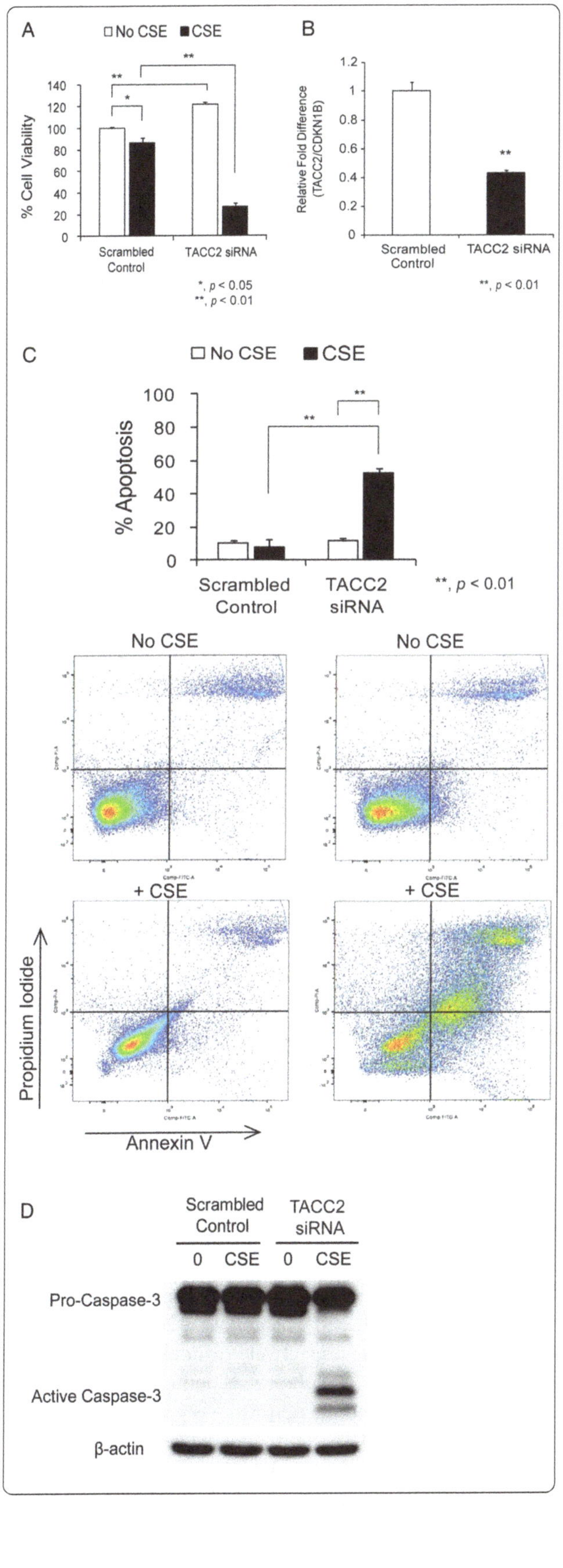

Fig. 2 Effects of TACC2 siRNA transfection on CSE-induced cytotoxicity and *TACC2* mRNA levels. **a** Forty-eight hours after transfection with either siRNA targeting *TACC2* (TACC2 siRNA) or the scrambled siRNA (scrambled control) as control, HBEC2 cells were incubated in the absence (no CSE) or presence of 2 % CSE (CSE) for 24 h. Cell viability was determined using the MTT assay at 24 h. Data are expressed as mean ± SEM for three independent experiments with triplicated samples (*$p < 0.05$; **$p < 0.01$). **b** Steady-state levels of *TACC2* mRNA were measured by RT-PCR and presented as relative fold difference compared with *CDKN1B* in HBEC2 cells after 48 h with either TACC2 siRNA or scrambled control. Data are expressed as mean ± SEM from two independent experiments with triplicated samples (**$p < 0.01$). **c** HBEC2 cells were treated as in **a**. Cell death was analyzed by Annexin V and propidium iodide (PI) staining 24 h after 2 % CSE exposure. The percentage of Annexin V positive cells/total cell number was expressed as percentage apoptosis. Data are expressed as mean ± SEM for three independent experiments (**$p < 0.01$). Representative flow cytometry data are shown. **d** HBEC2 cells were treated as in **a**. Immunoblot analysis of active caspase-3 was performed 24 h after 2 % CSE exposure. Immunoblotting data are representative of three experiments

identified by searching for COPD related keywords (chronic obstructive lung disease, chronic obstructive airway disease, chronic airflow limitation, chronic obstructive respiratory disease, chronic bronchitis, COAD, COPD) (FCBF tab in Additional file 2). These results suggest that the selected gene set contains a substantial number of known COPD genes and is possibly enriched in yet undescribed COPD and/or COPD-resistance associated genes. With this knowledge, we will be able to examine the role of gene-gene interaction and weak functionally deleterious mutations in the future as larger sample sizes for genetic studies are acquired and more sophisticated genome-editing technologies are employed in functional studies.

A total of 81 candidate COPD genes selected from our analysis were functionally tested using an in vitro gene knockdown model combined with exposure to CSE. We chose the 3-(4,5-dimethythiazol-2-yl)-2,5-diphenyl tetrazolium bromide (MTT) assay as a measure of cell viability because it is more sensitive in detecting CSE effects (such as cell growth inhibition) than the trypan blue exclusion or LDH release assay (unpublished data).

Table 5 Top five candidate genes whose gene silencing alone reduced cell viability

	Gene ID	Ratio (C/A)	Z-score	*p* value	BH FDR
1	*PLCH2**	0.490	−3.359	0.00039	0.00109
2	*KIAA1919**	0.492	−3.342	0.00041	0.00217
3	*MRPS34**	0.528	−3.030	0.00122	0.00326
4	*EPPK1*	0.607	−2.365	0.00901	0.00435
5	*NAT10*	0.612	−2.281	0.01127	0.00544

*p value < BH FDR

Notably, in addition to cigarette smoking, premature growth of the lung is also thought to be a risk factor for COPD [28]. Therefore, when analyzing the siRNA screening data, we considered two experimental states: (1) gene-environment interaction (both gene silencing and CSE exposure) and (2) siRNA-induced cytotoxicity (gene silencing alone). From the interaction analysis, we identified two candidate genes (*TACC2* and *MYO1E*) whose knockdown increased CSE-induced cytotoxicity. Both genes were identified in susceptible smokers. *TACC2* gene knockdown most dramatically augmented CSE-induced cytotoxicity. *TACC2*, a member of the transforming acidic coiled-coil-containing protein family, is involved in the regulation of centrosome and microtubule dynamics during cell cycling [29] but is not necessary for cell growth and mouse fertility and development [30]. The *TACC2* functions suggest that it may modulate both resistance (through a gain-of-function mutation) and susceptibility (through a loss-of-function mutation) to smoking-induced COPD. In fact, a recent human study identified a common *TACC2* variant as significantly associated with the resistant smoker phenotype [13]. *MYO1E* encodes a non-muscle class I myosin involved in cytoskeleton dynamics; homozygous nonsynonymous SNPs in *MYO1E* are associated with familial focal segmental glomerulosclerosis [31]. A recent in vitro study demonstrated that genetic deletion of *MYO1E* augments LPS-induced chemokine (C-C motif) ligand 2 secretion in inflammatory cells [32]. The effects of CS on the lungs of *Myo1e*$^{-/-}$mice are unknown, but CS exposure may enhance lung inflammation and possibly COPD in *Myo1e*$^{-/-}$mice.

We also identified three candidate genes (*PLCH2*, *KIAA1919*, and *MRPS34*), which significantly augmented cytotoxicity by siRNA knockdown alone. PLCH2 is a member of the phospholipase C superfamily of enzymes that regulate phosphoinositide turnover. *Plch2*$^{-/-}$ mice exhibit no obvious phenotype. However, effects on lung development have not been investigated [33]. *KIAA1919* and *MRPS34* encode a sodium-dependent glucose transporter [34] and mitochondrial ribosomal protein S34 [35], respectively. Future studies will be needed to determine whether these genes are involved in the development of respiratory systems using in vivo animal models.

Our in vitro siRNA screening further revealed a candidate gene whose knockdown protected against CSE-induced cytotoxicity (*SLC7A1*). *SLC7A1* was initially selected from the VAAST analysis of susceptible smokers. Since most rare disease-causing variants represent loss-of-function mutations, and *SLC7A1* was identified in susceptible smokers, we might expect knockdown of this gene to augment CSE-induced cytotoxicity. However, the opposite effect was observed in the siRNA screening. It may be that the specific mutations identified by the VAAST analysis were rare gain-of-function mutations. Further experiments examining the precise mutations are needed.

There are several limitations of this siRNA screening approach, including inability to model gain-of-function mutations or to assess epistasis, possible off-target effects, and considerable variability among the individual siRNA effects on gene silencing (e.g., 2 vs. 20 % of the scrambled control). Further studies using primary murine tracheobronchial cells are required to validate our in vitro findings. There are also other limitations when using a simplistic siRNA/cell viability assay to assess candidate genes. For example, inflammatory and endothelial cells were not evaluated, though both are important in disease progression [36, 37]. A number of genes are known to have high expression in inflammatory cells, such as alveolar macrophages and pulmonary endothelial cells, but not in airway epithelial cells [38, 39]. Based on our requirement that genes be expressed in HBECs, we may have excluded genes that may be important in COPD through mechanisms involving inflammatory cells. In future studies, we will include candidate genes expressed in inflammatory cells, such as alveolar macrophages or lymphocytes. Also note that knocking down genes is a rather blunt method. Recapitulating precise mutations of interest using genome-editing technologies, followed by in vitro functional studies, is likely much more precise and a future goal of ours.

Conclusions

In this study, we identified candidate COPD genes that augment spontaneous or CSE-induced cytotoxicity through genomic, transcriptomic analyses, and in vitro siRNA screening. In the future, we will determine the effects of candidate gene deletion on the development of spontaneous and/or CS-induced COPD in animal models.

Methods

Study subjects

Samples were collected from a cross-sectional case-control study design, with no therapeutic intervention or longitudinal follow-up information. All research involving human subjects was approved by the authors' institutional review board (University of Iowa, New Mexico VA Healthcare System (NMVAHCS), Lovelace Respiratory Research Institute (LRRI) protocols: #200612711, e#11-631, and #20031684, respectively). Informed consent was obtained from all study subjects. Blood or whole lung tissue samples were obtained from very advanced COPD patients ($n = 62$) and resistant smokers ($n = 30$) enrolled in the University of Iowa, NMVAHCS, LRRI, or the Lung Tissue Research Consortium (LTRC). Advanced stage COPD was defined as stages 3 or 4 COPD using the GOLD executive summary 2007 (forced expiratory volume in 1 s [FEV1]/forced vital capacity [FVC] <0.70 and percentage of predicted FEV1 <50) [40], except for three smokers who have more than 50 % predicted FEV1 but very low percentage of predicted diffusing capacity for carbon monoxide (DLCO) (<30). We define exceptionally resistant smokers as individuals who smoke at least 35 pack years, are >60 years of age, have normal spirometries, and absence of significant comorbidities, such as non-dermal malignancy or coronary artery disease. All study subjects were of self-reported non-Hispanic white ethnicity. We have excluded subjects with severe lung inflammation due to chronic infection, immunosuppression due to HIV infection, and known genetic diseases involving the lung (such as cystic fibrosis or A1ATD).

DNA extraction

DNA extraction was performed using the QIAamp Gentra Puregene Blood and Tissue kit according to manufacturer's instructions (Qiagen, Hilden, Germany). For isolation of DNA from lung tissues, cell lysis solution was added to 50 mg of minced lung tissues and then followed by addition of proteinase K and incubated for 3 h at 55 °C. For isolation of DNA from blood samples, 10 mL of whole blood was resuspended in Puregene Red Blood Cell Lysis Solution to lyse red blood cells; this was followed by incubation at room temperature for 5 min. Cell lysis solution was added to the white blood cell pellet and vortexed vigorously for 10 s. To the completely digested tissue and blood lysates, we added RNase A solution and protein precipitation solution according to manufacturer's instructions. The supernatant from the previous step was added to a tube containing isopropanol and mixed by inverting gently 50 times; 70 % ethanol was added and inverted several times to wash the DNA. After final centrifugation and removal of supernatant, the tubes were air-dried to remove residual ethanol for 5–10 min. DNA hydration solution was added and vortexed for 5 s and followed by incubation at 65 °C for 1 h to dissolve the DNA. Concentration and purity of the collected DNA were determined using the NanoDrop ND-1000 spectrophotometer.

Whole exome data acquisition, processing, and variant discovery

The whole exome sequencing library was constructed using the SureSelectXT Human All Exon V4+UTRs kit (Agilent, Santa Clara, CA) and sequenced using an Illumina HiSeq 2000 system. The sequencing was performed to achieve high-quality variant identification: overall ~60 million reads were sequenced for each individual, and an average read depth of at least 50× was achieved for all samples.

The variant discovery followed the Genome Analysis Toolkit (GATK) Best Practices recommendations [41]. Briefly, raw sequences were aligned to the human reference genome (version hg 19) using Burrows-Wheeler Aligner (BWA) [42]. The resulting raw alignments (in binary sequence alignment/map format) were processed by Picard (http://broadinstitute.github.io/picard) to remove PCR duplicates, followed by local realignment around indels and base quality score recalibration using GATK IndelRealigner and BaseRecalibrator respectively [43]. Variant discovery was performed using GATK UnifiedGenotyper using the joint calling function on all samples. Raw variants quality scored were recalibrated using GATK VariantRecalibrator and ApplyRecalibration was used to generate a Variant Quality Score logarithm of odds (VQSLOD). Finally, variants were filtered using the following criteria: (1) VQSLOD ≥2 and (2) read depth (DP) ≥6 in at least 80 % of samples. In addition, any individual genotype with DP < 6 or GQ < 30 was considered a no-call in the final dataset. The final dataset along with annotation by ANNOVAR [44] are available as a supplemental file on our website under Published Data (https://xinglab.genetics.rutgers.edu/PublishedData/COPD/AllVariants.zip).

VAAST Analysis

The variant data were analyzed using the Variant Annotation, Analysis, and Search Tool (VAAST 2.0) package [18]. The variants were annotated for their functional impact using Variant Annotation Tool (VAT) in the VAAST package. Then annotated variants in all susceptible individuals and all resistant individuals were combined into two condenser files using Variant Selection Tool (VST). The susceptible and resistant condenser files were analyzed using VAAST to prioritize candidate genes under both dominant and recessive modes of inheritance, allowing locus heterogeneity. We performed the analysis either assuming complete penetrance of a variant or allowing 5 % prevalence of a variant in the control population. Complete penetrance means that no individual in the background population can carry a certain disease-causing variant, while allowing 5 % prevalence means we estimate the expected disease allele frequency within the background population to be 5 % or lower.

The background population used for the analysis of COPD samples were composed of variants from the 10 Gen dataset [45], 1057 genomes from the 1000 Genome Project Phase I [46], 184 Danish exomes [47], 54 whole genomes from the Complete Genomics Diversity Panel [48], and resistant smoker sample exomes. The background files used for resistant smoker analysis are composed of all the genomes or exomes mentioned above (except the resistant smoker sample exomes) and COPD susceptible sample exomes. We did not include 6500

NHLBI exomes in the COPD sample background because COPD subjects are present in this data set [12].

VAAST candidate-gene prioritization was performed and variants in each gene were scored as a group. A rank list of candidate genes was generated based on their disease-causing probability. The significance level was assessed with individual permutation tests. VAAST analysis parameters used: "-m lrt -lh y –significance 1e-4 -d 1e5 [-r 0.05|-pnt c] -iht [d|r]".

Gene functional deficit score-based TTest and FCBF analysis

For the 92 individuals in our cohort, we extracted the set of ANNOVAR annotated exonic variants [44]. We mapped the RefSeq messenger RNA (mRNA) IDs identified by ANNOVAR to UniProt proteins. If a variant mapped to different RefSeq IDs, all affected UniProt IDs were included into the affected set. However, for each variant only the primary isoform of the protein was considered. For each non-synonymous variant, we computed the raw SNAP (screening for non-acceptable polymorphisms) score (range −100 to 100, any score ≥0 is classified as neutral/no functional change, and non-neutral otherwise) [49]. For each gene in every individual, we computed a *gene functional deficit score* as a sum over all gene-specific variant scores. An individual variant score was computed for each: (1) InDel or Stop loss/gain, score = 1, (2) SNAP identified non-neutral non-synonymous variant, score = SNAP score/100, (3) SNAP identified neutral variant score = 0.055, and (4) synonymous variant, score = 0.05. Further, individual variant scores of heterozygous variants were multiplied by a factor of 0.25 to approximately modulate the effects of heterozygosity. Gene scores computed in this fashion are 0 only for genes that have no variants at all. However, further comparison between gene scores for different genes is not possible, as the score is highly dependent on gene length and overall tolerance for variability (e.g., longer genes with more variable regions will tend to score higher while remaining relatively functional biochemically). Note that higher gene scores of statistically significant genes in resistant individuals in our study may indicate genes relevant for a resistance phenotype. We compared the distribution of gene scores for resistant smokers vs. the COPD-affected individuals using the *t* test metric. We also applied to the entire set of genes in leave one out cross-validation (92-fold) the fast correlation-based filter solution (FCBF) feature selection algorithm [50]. The symmetrical uncertainty attribute set evaluation in leave one out cross-validation (WEKA implementation [51], parameters = weka.attributeSelection.SymmetricalUncertAttrib uteSetEval -s "weka.attributeSelection.FCBFSearch -D false -T −1.7976931348623157E308 -N −1") was also combined to select a set of genes responsible for the observed phenotype as a group, as opposed to a set of genes individually contributing to the phenotype. The feature set evaluation algorithm [50] evaluates the worth of a set of attributes by measuring its symmetrical uncertainty with respect to another set of attributes. Performance of a particular attribute set was measured here by training J48 tree classifier [52] (parameters: weka.classifiers.meta.AttributeSelectedClassifier -E "weka.attributeSelection.SymmetricalUncertAttributeSetEval " -S "weka.attributeSelection.FCBFSearch -D false -T −1.7976931348623157E308 -N −1" -W weka.classifiers.trees.J48 – -C 0.25 -M 2).

Sanger sequencing validation of variant calls

Sanger sequencing was used to validate the seven variants in *TACC2* genes found from the WES. Primer pairs used to amplify PCR products harboring individual variants were designed using Primer3 (http://biotools.umass med.edu/bioapps/primer3_www.cgi). The detail information of primers used for sequencing, including the sequences, PCR product sizes, and annealing temperatures, are listed (Additional file 7: Table S1). PCR with those primer pairs was performed in 25-μl reaction with OneTaq DNA polymerase (New England Biolabs, Ipswich, MA). Half of the amount of each reaction was used for gel electrophoresis to confirm the presence of a single amplicon with the expected fragment size. After the confirmation, the remaining PCR product was purified by using Illustra ExoProStar according to the manufacturer's protocol (GE Healthcare Life Sciences, Buckinghamshire, UK). The purified PCR products were then used for Sanger sequencing (GenScript, Piscataway, NJ).

Cell culture

Primary HBECs (pHBECs) from human lung tissues were isolated from five nonsmoking donors and were cultured in conventional systems under a protocol approved by the LRRI Institutional Review Board as previously described [53]. HBEC2 cells (immortalized HBECs) were originally generated by Ramirez, et al. [54] and maintained as previously described [55]. Experiments were performed in 12-well Costar tissue culture plates at a starting cell density of $15 \times 10^3/cm^2$. Cell counts were performed by an electric particle counter (Beckman Coulter, Indianapolis, IN).

Preparation of cigarette smoke extract (CSE)

One hundred millimeter research cigarettes (3R4F) were purchased from the University of Kentucky. CSE were prepared as previously described [56].

RNA-Seq

Total RNA was extracted using trizol from pHBEC cultures of five non-smokers without COPD as previously described [53]. Sequence libraries were constructed using the Illumina TruSeq RNA prep kit (San Diego, CA). Multiplex sequencing was performed using the Illumina HiSeq 2000 system. We generated between 30 and 35 million 2×50 pair-end reads per sample. The raw sequences were aligned to the NCBI human genome reference build 37 using BWA, and the expression level of each transcript was determined using Tophat [57]. Gene expressions of approximately 32,500 gene transcripts, including multiple transcript variants, were analyzed (as the fragments per kilobase of transcript per million mapped reads (FPKM)). In total, 96.2 % of the gene transcripts have less than tenfold variability (the interquartile range of log10 expression equal to or less than 1.0). To control the quality of our computation of interquartile range, we have excluded all gene transcripts with 2 or more zero FPKM values among the five subjects.

Cell viability

Cell viability was determined by measuring the reduction of 3-(4,5-dimethythiazol-2-yl)-2,5-diphenyl tetrazolium bromide (MTT) as previously described [55]. The MTT absorbance was read at 570 nm.

RNA interference

Small interfering RNA (siRNA) for selected genes and the scrambled (control) siRNA were purchased from Applied Biosystems (Carlsbad, CA). Transfection of siRNA was performed using INTERFERin (Polyplus-tranfection Inc, New York, NY) according to the manufacturer's instructions. The targeted sequences for the selected genes are available (Additional file 8: Table S2). Twenty-four hours after plating, cells were transfected with the individual siRNA or the scrambled control for 48 h. The transfected cells were exposed to 2 % CSE for 24 h in the absence or presence of CSE as determined by the MTT assay.

Flow cytometry

To detect apoptosis, flow cytometry using Annexin V-FITC (BD Biosciences, Franklin Lakes, NJ) and propidium iodide (Sigma, St Louis, MO) was performed as previously described [58]. Early or late apoptosis was defined by Annexin V+/PI− or Annexin V+/PI+ staining, respectively.

Immunoblotting

Immunoblot analysis was performed for active caspase-3 as previously described [59]. Equivalent loading was verified by stripping the blot and reprobing with antibodies to β-actin.

Statistical analysis

The MTT absorption values from triplicated samples per group were averaged for the following four groups: (A) CSE-unexposed scrambled control cells, (B) CSE-exposed scrambled control cells, (C) CSE-unexposed siRNA-transfected cells, and (D) CSE-exposed siRNA-transfected cells. Every 2–10 genes per independent experiment were tested by siRNA transfection with a scrambled control. The interaction of CSE and siRNA effects on cell viability is quantified by the multiplicative interaction ratio of averaged MTT absorption values ([D/C] divided by [B/A]). This ratio is expected to be low or high if the targeted gene is protective or permissive for CSE-induced cytotoxicity. The mean and robust estimate of the standard deviation (nonparametric pseudo SD = interquartile range divided by 1.35) of these interactions are used to compute a Z-score. A one-side p value for each gene (by one-sample Z-test) was applied to detect both tails of genes. The effects of siRNA alone were also quantified by the ratio of C/A and further analyzed to detect other candidate genes among the 45 genes, whose siRNA reduced the MTT value compared with the scrambled control. The Benjamini and Hochberg (BH) false discovery rate (FDR) procedure is applied to the sorted list of p values (smallest to largest) using the formula $BH(j) = \alpha j/n$, where $\alpha = 0.05$ and j is the index number of the list of p values and n is the total number of candidate genes. The p values, $p(j)$, and the corresponding genes are considered significant for all genes whose $p(j) \leq BH(j)$.

Additional files

Additional file 1: COPD candidate genes prioritized by the VAAST analysis. Results from different VAAST run settings (Table 1) are shown in seperated tabs. The tab list and the column definition are shown in the tab "TableList".

Additional file 2: COPD candidate genes prioritized by the Gene Functional Deficit Score-based TTest and FCBF Analysis. Results from Gene Functional Deficit Score-based t test, FCBF analysis, and induced network modules analysis are shown in different tabs.

Additional file 3: The abundance of gene expression in primary human epithelial cell cultures using RNA-Seq. Expression level (FPKM) of 32,457 transcripts from five nonsmoking donors and the average expression level are shown.

Additional file 4: siRNA screening in vitro analysis for 81 COPD candidate genes. For the 81 candidate genes, the detailed gene information, selection method, and annotation are shown in the tab "Gene Selection Methods". The siRNA knockdown results in three trials are shown in the tab "siRNA summary".

Additional file 5: Annotated variants for the 81 COPD candidate genes. Whole exome sequencing variants identified in the 81 COPD candidate genes in the 62 susceptible smokers with COPD (tab "Susceptible") and 30 resistant smokers (tab "Resistant"). The variant list is in VCF format and annotated with ANNOVAR, except the first column which lists the gene id.

Additional file 6: Gene Ontology term enrichment analysis on the 81 candidate genes.

Additional file 7: Table S1. The information of Sanger sequencing validation for seven exonic variants of TACC2.

Additional file 8: Table S2. Table S2 siRNA sequences targeting 81 COPD candidate genes.

Abbreviations

A1ATD: alpha1 antitrypsin deficiency; BH: Benjamini and Hochberg; BWA: Burrows-Wheeler aligner; COPD: chronic obstructive pulmonary disease; CS: cigarette smoke; CSE: cigarette smoke extract; DP: read depth; FCBF: fast correlation-based filter solution; FDR: false discovery ratio; FEV1: forced expiratory volume in 1 second; FPKM: fragments per kilobase of transcript per million mapped reads; FVC: forced vital capacity; GATK: Genome Analysis Toolkit; GWAS: genome-wide association study; HBECs: human bronchial epithelial cells; MTT: 3-(4,5-dimethythiazol-2-yl)-2,5-diphenyl tetrazolium bromide; pHBECs: primary human bronchial epithelial cells; siRNA: Small interfering RNA; VAAST: Variant Annotation, Analysis, and Search Tool; VAT: Variant Annotation Tool; VQSLOD: variant quality score logarithm of odds; VST: Variant Selection Tool; WES: whole exome sequencing.

Competing interests

The authors declare that they have no competing interests.

Authors' contributions

TN, JX, SB, YB, MM, JZ, and JM helped in the study conception and design. TN, JX, SB, YB, MM, JJ, NW, HH, MP, RL, JKT, and SL helped in the acquisition of data. TN, JX, SB, YB, MM, NW, and CQ Analyzed and interpreted the data. TN, JX, SB, YB, and MM drafted of manuscript. SB, RL, and YL helped in the critical revision. All authors read and approved the final manuscript.

Acknowledgements

Primary human bronchial epithelial cell cultures were kindly provided by Dr. Yohannes Tesfaigzi's lab. This study was supported by Biomedical Research Institute of New Mexico, Clinical and Translational Science Center 003-2, VA merit review grant CX001048-02 (to TN), NIH CA097356 and P30 CA118100 (to SAB), JX was partly supported by NIH R00HG005846. YB was supported in part by an Informatics Research Starter grant from the PhRMA foundation and by NIH U01GM115486..

Author details

[1]Lovelace Respiratory Research Institute, 2425 Ridgecrest Drive SE, Albuquerque, NM 87108, USA. [2]Department of Genetics, Rutgers, The State University of New Jersey, 145 Bevier Road, Piscataway, NJ 08854, USA. [3]Department of Biochemistry and Microbiology, Rutgers, the State University of New Jersey, Piscataway, NJ, USA. [4]Department of Internal Medicine, University of New Mexico and New Mexico VA Health Care System, Albuquerque, NM, USA. [5]Biomedical Research Institute of New Mexico, Albuquerque, NM, USA. [6]Department of Medicine, University of Iowa, Iowa City, IA, USA.

References

1. Vos T, Flaxman AD, Naghavi M, Lozano R, Michaud C, Ezzati M, et al. Years lived with disability (YLDs) for 1160 sequelae of 289 diseases and injuries 1990–2010: a systematic analysis for the Global Burden of Disease Study 2010. Lancet. 2012;380(9859):2163–96. doi:10.1016/S0140-6736(12)61729-2.
2. Lozano R, Naghavi M, Foreman K, Lim S, Shibuya K, Aboyans V, et al. Global and regional mortality from 235 causes of death for 20 age groups in 1990 and 2010: a systematic analysis for the Global Burden of Disease Study 2010. Lancet. 2012;380(9859):2095–128. doi:10.1016/S0140-6736(12)61728-0.
3. Lokke A, Lange P, Scharling H, Fabricius P, Vestbo J. Developing COPD: a 25 year follow up study of the general population. Thorax. 2006;61(11): 935–9. doi:10.1136/thx.2006.062802.
4. Silverman EK. Progress in chronic obstructive pulmonary disease genetics. Proc Am Thorac Soc. 2006;3(5):405–8. doi:10.1513/pats.200603-092AW.
5. Janus ED, Phillips NT, Carrell RW. Smoking, lung function, and alpha 1-antitrypsin deficiency. Lancet. 1985;1(8421):152–4.
6. Thun GA, Ferrarotti I, Imboden M, Rochat T, Gerbase M, Kronenberg F, et al. SERPINA1 PiZ and PiS heterozygotes and lung function decline in the SAPALDIA cohort. PLoS One. 2012;7(8):e42728. doi:10.1371/journal.pone.0042728.
7. American Thoracic Society/European Respiratory Society statement: standards for the diagnosis and management of individuals with alpha-1 antitrypsin deficiency. Am J Respir Crit Care Med. 2003;168 (7):818–900. doi:10.1164/rccm.168.7.818
8. Cho MH, Boutaoui N, Klanderman BJ, Sylvia JS, Ziniti JP, Hersh CP, et al. Variants in FAM13A are associated with chronic obstructive pulmonary disease. Nat Genet. 2010;42(3):200–2. doi:10.1038/ng.535.
9. Wilk JB, Shrine NR, Loehr LR, Zhao JH, Manichaikul A, Lopez LM, et al. Genome-wide association studies identify CHRNA5/3 and HTR4 in the development of airflow obstruction. Am J Respir Crit Care Med. 2012;186(7): 622–32. doi:10.1164/rccm.201202-0366OC.
10. Cho MH, Castaldi PJ, Wan ES, Siedlinski M, Hersh CP, Demeo DL, et al. A genome-wide association study of COPD identifies a susceptibility locus on chromosome 19q13. Hum Mol Genet. 2012;21(4):947–57. doi:10.1093/hmg/ddr524.
11. Hersh CP, Pillai SG, Zhu G, Lomas DA, Bakke P, Gulsvik A, et al. Multistudy fine mapping of chromosome 2q identifies XRCC5 as a chronic obstructive pulmonary disease susceptibility gene. Am J Respir Crit Care Med. 2010; 182(5):605–13. doi:10.1164/rccm.200910-1586OC.
12. Fu W, O'Connor TD, Jun G, Kang HM, Abecasis G, Leal SM, Gabriel S, Rieder MJ, Altshuler D, Shendure J, Nickerson DA, Bamshad MJ, Project NES, Akey JM. Analysis of 6,515 exomes reveals the recent origin of most human protein-coding variants. Nature. 2013;493(7431):216–20. doi:10.1038/nature11690.
13. Wain LV, Sayers I, Soler Artigas M, Portelli MA, Zeggini E, Obeidat M, et al. Whole exome re-sequencing implicates CCDC38 and cilia structure and function in resistance to smoking related airflow obstruction. PLoS Genet. 2014;10(5):e1004314. doi:10.1371/journal.pgen.1004314.
14. Barnett IJ, Lee S, Lin X. Detecting rare variant effects using extreme phenotype sampling in sequencing association studies. Genet Epidemiol. 2013;37(2):142–51. doi:10.1002/gepi.21699.
15. Nyunoya T, Mebratu Y, Contreras A, Delgado M, Chand HS, Tesfaigzi Y. Molecular processes that drive cigarette smoke-induced epithelial cell fate of the lung. Am J Respir Cell Mol Biol. 2014;50(3):471–82. doi:10.1165/rcmb.2013-0348TR.
16. Petrache I, Natarajan V, Zhen L, Medler TR, Richter AT, Cho C, et al. Ceramide upregulation causes pulmonary cell apoptosis and emphysema-like disease in mice. Nat Med. 2005;11(5):491–8. doi:10.1038/nm1238.
17. Aoshiba K, Zhou F, Tsuji T, Nagai A. DNA damage as a molecular link in the pathogenesis of COPD in smokers. Eur Respir J. 2012. doi:10.1183/09031936.00050211
18. Hu H, Huff CD, Moore B, Flygare S, Reese MG, Yandell M. VAAST 2.0: improved variant classification and disease-gene identification using a conservation-controlled amino acid substitution matrix. Genet Epidemiol. 2013;37(6):622–34. doi:10.1002/gepi.21743.
19. Yandell M, Huff C, Hu H, Singleton M, Moore B, Xing J, et al. A probabilistic disease-gene finder for personal genomes. Genome Res. 2011;21(9):1529–42. doi:10.1101/gr.123158.111.
20. Kamburov A, Stelzl U, Lehrach H, Herwig R. The ConsensusPathDB interaction database: 2013 update. Nucleic Acids Res. 2013;41(Database issue):D793–800. doi:10.1093/nar/gks1055.
21. Kamburov A, Stelzl U, Herwig R. IntScore: a web tool for confidence scoring of biological interactions. Nucleic Acids Res. 2012;40(Web Server issue): W140–146. doi:10.1093/nar/gks492.
22. Treutlein B, Brownfield DG, Wu AR, Neff NF, Mantalas GL, Espinoza FH, et al. Reconstructing lineage hierarchies of the distal lung epithelium using single-cell RNA-seq. Nature. 2014;509(7500):371–5. doi:10.1038/nature13173.
23. Manolio TA, Collins FS, Cox NJ, Goldstein DB, Hindorff LA, Hunter DJ, et al. Finding the missing heritability of complex diseases. Nature. 2009;461(7265): 747–53. doi:10.1038/nature08494.
24. Eichler EE, Flint J, Gibson G, Kong A, Leal SM, Moore JH, et al. Missing heritability and strategies for finding the underlying causes of complex disease. Nat Rev Genet. 2010;11(6):446–50. doi:10.1038/nrg2809.
25. Gibson G. Rare and common variants: twenty arguments. Nat Rev Genet. 2011;13(2):135–45. doi:10.1038/nrg3118.

26. Kryukov GV, Pennacchio LA, Sunyaev SR. Most rare missense alleles are deleterious in humans: implications for complex disease and association studies. Am J Hum Genet. 2007;80(4):727–39. doi:10.1086/513473.
27. Schork NJ, Murray SS, Frazer KA, Topol EJ. Common vs. rare allele hypotheses for complex diseases. Curr Opin Genet Dev. 2009;19(3):212–9. doi:10.1016/j.gde.2009.04.010.
28. Eisner MD, Anthonisen N, Coultas D, Kuenzli N, Perez-Padilla R, Postma D, et al. An official American Thoracic Society public policy statement: novel risk factors and the global burden of chronic obstructive pulmonary disease. Am J Respir Crit Care Med. 2010;182(5):693–718. doi:10.1164/rccm.200811-1757ST.
29. Ha GH, Kim JL, Breuer EK. Transforming acidic coiled-coil proteins (TACCs) in human cancer. Cancer Lett. 2013;336(1):24–33. doi:10.1016/j.canlet.2013.04.022.
30. Schuendeln MM, Piekorz RP, Wichmann C, Lee Y, McKinnon PJ, Boyd K, et al. The centrosomal, putative tumor suppressor protein TACC2 is dispensable for normal development, and deficiency does not lead to cancer. Mol Cell Biol. 2004;24(14):6403–9. doi:10.1128/MCB.24.14.6403-6409.2004.
31. Mele C, Iatropoulos P, Donadelli R, Calabria A, Maranta R, Cassis P, et al. MYO1E mutations and childhood familial focal segmental glomerulosclerosis. N Engl J Med. 2011;365(4):295–306. doi:10.1056/NEJMoa1101273.
32. Wenzel J, Ouderkirk JL, Krendel M, Lang R. Class I myosin Myo1e regulates TLR4-triggered macrophage spreading, chemokine release, and antigen presentation via MHC class II. Eur J Immunol. 2015;45(1):225–37. doi:10.1002/eji.201444698.
33. Kanemaru K, Nakahara M, Nakamura Y, Hashiguchi Y, Kouchi Z, Yamaguchi H, et al. Phospholipase C-eta2 is highly expressed in the habenula and retina. Gene Expr Patterns. 2010;10(2–3):119–26. doi:10.1016/j.gep.2009.12.004.
34. Horiba N, Masuda S, Takeuchi A, Takeuchi D, Okuda M, Inui K. Cloning and characterization of a novel Na+-dependent glucose transporter (NaGLT1) in rat kidney. J Biol Chem. 2003;278(17):14669–76. doi:10.1074/jbc.M212240200.
35. Ogawa F, Adachi S, Kohu K, Shige K, Akiyama T. Binding of the human homolog of the Drosophila discs large tumor suppressor protein to the mitochondrial ribosomal protein MRP-S34. Biochem Biophys Res Commun. 2003;300(3):789–92.
36. Hautamaki RD, Kobayashi DK, Senior RM, Shapiro SD. Requirement for macrophage elastase for cigarette smoke-induced emphysema in mice. Science (New York, NY). 1997;277(5334):2002–4.
37. Nakanishi K, Takeda Y, Tetsumoto S, Iwasaki T, Tsujino K, Kuhara H, et al. Involvement of endothelial apoptosis underlying chronic obstructive pulmonary disease-like phenotype in adiponectin-null mice: implications for therapy. Am J Respir Crit Care Med. 2011;183(9):1164–75. doi:10.1164/rccm.201007-1091OC.
38. Woodruff PG, Koth LL, Yang YH, Rodriguez MW, Favoreto S, Dolganov GM, et al. A distinctive alveolar macrophage activation state induced by cigarette smoking. Am J Respir Crit Care Med. 2005;172(11):1383–92. doi:10.1164/rccm.200505-686OC.
39. Carolan BJ, Harvey BG, Hackett NR, O'Connor TP, Cassano PA, Crystal RG. Disparate oxidant gene expression of airway epithelium compared to alveolar macrophages in smokers. Respir Res. 2009;10:111. doi:10.1186/1465-9921-10-111.
40. Rabe KF, Hurd S, Anzueto A, Barnes PJ, Buist SA, Calverley P, et al. Global strategy for the diagnosis, management, and prevention of chronic obstructive pulmonary disease: GOLD executive summary. Am J Respir Crit Care Med. 2007;176(6):532–55.
41. DePristo MA, Banks E, Poplin R, Garimella KV, Maguire JR, Hartl C, et al. A framework for variation discovery and genotyping using next-generation DNA sequencing data. Nat Genet. 2011;43(5):491–8. doi:10.1038/ng.806.
42. Li H, Durbin R. Fast and accurate short read alignment with Burrows-Wheeler transform. Bioinformatics. 2009;25(14):1754–60. doi:10.1093/bioinformatics/btp324.
43. McKenna A, Hanna M, Banks E, Sivachenko A, Cibulskis K, Kernytsky A, et al. The Genome Analysis Toolkit: a MapReduce framework for analyzing next-generation DNA sequencing data. Genome Res. 2010;20(9):1297–303. doi:10.1101/gr.107524.110.
44. Wang K, Li M, Hakonarson H. ANNOVAR: functional annotation of genetic variants from high-throughput sequencing data. Nucleic Acids Res. 2010;38(16):e164. doi:10.1093/nar/gkq603.
45. Reese MG, Moore B, Batchelor C, Salas F, Cunningham F, Marth GT, et al. A standard variation file format for human genome sequences. Genome Biol. 2010;11(8):R88. doi:10.1186/gb-2010-11-8-r88.
46. Genomes Project C, Abecasis GR, Altshuler D, Auton A, Brooks LD, Durbin RM, Gibbs RA, Hurles ME, McVean GA. A map of human genome variation from population-scale sequencing. Nature. 2010;467(7319):1061–73. doi:10.1038/nature09534.
47. Li Y, Vinckenbosch N, Tian G, Huerta-Sanchez E, Jiang T, Jiang H, et al. Resequencing of 200 human exomes identifies an excess of low-frequency non-synonymous coding variants. Nat Genet. 2010;42(11):969–72. doi:10.1038/ng.680.
48. Drmanac R, Sparks AB, Callow MJ, Halpern AL, Burns NL, Kermani BG, et al. Human genome sequencing using unchained base reads on self-assembling DNA nanoarrays. Science. 2010;327(5961):78–81. doi:10.1126/science.1181498.
49. Bromberg Y, Rost B. SNAP: predict effect of non-synonymous polymorphisms on function. Nucleic Acids Res. 2007;35(11):3823–35. doi:10.1093/nar/gkm238.
50. Yu L, Liu H. Feature selection for high-dimensional data: a fast correlation-based filter solution. Proceedings of the twentieth international conference on machine learning. 2003. p. 856–63.
51. Hall M, Frank E, Holmes G, Pfahringer B, Reutemann P, Wittn I. The WEKA data mining software: an update. ACM SIGKDD Explor Newsl. 2009;11(1):10–8.
52. Quinlan JR. Book Review: C4.5: programs for machine learning. Mach Learn. 1993;16:235–40.
53. Jang JH, Bruse S, Liu Y, Duffy V, Zhang C, Oyamada N, et al. Aldehyde dehydrogenase 3A1 protects airway epithelial cells from cigarette smoke-induced DNA damage and cytotoxicity. Free Radic Biol Med. 2014;68:80–6. doi:10.1016/j.freeradbiomed.2013.11.028.
54. Ramirez RD, Sheridan S, Girard L, Sato M, Kim Y, Pollack J, et al. Immortalization of human bronchial epithelial cells in the absence of viral oncoproteins. Cancer Res. 2004;64(24):9027–34. doi:10.1158/0008-5472.CAN-04-3703.
55. Jang JH, Bruse S, Liu Y, Duffy V, Zhang C, Oyamada N, et al. Aldehyde dehydrogenase 3A1 protects airway epithelial cells from cigarette smoke-induced DNA damage and cytotoxicity. Free Radic Biol Med. 2013;68C:80–6. doi:10.1016/j.freeradbiomed.2013.11.028.
56. Nyunoya T, Monick MM, Klingelhutz AL, Glaser H, Cagley JR, Brown CO, et al. Cigarette smoke induces cellular senescence via Werner's syndrome protein down-regulation. Am J Respir Crit Care Med. 2009;179(4):279–87. doi:10.1164/rccm.200802-320OC.
57. Trapnell C, Roberts A, Goff L, Pertea G, Kim D, Kelley DR, et al. Differential gene and transcript expression analysis of RNA-seq experiments with TopHat and Cufflinks. Nat Protoc. 2012;7(3):562–78. doi:10.1038/nprot.2012.016.
58. Chand HS, Montano G, Huang X, Randell SH, Mebratu Y, Petersen H, et al. A genetic variant of p53 restricts the mucous secretory phenotype by regulating SPDEF and Bcl-2 expression. Nat Commun. 2014;5:5567. doi:10.1038/ncomms6567.
59. Shi S, Wang Q, Xu J, Jang JH, Padilla MT, Nyunoya T, et al. Synergistic anticancer effect of cisplatin and Chal-24 combination through IAP and c-FLIPL degradation, Ripoptosome formation and autophagy-mediated apoptosis. Oncotarget. 2015;6(3):1640–51.

3

Intragraft transcriptional profiling of renal transplant patients with tubular dysfunction reveals mechanisms underlying graft injury and recovery

Hátylas Azevedo[1], Paulo Guilherme Renesto[3], Rogério Chinen[3,4], Erika Naka[3], Ana Cristina Carvalho de Matos[3,4], Marcos Antônio Cenedeze[3], Carlos Alberto Moreira-Filho[1], Niels Olsen Saraiva Câmara[2,3*] and Alvaro Pacheco-Silva[3,4*]

Abstract

Background: Proximal tubular dysfunction (PTD) is associated with a decreased long-term graft survival in renal transplant patients and can be detected by the elevation of urinary tubular proteins. This study investigated transcriptional changes in biopsies from renal transplant patients with PTD to disclose molecular mechanisms underlying graft injury and functional recovery.

Methods: Thirty-three renal transplant patients with high urinary levels of retinol-binding protein, a biomarker of PTD, were enrolled in the study. The initial immunosuppressive scheme included azathioprine, cyclosporine, and steroids. After randomization, 18 patients (group 2) had their treatment modified by reducing cyclosporine dosage and substituting azathioprine for mycophenolate mofetil, while the other 15 patients (group 1) remained under the initial scheme. Patients were biopsied at enrollment and after 12 months of follow-up, and paired comparisons were performed between their intragraft gene expression profiles. The differential transcriptome profiles were analyzed by constructing gene co-expression networks and identifying enriched functions and central nodes in each network.

Results: Only the alternative immunosuppressive scheme used in group 2 ameliorated renal function and tubular proteinuria after 12 months of follow-up. Intragraft molecular changes observed in group 2 were linked to autophagy, extracellular matrix, and adaptive immunity. Conversely, gene expression changes in group 1 were related to fibrosis, endocytosis, ubiquitination, and endoplasmic reticulum stress.

Conclusion: These results suggest that molecular networks associated with the control of endocytosis, autophagy, protein overload, fibrosis, and adaptive immunity may be involved in improvement of graft function.

Keywords: Network analysis, Kidney transplantation, Genomics, Proximal tubular dysfunction, Transcriptional profiling

* Correspondence: niels@icb.usp.br; alvaro.pfilho@einstein.br
[2]Laboratory of Transplantation Immunobiology, Department of Immunology, Institute of Biomedical Sciences, Universidade de São Paulo (USP), São Paulo, Brazil
[3]Laboratory of Clinical and Experimental Immunology, Nephrology Division, Universidade Federal de São Paulo (UNIFESP), São Paulo, Brazil
Full list of author information is available at the end of the article

Introduction

Proximal tubular dysfunction (PTD) is characterized by proteinuria, aminoaciduria, and glucosuria. It is associated with a decreased long-term graft survival in renal transplant patients and can be detected by the elevation of urinary tubular proteins [1, 2].

Low-molecular-weight proteins (LMWP) are prominent urine biomarkers of PTD [3–7], as they are cleared by glomerular filtration and almost totally reabsorbed by tubular epithelial cells. Specifically, increased urinary levels of the LMWP urinary retinol-binding protein (uRBP) have been associated with tubular injury and fibrosis after renal transplantation [8, 9]. Therefore, measuring uRBP levels may contribute to the detection of patients at a higher risk for renal function loss [10, 11].

The early detection of transplanted patients with kidney dysfunction helps optimizing their immunosuppression protocols in the attempt to improve graft outcomes. For instance, PTD is found in heart transplant patients with a progressive worsening on renal function due to cyclosporine (CsA) nephrotoxicity [12]. In parallel, immunosuppression with mycophenolate mofetil (MYF) is associated with less acute rejection after kidney transplantation [13], offering a therapeutic alternative for patients with renal dysfunction caused by CsA. However, a large number of transplant patients in developing countries still receive maintenance immunosuppression regimens containing low-cost drugs, including CsA, azathioprine (AZA), and steroids [14]. Hence, the follow-up of renal transplant patients diagnosed with PTD allows the adoption of tailored immunosuppression regimens and provides a window of opportunity for investigating renal injury and recovery.

Here, we followed up renal transplant patients with elevated uRBP levels that were submitted to specific immunosuppression regimens for 12 months. Kidney biopsies were performed to investigate intragraft transcriptional profiles and to correlate their molecular changes with renal function outcomes. Through this approach, we identified some potential molecular mechanisms associated with graft function improvement in renal transplant patients.

Results

Assessment of renal dysfunction in transplanted patients and association between tubular proteinuria and glomerular function rates

Table 1 summarizes the patients' baseline clinical and laboratory features. None of these characteristics showed statistically significant differences ($p < 0.05$) between group 1 and group 2 patients. The analysis of renal histology in our cohort revealed no significant differences in the baseline Banff classification scores between the groups and also no significant differences in the Banff scores within the same group after 12 months of follow-up. At enrollment, the biopsies of three patients from group 1 showed no evidence of interstitial fibrosis and tubular atrophy (IF/TA), while nine, three, and zero biopsies showed mild, moderate, and severe IF/TA, respectively. At the end of the follow-up period, all biopsies from group 1 showed some evidence of IF/TA, with eight biopsies at mild, two at moderate, and two at severe IF/TA classification, respectively.

For group 2, the biopsies of two patients at enrollment showed no evidence of IF/TA, while nine, five, and two biopsies showed mild, moderate, and severe IF/TA,

Table 1 Baseline clinical and laboratory characteristics of the renal transplant patients enrolled in this study (*n* = 33)

Variables	Group 1[a] (*n* = 15)	Group 2[b] (*n* = 18)	*p* value
Recipient age ± SD, in years	44.9 ± 11.2	45.4 ± 9.9	0.79
Recipient weight ± SD, in kg	67.5 ± 13.8	69.6 ± 14.7	0.35
Recipient male, *N* (%)	11 (73.3 %)	10 (55.6 %)	0.34
Deceased donors, *N* (%)	6 (40 %)	6 (33.3 %)	0.69
Haploidentical HLA, *N* (%)	5 (33.3 %)	10 (55.6 %)	0.20
Time post-transplant ± SD, in months	93.3 ± 35.7	92.4 ± 33.1	0.94
Systemic arterial hypertension, *N* (%)	13 (86.7 %)	16 (88.9 %)	0.84
Diabetes mellitus, *N* (%)	2 (13.3 %)	1 (5.6 %)	0.44
Serum creatinine levels ± SD, in mg/dL	1.59 ± 0.28	1.67 ± 0.33	0.46
uRBP levels ± SD, in mg/L	2.79 ± 2.55	2.72 ± 2.24	0.88
Banff classification[c] at enrollment	No evidence (3), mild (9), moderate (3), severe (0)	No evidence (2), mild (9), moderate (5), severe (2)	0.64
Banff classification[c] after 12 months	No evidence (0), mild (8), moderate (2), severe (2)	No evidence (3), mild (7), moderate (6), severe (1)	0.25

[a]Group 1 patients: initial immunosuppression scheme with cyclosporine, corticosteroids, and azathioprine

[b]Group 2 patients: alternative immunosuppression scheme with reduced levels of cyclosporine and use of mycophenolate mofetil

[c]Banff classification: mild, moderate, or severe evidence of interstitial fibrosis and tubular atrophy (IF/TA)

respectively. At the end of the follow-up period, three biopsies from group 2 showed no evidence of IF/TA, while the biopsies of seven, six, and one patients displayed mild, moderate, and severe IF/TA, respectively.

Figure 1 illustrates serum creatinine levels, estimated creatinine clearance, and uRBP levels at 0 and 12 months of follow-up. Interestingly, only patients from group 2 significantly improved these renal function parameters. Serum creatinine levels increased in group 1 after 12 months of follow-up, indicating a worsening in renal function (Fig. 1a). In parallel, creatinine clearance calculated by either Cockcroft-Gault (Fig. 1b) or modification of diet in renal disease (MDRD) (Fig. 1c) equations was only significantly improved in patients submitted to the alternative immunosuppression protocol (group 2). uRBP concentration was also decreased in patients from group 2 after 12 months (Fig. 1d, e).

The distribution of uRBP values relative to estimated glomerular function rate (eGFR) equally showed a significant increase in RBP proteinuria in patients whose eGFR values ranged from 30 to 45 mL/min (with moderate kidney function reduction). uRBP levels in patients with eGFR <30 mL/min did not differ to those with eGFR values >45 mL/min (Fig. 2a).

A linear regression analysis was conducted to examine the association between delta values (t12–t0) for uRBP and eGFR over 12 months of follow-up. We found a small but significant correlation between delta values for uRBP and eGFR calculated via either Cockcroft-Gault ($r^2 = 0.2062$, $p = 0.0103$) or MDRD ($r^2 = 0.2880$, $p = 0.0019$) formulas, as depicted in Fig. 2b. This analysis showed that around 25 % of the variance in eGFR values was significantly explained by the variation in uRBP values.

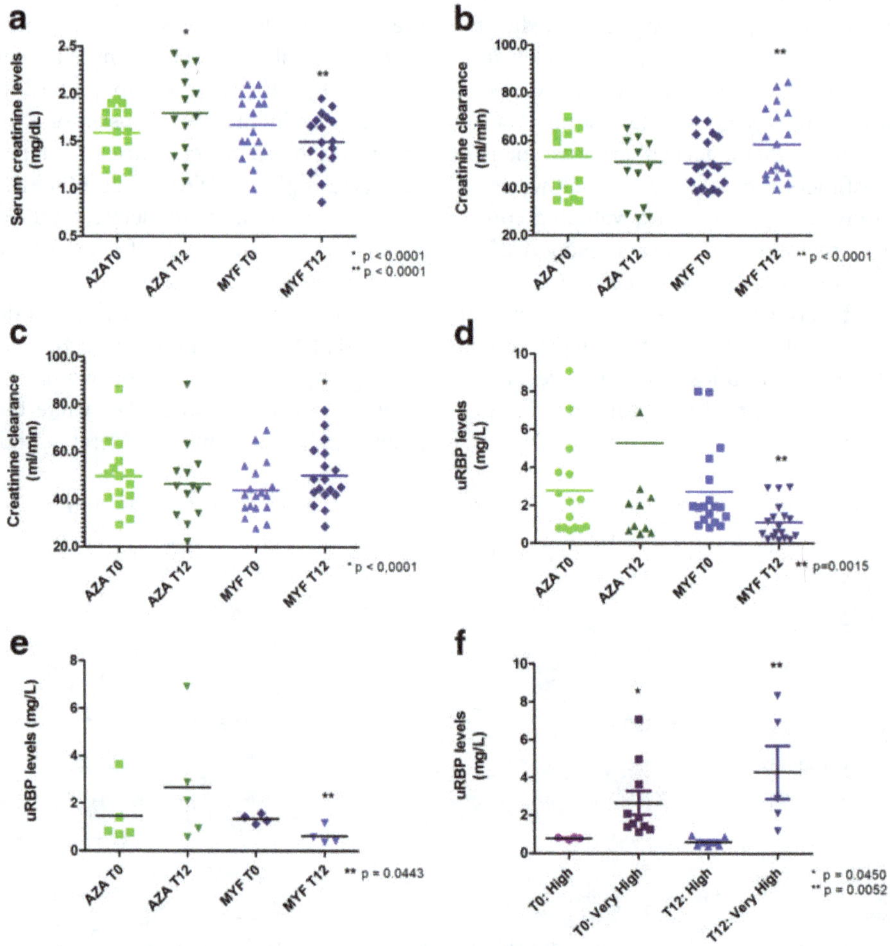

Fig. 1 Measurement of serum creatinine, creatinine clearance, and urinary retinol-binding protein (*uRBP*) levels. **a** Serum creatinine levels (mg/dL) at the beginning and at the end of the follow-up period of 12 months. **b** Creatinine clearance (mL/min) calculated using Cockcroft-Gault formula. **c** Creatinine clearance (mL/min) calculated using MDRD formula. **d** uRBP levels (mg/L) from all enrolled patients. **e** uRBP levels (mg/L) from patients used for intragraft gene expression profiling. **f** Comparison between very high and high uRBP levels at t0 and t12 post-enrollment periods. Statistical significance was assessed using paired Student's *t* test with $p < 0.05$. Renal function parameters were significantly altered in the groups treated with azathioprine (*AZA group comparison) or mycophenolate mofetil (**MYF group comparison) for 12 months

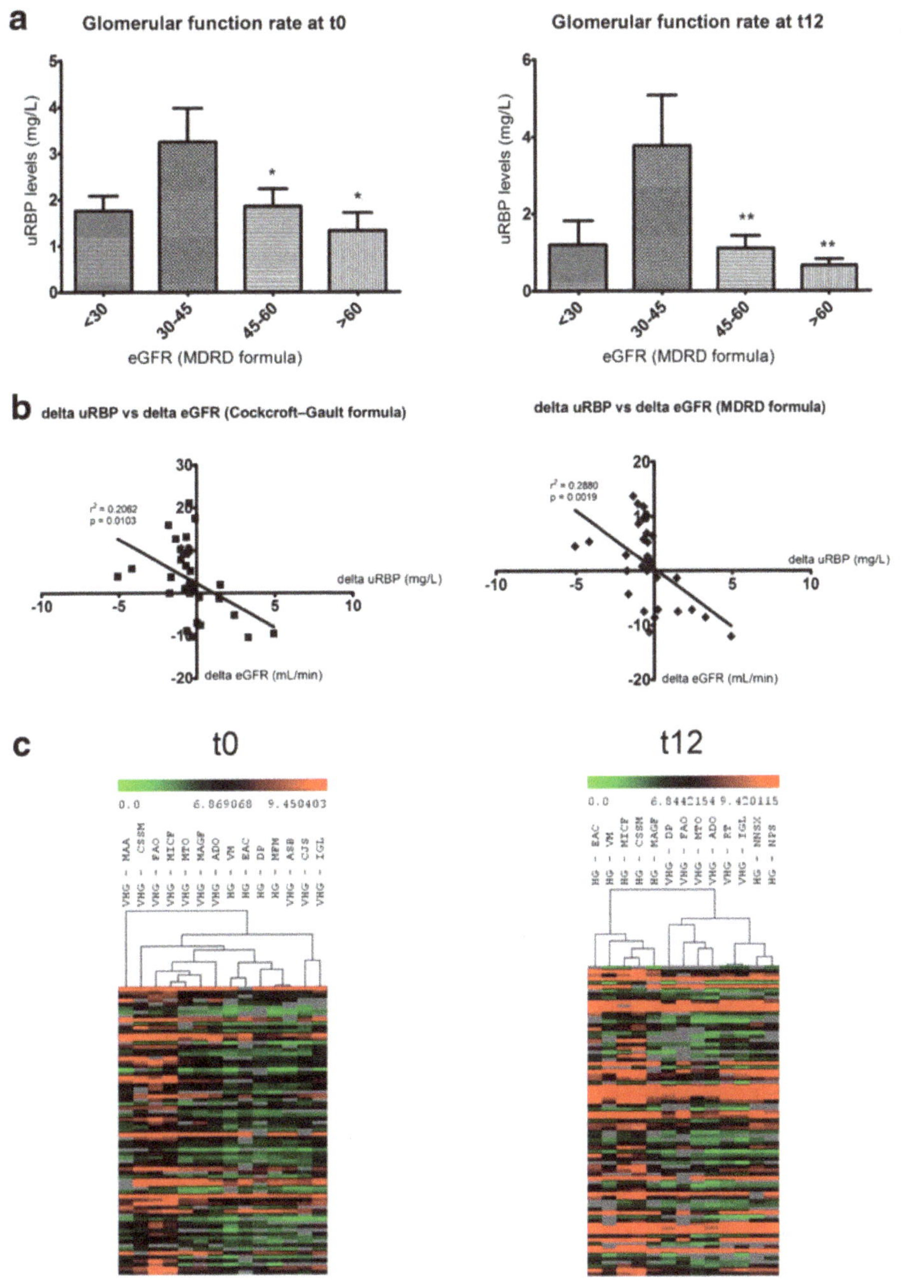

Fig. 2 Distribution of uRBP levels according to glomerular filtration rates and hierarchical clustering analysis of samples from different ranges of uRBP levels. **a** Histograms showing uRBP values according to eGFR levels at enrollment (t0) and after 12 months of follow-up (t12). uRBP levels are expressed as the mean ± SEM. The estimated glomerular filtration rates (*eGFR*) were calculated using the MDRD formula. $^{**}p < 0.05$, $^{**}p < 0.001$ compared to eGFR at 30–45 mL/min, using ANOVA followed by Tukey's correction. **b** Best-fit slope of the linear regression of the delta values from urinary retinol-binding protein (*uRBP*) levels and eGFR between t0 and t12. The estimated eGFR was calculated using Cockcroft-Gault or MDRD formulas. **c** Hierarchical clustering analysis of samples in the high group (*HG*, uRBP levels from 0.4 to 1 mg/L) and very high group (*VHG*, uRBP ≥ 1 mg/L) at t0 and t12

Network enrichment analysis reveals intragraft molecular changes associated with specific immunosuppression regimens

The lists of differentially expressed (DE) genes obtained in each comparison and their fold changes are displayed in Additional file 1. These DE genes were used to construct co-expression gene networks, and centrality measures were calculated for all nodes in order to determine the central nodes in each network. The centrality measures calculated were degree, which denotes the number of interactions a node has, and betweenness, which represents the fraction of shortest paths that passes through a specific node.

These central nodes were then categorized in three subclasses of hubs, high-hubs, and bottlenecks, using the information from the scatterplots in Figs. 3b, 4b, 5b, and 6b. High-hubs, hubs, and bottlenecks are displayed in the scatterplots respectively at the up-right, down-right, and up-left quadrants, due to their differences in degree and betweenness values. This classification is in accordance with previous studies showing that node subclasses exert important functions in the context of health and disease [15–17]. For instance, hub genes regulate many other genes and tend to be essential in biological networks [18] and their node degrees increase in tumor networks, suggesting their association with gain of regulatory control [19]. In addition, high-hubs were linked to mechanisms underlying refractory epilepsy [20] and bottleneck genes were correlated with essentiality [21] in biological networks. The high-hubs, hubs, and bottlenecks are described in Additional file 2.

Enriched biological functions were revealed among central nodes in the subnetworks. These biological functions are displayed in Figs. 3a, 4a, 5a, and 6a. Biological functions enriched by the genes from group 1 and group 2 comparisons are shown respectively in Figs. 3 and 4. Genes derived from the group 1 comparison (group 1 network) were related to endocytosis, ubiquitin-mediated proteolysis, endoplasmic reticulum stress, TGF-β pathway, and adherens junctions. Conversely, genes from the group 2 comparison (group 2 network) were linked to phagosome, antigen processing and presentation, cell adhesion, autoimmunity, allograft rejection, and graft-versus-host disease. The subnetworks containing the most central nodes in each group are displayed in Figs. 3c to 6c. Genes associated with specific keywords in each case were marked in a different color, as described in the legend of each figure.

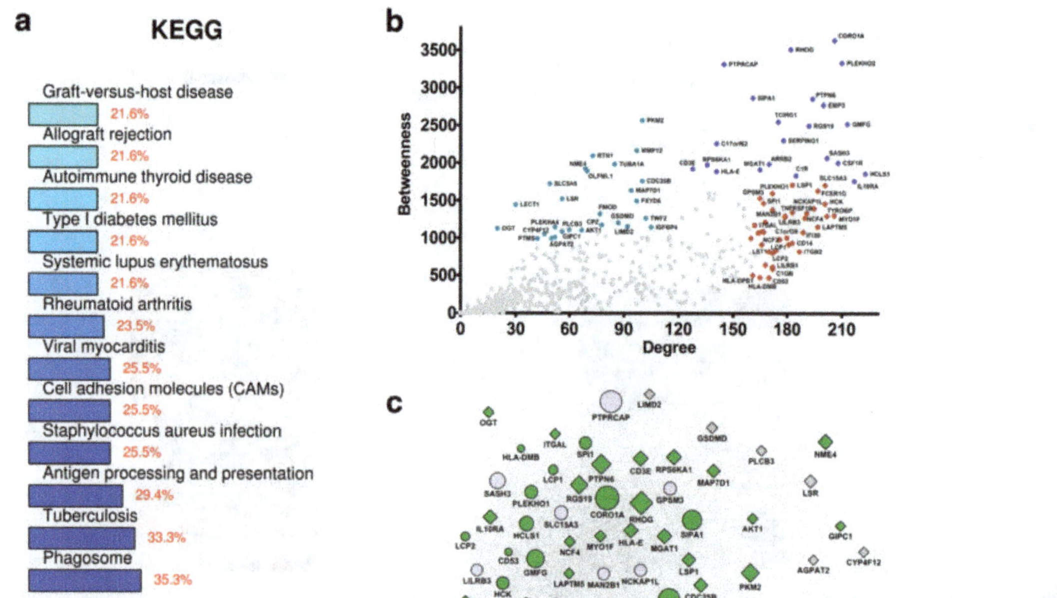

Fig. 3 Analysis of the intragraft transcriptional network related to the alternative immunosuppression scheme (group 2 patients) after 12 months of follow-up. Network genes were obtained by performing paired comparisons between the intragraft transcriptional profiles (t12 versus t0) of patients submitted to the alternative immunosuppression scheme and by subsequently inputting those genes in the GeneMania tool. **a** KEGG categories showing enrichment in functions for the network nodes. **b** Scatterplot of degree versus betweenness centralities for nodes obtained in the transcriptional network analysis. High-hubs, hubs, and bottlenecks are depicted respectively in *royal blue, red*, and *oil blue colors*. **c** Transcriptional subnetwork containing the most central nodes in the group 2 network. Differentially expressed (DE) genes and DE-related genes are represented respectively as *gray diamonds* and *blue circles* in the network. Node sizes are based on betweenness centrality measurements. Genes previously associated with the keywords "autophagy," "extracellular matrix," "cell adhesion," and "autoimmunity" are displayed in *green color*

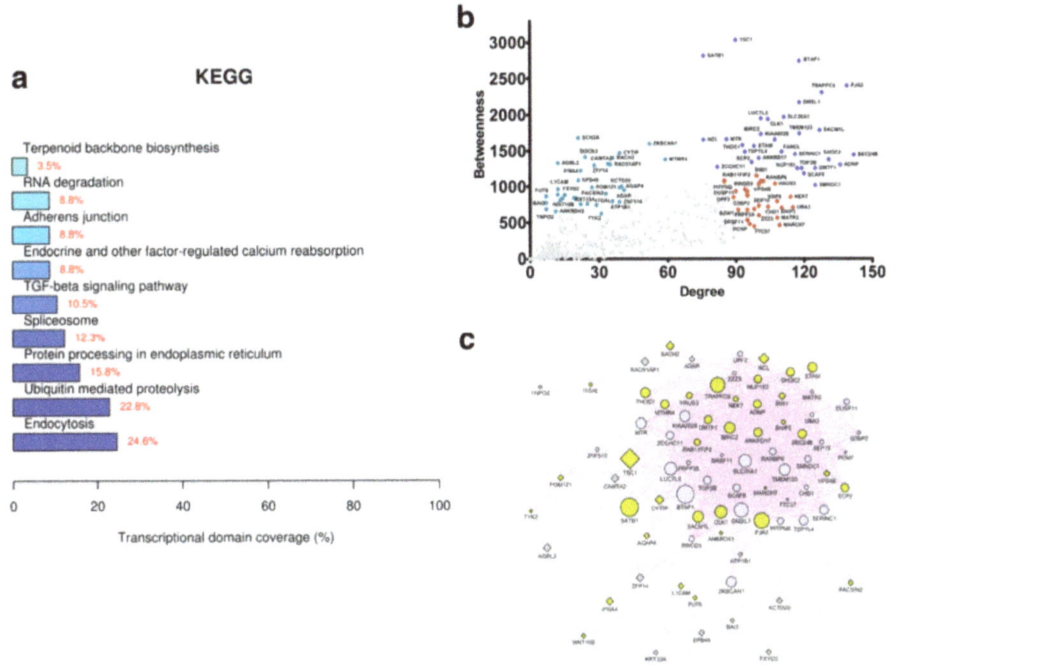

Fig. 4 Analysis of the intragraft transcriptional network related to the standard immunosuppression scheme (group 1 patients) after 12 months of follow-up. Network genes were obtained by performing paired comparisons between the intragraft transcriptional profiles (t12 versus t0) of patients submitted to the standard immunosuppression scheme and by subsequently inputting those genes in the GeneMania tool. **a** KEGG categories showing enrichment in functions for the network nodes. **b** Scatterplot of degree versus betweenness centralities for nodes obtained in the transcriptional network analysis. High-hubs, hubs, and bottlenecks are depicted respectively in *royal blue*, *red*, and *oil blue colors*. **c** Transcriptional subnetwork containing the most central nodes in the group 1 network. Differentially expressed (DE) genes and DE-related genes are represented respectively as *gray diamonds* and *blue circles* in the network. Node sizes are based on betweenness centrality measurements. Genes previously associated with the keywords "endocytosis," "fibrosis," and "inflammation" are displayed in *beige color*

Analysis of the intragraft molecular changes associated with higher retinol-binding proteinuria

We also associated the increased tubular proteinuria with a specific gene expression pattern, considering the differences in uRBP levels between the very high group (VHG) and high group (HG) (Fig. 1f). In addition, a hierarchical clustering analysis on the DE gene subsets was conducted to verify if the samples in each group showed similar gene expression patterns (Fig. 2c). With this approach, we confirmed that patients in the HG and VHG clustered together at t0 and t12, respectively.

Biological functions overrepresented by network genes from t0 and t12 comparisons are shown in Figs. 5 and 6. Genes derived from the former comparison (t0 network) are associated with osteoclast differentiation, protein digestion and absorption, and Toll-like, B cell, and T cell receptor pathways. On the other hand, genes derived from the latter comparison (t12 network) were linked to neuroactive ligand-receptor interaction, pathways in cancer, diabetes, and glycosaminoglycan degradation.

Discussion

Patients submitted to the alternative immunosuppression scheme show better renal function outcomes and a survival-related intragraft molecular profile

In the study cohort, patients that had their immunosuppression scheme altered by reducing the cyclosporine dosage and replacing AZA by MYF (group 2) improved serum creatinine and uRBP levels after 12 months of follow-up. Therefore, we hypothesized that distinct renal function outcomes could reflect intragraft molecular differences between patients from groups 1 and 2.

In the group 1 network (Fig. 3), the most overrepresented function was endocytosis. Therefore, endocytosis disturbance may be involved in proteinuria, as tubular injury is linked to the inhibition of LMWP receptor-mediated endocytosis [22]. Moreover, tubular atrophy biomarkers are associated with endocytosis in renal transplant patients [23]. Besides endocytosis, genes in the group 1 network were related to ubiquitination and endoplasmic reticulum stress. Endoplasmic reticulum stress is activated by unfolded protein accumulation and induces fibrosis in proteinuric kidney diseases [24, 25].

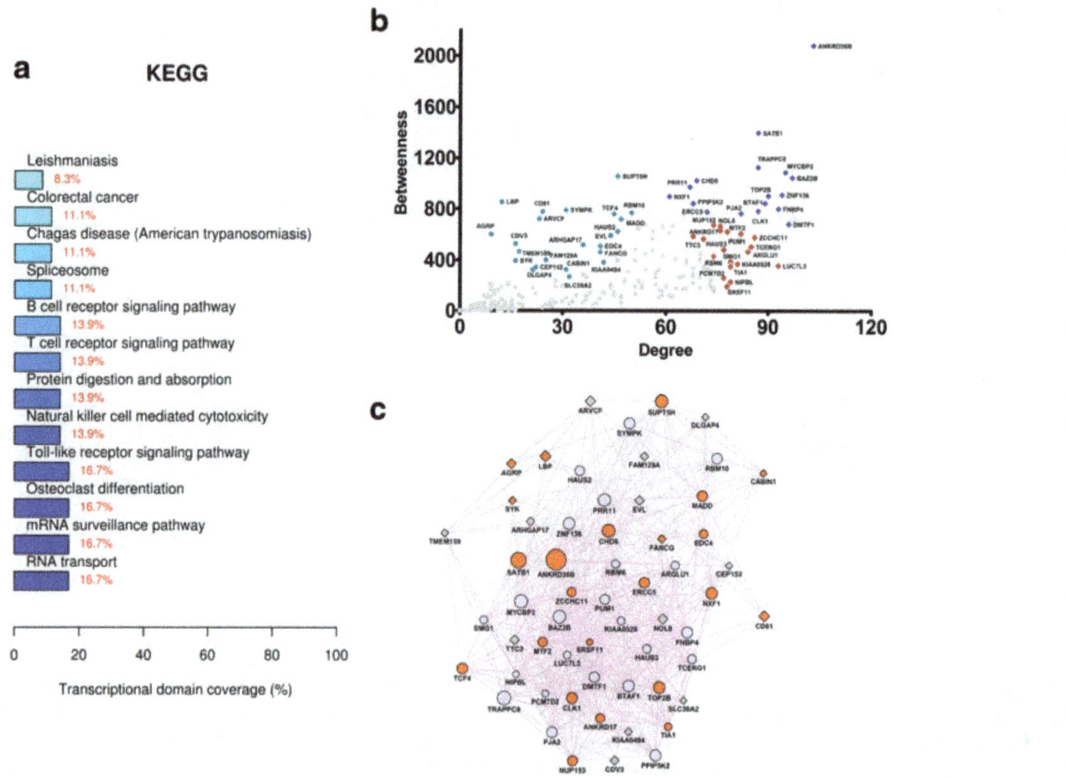

Fig. 5 Analysis of the intragraft transcriptional network related to proximal tubular dysfunction (PTD) right after enrollment (t0). Network genes were obtained by comparing the intragraft transcriptional profiles of patients with elevated uRBP levels and by subsequently inputting those genes in the GeneMania tool. **a** KEGG categories showing enrichment in functions for the network nodes. **b** Scatterplot of degree versus betweenness centralities for nodes obtained in the transcriptional network analysis. High-hubs, hubs, and bottlenecks are depicted respectively in *royal blue*, *red*, and *oil blue colors*. **c** Transcriptional subnetwork containing the most central nodes in the t0 network. Differentially expressed (DE) genes and DE-related genes are represented respectively as *gray diamonds* and *blue circles* in the network. Node sizes are based on betweenness centrality measurements. Genes previously associated with the keywords "immune response," "T regulatory cells," "autophagy," or "ubiquitin-proteasome" are displayed in *orange color*

In the group 2 network (Fig. 4), the most overrepresented function was phagosome, a type of vesicle formed during phagocytosis and autophagy. Interestingly, autophagy plays a role in tissue homeostasis, as damaged structures are degraded by autolysosomes after injury [26, 27]. Moreover, the role of autophagy as a protective mechanism against ischemic injury was investigated in tubular cells [28] and autophagy activation by MYF was able to prolong cell survival [29]. Central nodes in the group 2 network were also related to autophagy, such as Coro1a, Akt1, Ifi30, Cd14, and Ncf4. Coro1a, for example, inhibits autophagosomes in macrophages [30].

Other overrepresented functions in the group 2 network were antigen presentation, autoimmune diseases, and allograft rejection. Two of these genes, Igfpb4 and Il10ra, were already associated with graft function loss [31]. Moreover, the up-regulated gene Klrb1 disclosed the existence of tolerance mechanisms related to FoxP3+ T cells [32]. Genes coding for class I and II major histocompatibility molecules (MHC) were also found in the group 2 network, highlighting the role of HLA-G (MHC class I) for protecting renal transplants from autoimmunity [33] and the function of Hla-dpb1 (MHC class II) in renal allograft rejection [34].

Finally, cell adhesion and extracellular matrix functions were overrepresented by genes in the group 2 network. Serping1 and Mmp12, for example, were already associated with parenchyma deterioration in allografts undergoing rejection [35, 36]. Therefore, the alternative immunosuppression regimen may exert part of its protective effects by regulating the extracellular matrix, considering that MYF reduces fibroblast infiltration, collagen deposition, and ECM synthesis in kidney [37].

Patients with higher retinol-binding proteinuria have intragraft transcriptional changes associated with injury and repair mechanisms

Increased uRBP levels were significantly correlated with lower creatinine clearance in the cohort, corroborating the relevance of uRBP levels for assessing the extent of glomerular dysfunction after renal transplantation. Thus,

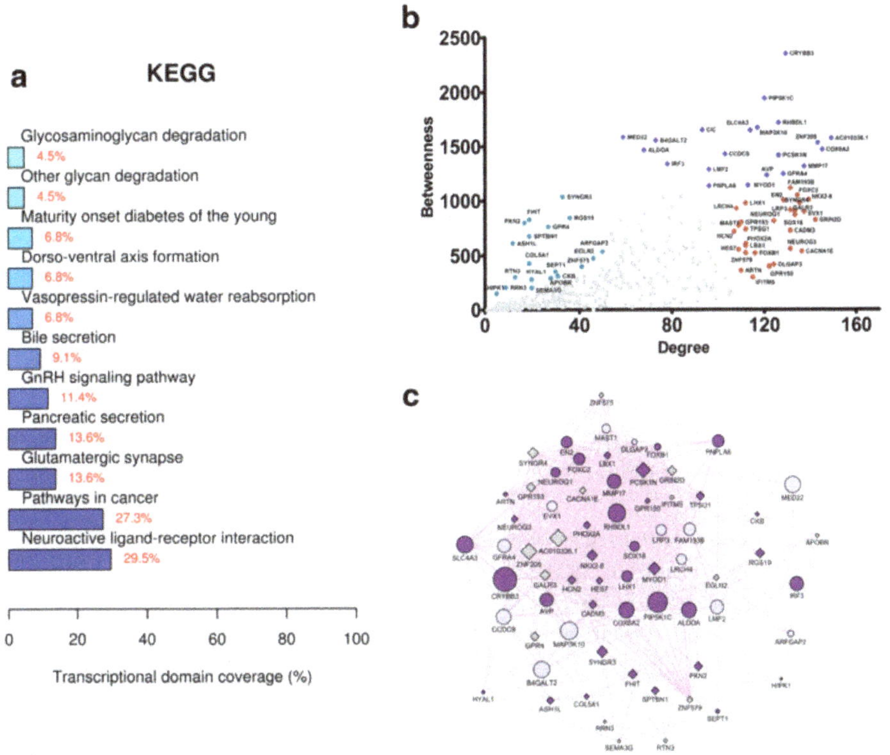

Fig. 6 Analysis of the intragraft transcriptional network related to proximal tubular dysfunction (PTD) after 12 months of follow-up (t12). Network genes were obtained by comparing the intragraft transcriptional profiles of patients with elevated uRBP levels and by subsequently inputting those genes in the GeneMania tool. **a** KEGG categories showing enrichment in functions for the network nodes. **b** Scatterplot of degree versus betweenness centralities for nodes obtained in the transcriptional network analysis. High-hubs, hubs, and bottlenecks are depicted respectively in *royal blue*, *red*, and *oil blue colors*. **c** Transcriptional subnetwork containing the most central nodes in the t12 network. Differentially expressed (DE) genes and DE-related genes are represented respectively as *gray diamonds* and *blue circles* in the network. Node sizes are based on betweenness centrality measurements. Genes previously associated with the keywords "fibrosis" and "tissue repair" are displayed in *purple color*

we conducted a network analysis to identify molecular events correlated with higher tubular proteinuria.

Intragraft transcriptional changes in the t0 network (Fig. 5) reflected molecular mechanisms of immune-mediated allograft injury. Indeed, intragraft gene expression changes during transplant rejection are associated with immune response and precede the onset of IF/TA [38, 39]. In parallel, a gene subset in the t0 network was associated with the Foxp3+ T regulatory cell phenotype. The high-hub Satb1, for example, inhibits the transcription factor Foxp3 [40], and Srrm1, the most up-regulated gene in the t0 comparison (fold change = 23.9), is part of the Foxp3 network [41].

Central nodes in the t0 network such as Trappc8, Agrp, Syk, and Baz2b were also linked to autophagy and ubiquitin-proteasome functions. As a matter of fact, autophagy components interplay with ubiquitination processes [42], and autophagy exerts protective functions by inhibiting endoplasmic reticulum stress and ubiquitin-proteasome system [43]. Hence, the imbalance of these two protein degradation systems may be involved in kidney graft dysfunction.

In the t12 network, central nodes were associated with fibrosis (Fig. 6), like Irf3, Myod1, and Crybb3 [44–46]. Moreover, genes encoding actin and myosin, biomarkers of fibrosis, were highly up-regulated in the VHG (fold change around 300). Notably, the most overrepresented function in the t12 network was "neuroactive ligand-receptor interaction", highlighting the role of neuronal genes for the response against ischemic renal injury [47] and maintenance of an epithelial phenotype in tubular cells [48].

Conclusions

This study showed that increased retinol-binding proteinuria is associated with kidney function worsening in renal transplant patients. Patients under the alternative immunosuppressive scheme (group 2) improved renal function after 12 months of follow-up. The examination of their intragraft transcriptional profiles revealed mechanisms linked to graft injury and recovery. These results can be useful for further studying the mechanisms underlying graft injury and functional recovery. In particular, the role of genes participating in ubiquitination, autophagy, protein overload, and neuroactive ligand-receptor interaction should be further investigated.

Methods

Patients

We studied a cohort of 33 subjects who underwent renal transplantation in the Kidney Hospital at the Federal University of São Paulo. Patients with at least 1 year of transplant surgery were checked for uRBP levels during a 3-month follow-up. Patients with uRBP >0.4 mg/L in three consecutive monthly measurements were diagnosed with PTD, according to studies establishing this upper normal limit [2, 12], and sequentially included in this study. The study was approved by the Research Ethics Committee at UNIFESP (number 0420/05).

Clinical study design

All patients enrolled were initially (t0) in use of CsA, prednisone, and AZA. They were randomized at enrollment for a prospective treatment with the standard protocol (group 1, $n = 15$) or with an alternative protocol with reduced CsA dosage and MYF introduction (group 2, $n = 18$). Patients were followed up for 12 months (t12) to determine if modifying their immunosuppression regimens could result in better renal function outcomes. All patients were submitted to graft biopsy at t0 and t12, and each biopsy was individually assessed in the transcriptomic profiling analysis.

Kidney histological analysis

Histological findings were evaluated by a pathologist using optical microscopy. Biopsies were interpreted following the Banff classification for kidney allograft pathology: each biopsy was categorized according to classification grades of interstitial fibrosis and tubular atrophy (IF/TA), which establishes the degree (I, II, and III) of interstitial fibrosis (mild, moderate, severe) and tubular atrophy (mild to moderate) for each evaluated sample [49]. Histological differences between the groups were determined using the χ^2 test for the categorical variables.

Renal function evaluation

Serum creatinine, creatinine clearance, and uRBP levels were measured in patients at enrollment and at the end of the follow-up period. Serum creatinine and uRBP levels were measured according to described elsewhere [10]. The glomerular function rate (eGFR) estimation was based on serum creatinine concentration and calculated using the Cockcroft-Gault and the modification of diet in renal disease (MDRD) formulas. The prediction of creatinine clearance by the Cockcroft-Gault equation was calculated as (140 – age) × body weight/plasma creatinine × 72 (×0.85 if female), and the MDRD estimate was determined as 175 × plasma creatinine – 1.154 × age – 0.203 (×0.742 if female; ×1.21 if black). Serum creatinine levels are represented in mg/dL, creatinine clearance in mL/min, and uRBP levels in mg/L. Statistical analysis (Student's t tests, with $p < 0.05$) was performed using GraphPad Prism 5.

Comparative analysis of the intragraft gene expression profiles

To investigate transcriptional changes related to specific immunosuppressive regimens, we performed paired comparisons (t12 versus t0) between biopsies from the same patients in group 1 (group 1 comparison) and in group 2 (group 2 comparison). A second analysis was performed to determine transcriptional changes associated with increased retinol-binding proteinuria. Patients were divided into subgroups: patients with uRBP values above 1.0 mg/L were classified in the very high group (VHG), whereas patients with uRBP levels above 0.4 mg/L but less than 1.0 mg/L were classified in the high group (HG). The uRBP cutoff level of 1.0 mg/L was selected because renal transplant patients with uRBP levels ≥1 mg/L are at a much higher risk for kidney function deterioration [10]. Moreover, the average uRBP levels in transplant patients with an estimated GFR less than 60 mL/min (i.e., with moderate to severe renal damage) were also higher than 1 mg/L [12]. The comparative analyses were done between the transcriptional profiles of patients from the VHG and HG at t0 (t0 comparison) and t12 (t12 comparison).

Oligonucleotide microarray experiments

Microarray experiments were performed as previously described [50]. Only samples with a RNA integrity number greater than 7 were employed in microarray experiments. Gene expression profiles were assessed using Agilent whole human genome 4x44K oligonucleotide microarrays. The R environment (http://www.r-project.org) was used to filter and pre-process the data. The mean of the probes for each gene was calculated, and the signal intensities were log2 transformed. These logarithmic values were input in the TmeV software to perform the statistical analyses [51]. No further normalization step was performed after visual inspection of the data distribution across the samples. The complete data set is available at the National Center for Biotechnology Information (NCBI) Gene Expression Omnibus, through the accession number GSE48250.

Bioinformatics workflow

The analyses described below were performed to obtain further information on the comparative gene expression profiles.

i) *Statistical testing.* Differentially expressed (DE) genes were found by paired (group 1 and group 2 comparisons) and unpaired (VHG × HG comparisons) t tests, with $p < 0.05$ as the significance threshold. The false discovery rate (FDR) control was initially applied to adjust the p values for multiple comparisons, but only few significant genes were obtained using this procedure. Therefore, a less stringent approach was used by analyzing the data with no FDR adjustment, with the understanding that false positives were not restrained a priori. In addition, the FDR corrections were not used in order to obtain a relevant number of genes for the network enrichment analysis.
ii) *Hierarchical clustering analysis.* Hierarchical clustering (HCL) analysis [52] was performed using TmeV. HCL was used to group DE genes based on their expression similarities across the samples. The average distance clustering method was employed, using sample tree selection and sample leaf order optimization. The distance metric used was the Pearson correlation, and HCL was performed only in the significant genes to reduce cluster noise.
iii) *Gene co-expression network analysis.* The differential transcriptomic datasets were used to generate the gene co-expression networks. The Cytoscape plug-in GeneMANIA [53] was used to predict DE gene interactions. Networks were generated using information from the co-expression category in GeneMANIA. Genes that co-express with DE genes (DE-related genes) were also included in the networks to study the interactions between DE genes and other co-expressed genes. To analyze the centrality of the nodes (genes) contained in the networks, the node centrality parameters "degree" and "betweenness" were calculated using the Cytoscape plug-in CentiScaPe [54]. Node degree is a local structure measure in networks that determines the number of edges linked to a node. Conversely, betweenness centrality is a global structure measure that defines the fraction of shortest paths passing through a node.
iv) *Identification of high-hubs, hubs, and bottlenecks.* Scatterplots were constructed using degree and betweenness values for each node in GraphPad Prism®5. These scatterplots allowed node hierarchy categorization in high-hubs, hubs, and bottlenecks. This categorization takes in account gene localization in different quadrants of the graph. High-hubs are placed in the up-right quadrant due to their higher degree and betweenness values. Conversely, hubs are located in the down-right quadrant, as they present high degree but lower betweenness values compared to high-hubs. Finally, bottlenecks are located in the up-left quadrant, as they show high betweenness but low degree values.
v) *Functional enrichment analysis.* Overrepresented biological functions were searched in the differential transcriptomic datasets using FunNet, a bioinformatics tool that performs functional profiling of gene expression data [55].
vi) *Analysis of subnetworks derived from central nodes.* Subnetworks were built with the aid of Cytoscape [56], using the central nodes identified in each comparison. Semantic relationships were identified between genes and keywords with the text mining tool GenClip [57]. We searched for relationships using the keywords "immune response", "T regulatory cells", "autophagy", "ubiquitin-proteasome", "endocytosis", "fibrosis", "inflammation", "extracellular matrix", "cell adhesion", and "autoimmunity". These relationships were highlighted in each subnetwork.

Additional files

Additional file 1: Differentially expressed genes obtained in each comparison. This table contains the differentially expressed (DE) genes obtained in each comparison and their fold changes. Statistical analysis using unpaired and paired t tests identified 250, 434, 417, and 593 DE genes according to the respective comparisons: t0 (very high versus high uRBP levels), t12 (very high versus high uRBP levels), group 1 (t12 × t0), and group 2 (t12 × t0).

Additional file 2: High-hub, hub, and bottleneck genes identified in each comparison. Hubs were defined as highly connected nodes according to node degree values. High-hubs are top-ranked hubs presenting also high betweenness centrality values. Bottleneck genes were classified as nodes with high betweenness centrality but low node degree values.

Abbreviations
AZA: azathioprine; CsA: cyclosporine; DE: differentially expressed; eGFR: estimated glomerular function rate; HCL: hierarchical clustering; HG: high group; IF/TA: interstitial fibrosis and tubular atrophy; LMWP: low-molecular-weight proteins; MDRD: modification of diet in renal diseases; MYF: mycophenolate mofetil; PTD: proximal tubular dysfunction; uRBP: urinary retinol-binding protein; VHG: very high group.

Competing interests
The authors declare that they have no competing interests.

Authors' contributions
PGR, RC, NOSC, and APS designed the study. RC, PGR, ACCM, and APS performed the clinical procedures. EN, MAC, PGR, and HA performed the RNA extraction and microarray experiments. HA, PGR, RC, CAMF, NOSC, and APS analyzed the data. HA, CAMF, NOSC, and APS drafted the manuscript. All authors read and approved the final manuscript.

Acknowledgements
We would like to thank the following funding agencies for supporting this study: Fundação de Amparo à Pesquisa do Estado de São Paulo-FAPESP (grants 2009/53443-1, 2011/50761-2, 2012/02270-2) and Conselho Nacional de Desenvolvimento Científico e Tecnológico-CNPq (grants 307626/2014-8 and INCT Complex Fluids) and NAP e-Science USP.

Author details
[1]Department of Pediatrics, Faculdade de Medicina da Universidade de São Paulo (FMUSP), São Paulo, Brazil. [2]Laboratory of Transplantation

Immunobiology, Department of Immunology, Institute of Biomedical Sciences, Universidade de São Paulo (USP), São Paulo, Brazil. [3]Laboratory of Clinical and Experimental Immunology, Nephrology Division, Universidade Federal de São Paulo (UNIFESP), São Paulo, Brazil. [4]Instituto Israelita de Ensino e Pesquisa Albert Einstein, Hospital Albert Einstein, São Paulo, Brazil.

References

1. Pinsky BW, Lentine KL, Ercole PR, Salvalaggio PR, Burroughs TE, Schnitzler MA. Predicting long-term graft survival in adult kidney transplant recipients. Saudi J Kidney Dis Transpl. 2012;23:693–700.
2. Camara NO, Silva MS, Nishida S, Pereira AB, Pacheco-Silva A. Proximal tubular dysfunction is associated with chronic allograft nephropathy and decreased long-term renal-graft survival. Transplantation. 2004;78:269–75.
3. Nauta FL, Bakker SJ, van Oeveren W, Navis G, van der Heide JJ, van Goor H, et al. Albuminuria, proteinuria, and novel urine biomarkers as predictors of long-term allograft outcomes in kidney transplant recipients. Am J Kidney Dis. 2011; 57:733–43.
4. Woitas RP, Stoffel-Wagner B, Poege U, Schiedermaier P, Spengler U, Sauerbruch T. Low-molecular weight proteins as markers for glomerular filtration rate. Clin Chem. 2001;47:2179–80.
5. van den Brand JA, Hofstra JM, Wetzels JF. Low-molecular-weight proteins as prognostic markers in idiopathic membranous nephropathy. Clin J Am Soc Nephrol. 2011;6:2846–53.
6. Ball ST, Lapsley M, Norden AG, Cairns TD, Palmer AB, Taube DH. Urinary retinol binding protein in Indo-Asian patients with idiopathic interstitial nephritis. QJM. 2003;96:363–7.
7. Jaconi S, Saurat JH, Siegenthaler G. Analysis of normal and truncated holo- and apo-retinol-binding protein (RBP) in human serum: altered ratios in chronic renal failure. Eur J Endocrinol. 1996;134:576–82.
8. Schaub S, Mayr M, Hönger G, Bestland J, Steiger J, Regeniter A, et al. Detection of subclinical tubular injury after renal transplantation: comparison of urine protein analysis with allograft histopathology. Transplantation. 2007;84:104–12.
9. Pallet N, Chauvet S, Chassé JF, Vincent M, Avillach P, Levi C, et al. Urinary retinol binding protein is a marker of the extent of interstitial kidney fibrosis. PLoS One. 2014;9:e84708.
10. de Matos AC, Câmara NO, de Oliveira AF, Franco MF, Moura LA, Nishida S, et al. Functional and morphologic evaluation of kidney proximal tubuli and correlation with renal allograft prognosis. Transpl Int. 2010;23:493–9.
11. Amer H, Lieske JC, Rule AD, Kremers WK, Larson TS, Franco Palacios CR, et al. Urine high and low molecular weight proteins one-year post-kidney transplant: relationship to histology and graft survival. Am J Transplant. 2013;13:676–84.
12. Chinen R, Câmara NO, Nishida S, Silva MS, Rodrigues DA, Pereira AB, et al. Determination of renal function in long-term heart transplant patients by measurement of urinary retinol-binding protein levels. Braz J MedBiol Res. 2006;39:1305–13.
13. Wüthrich RP, Weinreich T, Schwarzkopf AK, Candinas D, Binswanger U. Postmarketing evaluation of mycophenolate mofetil-based triple therapy immunosuppression compared with a conventional azathioprine-based regimen reveals enhanced efficacy and early pharmacoeconomic benefit after renal transplantation. Transplant Proc. 1998;30:4096–7.
14. Akoh JA. Renal transplantation in developing countries. Saudi J Kidney Dis Transpl. 2011;22:637–50.
15. Ntemka A, Iliadis F, Papanikolaou N, Grekas D. Network-centric analysis of genetic predisposition in diabetic nephropathy. Hippokratia. 2011;15:232–7.
16. McDermott JE, Taylor RC, Yoon H, Heffron F. Bottlenecks and hubs in inferred networks are important for virulence in Salmonella typhimurium. J Comput Biol. 2009;16:169–80.
17. Hsu CW, Juan HF, Huang HC. Characterization of microRNA-regulated protein-protein interaction network. Proteomics. 2008;8:1975–9.
18. Benson M, Breitling R. Network theory to understand microarray studies of complex diseases. Curr Mol Med. 2006;6:695–701.
19. Reddy A, Huang CC, Liu H, Delisi C, Nevalainen MT, Szalma S, et al. Robust gene network analysis reveals alteration of the Stat5a network as a hallmark of prostate cancer. Genome Inform. 2010;24:139–53.
20. Bando SY, Silva FN, Costa Lda F, Silva AV, Pimentel-Silva LR, Castro LH, et al. Complex network analysis of CA3 transcriptome reveals pathogenic and compensatory pathways in refractory temporal lobe epilepsy. Plos One. 2013;8:e79913.
21. Yu H, Kim PM, Sprecher E, Trifonov V, Gerstein M. The importance of bottlenecks in protein networks: correlation with gene essentiality and expression dynamics. PLoS Comput Biol. 2007;3:e59.
22. Lebeau C, Debelle FD, Arlt VM, Pozdzik A, De Prez EG, Phillips DH, et al. Early proximal tubule injury in experimental aristolochic acid nephropathy: functional and histological studies. Nephrol Dial Transplant. 2005;20:2321–32.
23. Le Roux S, Devys A, Girard C, Harb J, Hourmant M. Biomarkers for the diagnosis of the stable kidney transplant and chronic transplant injury using the ProtoArray® technology. Transplant Proc. 2010;42:3475–81.
24. Tanjore H, Lawson WE, Blackwell TS. Endoplasmic reticulum stress as a pro-fibrotic stimulus. Biochim Biophys Acta. 1832;2013:940–7.
25. Cybulsky AV. The intersecting roles of endoplasmic reticulum stress, ubiquitin- proteasome system, and autophagy in the pathogenesis of proteinuric kidney disease. Kidney Int. 2013;84:25–33.
26. Heymann D. Autophagy: a protective mechanism in response to stress and inflammation. Curr Opin Investig Drugs. 2006;7:443–50.
27. Deretic V. Autophagosome and phagosome. Methods Mol Biol. 2008;445:1–10.
28. Jiang M, Liu K, Luo J, Dong Z. Autophagy is a renoprotective mechanism during in vitro hypoxia and in vivo ischemia-reperfusion injury. Am J Pathol. 2010;176:1181–92.
29. Dohi E, Tanaka S, Seki T, Miyagi T, Hide I, Takahashi T, et al. Hypoxic stress activates chaperone-mediated autophagy and modulates neuronal cell survival. Neurochem Int. 2012;60:431–42.
30. Seto S, Tsujimura K, Koide Y. Coronin-1a inhibits autophagosome formation around Mycobacterium tuberculosis-containing phagosomes and assists mycobacterial survival in macrophages. Cell Microbiol. 2012;14:710–27.
31. Maluf DG, Mas VR, Archer KJ, Yanek K, Gibney EM, King AL, et al. Molecular pathways involved in loss of kidney graft function with tubular atrophy and interstitial fibrosis. Mol Med. 2008;14:276–85.
32. Pesenacker AM, Bending D, Ursu S, Wu Q, Nistala K, Wedderburn LR. CD161 defines the subset of FoxP3+ T cells capable of producing proinflammatory cytokines. Blood. 2013;121:2647–58.
33. Rouas-Freiss N, LeMaoult J, Moreau P, Dausset J, Carosella ED. HLA-G in transplantation: a relevant molecule for inhibition of graft rejection? Am J Transplant. 2003;3:11–6.
34. Jolly EC, Key T, Rasheed H, Morgan H, Butler A, Pritchard N, et al. Preformed donor HLA-DP-specific antibodies mediate acute and chronic antibody-mediated rejection following renal transplantation. Am J Transplant. 2012; 12:2845–8.
35. Ling XB, Sigdel TK, Lau K, Ying L, Lau I, Schilling J, et al. Integrative urinary peptidomics in renal transplantation identifies biomarkers for acute rejection. J Am Soc Nephrol. 2010;21:646–53.
36. Famulski KS, Kayser D, Einecke G, Allanach K, Badr D, Venner J, et al. Alternative macrophage activation-associated transcripts in T-cell-mediated rejection of mouse kidney allografts. Am J Transplant. 2010;10:490–7.
37. Badid C, Vincent M, McGregor B, Melin M, Hadj-Aissa A, Veysseyre C, et al. Mycophenolate mofetil reduces myofibroblast infiltration and collagen III deposition in rat remnant kidney. Kidney Int. 2000;58:51–61.
38. Bradley SP, Pahari M, Elias G, Uknis ME, Misra MV, Rastellini C, et al. Intragraft gene expression profile during acute cellular rejection in clinical small bowel transplantation: a case report. Transplant Proc. 2006;38:1740–1.
39. Scherer A, Gwinner W, Mengel M, Kirsch T, Raulf F, Szustakowski JD, et al. Transcriptome changes in renal allograft protocol biopsies at 3 months precede the onset of interstitial fibrosis/tubular atrophy (IF/TA) at 6 months. Nephrol Dial Transplant. 2009;24:2567–75.
40. Grzanka J, Leveson-Gower D, Golab K, Wang XJ, Marek-Trzonkowska N, Krzystyniak A, et al. FoxP3, Helios, and SATB1: roles and relationships in regulatory T cells. Int Immunopharmacol. 2013;16:343–7.
41. Rudra D, deRoos P, Chaudhry A, Niec RE, Arvey A, Samstein RM, et al. Transcription factor Foxp3 and its protein partners form a complex regulatory network. Nat Immunol. 2012;13:1009–10.
42. Behrends C, Sowa ME, Gygi SP, Harper JW. Network organization of the human autophagy system. Nature. 2010;466:68–76.
43. Ding WX, Ni HM, Gao W, Yoshimori T, Stolz DB, Ron D, et al. Linking of autophagy to ubiquitin-proteasome system is important for the regulation of endoplasmic reticulum stress and cell viability. Am J Pathol. 2007;171:513–24.
44. Tsushima K, Osawa T, Yanai H, Nakajima A, Takaoka A, Manabe I, et al. IRF3 regulates cardiac fibrosis but not hypertrophy in mice during angiotensin II-induced hypertension. FASEB J. 2011;25:1531–43.

45. Walker JL, Zhai N, Zhang L, Bleaken BM, Wolff I, Gerhart J, et al. Unique precursors for the mesenchymal cells involved in injury response and fibrosis. Proc Natl Acad Sci U S A. 2010;107:13730–5.
46. Stacchiotti A, Rezzani R, Angoscini P, Rodella L, Bianchi R. Small heat shock proteins expression in rat kidneys treated with cyclosporine A alone and combined with melatonin. Histochem J. 2002;34:305–12.
47. Abbate M, Brown D, Bonventre JV. Expression of NCAM recapitulates tubulogenic development in kidneys recovering from acute ischemia. Am J Physiol. 1999;277:F454–63.
48. Bozic M, de Rooij J, Parisi E, Ortega MR, Fernandez E, Valdivielso JM. Glutamatergic signaling maintains the epithelial phenotype of proximal tubular cells. J Am Soc Nephrol. 2011;22:1099–111.
49. Solez K, Colvin RB, Racusen LC, Sis B, Halloran PF, Birk PE, et al. Banff '05 meeting report: differential diagnosis of chronic allograft injury and elimination of chronic allograft nephropathy. Am J Transplant. 2007;7:518–26.
50. Correa-Costa M, Azevedo H, Amano MT, Gonçalves GM, Hyane MI, Cenedeze MA, et al. Transcriptome analysis of renal ischemia/reperfusion injury and its modulation by ischemic pre-conditioning or hemin treatment. PLoS One. 2012;7:e49569.
51. Saeed AI, Sharov V, White J, Li J, Liang W, Bhagabati N, et al. TM4, a free, open-source system for microarray data management and analysis. Biotechniques. 2003;34:374–8.
52. Eisen MB, Spellman PT, Brown PO, Botstein D. Cluster analysis and display of genome-wide expression patterns. Proc Natl Acad Sci U S A. 1998;95:14863–8.
53. Montojo J, Zuberi K, Rodriguez H, Kazi F, Wright G, Donaldson SL, et al. GeneMANIA Cytoscape plugin: fast gene function predictions on the desktop. Bioinformatics. 2010;26:2927–8.
54. Scardoni G, Petterlini M, Laudanna C. Analyzing biological network parameters with CentiScaPe. Bioinformatics. 2009;25:2857–9.
55. Prifti E, Zucker JD, Clement K, Henegar C. FunNet: an integrative tool for exploring transcriptional interactions. Bioinformatics. 2008;24:2636–8.
56. Smoot M, Keiichiro O, Ruscheinski J, Wang PL, Ideker T. Cytoscape 2.8: new features for data integration and network visualization. Bioinformatics. 2011;27:431–2.
57. Huang ZX, Tian HY, Hu ZF, Zhou YB, Zhao J, Yao KT. GenCLiP: a software program for clustering gene lists by literature profiling and constructing gene co-occurrence networks related to custom keywords. BMC Bioinformatics. 2008;9:308.

4

The clinical trial landscape in oncology and connectivity of somatic mutational profiles to targeted therapies

Sara E. Patterson, Rangjiao Liu, Cara M. Statz, Daniel Durkin, Anuradha Lakshminarayana and Susan M. Mockus*

Abstract

Background: Precision medicine in oncology relies on rapid associations between patient-specific variations and targeted therapeutic efficacy. Due to the advancement of genomic analysis, a vast literature characterizing cancer-associated molecular aberrations and relative therapeutic relevance has been published. However, data are not uniformly reported or readily available, and accessing relevant information in a clinically acceptable time-frame is a daunting proposition, hampering connections between patients and appropriate therapeutic options. One important therapeutic avenue for oncology patients is through clinical trials. Accordingly, a global view into the availability of targeted clinical trials would provide insight into strengths and weaknesses and potentially enable research focus. However, data regarding the landscape of clinical trials in oncology is not readily available, and as a result, a comprehensive understanding of clinical trial availability is difficult.

Results: To support clinical decision-making, we have developed a data loader and mapper that connects sequence information from oncology patients to data stored in an in-house database, the JAX Clinical Knowledgebase (JAX-CKB), which can be queried readily to access comprehensive data for clinical reporting via customized reporting queries. JAX-CKB functions as a repository to house expertly curated clinically relevant data surrounding our 358-gene panel, the JAX Cancer Treatment Profile (JAX CTP), and supports annotation of functional significance of molecular variants. Through queries of data housed in JAX-CKB, we have analyzed the landscape of clinical trials relevant to our 358-gene targeted sequencing panel to evaluate strengths and weaknesses in current molecular targeting in oncology. Through this analysis, we have identified patient indications, molecular aberrations, and targeted therapy classes that have strong or weak representation in clinical trials.

Conclusions: Here, we describe the development and disseminate system methods for associating patient genomic sequence data with clinically relevant information, facilitating interpretation and providing a mechanism for informing therapeutic decision-making. Additionally, through customized queries, we have the capability to rapidly analyze the landscape of targeted therapies in clinical trials, enabling a unique view into current therapeutic availability in oncology.

Keywords: Cancer, Precision medicine, Actionability, Clinical trials, Curation

* Correspondence: susan.mockus@jax.org
The Jackson Laboratory for Genomic Medicine, 10 Discovery Dr., Farmington, CT 06032, USA

Introduction

The advent of the genomic era has provided clinicians and researchers the ability to analyze molecular data from patients and identify genetic variants that may have an impact on their clinical outcome and treatment options. Cancer research has additionally identified a myriad of genetic variations that impact protein function, the pathology of tumor cells, and potential response to targeted therapies. Connecting this information to clinical patient data is critical for the implementation of precision medicine. However, this information is vast and disparate, which hampers the ability to access potentially crucial information in a clinically acceptable time frame. Access to this data requires several key components: a structured and well-organized database for deposition of clinically relevant data, accurate manual curation of data with limited variability, accessibility of connections between data elements via well-defined relationships, and a system for routinely and automatically mapping clinical sample data to the database. A number of publicly available databases exist that catalog cancer-related genomic variations or that connect variations to potentially relevant therapies, but none complement the need for connecting patient aberrations to targeted therapy—either through clinical trials or approved drugs, while incorporating supporting efficacy information. For instance, the COSMIC database provides an invaluable catalog of cancer-related somatic genetic aberrations but does not assess relationships between those variants and therapies [1]. The My Cancer Genome database from Vanderbilt incorporates efficacy data for well-studied molecular aberrations that could prove useful in clinical interpretation [2]. However, the content is confined to a small variant list and is not routinely updated and as a result, the depth and breadth of the coverage of molecular targets and targeted therapies, as well as patient indications and clinical trials curated is limited, effectively hindering its utility. In addition to the scarcity of databases populated with comprehensive targeted oncology clinical data, a system that can directly link patient sequence data to clinical information is lacking, and thus, the speed at which these data can be related to targetable mutations in tumor samples is greatly reduced.

To enable this process, we have developed a clinical bioinformatics and curation pipeline that operates within a Clinical Laboratory Improvements Amendment (CLIA) and College of American Pathologists (CAP)-accredited environment, the JAX Clinical Genome Analytics (CGA) system. This system enables systematic identification and annotation of clinically relevant cancer variants and facilitates connections to therapeutic interventions. JAX-CGA comprises several components, including an automated data loader and mapper, which loads called variants from clinical samples and transforms them to Human Genome Variation Society (HGVS) nomenclature using Human Genome Organisation (HUGO) gene symbols and subsequently maps the variants to the database. In addition, our in-house curation database, the JAX Clinical Knowledgebase (JAX-CKB) enables dynamic curation of data connecting genetic variant to phenotype and protein effect, as well as therapeutic relevance and potential treatment approaches, which can be queried readily for clinical reporting. To facilitate interoperability between databases, the JAX-CKB utilizes standardized variant nomenclature and incorporates specific ontologies. HUGO, through the HUGO Gene Nomenclature Committee (HGNC), maintains a catalog of unique approved gene names, which we have incorporated into the JAX-CKB, reducing ambiguity and enabling interoperability between databases [3]. Additionally, the Human Genome Variation Society (HGVS) provides guidelines for the standardization of variant nomenclature, which are actively maintained and updated, facilitating unambiguous variant naming [4].

A fundamental challenge to precision medicine is the ability to easily connect a patient's genetic variants with a therapeutic approach, which could include either FDA-approved targeted therapies or targeted therapies in recruiting clinical trials. Recent studies have demonstrated that linking patients to relevant targeted therapies based on genomic data has the potential to achieve higher clinical success relative to recruitment solely on histological subtype [5]. Leveraging the power of an integrated knowledgebase, we can identify opportunities for clinical intervention, including available clinical trials for targeted therapies. A comprehensive database can also provide a unique view into the landscape of molecular targets and therapies, uncovering potential opportunities for new research and development. In addition to supporting clinical reporting, the JAX-CKB allows visibility into deficiencies in the characterization of potentially actionable mutations and therapeutic interventions, exposing opportunities for research that have the potential to advance cancer treatment. Here, we detail and share the processes by which we map patient data to the knowledgebase and provide a detailed view of the clinical trial landscape in solid tumors.

Results and discussion

The organization of the CGA system (Fig. 1) incorporates several components, including the bioinformatics pipeline, the data loader and mapper, the JAX Clinical Knowledgebase, and reporting tools. The bioinformatics pipeline portion of the CGA system, including its validation, is discussed elsewhere [6].

The flow of data through the CGA-CKB system begins with the transformation and automated mapping of patient variants to JAX-CKB. Mapped variants are then filtered for actionability, which is defined as those with a

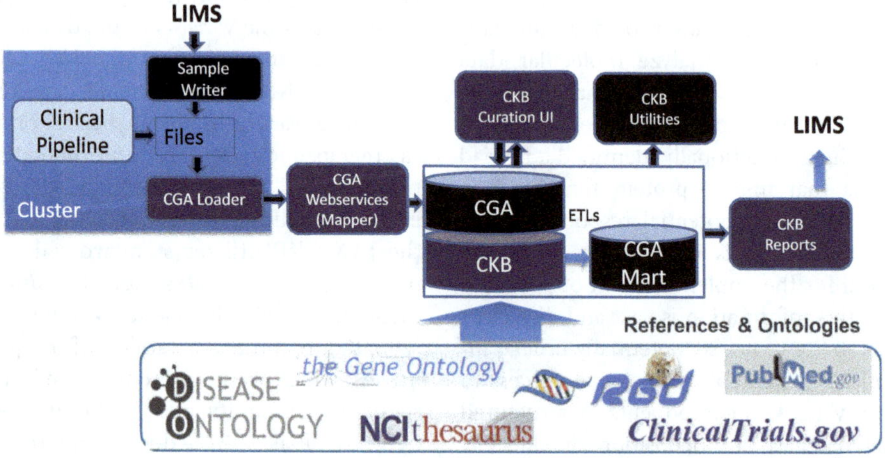

Fig. 1 Flow of data through the clinical pipeline. Patient sample sequence data from the clinical genome analytics (CGA) pipeline is integrated with sample meta-data from the Laboratory Information Management System (LIMS), and called variants are mapped to the JAX-Clinical Knowledgebase (JAX-CKB) database. The JAX-CKB incorporates information populated from various incorporated ontologies and databases, and via the curation User Interface (UI), and is maintained using the CKB-Utilities tools. Using information from the JAX-CGA/CKB and the CKB reports, tools are used to generate datasets to enable clinical report generation

related FDA-approved therapy or targeted therapy in clinical trials. Filtered actionable variants are then linked to treatment approaches and relevant recruiting clinical trials using unique clinical reporting tools through drug class or therapy, as well as by molecular criteria. This process is facilitated by the design of the JAX-CKB database, which incorporates connectivity of various database attributes, through the use of standard language and ontologies. The user interface allows for annotation of the most current data surrounding genetic variants and targeted therapies, on top of a system that incorporates standard ontologies and controlled vocabularies. This data is then readily accessible via reporting and analysis queries for both patient clinical reporting applications and for comprehensive analysis of data, including data regarding clinical trials. To enable both clinical reporting and analysis of clinically relevant data surrounding the JAX-CTP-targeted gene panel, we have implemented a unique framework for managing and connecting patient molecular data to efficacy information and relevant treatment approaches in real-time.

Data loader and mapper

The data loader and mapper play a critical role in routine processing of large scale variation results from the CGA pipeline. The roles of the data loader and mapper are to upload genomic sequence data and integrate that sequence data with sample meta-data and to automatically map patient variants to JAX-CKB, enabling further queries for clinical interpretation.

The CGA loader (Fig. 2) runs as a daemon program, which intermittently scans the file system in a cluster server searching for new runs from the clinical analytics (CGA) pipeline. New runs consist of either analysis results of a new sample or repeat analysis results from a previously mapped sample. The CGA loader program additionally extracts sample meta-data, such as patient diagnosis, from our Laboratory Information Management System (LIMS) and submits these data along with the variant call format (VCF) files to a Restful web service.

The web service, running in a Tomcat application server, saves all data files in the database with an initial "pending" file status and manages the workflow of all computational jobs in a queue. The format of each data file is defined as a template in the database allowing an integrated Java program, using Reflection, to dynamically parse, transform, and map file columns to database tables and columns. During this workflow, additional filtering or validation steps can be integrated for quality or biological reasons. If an error occurs in the loading process, the file status will be updated as "error" and a tracking message will be logged. Following successful completion of a job, the file status will be updated as "loaded."

The Mapper (Fig. 3) is the nickname of the web service that manages the automated data loading and mapping jobs. During the data loading workflow, sequence variants, which include point mutations or small insertions or deletions (InDels), are transformed to standard Human Genome Variation Society (HGVS) nomenclature [4], which is consistent with the regular expression (regex) vocabulary of variants in the JAX-CKB. For example, the SnpEff and SnpSift programs annotate an amino acid from nucleotide sequence information using a three-letter amino acid code (e.g., "Gly" for Glycine). This three-letter code is then transformed to the single letter code for that amino

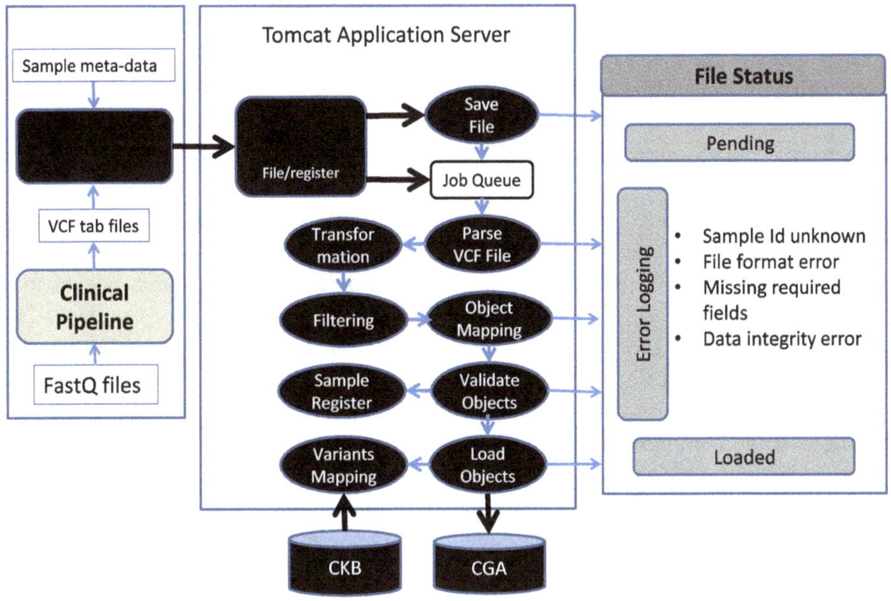

Fig. 2 Automated data loading process. The data loader scans the file system hourly to find any new runs from the clinical pipeline and submits the variant call files (VCF) along with their sample meta-data to a web service. The web service manages the data loading jobs in a queue, to transform, filter, and upload the datasets to our CGA database and map the variants to the JAX-CKB

acid, using the appropriate syntax (e.g., "G" for "Gly"). For example, p.Gly749Glu/c.2246G > A will be transformed to "G749E," and the frameshift Arg34_Val35fs/c.102_103insG will be transformed to "R34fs." Subsequent to transformation, several filtering steps have been implemented to ensure the use of only high-quality somatic variations and reduce false-positives. Filters include removal of variants with low coverage, variants with allele frequencies that do not meet minimum requirements for sensitivity, low impact variants including silent mutations, variants that are likely germline, and those that are likely false-positive or outside of coding regions. Simultaneously, Copy Number Variation

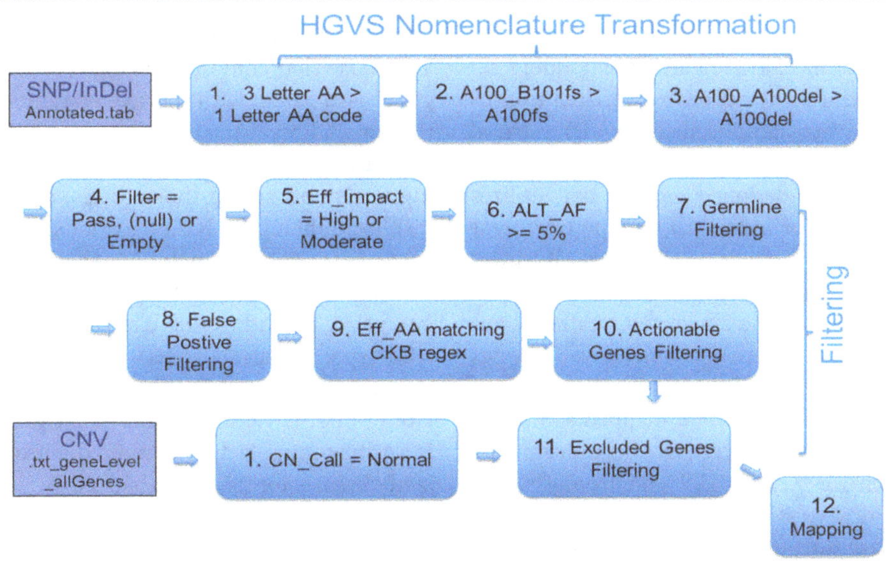

Fig. 3 Automated variant mapping process. The genomic datasets are transformed to standard Human Genome Variation Society (HGVS) nomenclature, using a standardized Regular Expressions (regex) vocabulary incorporated into the JAX-CKB. This enables mapping and identification of clinically meaningful variants in a scalable and reproducible manner. The variant mapping program also includes several filtering steps, including removal of low-impact variants, those that did not pass metrics, and those that are likely germline or false-positive, to ensure reporting of only high-quality somatic variations

(CNV) files are filtered to remove genes with fewer than six copies, based on the analytical validation of the assay [6]. Following filtering, CNV files are merged with the filtered VCFs. The analytics program will then automatically recognize and map the variants in the JAX-CKB, enabled by the transformation to the HGVS syntax used by JAX-CKB.

This process allows for direct connectivity between clinically actionable somatic variants present in patient tumor samples with applicable clinical data.

Database queries and clinical reporting tools

While compiling and organizing clinically relevant data has utility, the power to utilize these data lies in the ability to easily access and leverage complex data to inform clinical decision-making. To that end, we have designed and implemented several database queries, which both facilitate reporting and enable unique views into the data for analysis and quality assurance.

Following the mapping of samples to the JAX-CKB, mapped variants are filtered and displayed in a JAX-CKB reporting application. From the sample landing page, a user can select the relevant sample and launch queries directly from the sample results, allowing the user to tailor the queries to sample-specific molecular aberrations while providing traceability. Several queries have been built to generate result sets that support clinical reporting. The queries are designed to run based on all clinically relevant variants in a single sample simultaneously; however, the search page for each query incorporates a selection tool from which a user can deselect any variant in the sample to exclude as search parameters. This enables the user to run the queries on only variants that have clinical relevance, as well as limiting query results from samples with many actionable variants to produce multiple, more manageable, datasets. To retrieve the list of targeted therapies associated with sample-specific variants, we have designed a query that utilizes the associated drug class or individual therapy treatment approaches specific to variants to retrieve all drugs assigned to those treatment approaches. For example, the variant KIT D816V has the associated treatment approach "KIT Inhibitors." There are currently 31 drugs curated into JAX-CKB that have been assigned the drug class "KIT Inhibitors." If present in a mapped sample in CKB, launching the drug query retrieves the drug class treatment approach for KIT D816V and from this retrieves the list of drugs associated with this drug class. In addition, there are three separate queries that leverage different aspects of curated efficacy evidence. The approval status query retrieves data from the "approval status" attribute of curated efficacy evidence for all drugs related to treatment approaches for given variants. This enables identification of targeted therapies that are FDA-approved versus investigational therapies, related to a treatment approach for the patient variant(s). The drug resistance query retrieves efficacy evidence entries for all molecular profile responses with response type of "resistant" that contains the same parent gene of sample-specific variants. For example, for a sample containing a KRAS G12C variant, the resistance query would retrieve all resistance efficacy evidence associated with any molecular profile response containing a variant in KRAS. This would include the efficacy evidence related to KRAS G12C, as well as category variants "KRAS G12X" and closely related variants KRAS G12D and KRAS G13C. Relating to parent gene, rather than specific variant, enables visibility into all potential resistance mechanisms without requiring duplication of data in CKB. A third query pulls all evidence with sensitive response type related to the treatment approach for patient variants, to support rationale for therapies included on patient reports.

Clinical trial query

Clinical trials function as an avenue toward clinical intervention and constitute an important component of the therapeutic landscape for oncology patients. Thus, we have incorporated the identification of clinical trials relevant to patient variants into clinical reporting to provide multiple potential options for clinicians to direct patient treatment. Clinical trials can be relevant to patient variants via inclusion of drugs contained in variant treatment approaches, or through molecular inclusion criteria. To retrieve a list of open clinical trials related to sample variants, we have implemented a complex query that retrieves clinical trials both for drugs in treatment approaches and those that contain molecular criteria for the parent gene associated with sample variants, irrespective of therapy. Clinical trials are queried first by drugs contained in variant treatment approaches, and for resulting trials containing those drugs, data is retrieved and displayed including NCTID, sponsor, title, and therapy, as well as a column including the variant for which the trial was pulled. Clinical trials are also queried on parent gene, excluding all trials with a requirement for wild-type. Clinical trials can be further restricted through the selection of Disease Ontology terms, related to patient indication. The Disease Ontology (DO) is a robust, regularly maintained, ontology of standardized disease terms, each backed by individual unique identifiers (DOIDs), which facilitates precise mapping to patient diagnosis [7]. When selected, returned trials will include only those that are related to the selected DOID and all child terms. This process is enabled by the integration of the DO tree in clinical trial curation, which can be employed for retrieval of only those trials relevant to the patient's indication.

Clinical trial landscape

Outside of CKB, there are currently no publicly available comprehensive databases that curate information on molecular eligibility for clinical trials. Access of clinical trials recruiting on molecular criteria is not readily accessible through clinical trial registries, such as clinicaltrials.gov, which is primarily the result of a lack of standardization of nomenclature and syntax for molecular eligibility criteria in clinical trial registries [8]. One significant potential outcome of this is an inadequate representation of potentially relevant clinical trials for clinicians performing a cursory keyword search. In addition, the lack of molecular criteria accessibility likely increases the time necessary for identifying potentially relevant trials, as a clinician must sort through the trial record to identify molecular inclusion criteria. To expedite retrieval of relevant clinical trials and streamline the path toward clinical intervention based on molecular variant, we have incorporated manual curation of molecular eligibility into the JAX-CKB database to easily connect patients with trials that are relevant to findings from next-generation sequencing.

In addition to providing a path to intervention in clinical reporting, the organization and design of the JAX-CKB (discussed below) allows us to readily analyze the current landscape of clinical trials. This analysis of the relative number of clinical trials for drug class or molecular criteria enables a view into the strengths and weaknesses in targeting actionable variants in clinical trials. Previous analysis of the landscape of cancer clinical trials in the USA, in the absence of a structured curated database, has required significant effort using data extracted manually via keyword search of clinicaltrials.gov [9]. The organization of data elements curated into the JAX-CKB database facilitates rapid analysis of data from clinical trials through customized queries of the database. Using these queries, we have analyzed the landscape of trials restricted to those currently open (defined as those with the recruitment status of "not yet recruiting," "recruiting," or "available") related to our 358-gene targeted panel.

When broken down by patient indication (Fig. 4), 19.1 % of open trials curated into the JAX-CKB are recruiting patients with advanced solid tumors with

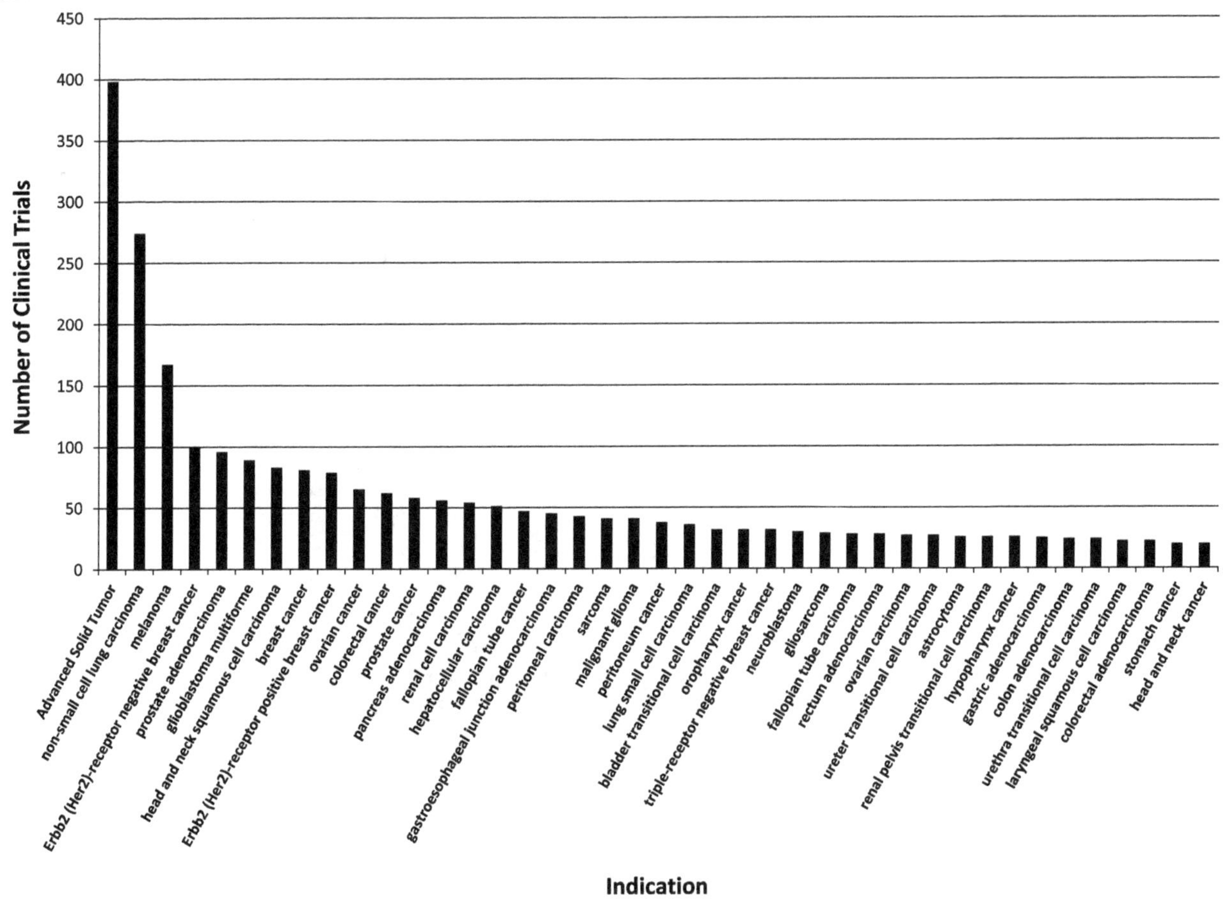

Fig. 4 Analysis of JAX-CKB curated clinical trials by solid tumor indication. Disease Ontology terms selected as patient indication for greater than 20 clinical trials plotted relative to the number of open clinical trials recruiting on each indication

unspecified histology (398/2058; Fig. 4). This comprises the largest proportion of curated clinical trials and likely represents early stage trials for therapeutics that have not demonstrated clinical utility in specific tumor types. Other well-studied indications are highly represented in clinical trials (Fig. 4), particularly those for which a recurring molecular target has been identified. These include non-small cell lung cancer (13.23 %; 274/2058), melanoma (8.0 %; 167/2058), ERBB2 (Her2)-receptor negative breast cancer (4.8 %; 100/2058), prostate adenocarcinoma (4.7 %; 96/2058), glioblastoma multiforme (4.3 %; 89/2058), and head and neck squamous cell carcinoma (4.0 %; 83/2058) (Fig. 4). There are 96 indications present as inclusion criteria in only a single open clinical trial for a targeted therapy, representing an unmet need for these patients. These indications include rare cancers, such as chordoma (NCT01407198) and hemangiopericytoma (NCT01396408), which have very little data surrounding targeted therapeutic efficacy, and accordingly very few promising treatment options [10, 11].

Independent of histology, patients may enter clinical trials based on molecular eligibility, either through a molecular aberration that is targeted by an agent in the trial, or a trial that is directly recruiting on the presence of a specific molecular criterion, irrespective if the investigated therapy has demonstrated a connection to the molecular target. Drug classes are a proxy for treatment approach, and identifying the number of clinical trials investigating drugs within specific drug classes provides insight into the availability of targeted therapies for molecular aberrations. In an analysis of the most frequent drug classes containing drugs currently under investigation in open clinical trials, we have determined that Pan-VEGFR Inhibitors (378/2168; 17.4 %), KIT Inhibitors (312/2168; 14.4 %), Pan-PDGFR Inhibitors (208/2168; 9.6 %), and RET Inhibitors (205/2168; 9.5 %) are most highly represented (Fig. 5). This suggests that patients with activating mutations in KDR (VEGFR2), KIT, PDGFR, and RET, depending on indication, may have several options for potential therapeutic intervention through recruiting clinical trials. In contrast, the drug classes DNMT1 inhibitors, EZH2 inhibitors, and FGFR3 antibody are among those represented in a single open clinical trial, which suggests that patients harboring genetic aberrations targeted by these drug classes may have fewer options.

Clinical trials may also recruit patients based directly on molecular eligibility, which may be restricted to specific molecular aberrations, or a wider category of molecular aberrations. For instance, a patient with an exon

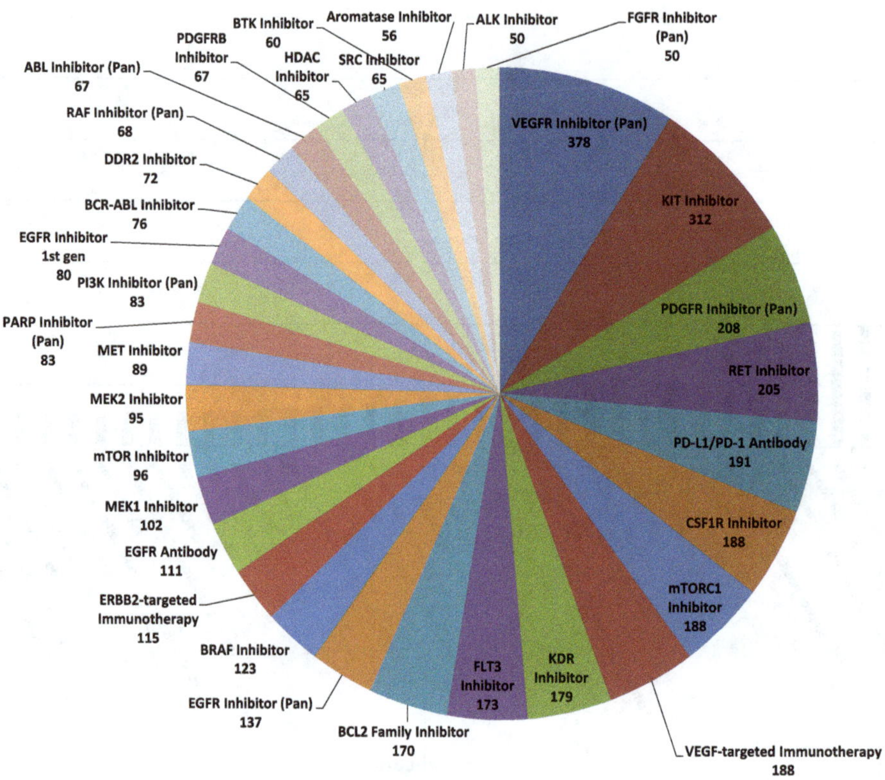

Fig. 5 Analysis of open clinical trials related to drug class. Chart displays the number of clinical trials investigating targeted drugs within specified drug classes, for all drug classes represented in greater than 50 open clinical trials

19 deletion in EGFR could potentially enroll in clinical trials recruiting patients with "EGFR mutations," "EGFR activating mutations," "EGFR exon 19 deletions," or the specific deletion harbored by this patient. Currently, the top 5 molecular criteria in open clinical trials are related to the expression of various hormone and growth factor receptors including ERBB2, PGR, and ESR1 (Fig. 6), which is likely related to their long-standing clinical association with breast cancer subtyping and therapeutic relevance [12]. Mutations in BRAF V600 have demonstrated clinical utility as targetable aberrations [13] and are highly represented as molecular eligibility criteria in clinical trials, with 36 open trials recruiting on BRAF V600 mutations at the time of this manuscript (Fig. 6b). BRCA1 and BRCA2 mutations are also frequently included as molecular eligibility criteria, with 35 and 33 open trials recruiting on these variants, respectively (Fig. 6b).

EGFR aberrations are a highly represented target for investigational therapies in clinical trials. EGFR is the fifth most common gene associated with molecular inclusion criteria for open clinical trials curated into CKB, with 86 clinical trials including EGFR-related molecular inclusion criteria (Fig. 6a). Most of these trials are related to the well-characterized EGFR L858R and exon 19 deletion mutations (Fig. 6c), which have demonstrated sensitivity to EGFR tyrosine kinase inhibitors (TKIs) [14]. In contrast, EGFR exon 20 insertion mutations demonstrate de novo resistance to EGFR TKIs and have few options for clinical intervention [15, 16]. At the time of this manuscript, EGFR exon 20 insertions are included as eligibility criteria for two open clinical trials.

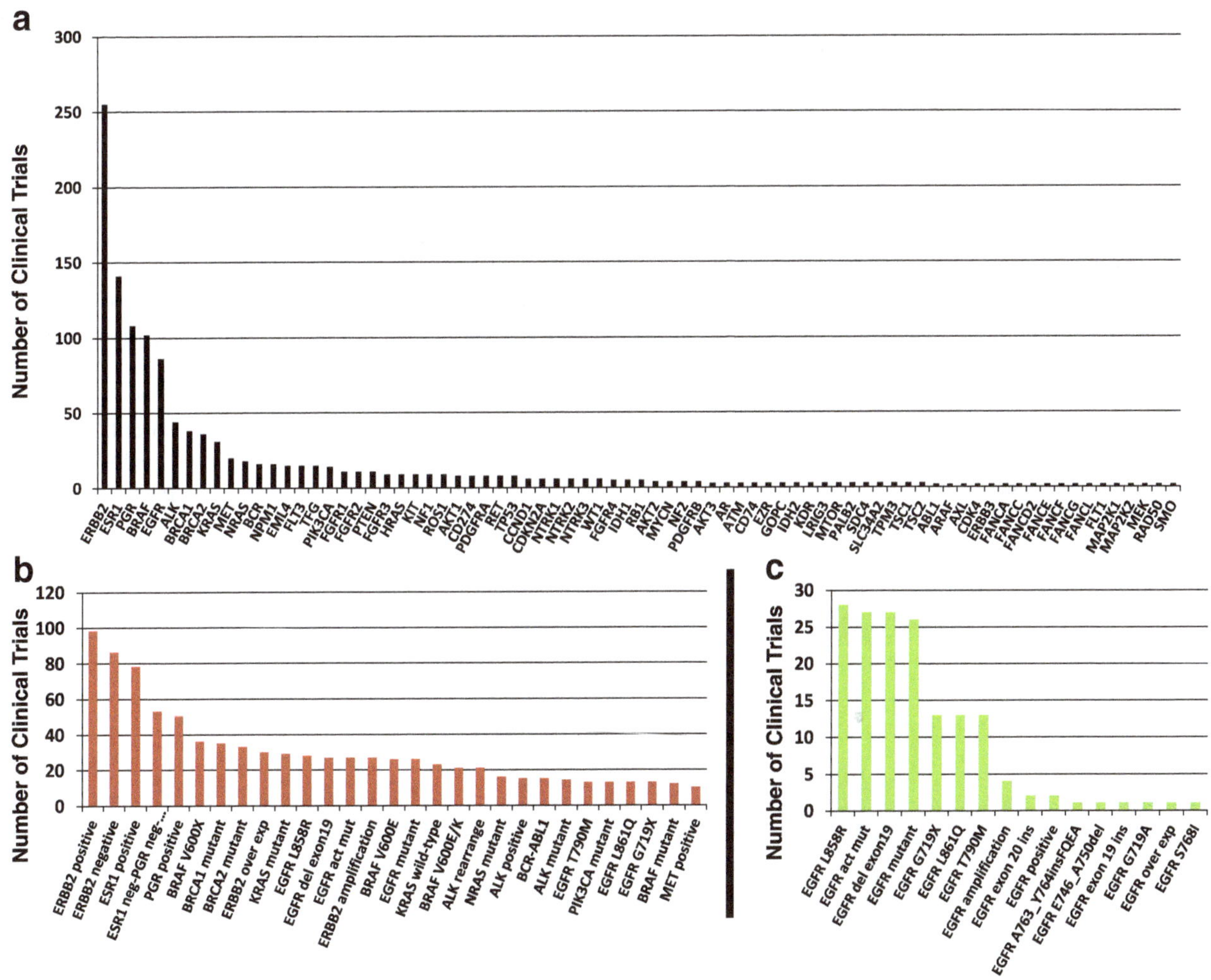

Fig. 6 Analysis of number of open clinical trials relative to gene and gene variant as molecular criteria. **a** Combined total of clinical trials with eligibility criteria related to individual genes. Graph displays genes that are represented by molecular eligibility criteria in two or greater clinical trials. **b** Number of clinical trials for the top specific or category molecular criteria represented in clinical trials, limited to those in 10 or more clinical trials. **c** Number of clinical trials for each type of EGFR molecular criteria present in open clinical trials

Both of these trials are investigating the HSP90 inhibitor AUY922 and are recruiting either specifically on exon 20 insertions or for EGFR mutations, including exon 20 insertions. HSP90 inhibitors have not demonstrated full clinical utility in the context of EGFR mutations but have preclinical support [17]. In this context, HSP90 inhibitors do not meet the stringency for a JAX-CKB treatment approach, and thus would not be an entry point for clinical trial association for patients with EGFR exon 20 insertion mutations. Therefore, the inclusion of molecular criteria in this instance represents a mechanism for therapeutic intervention that would not otherwise be available.

The JAX clinical knowledgebase and curation user interface

A critical component to accessing data for reporting or analysis is a comprehensive structured database. Thus, we have designed and implemented an in-house database to support clinical reporting and analysis of clinically relevant data. The JAX-CKB serves as a repository for data regarding oncology-relevant genetic variants and therapies and incorporates interconnectivity between protein effect, therapeutic efficacy, treatment approaches, and recruiting clinical trials. CKB features a structured database framework for managing clinical and biological data, into which clinical data content is entered through a curation user interface (Fig. 7). Data is dynamically curated by experts to populate comprehensive content related to oncology, to support clinical reporting, and to maintain the most up-to-date information on therapies, response, and relevant clinical trials. While JAX-CKB content is currently tailored toward solid tumor clinical reporting, the framework of the JAX-CKB is versatile and can support additional applications. The organization of the JAX-CKB is essential to its utility in supporting clinical reporting and includes several key elements (Fig. 3), which are outlined below.

Data elements

The JAX-CKB database incorporates several data elements, which are connected in various ways in curation of cancer-related data.

Genes and gene variants

To support standardization and mapping of genetic variants to JAX-CKB and facilitate interoperability with other databases, we have incorporated HUGO Gene Nomenclature Committee (HGNC) nomenclature through the integration of approved symbols and gene names from the NCBI Gene database into the JAX-CKB [3, 18]. By incorporating gene names from the Gene database, genes are assigned approved symbols, removing variability, and are additionally backed by unique numerical gene identifiers (NCBI GeneID). Although our current JAX Cancer Treatment Panel (JAX CTP) is limited to 358 cancer-related genes, the JAX-CKB incorporates all known annotated human genes. This supports the rapid pace of advancement in cancer research, as well as allowing for the flexibility to support additional assays, which would include sequencing assays for other disease types.

Naming of genetic variants in the JAX-CKB utilizes a Regular Expression (regex) system that conforms to Human Genome Variation Society (HGVS) nomenclature guidelines, restricting the entry of genetic variants to a predefined syntax and enabling mapping of variants to JAX-CKB. HGVS maintains standards for variant naming, which serves to eliminate ambiguity and ensure precision when reporting data related to genetic variants [4]. Conforming to these standards additionally provides a mechanism to promote interoperability between databases.

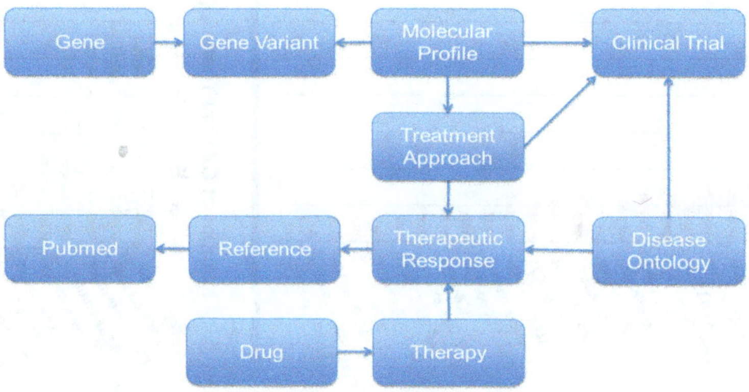

Fig. 7 Connections between various data elements in the JAX Clinical Knowledgebase (JAX-CKB). Molecular profiles consist of one of more molecular entities, linked to gene variants and parent genes. Molecular profiles are assigned relevant treatment approaches based on available data from the literature on efficacy, as a therapeutic response. The therapeutic response additionally takes into account patient indication, which is represented by an integrated disease ontology. Treatment approaches contain therapies, which consist of a single drug or combination of drugs. These drugs can be either FDA-approved, or under investigation in clinical trials, which are additionally curated into the JAX-CKB

Each variant type is assigned a specified syntax. For example, a mutation resulting in a single amino acid change, such as the gene variant KRAS G12C, would follow the syntax [ACDEFGHIKLMNPQRSTVWY][0-9] + [ACDEFGHIKLMNPQRSTVWY]. The regex syntax is periodically updated to maintain compliance with HGVS guidelines. The JAX-CKB additionally incorporates controlled vocabularies regarding protein effect, to annotate functional relevance of specific variants. This aids in the interpretation of actionability and provides a further mechanism for filtering gene variants. Gene variants are linked to a parent gene and can include any alteration to the gene, supporting multi-omic data curation. Examples of variants include mutations, changes to gene copy number, and expression changes. In addition, JAX-CKB supports annotation of category variants (e.g., activating mutation, inactivating mutation), which allows molecular profile responses and clinical trials to be curated to conditions where they are applicable to multiple or unknown/unspecified aberrations. For example, the category variant "EGFR activating mutation" represents any activating mutation in EGFR. There are also less broad category variants, such as "EGFR del exon 19," which represents a non-specified deletion within exon 19 of the EGFR gene. These variants are also subject to regex syntax and must meet predefined specifications. The inclusion of category variants enables capture of information that is relevant to bins of variants, rather than individual variants, or in conditions where the applicable variants are not specified.

Within the Gene Variant Curation User Interface (UI), there are multiple fields for data capture, each utilizing their own specific controlled vocabularies to minimize variability and allow for efficient analysis of curated information. The curation user interface allows users to capture information around genetic variants related to amino acid change, mutational impact to the protein (e.g., missense, deletion, and frameshift), and functional effect on the protein (e.g., gain-of function, and loss-of-function). In addition, gene variants can be linked to their impact on biological pathways, through the integration of a human pathway ontology from the Rat Genome Database [19]. The gene variant UI also incorporates a free-text annotation field for capture of specific information on the relationship of the variant to oncogenic transformation and protein function, which is supported by curated references from primary literature.

Molecular profiles

A molecular profile in the JAX-CKB contains one or more genetic aberrations, which can include mutations, such as single nucleotide polymorphisms (SNPs), frameshifts, insertions, and deletions, as well as gene fusions, copy number variations and/or changes in expression levels. The flexibility of the molecular profile and the ability to build complex molecular profiles allow for the curation of therapeutic efficacy to multiple aberrations simultaneously, rather than each variant in isolation. For example, a molecular profile may contain a mutation, such as the EGFR activating mutation EGFR L858R, as well as a copy number variation, such as amplification of the MET gene. MET amplification is a recognized mechanism for resistance to EGFR tyrosine kinase inhibitor therapy [20–22], and the ability to create a complex profile enables the curation of resistance data to the appropriate complex molecular conditions.

Current research supports the idea that targeted treatment of tumor driver mutations alone often leads to the upregulation of compensatory mechanisms and/or acquisition of secondary resistance mutations and that utilizing a multi-target approach may be beneficial in reducing the potential for resistance [23]. For example, the ability to assign treatment approaches and curate efficacy evidence to multiple aberrations simultaneously enables JAX-CKB and downstream clinical reporting to stay on-pace with clinical progress in the use of combination therapies in complex multi-omic conditions.

Drugs and therapies

A key component to capture response data related to molecular targets is a comprehensive database of relevant drugs, which are incorporated as data elements into JAX-CKB. Attributes associated with drugs include drug name, trade name, and synonyms, in which each are required to be unique terms within a drug record, and among other records, to effectively eliminate redundancy. CKB currently contains 1108 unique targeted therapies relevant to treatment approaches or clinical trials for variants related to our 358-gene panel. When appropriate, drugs are assigned a "Drug Class," as a selectable field within the Therapy UI using an in-house controlled vocabulary. Drug class represents a rational group of drugs with a unified target, such as "PIK3CA Inhibitor" for drugs that target PIK3CA. These drug classes are then utilized as "Treatment Approaches" for a molecular profile, which provides a link between molecular aberrations and relevant targeted therapies.

Therapies, similar to molecular profiles, are comprised of one or more drugs. The ability to build complex therapies supports the therapeutic trend toward using targeted therapies in combination with other targeted therapies or chemotherapies and enables precision and accuracy in efficacy curation.

Treatment approaches

The current utility of CKB is to support clinical reporting for our JAX CTP panel. To this end, we have incorporated the selection of potential treatment approaches for molecular profiles, which designates a relationship between a selected molecular profile and either a drug class or individual therapy. Treatment approaches are selected following extensive review of the literature supporting preclinical and/or clinical efficacy, which is additionally curated via molecular profile response. For example, the gene variant PIK3CA H1047R is a gain-of-function variant in PIK3CA [24]. Due to several lines of evidence, PIK3CA H1047R has been assigned the drug class treatment approaches "mTOR inhibitors," "PI3K inhibitors," and "AKT inhibitors." Each of these drug classes is linked to a milieu of drugs with the target mTOR, PI3K, and/or AKT, respectively. These treatment approaches are used in the reporting queries to identify relevant therapies and clinical trials.

Molecular profile response and efficacy evidence

An essential component of a clinical knowledgebase is the inclusion of evidence from primary literature sources supporting therapeutic efficacy in the context of specific molecular variants. The "Molecular Profile Response" element of the JAX-CKB represents the relationship between a molecular profile and indication, therapy, and related response type. The patient indication attribute integrates the cancer terms from the Disease Ontology [7], which are backed by unique identifiers (DOID) and represented in a navigable tree.

Associated with the molecular profile response is efficacy evidence, which is the point of entry for literature-based data supporting the molecular profile response. A molecular profile response can be associated with one or more lines of efficacy evidence. Efficacy evidence additionally includes information regarding the approval status of the associated therapy in relation to the curated evidence, using a controlled vocabulary to designate whether the study was preclinical, phase I, phase Ib/II, phase III, or FDA-approved, or if the therapy-molecular profile association is in National Comprehensive Cancer Network (NCCN), FDA, or European Society for Medical Oncology (ESMO) Guidelines. Curation of FDA-approval studies links tumor type to approval status, allowing for simultaneous identification of targeted therapies that have achieved FDA-approval in a patient indication. Efficacy evidence can be categorized, using controlled vocabulary, as actionable, which is characterized as data relating the efficacy of the selected therapy in the context of the selected profile response and is prioritized based on an in-house 25-tier ranking system. Additionally, the JAX-CKB supports the entry of prognostic and diagnostic evidence as well as emerging evidence, which supports potential emerging therapeutic targets.

Clinical trials

To facilitate the inclusion of relevant clinical trials in patient reporting to provide an additional potential avenue for therapeutic intervention, we have incorporated clinical trial content into the JAX-CKB. This additionally provides a unique mechanism for the analysis of clinical trial data, which is cumbersome through direct extraction from clinical trial databases. For the import of clinical trials into CKB, we have built a utility that searches clinical trials from clinicaltrials.gov using a predefined list of targeted therapies, restricted to trials in the USA or Canada in "cancer" that are "open," which are defined as those trials that have a recruitment status of "recruiting," "not yet recruiting," or "available." Returned trials are matched against those in the JAX-CKB to produce a unique list of new clinical trials. Information including NCTID, title, and recruitment status is imported, and remaining information from the trial record that cannot be automatically retrieved from clinicaltrials.gov is manually curated. The NCTID serves as a unique identifier for each clinical trial. The elements in the clinical trial user interface enable connectivity to molecular aberrations in two ways. First, an intervention attribute enables curation of therapies being tested in the clinical trial. These therapies are selected from a drop-down list of therapies that is populated from the therapy table. The drugs contained in those therapies can then be linked to a treatment approach via a drug class or individual therapy. Additionally, clinical trial curation integrates molecular eligibility criteria, so trials recruiting based on the presence or absence of specific molecular criteria can be easily identified. Therefore, a patient sample with an EGFR mutation would be linked to a trial testing erlotinib that is recruiting on EGFR activating mutations both by the targeted therapy and the required molecular criteria.

To ensure that we have the most up-to-date information in the JAX-CKB regarding clinical trial status, we have built a cron job that performs a nightly search of clinicaltrials.gov for any change to recruitment status for trials in the JAX-CKB and automatically updates any changes to the database. Thus, only clinical trials that have open recruitment statuses are included on clinical reports.

The connectivity of data in JAX-CKB and the associated reporting tools enables a direct connection between molecular aberration and targeted therapy and connects patients to trials investigating a relevant targeted therapy, but which may not be specifically recruiting on a molecular aberration. This may have the unintentional, beneficial consequence of increasing success and development of targeted agents by prescreening patients outside of the clinical trial protocol for the presence of relevant genomic targets.

Conclusions

Progress in patient treatment in oncology is dependent on both a comprehensive understanding of the relationships between molecular variants in tumors and efficacy, and the power to access clinically relevant information in a timely manner. Here, we share methods and demonstrate the implementation of a data mapper and loader and queries for rapid retrieval of data related to clinical efficacy to inform clinical interpretation of molecular aberrations, supported by the structured and organized design of the JAX Clinical Knowledgebase. In addition, global analysis of the clinical trial landscape provides insight into the types of targeted agents and molecular targets that are focal points, as well as those that could be served by additional investigation.

Methods

JAX-CKB mapper and JAX-CKB schema

The following versions of the JAX-CKB mapper and database were used in analysis:

JAX-CKB mapper: Version 1.05.
JAX-CKB Schema: Version 1.4. JAX-CKB; 81 Tables.

Data analysis

For data analysis, data was retrieved from the stage version of the JAX-CKB database using SQL queries. Due to the dynamic nature of the data in the production database due to continual curation, the stage database, which is not subject to continual update, was used to eliminate variability in the data. The stage database was last updated on September 18, 2015, from a snapshot of the production version of the JAX-CKB database.

Clinical trial recruitment status updating

A nightly cron job is implemented to query https://clinicaltrials.gov/ct2/results. It parses the HTML contents to obtain the trials that match the search criteria [cancer-related clinical trials in the United States and Canada] and check the recruitment status of all existing trials in the JAX-CKB database. Following comparison with the recruitment status of clinical trials in the JAX-CKB database, it updates the database and sends an email notification to data curators if any recruitment status is changed.

Abbreviations

CKB: Clinical Knowledgebase; CGA: Clinical Genome Analytics; CTP: Cancer Treatment Profile; DO: Disease Ontology; HGNC: HUGO Gene Nomenclature Committee; HGVS: Human Genome Variation Society; HUGO: Human Genome Organisation; regex: Regular Expression; VCF: variant call format.

Competing interests

The authors declare that they have no competing interests.

Authors' contributions

SEP prepared the manuscript and contributed to data analysis and interpretation and figure preparation. RL is the primary designer and developer of the JAX-CKB database and data loader and mapper and contributed to data analysis. SMM conceived of and participated in design of the database and contributed to data analysis. AL and DD designed queries and contributed to the database development. CMS participated in data analysis. All authors read and approved the final manuscript.

Acknowledgements

We would like to thank Taofei Yin, Sean Eddy, Guruprasad Ananda, and Danielle Ruby for their contributions to data curation and database support.

References

1. Forbes SA, Beare D, Gunasekaran P, Leung K, Bindal N, Boutselakis H, et al. COSMIC: exploring the world's knowledge of somatic mutations in human cancer. Nucleic Acids Res. 2015;43(Database issue):D805–811.
2. My Cancer Genome. Vanderbilt-Ingram Cancer Center. http://www.mycancergenome.org. Accessed 20 Oct 2015.
3. Gray KA, Yates B, Seal RL, Wright MW, Bruford EA. Genenames.org: the HGNC resources in. Nucleic Acids Res 2015. 2015;43(Database issue):D1079–1085.
4. den Dunnen JT, Antonarakis SE. Mutation nomenclature extensions and suggestions to describe complex mutations: a discussion. Hum Mutat. 2000;15:7–12.
5. Yuan Y, Van Allen EM, Omberg L, Wagle N, Amin-Mansour A, Sokolov A, et al. Assessing the clinical utility of cancer genomic and proteomic data across tumor types. Nat Biotechnol. 2014;32:644–52.
6. Ananda G, Mockus S, Lundquist M, Spotlow V, Simons A, Mitchell T, et al. Development and validation of the JAX Cancer Treatment Profile™ for detection of clinically actionable mutations in solid tumors. Exp Mol Pathol. 2015;98:106–12.
7. Schriml LM, Mitraka E. The Disease Ontology: fostering interoperability between biological and clinical human disease-related data. Mamm Genome Off J Int Mamm Genome Soc. 2015;26:584–9.
8. Mockus S, Patterson S, Statz C, Bult C, Tsongalis G. Clinical trials in precision oncology. Clin Chem. 2015.247437. [Epub ahead of print]
9. Roper N, Stensland KD, Hendricks R, Galsky MD. The landscape of precision cancer medicine clinical trials in the United States. Cancer Treat Rev. 2015;41:385–90.
10. Di Maio S, Yip S, Al Zhrani GA, Alotaibi FE, Al Turki A, Kong E, et al. Novel targeted therapies in chordoma: an update. Ther Clin Risk Manag. 2015;11:873–83.
11. Ramakrishna R, Rostomily R, Sekhar L, Rockhill J, Ferreira M. Hemangiopericytoma: radical resection remains the cornerstone of therapy. J Clin Neurosci Off J Neurosurg Soc Australas. 2014;21:612–5.
12. Sana M, Malik HJ. Current and emerging breast cancer biomarkers. J Cancer Res Ther. 2015;11:508–13.
13. Spagnolo F, Ghiorzo P, Orgiano L, Pastorino L, Picasso V, Tornari E, et al. BRAF-mutant melanoma: treatment approaches, resistance mechanisms, and diagnostic strategies. OncoTargets Ther. 2015;8:157–68.
14. Kuan F-C, Kuo L-T, Chen M-C, Yang C-T, Shi C-S, Teng D, et al. Overall survival benefits of first-line EGFR tyrosine kinase inhibitors in EGFR-mutated non-small-cell lung cancers: a systematic review and meta-analysis. Br J Cancer. 2015;113(10):1519-28.
15. Naidoo J, Sima CS, Rodriguez K, Busby N, Nafa K, Ladanyi M, et al. Epidermal growth factor receptor exon 20 insertions in advanced lung adenocarcinomas: clinical outcomes and response to erlotinib. Cancer. 2015; 121(18):3212-20.
16. Yasuda H, Park E, Yun C-H, Sng NJ, Lucena-Araujo AR, Yeo W-L, et al. Structural, biochemical, and clinical characterization of epidermal growth factor receptor (EGFR) exon 20 insertion mutations in lung cancer. Sci Transl Med. 2013;5:216ra177.
17. Choi YJ, Kim SY, So KS, Baek I-J, Kim WS, Choi SH, et al. AUY922 effectively overcomes MET- and AXL-mediated resistance to EGFR-TKI in lung cancer cells. PloS One. 2015;10:e0119832.
18. Brown GR, Hem V, Katz KS, Ovetsky M, Wallin C, Ermolaeva O, et al. Gene: a gene-centered information resource at NCBI. Nucleic Acids Res. 2015; 43(Database issue):D36–42.
19. Petri V, Jayaraman P, Tutaj M, Hayman GT, Smith JR, De Pons J, et al. The pathway ontology—updates and applications. J Biomed Semant. 2014;5:7.
20. Costa DB, Nguyen K-SH, Cho BC, Sequist LV, Jackman DM, Riely GJ, et al. Effects of erlotinib in EGFR mutated non-small cell lung cancers with resistance to gefitinib. Clin Cancer Res Off J Am Assoc Cancer Res. 2008;14:7060–7.
21. Robinson KW, Sandler AB. The role of MET receptor tyrosine kinase in non-small cell lung cancer and clinical development of targeted anti-MET agents. The Oncologist. 2013;18:115–22.
22. Turke AB, Zejnullahu K, Wu Y-L, Song Y, Dias-Santagata D, Lifshits E, et al. Preexistence and clonal selection of MET amplification in EGFR mutant NSCLC. Cancer Cell. 2010;17:77–88.

5

A first-line diagnostic assay for limb-girdle muscular dystrophy and other myopathies

Dorota Monies[1,2*], Hindi N. Alhindi[4], Mohamed A. Almuhaizea[3], Mohamed Abouelhoda[1,2], Anas M. Alazami[1,2], Ewa Goljan[1,2], Banan Alyounes[1,2], Dyala Jaroudi[1,2], Abdulelah AlIssa[1,2], Khalid Alabdulrahman[1,2], Shazia Subhani[1,2], Mohamed El-Kalioby[1,2], Tariq Faquih[1,2], Salma M. Wakil[1,2], Nada A. Altassan[1,2], Brian F. Meyer[1,2] and Saeed Bohlega[2,3]

Abstract

Background: Fifty random genetically unstudied families (limb-girdle muscular dystrophy (LGMD)/myopathy) were screened with a gene panel incorporating 759 OMIM genes associated with neurological disorders. Average coverage of the CDS and 10 bp flanking regions of genes was 99 %. All families were referred to the Neurosciences Clinic of King Faisal Specialist Hospital and Research Centre, Saudi Arabia. Patients presented with muscle weakness affecting the pelvic and shoulder girdle. Muscle biopsy in all cases showed dystrophic or myopathic changes. Our main objective was to evaluate a neurological gene panel as a first-line diagnostic test for LGMD/myopathies.

Results: Our panel identified the mutation in 76 % of families (38/50; 11 novel). Thirty-four families had mutations in LGMD-related genes with four others having variants not typically associated with LGMD. The majority of cases had recessive inheritance with homoallelic pathogenic variants (97.4 %, 37/38), as expected considering the high rate of consanguinity in the study population. In one case, we detected a heterozygous mutation in *DNAJB* responsible for LGMD-1E. Our cohort included seven different subtypes of LGMD2. Mutations of *DYSF* were the most commonly identified cause of disease followed by that in *CAPN3* and *FKRP*. Non-LGMD myopathies were due to mutations in genes associated with congenital disorder of glycosylation (*ALG2*), rigid spine muscular dystrophy 1 (*SEPN1*), inclusion body myopathy2/Nonaka myopathy (*GNE*), and neuropathy (*WNK1*). Whole exome sequencing (WES) of patients who remained undiagnosed with the neurological panel did not improve our diagnostic yield.

Conclusions: Our neurological panel achieved a high clinical sensitivity (76 %) and is an effective first-line laboratory test in patients with LGMD and other myopathies. This sensitive, cost-effective, and rapid assay significantly assists clinical practice especially in these phenotypically and genetically heterogeneous disorders. Moreover, the application of the American College of Medical Genetics (ACMG) and Association for Molecular Pathology (AMP) guidelines applied in the classification of variant pathogenecity provides a clear interpretation for physicians on the relevance of such findings.

Keywords: Neurological panel, Targeted resequencing, LGMD, Myopaties

* Correspondence: dmonies@kfshrc.edu.sa
[1]Department of Genetics, Research Centre, King Faisal Specialist Hospital and Research Centre, PO Box 3354, Riyadh 11211, Kingdom of Saudi Arabia
[2]Saudi Human Genome Program, King Abdulaziz City for Science and Technology, Riyadh, Saudi Arabia
Full list of author information is available at the end of the article

Background

The original definition of limb-girdle muscular dystrophy (LGMD) as non-Duchenne with autosomal recessive inheritance was introduced in 1953 [1]. LGMD classification has rapidly evolved since that time. A major advance in neuromuscular disorders over the last three decades has been the identification of many genes underlying this group of heterogeneous diseases [2–4]. The LGMDs now vary widely in their genetics and also in clinical features ranging from very mild forms which allow patients to maintain a fairly normal life to much more severe deterioration of proximal limb muscles that causes dramatic physical weakness along with a shortened life-span [4, 5]. There are two major groups: LGMD1 and LGMD2 with autosomal dominant and autosomal recessive patterns of inheritance, respectively [6]. To date, there are at least 8 genes associated with LGMD1 (LGMD1A-1H) and 23 genes in which mutations lead to different subtypes of LGMD2 (LGMD2A-2W) [4] that continues to expand. This growing genetic heterogeneity highlights the problem of a very complex clinical diagnosis [7]. Current recommendations for diagnosis and management of LGMD are very complex and require access to multiple specialties including thorough clinical examination, laboratory testing, muscle imaging, histological appearances of muscle, and more [5]. Since there are overlapping phenotypes in LGMDs and a number of other myopathic disorders, a precise diagnosis without genetic testing is very difficult. As a consequence of this, many patients remain undiagnosed.

Next-generation sequencing (NGS) also known as massively parallel sequencing has ushered in a new era in molecular diagnostics. The availability of exome sequencing has been rapidly applied to clinical settings [8, 9]. Whiles the cost of a clinical-grade whole exome sequencing (WES) is still high and its interpretation very complex, an alternative approach to WES is the use of a NGS-based gene panel. This approach is particularly well suited for genetically and clinically heterogeneous conditions where the number and/or size of genes is too large and expensive to sequence one gene at a time [10]. In this study, we aimed to determine clinical sensitivity of a neurological gene panel for diagnosis of LGMD/myopathies on the basis that this would be a cheaper, more practical, and effective approach.

Results

A total of 50 families were included in this study, of which 36 had an autosomal recessive pattern of inheritance with parental consanguinity. Mean age of disease onset for our patients was 10.6 years. Distribution of muscle weakness, age of onset, creatine kinase level, biopsy findings, and other clinical characteristics are presented in the supplementary data (see Additional file 1). Disease-causing mutations classified based on American College of Medical Genetics (ACMG)/Association for Molecular Pathology (AMP) guidelines were identified in 76 % of the study cohort (38/50 families). Novel mutations were present in 11 of these families (Table 1). The neurological panel revealed pathogenic variants in LGMD and non-LGMD myopathy related genes in 34 and 4 of our families, respectively (Table 2). The majority of our cases with pathogenic causal variants were homoallelic (97.4 %, 37/38), consistent with an autosomal recessive pattern of inheritance and as expected within a highly consanguineous population [11, 12]. We identified only one family with autosomal dominant inheritance, the result of a heterozygous mutation in *DNAJB* (LGMD type 1E), present in two affected memebers (father and his son). The neurological panel detected seven subtypes of LGMD2, mutations of *DYSF* (LGMD2???) being the most common cause of the disease. *CAPN3* (LGMD2??) and *FKRP*(LGMD2??) were two other commonly mutated genes in our study (Table 2).

Three affected members of family 25 had a homozygous mutation in *FKRP* (c.C941T, p.T314M) that segregated with disease with the exception of a 3-year-old asymptomatic sister also homozygous for this allele. She probably has not developed the phenotype yet as the first symptoms associated with LGMD2I usually occur between the first and third decade of life.

In several families, mutations in genes related to non-LGMD myopathies were identified. Family 3 had a mutation in *ALG2* associated with a congenital disorder of glycosylation. The patients presented with a congenital limb-girdle pattern of weakness with no ocular or bulbar involvement. Muscle biopsies showed myopathic features, ragged red fibers, and a sub-sarcolemmal accumulation of structurally normal mitochondria. Family 16 had a mutation in *SEPN1* associated with rigid spine muscular dystrophy 1. The proband displayed major weakness in lower extremities that progressed and affected the upper proximal shoulder and girdle muscles. He suffered from respiratory failure and kyphoscoliosis. Needle EMG examination and nerve conduction studies were consistent with myopathy.

The index patient from family 24 presented with proximal muscle weakness of the lower extremities that started after the delivery of her first child. The condition progressed in severity and she started to note frequent falls and tripping after delivery of her second child. Nerve conduction studies as well as the needle EMG were consistent with a clinical diagnosis of progressive distal myopathy. In this patient, a mutation in *GNE* (UDP-*N*-acetyloglucosamine 2-epimerase/*N*-acetylmannosamine kinase) associated with inclusion body myopathy2/Nonaka myopathy was identified. Neuropathy due to mutation in *WNK1* was found to underlie disease

Table 1 Summary of findings in positive cases and mutation classification based on ACMG/AMP guidelines

Index ID	Gene	Mutation	Novel	Effect	Evidence of pathogenicity
10R-00963 10R-00534	*CAPN3*	NM_000070:c.801+1G>A	No	Pathogenic	PVS1, PP1-S, PM2, PM3, PP1-M, PP1, PP3, PP4
10R-00309 10R-00658 11R-02841 11R-03100	*FKRP*	NM_001039885:c.941C>T;p.T314M	No	Pathogenic	PP1-S, PM2, PM3, PP1-M, PP1, PP3, PP2, PP4
11R-00685	*ALG2*	NM_033087.3:indel:c.214_224delGGGGACTGGCT delinsAGTCCCCG;p.72_75delGDWLinsSPR	No	Pathogenic	PVS1, PP1-S, PM2, PM3, PM4, PP1, PP3, PP4
10R-00500 10R-00405 10R-00538 14R-02300 11R-01506 12R-01190	*DYSF*	NM_003494:c.164_165insA;p.I57Hfs*8	No	Pathogenic	PVS1, PP1-S, PM2, PM3, PM4, PP1-M, PP1, PP3, PP4
11R-00680 10R-00857	*SGCA*	NM_001135697:c.608_609insCG;p.S203fs	Yes	Pathogenic	PVS1, PP1-S, PM2, PM3, PM4, PP1-M, PP1, PP3, PP4
794310 502098 11R-00463	*SGCA*	NM_001135697:c.101G>A;p.R34H	No	Pathogenic	PM2, PP1-S, PM3, PP1-M, PP1, PP2, PP3, PP4
10R-00739	*CAPN3*	NM_000070:cA2329A>G;p.I777V	Yes	Pathogenic	PP1-S, PM2, PM3, PP1-M, PP1, PP2, PP3, PP4
10R-00751	*SEPN1*	NM_206926:c.1270G>A;p.D424N	Yes	Pathogenic	PP1-S, PM2, PM3, PP1-M, PP1, PP2, PP3, PP4
10R-00779	*CAPN3*	NM_000070:c.146G>A;p.R49H	No	Pathogenic	PP1-S, PM2, PM3, PP1-M, PP1, PP2, PP3, PP4
10R-00926	*SGCB*	NM_000232:c.355A>T;p.I119F	No	Pathogenic	PP1-S, PM2, PM3, PP1-M, PP1, PP2, PP3, PP4
10R-00973	*FKTN*	NM_006731:c.314G>T;p.C105F	No	Pathogenic	PP1-S, PM2, PM3, PP1-M, PP1, PP2, PP3, PP4
11R-00018 13R-01080	*CAPN3*	NM_000070:c.1699G>A;p.G567R	Yes	Likely pathogenic	PM2, PM3, PP2, PP3, PP4
11R-00031	*GNE*	NM_001190384:c.1805T>C;p.M602T	No	Pathogenic	PP1-S, PM2, PM3, PP1-M, PP1, PP2, PP3, PP4
11R-00230 12R-01188	*FKRP*	NM_001039885:c.941C>T;p.T314M	No	Pathogenic	PP1-S, PM2, PM3, PP1-M, PP1, PP2, PP3, PP4
11R-00232	*WNK1*	NM_213655:c.2152C>T;p.R718C	Yes	Pathogenic	PP1-S, PM2, PM3, PP1-M, PP1, PP2, PP3, PP4
11R-00308 12R-01189	*DYSF*	NM_001130976:c.1433delC;p.T478fs	Yes	Pathogenic	PVS1, PP1-S, PM2, PM3, PM4, PP1-M, PP1, PP4, PP3
11R-00337	*FKRP*	NM_001039885:c.1012G>T;p.V338L	Yes	Pathogenic	PP1-S, PM2, PM3, PP1-M, PP1, PP2, PP3, PP4
11R-00745	*DYSF*	NM_001130976:c.A5201T;p.E1734V	No	Pathogenic	PP1-S, PM2, PM3, PP1-M, PP1, PP2, PP3, PP4
11R-02618	*CAPN3*	NM_000070:c.310G>T;p.E104X	Yes	Pathogenic	PVS1, PM2, PM3, PM4, PP3, PP4
12R-00001	*ANO5*	NM_001142649, c.169C>T;p.R57W	No	Pathogenic	PP1-S, PM2, PM3, PP1-M, PP1, PP2, PP3, PP4
12R-00316	*CAPN3*	NM_000070:c.2381-1G>A	Yes	Pathogenic	PVS1, PM2, PM3, PM4, PP1-M, PP1-S, PP1, PP3, PP4
13R-01177	*DYSF*	NM_001130976:c.89-1G>A	Yes	Pathogenic	PVS1, PP1-S, PM2, PM3, PM4, PP1-M, PP1, PP3, PP4
14R-00183	*DNAJB6*	NM_005494:c.C287T;p.P96L	Yes	Pathogenic	PP1-S, PM2, PP1-M, PP1, PP2, PP3, PP4

ACMG/AMP American College of Medical Genetics and Genomics/Association for Molecular Pathology, *PVS1* pathogenic very strong, *PM2-4* pathogenic moderate, *PP1-4* pathogenic supporting, *PP1-M* pathogenic supporting (moderate), *PP1-S* pathogenic supporting (strong)

in family 26. The female patient was normal until the age of 9, when she had a problem with standing and walking. She developed full Gower's sign and had frequent falls. The girl had a global proximal weakness affecting, lower more than the upper, extremities. There was also bilateral mild hypertrophy of the calf muscles with a marked lordotic gait. Muscle biopsy showed clear dystrophic changes. She had a similarly affected sister.

We performed WES on nine neurological panel "negative" consanguineous multiplex families with autosomal recessive disease inheritance. Accordingly, we concentrated on finding pathogenic homozygous variants within defined ROHs (shared by affected individuals only). WES did not reveal any such variants, in particular among neuromuscular genes within these regions.

Discussion

High-throughput solutions such as NGS have revolutionized the genetic approach for molecular diagnosis of Mendelian disorders. Utility of WES in the clinic has

Table 2 Overview of the genetic diagnosis based on the neurological panel results

Genetic diagnosis	Number of families ($n = 50$)	Inheritance
Dyspherlinopathy	10 (20 %)	AR
Calpainopathy	8 (16 %)	AR
Dystroglycenopathy (type C)	7 (14 %)	AR
α-Sarcoglyconapthy	5 (10 %)	AR
β-Sarcoglyconapthy	1 (2 %)	AR
Dystroglycenopathy (type B)	1 (2 %)	AR
Anoctaminopathy	1 (2 %)	AR
Congenital disorder of glycosylation	1 (2 %)	AR
Rigid spine muscular dystrophy	1 (2 %)	AR
Nonaka myopathy	1 (2 %)	AR
Neuropathy	1 (2 %)	AR
Limb-girdle muscular dystrophy, type 1E	1 (2 %)	AD
Undiagnosed	12 (24 %)	

AR autosomal recessive, *AD* autosomal dominant

been relatively successful, mainly for neurological conditions, with a yield of ~25 % [9, 13]. Due to high cost, a long turnaround time and challenging interpretation, arguably, WES should be considered mainly for the purpose of novel gene discovery. An alternative solution that substantially reduces the above limitations is an approach which targets known disease genes grouped in panels [14]. Several approaches have been taken to design and sequence NGS-based gene panels for LGMD patients [7, 15–19]. In our study, the neurological gene panel comprised 759 genes known to cause Mendelian neurological diseases (not only LGMD) as annotated by OMIM up to August 2013. The neurological panel is a part of our Mendeliome assay that includes 13 symptom/sign-based gene panels [14]. Minimal expertise is required by clinicians to choose the neurological panel from the 13 present in the "Mendeliome," e.g., any muscle weakness or movement problem will trigger testing for this panel. This comprehensive assay simplifies the molecular diagnostic process and takes into consideration the remarkable phenotypic variability between many neuromuscular disorders. In this way, we significantly reduce the chance of missing a genetic diagnosis due to atypical presentation.

In our cohort of 50 LGMD families, the neurological panel was able to resolve 76 % of cases (38/50 families), an exceptionally high diagnostic yield relative to previously published data (Table 3). In light of ambiguity associated with clinical significance of NGS variants, assigning an objective assessment of pathogenicity is crucial in making a molecular diagnosis. We interpreted the clinical significance of our findings (those which survived our filtration process) using ACMG/AMP

Table 3 Diagnostic yield of gene panels in LGMD/myopathic patients

Gene Panel	Number of genes in the gene panel	Number of families studied	Number of genetically diagnosed families
Ghosh at al. [15]	9	27	10 (37 %)
Savarese et al. [7]	93	177	108 (61 %)
Seong et al. [17]	18	35	20 (57 %)
Ankala et al. [18]	11	96	25 (26 %)
Dai et al. [19]	399	55	36 (65 %)
Kuhn et al. [20]	38	58	19 (33 %)
Our study	759	50	38 (76 %)

standards and guidelines. This objective and standardized approach to variant classification, while clearly indicated in clinical situations, has not been applied by previous studies evaluating NGS for diagnosis of myopathies [7, 15–21]. Given the nature and number of variants identified using NGS, we feel that classification in this manner is not only useful but also essential to clear interpretation by referring physicians. All but one disease associated variant identified in our cohort were classified as pathogenic. The one exception was classified as a likely pathogenic, consequent to there being only one affected member in the family, thus preventing demonstration of co-segregation with multiple affected individuals (Table 1).

The panel covers all types of muscular dystrophies, myopathies, and other neuromuscular disorders encompassing more than 300 diseases to date [22]. Unfortunately, most of them have very similar clinical presentations and even with thorough clinical evaluation and muscle pathology, a correct diagnosis without genetic testing remains difficult. A strong case may be made to test all muscle genes in analysis of patients with suspected LGMD [20]. In similar studies, a diagnostic yield of LGMD patients using a gene panel approach varies from 16 to 65 % [7, 20]. In these studies, the number and composition of genes sequenced were associated with diagnostic yield. Ankala and colleagues, by expanding their LGMD panel with 11 genes to a more comprehensive neuromuscular disease (NMD) panel containing 41 genes, achieved a threefold greater diagnostic rate. Their NMD panel also covered other non-LGMD movement disorders and increased the yield from approximately 15 to 46 % [18]. Dai et al. claimed to have designed a panel of 44 known genes underlying muscular dystrophies and congenital myopathies. In fact, their libraries incorporated 399 genes covering common inherited disorders including at least 55 genes associated with myopathies. They were able to find causative mutations in 65 % of patients [19]. The MotorPlex assay comprising 93 muscle disease loci identified pathogenic or potential causative variants in 61 % of patients tested [7]. Better performance of our panel (diagnostic yield = 76 %) versus others described may be associated with it being more comprehensive (759 genes) and its application in an inbred population. Our neurological panel was able to diagnose patients due to mutations in genes absent in assays described in other studies. Those gene panels would miss at least 8 to 10 % of pathogenic changes in our cohort. The Athena Diagnostics LGMD gene panel would not detect a mutation in seven of our cases: *FKTN, ANO5, DNAJB6, SEPN1, GNE, WNK1*, and *ALG2* [15]. Both assays used by Ankala et al. and Seong et al. would miss diagnosis due to changes in four to five genes (*DNAJB6, GNE, WNK1, ALG2*, and *SEPN1*) [17, 18]. Two other panels would fail to diagnose patients from our cohort with disorders associated with *DNAJB6, WNK1*, and *ALG2* [19] and *SEPN1, WNK1*, and *ALG2* [20]. While our neurological panel could be focused further, there is little to be gained from doing so other than to perhaps reduce incidental findings. Ghaoui and colleagues tested 60 LGMD families (undiagnosed by conventional candidate gene sequencing) using WES with a diagnostic success rate of 45 %. The identified mutations were present only in known myopathy genes and WES did not reveal any finding in novel genes. The group obtained identical results by retesting the samples using a neuromuscular panel which contained 336 neuromuscular disease-related genes. The panel confirmed all variants identified by WES highlighting the benefit of a panel-based approach relative to candidate gene sequencing or WES [21]. Comprehensive gene panels should be viewed as first-tier tests before considering WES. Based on our results and reports in the literature cited above, we consider application of gene panels to be a more effective approach for diagnosis of myopathies.

Considering very high genetic heterogeneity (similar phenotype associated with multiple genes) and phenotypic heterogeneity (a single gene associated with multiple phenotypes) in LGMD and other myopathies, applying a gene panel (such as our neurological panel), incorporating a comprehensive list of genes associated with neuromuscular disorders, that can be tested together at the same time, provides a very powerful and practical diagnostic tool. Besides the high diagnostic efficiency, our neurological panel is cost-effective. A multiplexing strategy running 24 samples per run dramatically reduces the sample processing cost (~$150) and time. A similar approach was taken by Savarese and colleagues who designed the MotorPlex panel (93 genes) covering all known forms of non-syndromic muscle disorders. They also applied a cost-effective pooled sequencing strategy with 100 % specificity and sensitivity of the assay in 20 LGMD or congenital myopathy patients [7].

Molecular diagnosis is crucial for genetic counseling and prognosis [23, 24]. An earlier genetic diagnosis provides better disease management and also protects patients from more invasive clinical evaluation [25].

WES applied to the remaining undiagnosed multiplex families ($n = 9$) did not detect any disease-causing mutations. The use of NGS technology, as applied in this study, has limitations including incomplete coverage of target sequences in PCR based libraries, amplification bias resulting non-uniform coverage of library amplicons, inability to detect structural changes, and poor sensitivity for copy number variation. We cannot exclude possibility of missing mutations in known genes which are not fully covered or those present in intronic regions which are not covered by WES [26, 27]. Another NGS limitation is copy number variation (CNV) (gross deletions and insertions) which are poorly detected so far by this approach [20]. WES results of our and other studies [21] also suggest that the majority of genes underlying LGMD and other myopathies have probably been identified with limited scope for novel discovery.

Conclusions

We have demonstrated that our neurological panel assay covering 759 neurological genes cited by OMIM has a high diagnostic yield (76 %) for LGMD and other myopathies. In addition, it is a rapid and cost-effective assay. We believe that the majority of LGMD patients can be diagnosed using this new very powerful genomic tool and WES should be reserved only for negative cases as an opportunity to discover novel candidate genes. Classification of disease-associated variants with respect to pathogenicity, using guidelines of the ACMG/AMP as applied in our study further, adds to the power and utility of NGS panels for the clinical diagnosis of LGMD and other myopathies.

Methods

Participants

A total of 50 random genetically unstudied families were included in this study; 36 and 14 of these were multiplex and simplex families, respectively. All were collected through neurosciences clinic at KFSHRC between 2010 and 2015. The age of disease onset varied from 1 to 35 years. All patients presented with muscle weakness affecting the pelvic and shoulder girdle. Muscle biopsy in all individuals showed dystrophic or myopathic changes. DNA was extracted from peripheral blood samples using standard procedures (Flexi Gene DNA Handbook, Qiagen). Samples were quantitated spectrophotometrically and stored at −20 °C.

Neurological panel assay and bioinformatics analysis

This gene panel was a part of the NGS targeted resequencing "Mendeliome" assay that consists of 13 symptom-based gene panels which cover all inherited disease associated genes in OMIM as of August 2013 [14]. The neurological panel included 759 OMIM genes associated with neurological disorders (see Additional file 2). Genes were amplified and a library constructed using an AmpliSeq HiFi mix, proprietary primers (see Additional file 3) and library kit (Thermo Fisher, Carlsband, CA, USA) followed by sequencing on the Ion Proton platform according to the manufacturer's protocol (Thermo Fisher, Carlsband, CA, USA). Variants were called and annotated using the Saudi Human Genome pipeline [14]. Briefly, only regions of the reads with high quality (Ion Torrent base calling algorithm, Thermo Fisher, Carlsband, CA, USA) were aligned to the UCSC hg19 (http://genome.ucsc.edu/) reference sequence and processed for variant calling using the Torrent Suite Variant Caller (TVC) program (Thermo Fisher, Carlsband, CA, USA). Performance of the neurological panel in this study resulted in >95 % of reads at Q17 with an average read depth of 166 X. Variants were annotated using in-house programs that extend the public Annovar package with other licensed commercial data sets such as the professional version of HGMD [28] and in-house databases made up of a collection of disease-causing and polymorphic variants observed in individuals of Arab ethnicity. As a final step, non-relevant variants were filtered out based on their quality, functional characteristics, and their frequency in our datasets. Intronic variants, synonymous variants, and those present in population databases (specifically those that are in 1000 Genome database with MAF >1 %) were also filtered out. Furthermore, variants that were frequent (MAF >1 %) in our population specific in-house variant database were also filtered out. After applying filtration criteria, all nonsense, frameshift, and canonical splice site variants were considered pathogenic. For interpretation and classification of the remaining SNVs, nucleotide and amino acid conservation and effects on protein sequence were analyzed with PolyPhen-2 (http://genetics.bwh.harvard.edu/pph2/) and SIFT (http://sift.jcvi.org/). Potential causative variants were validated by Sanger sequencing and further vetted for familial segregation. Finally, the remaining variants were identified as pathogenic or likely pathogenic following guidelines of the American College of Medical Genetics (ACMG) and Association for Molecular Pathology (AMP) [29].

DNA Sanger sequencing

Coding regions of candidate genes were sequenced using a BigDye Terminator kit (Applied Biosystems, Foster City, CA) and run on an ABI 3730xl automated sequencer (Applied Biosystems, Foster City, CA). SeqScape v.2.6 software (Applied Biosystems, Foster City, CA) was used to align sequence data against the relevant reference.

Genetic variant interpretation, ACMG/AMP guidelines

The clinical significance of NGS variants was classified using an openly available online tool for implementing the ACMG/AMP standards and guidelines: http://medschool.umaryland.edu/Genetic_Variant_Interpretation_Tool1.html. [30].

Genotyping and homozygosity mapping

All participating individuals (affected and unaffected) in cases where the neurological panel failed to identify a likely casual mutation were genotyped using the Affymetrix Axiom array (Affymetrix, Santa Clara, CA, USA) following the manufacturer's protocol (http://www.affymetrix.com/support/technical/manuals.affx). Resulting genotypes were analyzed for shared runs of homozygosity (ROH) using autoSNPa (http://dna.leeds.ac.uk/autosnpa/).

Whole exome sequencing and analysis

Multiplex families that tested negative on the neurological panel underwent WES. Briefly, 100 ng of each DNA was amplified in 12 separate wells using Exome Primer Pools, AmpliSeq HiFi mix (Thermo Fisher, Carlsbad, CA, USA) and 10 amplification cycles. All 12 PCR pools were combined in one well and subjected to primer digestion by incubation with FuPa reagent (Thermo Fisher, Carlsbad, CA, USA). Amplified Exome targets were ligated with Ion P1 and Ion Xpress Barcode adapters. After purification, libraries were quantified using qPCR with the Ion Library Quantification Kit (Thermo Fisher, Carlsbad, CA, USA). The prepared exome library was further used for emulsion PCR on an Ion OneTouch System and templated Ion Sphere particles were enriched using Ion OneTouch ES, both procedures following the manufacturer's instructions (Thermo Fisher, Carlsbad, CA, USA). The template-positive Ion PI Ion Sphere particles were processed for sequencing on the Ion Proton instrument (Thermo Fisher, Carlsbad, CA, USA). Reads were mapped to UCSC hg19 (http://genome.ucsc.edu/) and variants identified using the Saudi Human Genome Program (SHGP) pipeline [14].

Acknowledgements

The authors would like to thank the family members for their participation in this study. We thank the Genotyping and Sequencing Core Facilities of King Faisal Specialist Hospital and Research Centre for their contributions. We would also like to acknowledge the support of KACST through the Saudi Human Genome Project (SHGP).

Funding
Work described was supported by a grant from King Abdulaziz City for Science and Technology (KACST) (KACST#2070005).

Authors' contributions
DM, SB, BFM participated in the conception and design, acquisition and analysis of data, and drafting of the manuscript. HNA and MAM participated in the acquisition and analysis of data and drafting of the manuscript. MA, AMA, EG, BA, DJ, AA, KA, SS, ME, TF, SMW, and NAA participated in the acquisition and analysis of data. All authors read and approved the final manuscript.

Competing interests
The authors declare that they have no competing interests.

Consent for publication
Not applicable.

Author details
[1]Department of Genetics, Research Centre, King Faisal Specialist Hospital and Research Centre, PO Box 3354, Riyadh 11211, Kingdom of Saudi Arabia. [2]Saudi Human Genome Program, King Abdulaziz City for Science and Technology, Riyadh, Saudi Arabia. [3]Department of Neuroscience, King Faisal Specialist Hospital and Research Centre, Riyadh, Saudi Arabia. [4]Department of Pathology and Laboratory Medicine, King Faisal Specialist Hospital and Research Centre, Riyadh, Saudi Arabia.

References

1. Khadilkar SV, Singh RK. Limb girdle muscular dystrophies in India. Neurol India. 2008;56:281–8.
2. Laing NG. Genetics of neuromuscular disorders. Crit Rev Clin Lab Sci. 2012;49:33–48.
3. Murphy A, Kelly RJ. From molecular classification to targeted therapeutics: the changing face of systemic therapy in metastatic gastroesophageal cancer. Gastroenterol Res Pract. 2015;2015:896560.
4. Nigro V, Savarese M. Genetic basis of limb-girdle muscular dystrophies: the 2014 update. Acta Myol. 2014;33:1–12.
5. Narayanaswami P, Weiss M, Selcen D, David W, Raynor E, Carter G, Wicklund M, Barohn RJ, Ensrud E, Griggs RC, et al. Evidence-based guideline summary: diagnosis and treatment of limb-girdle and distal dystrophies: report of the guideline development subcommittee of the American Academy of Neurology and the practice issues review panel of the American Association of Neuromuscular & Electrodiagnostic Medicine. Neurology. 2014;83:1453–63.
6. Guglieri M, Straub V, Bushby K, Lochmuller H. Limb-girdle muscular dystrophies. Curr Opin Neurol. 2008;21:576–84.
7. Savarese M, Di Fruscio G, Mutarelli M, Torella A, Magri F, Santorelli FM, Comi GP, Bruno C, Nigro V. MotorPlex provides accurate variant detection across large muscle genes both in single myopathic patients and in pools of DNA samples. Acta Neuropathol Commun. 2014;2:100.
8. Fogel BL, Lee H, Deignan JL, Strom SP, Kantarci S, Wang X, Quintero-Rivera F, Vilain E, Grody WW, Perlman S, et al. Exome sequencing in the clinical diagnosis of sporadic or familial cerebellar ataxia. JAMA Neurol. 2014;71:1237–46.
9. Lee H, Deignan JL, Dorrani N, Strom SP, Kantarci S, Quintero-Rivera F, Das K, Toy T, Harry B, Yourshaw M, et al. Clinical exome sequencing for genetic identification of rare Mendelian disorders. JAMA. 2014;312:1880–7.
10. Vasli N, Bohm J, Le Gras S, Muller J, Pizot C, Jost B, Echaniz-Laguna A, Laugel V, Tranchant C, Bernard R, et al. Next generation sequencing for molecular diagnosis of neuromuscular diseases. Acta Neuropathol. 2012;124:273–83.
11. Alkuraya FS. Genetics and genomic medicine in Saudi Arabia. Mol Genet Genomic Med. 2014;2:369–78.
12. Abu-Safieh L, Alrashed M, Anazi S, Alkuraya H, Khan AO, Al-Owain M, Al-Zahrani J, Al-Abdi L, Hashem M, Al-Tarimi S, et al. Autozygome-guided exome sequencing in retinal dystrophy patients reveals pathogenetic mutations and novel candidate disease genes. Genome Res. 2013;23:236–47.
13. Yang Y, Muzny DM, Reid JG, Bainbridge MN, Willis A, Ward PA, Braxton A, Beuten J, Xia F, Niu Z, et al. Clinical whole-exome sequencing for the diagnosis of mendelian disorders. N Engl J Med. 2013;369:1502–11.
14. Saudi Mendeliome G. Comprehensive gene panels provide advantages over clinical exome sequencing for Mendelian diseases. Genome Biol. 2015;16:134.
15. Ghosh PS, Zhou L. The diagnostic utility of a commercial limb-girdle muscular dystrophy gene test panel. J Clin Neuromuscul Dis. 2012;14:86–7.
16. Bartoli M, Desvignes JP, Nicolas L, Martin K. Exome sequencing as a second-tier diagnostic approach for clinically suspected dysferlinopathy patients. Muscle Nerve. 2014;50:1007–10.
17. Seong MW, Cho A, Park HW, Seo SH, Lim BC, Seol D, Cho SI, Park SS, Chae JH: Clinical applications of next-generation sequencing-based gene panel in patients with muscular dystrophy: Korean experience. Clin Genet 2015. (Epub ahead of print).
18. Ankala A, da Silva C, Gualandi F, Ferlini A, Bean LJ, Collins C, Tanner AK, Hegde MR. A comprehensive genomic approach for neuromuscular diseases gives a high diagnostic yield. Ann Neurol. 2015;77:206–14.
19. Dai Y, Wei X, Zhao Y, Ren H, Lan Z, Yang Y, Chen L, Cui L. A comprehensive genetic diagnosis of Chinese muscular dystrophy and congenital myopathy patients by targeted next-generation sequencing. Neuromuscul Disord. 2015;25:617–24.
20. Kuhn M, Glaser D, Joshi PR, Zierz S, Wenninger S, Schoser B, Deschauer M. Utility of a next-generation sequencing-based gene panel investigation in German patients with genetically unclassified limb-girdle muscular dystrophy. J Neurol. 2016;263:743–50.
21. Ghaoui R, Cooper ST, Lek M, Jones K, Corbett A, Reddel SW, Needham M, Liang C, Waddell LB, Nicholson G, et al: Use of Whole-Exome Sequencing for Diagnosis of Limb-Girdle Muscular Dystrophy: Outcomes and Lessons Learned. JAMA Neurol. 2015;72:11424-1432.
22. Kaplan JC, Hamroun D. The 2015 version of the gene table of monogenic neuromuscular disorders (nuclear genome). Neuromuscul Disord. 2014;24: 1123–53.
23. Leung DG, Wagner KR. Therapeutic advances in muscular dystrophy. Ann Neurol. 2013;74:404–11.
24. Mercuri E, Muntoni F. Muscular dystrophies. Lancet. 2013;381:845–60.
25. Vasli N, Laporte J. Impacts of massively parallel sequencing for genetic diagnosis of neuromuscular disorders. Acta Neuropathol. 2013;125:173–85.
26. Biesecker LG, Biesecker BB. An approach to pediatric exome and genome sequencing. Curr Opin Pediatr. 2014;26:639–45.
27. Biesecker LG, Green RC. Diagnostic clinical genome and exome sequencing. N Engl J Med. 2014;371:1170.
28. Wang K, Li M, Hakonarson H. ANNOVAR: functional annotation of genetic variants from high-throughput sequencing data. Nucleic Acids Res. 2010;38:e164.
29. Richards S, Aziz N, Bale S, Bick D, Das S, Gastier-Foster J, Grody WW, Hegde M, Lyon E, Spector E, et al. Standards and guidelines for the interpretation of sequence variants: a joint consensus recommendation of the American College of Medical Genetics and Genomics and the Association for Molecular Pathology. Genet Med. 2015;17:405–24.
30. Kleinberger J, Maloney KA, Pollin TI, Jeng LJ: An openly available online tool for implementing the ACMG/AMP standards and guidelines for the interpretation of sequence variants. Genet Med 2016. (Epub ahead of print).

6

Head-and-neck squamous cell carcinoma risk in smokers: no association detected between phenotype and *AHR*, *CYP1A1*, *CYP1A2*, or *CYP1B1* genotype

Lucia F. Jorge-Nebert[1], Ge Zhang[1,2], Keith M. Wilson[3], Zhengwen Jiang[1,5], Randall Butler[4], Jack L. Gluckman[3], Susan M. Pinney[1] and Daniel W. Nebert[1*]

Abstract

Background: Head-and-neck squamous cell carcinoma (HNSCC) differs between smokers and nonsmokers in etiology and clinical presentation. Because of demonstrated unequivocal involvement in smoking-induced cancer in laboratory animals, four candidate genes--*AHR*, *CYP1A1*, *CYP1A2*, and *CYP1B1*--were selected for a clinical genotype-phenotype association study of HNSCC risk in smokers. Thirty-six single-nucleotide variants (mostly tag-SNPs) within and near these four genes [16 (*AHR*), 4 (*CYP1A1*), 4 (*CYP1A2*), and 12 (*CYP1B1*)] were chosen.

Methods: Extreme discordant phenotype (EDP) method of analysis was used to increase statistical power. HNSCC patients--having smoked 1–40 cigarette pack-years--represented the "highly-sensitive" (HS) population; heavy smokers having smoked ≥80 cigarette-pack-years without any type of cancer comprised the "highly-resistant" (HR) group. The vast majority of smokers were intermediate and discarded from consideration. Statistical tests were performed on $N = 112$ HS and $N = 99$ HR DNA samples from whole blood.

Conclusions: Among the four genes and flanking regions--one haploblock, ACTTGATC in the 5′ portion of *CYP1B1*, retained statistical significance after 100,000 permutations ($P = 0.0042$); among our study population, this haploblock was found in 36.4% of African-American, but only 1.49% of Caucasian, HNSCC chromosomes. Interestingly, in the 1000 Genomes Project database, frequency of this haplotype (in 1322 African and 1006 Caucasian chromosomes) is 0.356 and 0.003, respectively. This study represents an excellent example of "spurious association by population stratification". Considering the cohort size, we therefore conclude that the variant alleles chosen for these four genes, alone or in combinations, are not statistically significantly associated with risk of cigarette-smoking-induced HNSCC.

Keywords: *AHR* gene, *CYP1A1*, *CYP1A2*, *CYP1B1* genes, Tag-SNPs (single nucleotide polymorphisms), Head-and-neck squamous cell carcinoma (HNSCC), Cigarette smoking, Extreme discordant phenotype method, Population stratification, Candidate-gene approach to genotype-phenotype association

* Correspondence: dan.nebert@uc.edu
[1]Department of Environmental Health and Center for Environmental Genetics, University of Cincinnati Medical Center, Cincinnati, OH 45267-0056, USA
Full list of author information is available at the end of the article

Introduction

Worldwide, head-and-neck squamous cell carcinoma (HNSCC) is the sixth most common cancer. An increased risk of HNSCC among cigarette smokers is well known. In addition, high-risk types of human papilloma virus (HPV) are associated with certain HNSCCs, specifically a subset arising in the oropharynx. It also appears very likely that there exists a genetic predisposition to smoking-induced HNSCC risk. Clearly, "cancer" represents a multifactorial trait involving hundreds, if not thousands, of genes, plus epigenetic and environmental effects. It remains possible, however, that—if a candidate-gene approach that embraces a method having sufficient statistical power is applied to a sufficiently large cohort—then a genotype-phenotype association might be demonstrated for one or more "small-effect" genes. This study describes an attempt to establish such an association.

Hundreds of polycyclic aromatic hydrocarbons (PAHs) are present in cigarette smoke. Many laboratory animal studies have demonstrated that aryl hydrocarbon receptor (AHR)-regulated cytochrome P450 family-1 (CYP1) enzymes (CYP1A1, CYP1A2, and CYP1B1) metabolize PAHs to reactive oxygenated intermediates. When cancer initiation occurs via "direct contact" with a carcinogen, e.g., cigarette smoke, we believe that HNSCC will more likely be associated with CYP1-mediated metabolic activation [5, 27, 38], compared with a distal cancer site such as kidney [27].

Description of the four candidate genes

AHR codes for a ligand-activated transcription factor controlling numerous genes and critical cell pathways [40], including up-regulation of *CYP1A1*, *CYP1A2*, and *CYP1B1* genes [33]. AHR foreign ligands include chemicals such as PAHs; polyhalogenated dibenzo-*p*-dioxins, dibenzofurans and biphenyls; and benzoflavones found especially in cruciferous plants [26]. AHR endogenous ligands include indoles and tryptophan-derived moieties and an unknown number of the >150 members of the lipid mediator second-messenger family [7, 32]. The highly conserved AHR exists in all vertebrates and has also been reported to exist—without ligand-binding properties—in mollusk, *Caenorhabditis elegans*, and *Drosophila* [1].

CYP1A1 encodes the P450 monoxygenase that metabolizes planar substrates, many of which are PAHs and biphenyls. CYP1A1 metabolizes few, if any, drugs. Decades of PAH-treated lab animal studies have shown strong correlations of inducible CYP1A1 with various types of cancer—in tissues in contact with the administered PAH [27]. Although basal CYP1A1 expression in animal and human tissues is nearly always nil, inducible CYP1A1 activity is ubiquitous, located in virtually every tissue and cell type of the body. For example, inducible CYP1A1 is found in white blood cells, endothelial cells of blood vessels, lung, kidney, skin, and epithelial lining of the head and neck and upper and lower gastrointestinal (GI) tract. Inducible CYP1A1 also is seen early in embryogenesis [30].

CYP1A2 codes for the CYP1A2 monooxygenase that metabolizes about two dozen drugs—including caffeine and theophylline—plus many environmental aromatic amines. Substantial basal (constitutive) CYP1A2 activity occurs in mammalian liver. Whereas >60-fold differences in human hepatic CYP1A2 (mRNA, protein, and activity) exist between individuals in any population studied, etiology remains unknown. Human *CYP1A2* gene expression is not detectable in embryo, fetus, or kidney but is inducible by PAHs mostly in the liver, GI tract, pancreas, nasal epithelium, brain, and lung [30].

CYP1B1 encodes the CYP1B1 monooxygenase which, like CYP1A1, metabolizes numerous PAHs and biphenyls, *N*-heterocyclic amines, arylamines and amino azo dyes, and other carcinogenic and toxic environmental chemicals. Also, like CYP1A1, CYP1B1 metabolizes few, if any, drugs. Unlike CYP1A1, CYP1B1 often exhibits substantial basal levels (e.g., endocrine tissues, tumors). CYP1B1 expression is induced in vascular endothelial cells, thymus/marrow and immune cells, breast, prostate, uterus, epithelial lining of the head and neck and upper GI tract, and various types of cancers [30].

Mouse *Ahr* knockouts and all three *Cyp1* single-, plus all three double- and the triple-knockout lines are viable and able to reproduce––although serious problems occur in *Ahr(–/–)* knockout [10, 20] and in *Cyp1a1/1a2/1b1(–/–)* triple-knockout mice [8, 31]. Whereas no human "knockout" equivalent has been found for *AHR*, *CYP1A1*, or *CYP1A2*, null mutations in *CYP1B1* are associated with primary congenital glaucoma [44], suggesting that, during embryogenesis, development of the eye's anterior chamber requires metabolism of a critical endogenous CYP1B1 substrate, most likely a lipid mediator [34].

EDP method

Given any gradient for a multifactorial trait, if one selects the two extremes of the phenotypic gradient [28] and disregards intermediate responders in whom genes contributing to phenotype are likely to overlap—then statistical power can be increased [46]. Among HNSCC patients, we selected those with a history of 1–40 cigarette pack-years (Cig-Pk-Yrs) as "highly sensitive" (HS). In the same clinic, we selected heavy smoker volunteers with ≥80 Cig-Pk-Yrs having no types of cancer, as "highly resistant" (HR) controls. All nonsmokers, smokers intermediate between HS and HR criteria, and other patients were excluded from our study.

SNP typing of the four genes

The human *AHR* gene is located at chromosome (Chr) 7p15 and *CYP1B1* gene at Chr 2p22.2. The *CYP1A2_-CYP1A1* locus, on Chr 15q24.1, contains the two genes oriented head-to-head with a bidirectional promoter [6]. The purpose of this study was to search for single single-nucleotide polymorphism (SNP) marker and haplotype associations in these four selected genes that might be statistically significantly correlated with greater risk of HNSCC in HS cancer patients, compared with HR heavy smokers having no cancer.

Methods

Clinical screening and patient/volunteer recruitment

HS patients were identified at the Barrett Cancer Center (Department of Otolaryngology-Head and Neck Surgery, University of Cincinnati College of Medicine) and the Cincinnati Veterans' Association (VA) Hospital. Additional patients were identified within the Fernald Community Project. HR volunteers were recruited at the Barrett Cancer Center, Cincinnati VA Hospital, Fernald Project, and throughout the community in response to flyers. A questionnaire was used to identify HNSCC patients—who qualified as HS because they had smoked 1–40 Cig-Pk-Yrs (e.g., "20 Cig-Pk-Yr" denotes someone who might have smoked one pack per day for 20 years or one-half pack per day for 40 years). Volunteers, having no cancer of any kind (with exception of UV-caused skin cancers), despite having smoked ≥80 Cig-Pk-Yrs, qualified for the HR group and were also identified by questionnaire.

The Fernald Community blood samples originated from an earlier independent study, as described [39]. At all times, we followed the clinical protocol—titled "Human Cancer and *AHR/CYP1A1/1A2/1B1* Gene Polymorphisms" (#03-08-07-01)—annually approved for the entire study, without any HIPAA issues or concerns, by the University of Cincinnati Medical Center Institutional Review Board (IRB).

Questionnaires

Questions asked of all participants included self-identified ethnicity, pipe or cigar smoking, second-hand smoke, high-tar vs. low-tar cigarettes, alcohol consumption, use of snuff/chewing tobacco, amount of coffee consumed, occupational hazards, consumption and amount of daily antioxidants, frequency of eating grilled meat, and family history. Staff administering the questionnaire ranked these eleven caveats as zero (little or negligible contact), one (intermediate contact or exposure), or two (high exposure). A copy of this questionnaire is included as Additional file 1.

Clinical samples

Whole blood for DNA isolation was collected from HS and HR subjects via antecubital venipuncture. BD Vacutainer® systems (Becton, Dickinson and Co.; Franklin Lakes, NJ) were used. We collected two 13 × 100 mm Vacutainer® K_2EDTA tubes (lavender caps) for a total volume of 10 mL. Freshly collected whole blood was refrigerated at 4 °C and kept not longer than 24 h at that temperature. If genomic DNA (gDNA) was not isolated from blood within 24 h of collection, the blood was stored at −80 °C usually for not longer than 8 weeks and then thawed the same day the isolation procedure was performed.

DNA preparation

For the 10 mL of blood—collected at the Barrett Cancer Center and Cincinnati VA Hospital—gDNA was isolated using the QIAamp® DNA Blood Maxi kit (QIAGEN, Hilden, Germany) by applying the spin protocol. Whole-blood samples from the Fernald Community cohort, which had been stored at −70 °C, were thawed on the day of isolation. Average volume of these samples was 2 mL; gDNA was therefore isolated using the QIAamp® DNA Blood Midi kit (QIAGEN) and spin protocol. Both QIAGEN kits employed special patented columns for purification of gDNA present in blood pretreated with proteases, in a highly denaturing medium containing guanidine hydrochloride.

Highly purified gDNA isolated in this fashion was quantified by UV absorption at 260 nm using a mini-spectrometer (NanoDrop® ND-1000, Thermo Scientific; Waltham, MA). Purity was evaluated using the 260/280 and 260/230 absorbance ratios.

Selection of common tag-SNPs

Tag-SNPs and other variants of interest were chosen in order to study variation within each of the genetic loci and to correlate these variations within each person in the HS and HR cohorts. Tag-SNPs were selected by the HaploView program [3], using a MAF cutoff of 10.0% and an r^2 threshold of 0.8. We used previous sequencing data of the *CYP1* locus (~40 kb) which includes the 23.3-kb bidirectional promoter [14, 15]. For *AHR* and *CYP1B1*, we selected tag-SNPs from targeted sequencing, including 10 kb of 5′ flanking regions in 24 Caucasian DNA samples (Z.J., *data not shown*). The ultimate total was 36 tag-SNPs: 16 for *AHR*, 4 for *CYP1A1*, 4 for *CYP1A2*, and 12 for *CYP1B1*.

These tag-SNPs are in linkage disequilibrium (LD) with other variants in their vicinity. Thus, their selection and genotyping were expected to simplify studies of the four candidate genes—by reducing the number of variants needed for a comprehensive study. Whole-genome sequencing was not yet an option at the time of this study.

Genotyping of tag-SNPs

Tag-SNP genotypes for each HS and HR individual were determined using the ABI Prism® SnaPshot™ Multiplex system, which allows for the typing of up to ten SNPs simultaneously. Chemistry of the applied kit is based on the dideoxy single-base extension of unlabeled primer or primers, catalyzed by AmpliTaq DNA polymerase, FS; this permits incorporation of a single fluorescently-labeled ddNTP on the 3′-end of the primer.

Statistical analyses of genotype-phenotype association

Bioinformatics software HaploView (v. 4.2, Mark Daly's lab; MIT/Harvard Broad Institute) was used to determine associations between the markers—variation in SNPs—and phenotype in HS vs. HR samples. This software also analyzes LD patterns, generating haploblocks, and inferring haplotypes. Criteria applied to determine haploblocks within genes was the "solid spine of LD". Associations between these haplotypes and phenotypes were also calculated via the chi-square association test of every haplotype (exhaustive haplotype test). In addition, we applied the permutations test (results randomized 100,000 times) to adjust for multiple tests of multiple SNPs or haplotypes. Haplotype estimates for each individual HS and HR were obtained through the software program phase (v. 2.1.1, Matthew Stephens's lab; University of Chicago). Differences in frequencies of haplotypes between populations were examined for significance using Fisher's exact test.

Results

HNSCC in the present study refers exclusively to squamous cell carcinoma of the oral cavity, larynx, oropharynx, and hypopharynx. Tumors of the face, salivary glands, nasopharynx, and brain were excluded. A substantial number (~15%) of HNSCC patients have no cigarette smoking history. Clinical and genetic features of HNSCC in nonsmokers, former smokers, and current smokers are known to be distinctly different [17]. Compared with malignancies in nonsmokers, smokers exhibit more tumors of the larynx, hypopharynx, and floor of the mouth; a much greater *TP53* mutation rate; a substantially higher percent of loss of heterozygosity at Chr 3p, 4q, and 11q13; and a greater overall average number of chromosomal losses. In contrast, the percentage with HPV infection was marginally, but not statistically, lower in nonsmoker malignancies [17].

We excluded all nonsmokers (i.e., those having smoked no more than one Cig-Pk-Yr, i.e., <7300 cigarettes, <365 packs, over a lifetime) because the mechanism of HNSCC tumorigenesis is likely not to be relevant to activation by cigarette smoke PAHs involving any of the four genes under study. It should be noted that more than 85% of all HNSCC smokers did not qualify because they had smoked >40 Cig-Pk-Yrs and <80 Cig-Pk-Yrs; this selective stringency creates a more robust statistical power [46], but unfortunately fewer numbers of qualified individuals.

Over a 5-year period, we collected gDNA from a total of 149 blood samples from the Barrett Cancer Center and Cincinnati VA Hospital, plus 62 samples from the Fernald Project. Of the 211 total samples, there were 94 and 18 HS patients from the Barrett Center/VA Hospital and Fernald Project, respectively, and 55 and 44 HR samples from the Barrett Center/VA Hospital and Fernald Project, respectively. Total subjects having high-quality gDNA ultimately included 112 HS and 99 HR for this study (Table 1).

Statistical analyses of covariates

With regard to the 11 covariates included in the questionnaire (Appendix 1; *described above*), we specifically focused on occupational history, amount of dietary grilled meat, and family history of cancer (ranked as "0," "1," or "2" for each individual); the other 8 parameters did not have sufficient information for a substantial number of participants. We used logistic regression and random-forests models for statistical analysis of possible associations between the selected best SNP markers and these three covariates. No statistically significant associations or trends of association ($P < 0.05$) were found between any of these covariates and phenotype (*data not shown*). The "self-identified ethnicity" was important and is discussed in great detail below.

Genotype-phenotype association analysis

The four genes studied, with *arrows* displaying locations of each selected tag-SNP, are illustrated in Fig. 1. It is

Table 1 Demographics of entire cohort studied

Ethnicity (self-identified)					
	HS	HR	Total	Percent	
Caucasian-American	101	95	196	92.9	
African-American	11	3	14	6.6	
Latino-American	0	1	1	0.5	
Total	112	99	211	100	
By site of collection					
Ethnicity (self-identified)	Barrett/VA		Fernald		Total
	HS	HR	HS	HR	
Caucasian-American	83	51	18	44	196
African-American	11	3	0	0	14
Latino-American	0	1	0	0	1
Total	94	55	18	44	211

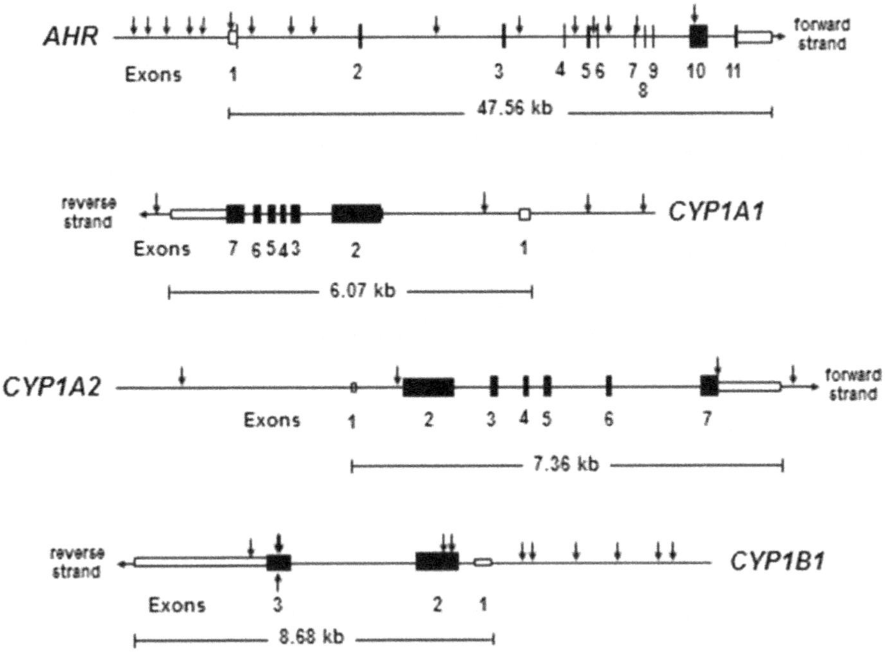

Fig. 1 Diagram of locations of all tag-SNPs and other SNPs in the four genes chosen for study. The *AHR* gene, having 11 exons, is located on chromosome 7p15, and the transcribed region spans 47.53 kb; 16 tag-SNPs inside or near the gene were selected. *CYP1A1* (spanning 6.07 kb) and *CYP1A2* (7.36 kb) on chromosome 15q24.1 each have seven exons and are arranged in tandem, head-to-head, with a 23.3-kb bidirectional promoter between them; four tag-SNPs inside and near each gene were chosen. Human *CYP1B1* on Chr 2p22.2 has three exons and spans 8.68 kb; 12 tag-SNPs inside and near the gene were selected. Note that three SNPs (#2, 3 and 4) are located in *CYP1B1* exon 3, within four nucleotides of one another. *Closed rectangles* of the exons denote the translated region and *open rectangles* the 5'- and 3'-untranslated regions. Whereas translation of *AHR* is initiated at the 3' end of exon 1, all three *CYP1* genes have noncoding first exons. Note that *CYP1A1* and *CYP1B1* are on reverse strand of gDNA, meaning that the chromosomal location of each SNP for these two genes in the Genome Assembly is numbered from the 3'- to 5'-end, whereas *AHR* and *CYP1A2* are on the positive strand and therefore each SNP for these two genes is numbered from the 5'- to 3'-end.

noteworthy that two of the four genes are on the reverse strand; by convention, all marker alleles were converted to the positive strand. Moreover, *CYP1A1* lies 5′-ward of *CYP1A2*, with the two genes situated head-to-head and a 23.3-kb bidirectional promoter between them; thus, two tag-SNPs located 5′-ward of *CYP1A1* and one tag-SNP 5′-ward of *CYP1A2* exon 1 are located within the bidirectional promoter—which has well-known regulatory elements, i.e., AHR-binding sites [6].

Table 2 lists the 36 selected tag-SNPs, dbSNP rs numbers, chromosomal positions, locations in or near each gene, ancestral alleles (i.e., the "phylogenetic root" based on sequence alignment of multiple ($N = 6$) primates), reference/alternative alleles, alternative allele frequencies [q(alt)] in Caucasian (EUR) samples from the 1000 Genomes Project, q(alt) in African (AFR) samples from the 1000 Genomes Project, and q(alt) found in the cohort studied (HR and HS combined). We used Genome Assembly GRCh38.p2 (Annotation Release 107) for this analysis; data were calculated from the genomic sequences of Ensembl transcripts AHR-002 (ENST00000242057), CYP1A1-001 (ENST00000379727), CYP1A2-001 (ENST00000343932), and CYP1B1-001 (ENST00000610745).

Single-marker association study (allele test)

The entire cohort (HS and HR combined, $N = 211$) included 196 self-identified Caucasian-Americans, 14 African-Americans, and 1 Latino (Table 1). Using the chi-square test for allele-frequency difference of each tag-SNP between HS and HR samples (Table 3), we found three P values significant at the $P < 0.05$ level; however, these associations did not retain statistical significance in the 100,000 permutations test. This is a very common observation and often not appreciated—when comparing standard statistical tests with a permutation test that is mandatory for association studies with regard to multiple markers throughout all chromosomes [4].

Case-control association tests in Caucasians ($N = 196$) were achieved by removing the 1 Latino and 14 African-American subjects. Table 4 displays the HaploView analysis—examining individual markers or SNPs. From chi-square analysis, there is only one SNP that appears to show a significant P value in the *AHR* gene: rs4719497 ($P = 0.0134$). All other SNPs have P values >0.05. However, following permutation testing, rs4719497 did not retain significance ($P = 0.1048$), indicating that this SNP did not survive the mandatory correction for multi-testing.

Table 2 List of tag-SNPs, chromosomal location, SNP identifier number, chromosomal position, location in or near gene, ancestral gene, reference/alternative alleles, alternative allele frequency q(alt) in Caucasian (EUR) and African (AFR) populations from the 1000 Genomes Project, and q(alt) in cohort studied

SNP #	Gene	Chr	SNP ID	Chromosomal position[a]	Location in or near the gene[b]	Ancestral allele[c]	Reference /alternative allele	q(alt) EUR	q(alt) AFR	q(alt) in present cohort
1	*AHR*	7	rs62444550	17290494	−8128C > A, promoter	A	C/A	0.4264	0.6407	0.4265
2	*AHR*	7	rs77821156	17291826	−6796A > G, promoter	A	A/G	0.0815	0.0045	0.0900
3	*AHR*	7	rs10250822	17293365	−5257 T > C, promoter	T	T/C	0.2217	0.0908	0.2370
4	*AHR*	7	rs4719497	17295275	−3347 T > C, promoter	T	T/C	0.1243	0.0083	0.1374
5	*AHR*	7	rs3757824	17296411	−2211 T > C, promoter	T	T/C	0.2127	0.1490	0.1706
6	*AHR*	7	rs7796976	17298806	+185G > A, 5′-UTR, exon 1	G	G/A	0.2217	0.0840	0.00
7	*AHR*	7	rs713150	17300533	+1912C > G, intron 1	C	C/G	0.2177	0.1354	0.2322
8	*AHR*	7	rs17722841	17303970	+5349G > A, intron 1	G	G/A	0.1690	0.0061	0.1919
9	*AHR*	7	rs2282885	17305990	+7369A > G, intron 1	A	A/G	0.3956	0.0189	0.3697
10	*AHR*	7	rs2282883	17316723	+18102C > T, intron 2	T	C/T	0.3469	0.6293	0.3720
11	*AHR*	7	rs4236290	17323944	+253233 T > C, intron 3	T	T/C	0.1064	0.0061	0.1114
12	*AHR*	7	rs2158041	17328796	+30175C > T, intron 4	C	C/T	0.2187	0.0340	0.2322
13	*AHR*	7	rs3802082	17330557	+31936A > T, intron 5	A	A/T	0.1769	0.1513	0.1374
14	*AHR*	7	rs7811989	17331739	+33118G > A, intron 6	G	G/A	0.2396	0.1135	0.2749
15	*AHR*	7	rs2074113	17334147	+35526G > T, intron 7	G	G/T	0.1043	0.1520	0.0711
16	*AHR*	7	rs2066853	17339486	+40865G > A, exon 10, R554K	A	G/A	0.1083	0.4622	0.0995
1	*CYP1A1*	15 (rev)	rs4646903	74719300	*CYP1A1*2A*, *Msp* I RFLP Site, +6311A > G, downstream 3′	A	A/G	0.1074	0.2345	0.1066
2	*CYP1A1*	15 (rev)	rs2606345	74724835	+776A > C, intron 1	C	A/C	0.3370	0.9501	0.3768
3	*CYP1A1*	15 (rev)	rs3826042	74726564	−954C > T, promoter	C	C/T	0.0417	0.0053	0.0498
4	*CYP1A1*	15 (rev)	rs7495708	74727502	−1892 T > C, promoter	C	T/C	0.1541	0.5696	0.1919
1	*CYP1A2*	15	rs2069514	74745879	*CYP1A2*1C*, −2965G > A, promoter	G	G/A	0.0199	0.3132	0.0616
2	*CYP1A2*	15	rs762551	74749576	*CYP1A2*1 F*, +733A > C, intron 1	A	A/C	0.3201	0.4380	0.2796
3	*CYP1A2*	15	rs2470890	74755085	*CYP1A2*1B*, +6242 T > C, exon 7, N516N (syn)	C	T/C	0.4036	0.9720	0.3981
4	*CYP1A2*	15	rs17861162	74756412	+7569C > G, downstream 3′	C	C/G	0.0626	0.1604	0.0592
1	*CYP1B1*	2 (rev)	rs162562	38070372	+5810 T > G, 3′UTR	T	T/G	0.2187	0.6694	0.2701
2	*CYP1B1*	2 (rev)	rs1800440	38070996	*CYP1B1*4*, +5186 T > C, exon 3, N453S	T	T/C	0.1958	0.0068	0.1588
3	*CYP1B1*	2 (rev)	rs1056837	38071007	+5175G > A, exon 3, D449D (syn)	G	G/A	0.3976	0.7988	0.4692
4	*CYP1B1*	2 (rev)	rs1056836	38071060	*CYP1B1*3*, +5122C > G, exon 3, L432V	C	C/G	0.3976	0.8169	0.4716
5	*CYP1B1*	2 (rev)	rs1056827	38075034	*CYP1B1*2*, +1148C > A, exon 2, A119S	A	C/A	0.2892	0.5234	0.2962
6	*CYP1B1*	2 (rev)	rs10012	38075247	*CYP1B1*2*, +935G > C, exon 2, R48G	C	G/C	0.2903	0.5741	0.2986
7	*CYP1B1*	2 (rev)	rs162558	38076937	−756 T > C, promoter	T	T/C	0.1720	0.1846	0.2085
8	*CYP1B1*	2 (rev)	rs2855655	38077346	−1165C > T, promoter	C	C/T	0.3658	0.7345	0.4431
9	*CYP1B1*	2 (rev)	rs162557	38078309	−2128G > A, promoter	G	G/A	0.2376	0.2224	0.2536
10	*CYP1B1*	2 (rev)	rs162556	38079312	−3131A > G, promoter	A	A/G	0.5000	0.0348	0.4526
11	*CYP1B1*	2 (rev)	rs162555	38080367	−4186 T > C, promoter	T	T/C	0.1750	0.1664	0.2014
12	*CYP1B1*	2 (rev)	rs10175368	38080719	−4538C > T, promoter	C	C/T	0.2853	0.0613	0.2559

Chr chromosome, *rev* reverse strand, *syn* synonymous, *q(alt)* frequency of alternative allele

[a]Genome Assembly GRCh38.p2 (Annotation Release 107)

[b]Calculated from the genomic sequences of Ensembl transcripts AHR-002 (ENST00000242057), CYP1A1-001(ENST00000379727), CYP1A2-001 (ENST00000343932), and CYP1B1-001 (ENST00000610745)

[c]Ancestral allele refers to the "phylogenetic root" based on sequence alignment of multiple ($N = 6$) primates, as reported in dbSNP

Table 3 Comparison of allele frequencies (additive model), as tested by chi-square analysis and permutations test, of individual tag-SNPs, HS vs. HR samples, entire cohort ($N = 211$)

Gene	Tag-SNP #	SNP ID	Asso-ciated allele[a]	Cases (HS) allele frequencies	Controls (HR) allele frequencies	*P* value (chi-square analysis)	*P* value, following 100 K permutations
AHR	1	rs62444550	A	0.433	0.419	0.7741	1.00
AHR	2	rs77821156	A	0.929	0.889	0.1553	0.7627
AHR	3	rs10250822	C	0.272	0.197	0.0693	0.4745
AHR	4	rs4719497	C	0.174	0.096	0.0200*	0.1727
AHR	5	rs3757824	T	0.862	0.793	0.0613	0.4203
AHR	7	rs713150	G	0.263	0.197	0.1068	0.6212
AHR	8	rs17722841	G	0.817	0.798	0.6212	0.9998
AHR	9	rs2282885	A	0.665	0.591	0.1147	0.6360
AHR	10	rs2282883	T	0.397	0.343	0.2531	0.9128
AHR	11	rs4236290	T	0.902	0.874	0.3607	0.9774
AHR	12	rs2158041	T	0.250	0.212	0.3578	0.9742
AHR	13	rs3802082	A	0.884	0.838	0.1751	0.8056
AHR	14	rs7811989	A	0.295	0.253	0.3335	0.9620
AHR	15	rs2074113	G	0.942	0.914	0.2670	0.9201
AHR	16	rs2066853	A	0.107	0.091	0.5783	0.9996
CYP1A1	1	rs4646903	G	0.125	0.086	0.1936	0.7207
CYP1A1	2	rs2606345	C	0.397	0.354	0.3543	0.9355
CYP1A1	3	rs3826042	C	0.960	0.939	0.3355	0.9121
CYP1A1	4	rs7495708	C	0.205	0.177	0.4567	0.9711
CYP1A2	1	rs2069514	A	0.089	0.030	0.0119*	0.0896
CYP1A2	2	rs762551	A	0.723	0.717	0.8902	1.0000
CYP1A2	3	rs2470890	C	0.438	0.354	0.0787	0.4071
CYP1A2	4	rs17861162	G	0.076	0.040	0.1233	0.5885
CYP1B1	1	rs162562	G	0.317	0.217	0.0212*	0.1262
CYP1B1	2	rs1800440	T	0.862	0.818	0.2232	0.7441
CYP1B1	3	rs1056837	A	0.478	0.460	0.7103	0.9998
CYP1B1	4	rs1056836	C	0.482	0.460	0.6433	0.9979
CYP1B1	5	rs1056827	A	0.317	0.273	0.3206	0.8683
CYP1B1	6	rs10012	C	0.326	0.268	0.1922	0.6692
CYP1B1	7	rs162558	T	0.821	0.758	0.1071	0.4914
CYP1B1	8	rs2855655	C	0.562	0.551	0.8045	1.0000
CYP1B1	9	rs162557	A	0.272	0.232	0.3459	0.9051
CYP1B1	10	rs162556	A	0.567	0.525	0.3903	0.9346
CYP1B1	11	rs162555	T	0.826	0.768	0.1367	0.5529
CYP1B1	12	rs10175368	C	0.746	0.742	0.9417	1.0000

*Statistically significant ($P < 0.05$)
[a]Associated allele, as selected by the HaploView program

Haplotype association study

Figure 2 shows the LD heat map for *CYP1B1* SNPs, using r^2 (correlation between pairs of loci) as a measure of linkage disequilibrium. Haploblocks 1 (tag-SNPs #1, 2, 3, and 4) and 2 (tag-SNPs #5, 6, 7, 8, 9, 10, 11, and 12) were determined by the program HaploView because they showed the best continuous solid spine of LDs. We defined haploblocks by applying this nonstringent criterion—since no strong correlations were expected between tag-SNP pairs, i.e., by definition, tag-SNPs are selected because they should capture as much information as possible about independent regions of the gene.

Table 4 Comparison of allele frequencies (additive model), as tested by chi-square analysis and permutations test, of individual tag-SNPs, HS vs. HR Caucasian-only sample ($N = 196$)

Gene	Tag-SNP #	SNP ID	Asso-ciated allele[a]	Cases (HS) allele frequencies	Controls (HR) allele frequencies	*P* value (chi-square analysis)	*P* value, following 100 K permutations
AHR	1	rs62444550	A	0.926	0.884	0.1598	0.741
AHR	2	AHR10SNP	A	0.431	0.416	0.7654	1.00
AHR	3	rs10250822	C	0.282	0.205	0.0768	0.4705
AHR	4	rs4719497	C	0.188	0.100	0.0134*	0.1048
AHR	5	rs3757824	T	0.856	0.795	0.1069	0.6213
AHR	7	rs713150	G	0.272	0.205	0.1204	0.6554
AHR	8	rs17722841	G	0.797	0.795	0.9551	1.00
AHR	9	rs2282885	A	0.629	0.579	0.3139	0.9577
AHR	10	rs2282883	T	0.386	0.337	0.3102	0.9496
AHR	11	rs4236290	T	0.896	0.868	0.3958	0.9802
AHR	12	rs2158041	T	0.267	0.221	0.287	0.9119
AHR	13	rs3802082	A	0.876	0.842	0.3308	0.9603
AHR	14	rs7811989	A	0.302	0.263	0.3939	0.9776
AHR	15	rs2074113	G	0.941	0.921	0.4452	0.992
AHR	16	rs2066853	A	0.084	0.079	0.8506	1.00
CYP1A1	1	rs4646903	G	0.104	0.084	0.5039	0.9866
CYP1A1	2	rs2606345	C	0.351	0.337	0.7604	0.9999
CYP1A1	3	rs3826042	C	0.955	0.937	0.4136	0.9446
CYP1A1	4	rs7495708	C	0.158	0.153	0.8745	1.00
CYP1A2	1	rs2069514	A	0.035	0.016	0.2365	0.8158
CYP1A2	2	rs762551	C	0.292	0.279	0.7736	1.00
CYP1A2	3	rs2470890	C	0.386	0.332	0.2606	0.8355
CYP1A2	4	rs17861162	G	0.064	0.037	0.216	0.7605
CYP1B1	1	rs162562	G	0.282	0.221	0.1639	0.6189
CYP1B1	2	rs1800440	T	0.856	0.821	0.3404	0.8777
CYP1B1	3	rs1056837	G	0.545	0.542	0.9612	1.00
CYP1B1	4	rs1056836	C	0.460	0.458	0.9604	1.00
CYP1B1	5	rs1056827	A	0.297	0.268	0.5298	0.9853
CYP1B1	6	rs10012	C	0.302	0.263	0.3939	0.9263
CYP1B1	7	rs162558	T	0.812	0.763	0.238	0.751
CYP1B1	8	rs2855655	C	0.589	0.547	0.4043	0.9302
CYP1B1	9	rs162557	A	0.277	0.242	0.4283	0.9473
CYP1B1	10	rs162556	A	0.540	0.516	0.6369	0.9954
CYP1B1	11	rs162555	T	0.817	0.774	0.2895	0.8232
CYP1B1	12	rs10175368	T	0.277	0.258	0.6658	0.9985

* Statistically significant ($P < 0.05$)
[a]Associated allele, as selected by the HaploView program

However, some of the SNPs selected for this study are in very close proximity to one other and did show substantially strong correlations (depicted as *darkest squares* in the heat map, e.g., SNPs #5 and 6). We included these SNPs in our study because of functional relevance in cancer studies or known to alter enzyme activity levels. LD heat maps and haploblocks were similarly generated for the other three genes studied and were unremarkable (*data not shown*).

Of the 38 haplotypes inferred in the six defined haploblocks among our four genes in the entire cohort (Table 5, $N = 211$), six were statistically significant ($P < 0.05$) by chi-square analysis. After 100,000 permutations, however, only one *CYP1B1* haplotype in haploblock 2 (ACTTGATC)

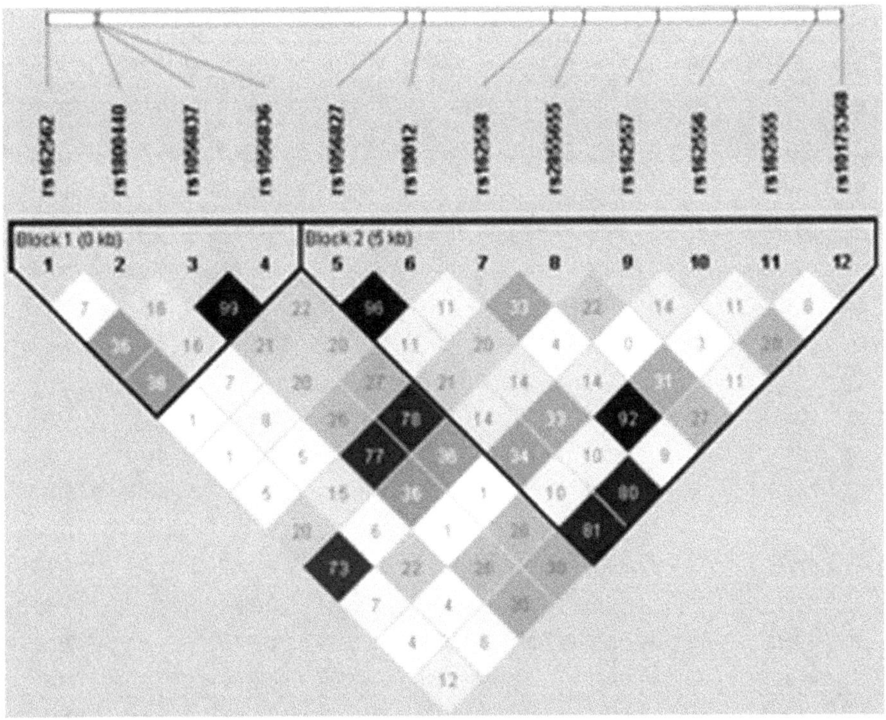

Fig. 2 LD heat map of *CYP1B1* SNPs from the entire cohort ($N = 211$), using r^2 (correlation between pair of loci) as a measure of LD. Haploblock 2—which we had found initially as apparently the only haplotype (ACTTGATC) significantly associated with HNSCC risk in smokers—contains two SNPs in the 5′ half of exon 2 plus six SNPs in the 5′-flanking region (reverse strand), extending as far 5′-ward as 4538 bp upstream of the transcription initiation start-site. *Shading* denotes the following: *white* ($r^2 = 0$), black ($r^2 = 1$), and shades of *gray* ($0 < r^2 < 1$), with deepening *gray colors* depicting increasing r^2 values (correlation) between SNP pairs. *Numbers* in each *square* indicate percentage of correlation ($r^2 \times 100$).

retained a statistically significant ($P = 0.0042$) association with HNSCC risk in smokers.

The same analysis was applied to the Caucasian-only sample ($N = 196$) and the results are shown in Table 6. Only two haplotypes revealed significant P values: one in *AHR* Block 1 ($P = 0.0259$) and one in *CYP1B1* Block 2 ($P = 0.0392$). Following correction for multi-testing (100K permutations), however, the appearance of statistical significance was lost: $P = 0.142$ and 0.1666, respectively. Note that the inferred haplotypes, and order of their ranking by frequency (*left-most columns*), differs between Table 5 (entire cohort) and Table 6 (Caucasian-only sample).

Further analysis of *CYP1B1* haplotype ACTTGATC

In our initial approach, all HS and HR subjects ($N = 211$, i.e., 422 chromosomes, including one self-identified Latino) were included in the same association study of haplotype ACTTGATC with HNSCC risk among smokers, irrespective of racial origin (Table 5); and it was initially exciting to find the ACTTGATC haplotype (apparently) statistically significantly associated ($P = 0.0042$) with cigarette smoking-induced risk of HNSCC. When the Latino subject was excluded from the analysis, leaving only African-American and Caucasian-Americans (Table 7), Fisher's exact test indicated a significant difference in the frequency of this haplotype between HS and HR ($P = 0.0011$).

However, when individuals carrying this haplotype were identified by race (Table 8), it became apparent that the haplotype was tracking extensively with African origin, rather than HNSCC risk; in a total of 112 HS subjects (224 chromosomes and excluding the one Latino), this haplotype appeared 11 times—8 times in heterozygous state in African-Americans and 3 times in heterozygous state among Caucasian-Americans. This distribution is disproportionately larger for African-Americans, considering we had recruited only 11 African-American HS vs. 101 Caucasian-American HS patients. The Fisher exact test demonstrated a significant difference ($P < 0.0001$) in this haplotype frequency between African-American and Caucasian-American HS chromosomes (Table 8).

From a total of 28 African-American chromosomes of both HS and HR studied (Table 9), 8 carried the ACTTGATC haplotype (allele frequency = 0.286). However, when we compared this haplotype and HNSCC risk in African-Americans, no statistically significant conclusion could be reached ($P = 0.141$), perhaps because our study is under-powered due to small sample sizes—especially a control HR group comprising only six chromosomes.

Table 5 Comparison of haplotype frequencies (additive model), as tested by chi-square analysis and permutations test, HS vs. HR samples, entire cohort ($N = 211$)

Haploblock and inferred haplotypes	Haplotype total frequency	Cases (HS) frequencies	Controls (HR) frequencies	*P* value (chi-square analysis)	*P* value, following 100 K permutations
AHR Block 1					
ACTTTC	0.479	0.487	0.470	0.7288	1.00
AATTCC	0.171	0.138	0.207	0.0613	0.5326
AACCTG	0.133	0.165	0.096	0.0365 *	0.3665
AACTTG	0.097	0.094	0.101	0.8016	1.0000
GCTTTC	0.090	0.071	0.111	0.1553	0.8236
AATTTC	0.024	0.031	0.015	0.2781	0.9680
AHR Block 2					
GACTCAGGG	0.243	0.249	0.237	0.7678	1.00
AGCTCAGGG	0.187	0.174	0.201	0.464	1.00
GATTTAAGG	0.152	0.175	0.126	0.17	0.8445
GGCTCAGGG	0.119	0.108	0.132	0.4426	1.0000
GATCTAAGG	0.075	0.066	0.085	0.4563	1.0000
GATTCTGTA	0.071	0.058	0.086	0.267	0.9607
GGCTCTGGG	0.061	0.049	0.075	0.2578	0.9456
GATCCAAGG	0.033	0.031	0.035	0.8143	1.00
GATTCAGGA	0.026	0.045	0.005	0.0109 *	0.0786
GACTCAAGG	0.012	0.018	0.005	0.225	0.9396
CYP1A1					
AACT	0.623	0.603	0.646	0.3543	0.9613
ACCT	0.185	0.192	0.177	0.6882	0.9998
GCCC	0.107	0.125	0.086	0.1936	0.7935
ACTC	0.050	0.040	0.061	0.3355	0.9493
ACCC	0.036	0.040	0.030	0.5845	0.9975
CYP1A2					
GAT	0.602	0.562	0.646	0.0787	0.4736
GCC	0.280	0.277	0.283	0.8902	1.00
AAC	0.062	0.089	0.030	0.0119 *	0.1239
GAC	0.057	0.071	0.040	0.1697	0.7649
CYP1B1 Block 1					
TTGG	0.360	0.364	0.357	0.8781	1.00
GTAC	0.261	0.301	0.215	0.0444 *	0.2237
TTAC	0.208	0.176	0.244	0.086	0.5013
TCGG	0.159	0.138	0.182	0.2232	0.8763
CYP1B1 Block 2					
ACTCGATT	0.256	0.254	0.258	0.9417	1.00
CGTCGGTC	0.251	0.237	0.268	0.4627	1.00
CGCTGACC	0.185	0.156	0.217	0.1078	0.6369
CGTTAGTC	0.182	0.174	0.192	0.6369	1.00
CGTCAATC	0.037	0.057	0.015	0.0248	0.1204

Table 5 Comparison of haplotype frequencies (additive model), as tested by chi-square analysis and permutations test, HS vs. HR samples, entire cohort ($N = 211$) *(Continued)*

ACTTGATC	0.028	0.052	0.000	0.0011 **	0.0042 **
CGTTAATC	0.020	0.024	0.015	0.5256	1.00
CGCTAGCC	0.012	0.013	0.010	0.7551	1.00
ACTCGATC	0.010	0.010	0.010	0.9756	1.00

AHR Block 1 SNPs 1, 2, 3, 4, 5, 7, *AHR Block 2* SNPs 8, 9, 10, 11, 12, 13, 14, 15, 16, *CYP1A1* SNPs 1, 2, 3, 4, *CYP1A2* SNPs 1, 2, 3, *CYP1B1 Block 1* SNPs 1, 2, 3, 4, *CYP1B1 Block 2* SNPs 5, 6, 7, 8, 9, 10, 11, 12

*Statistically significant ($P < 0.05$)

**Statistically significant ($P < 0.005$)

To investigate the possibility that haplotype ACTTGATC in our African-American cohort simply represented a high-frequency haplotype among Africans, we turned to the 1000 Genomes Project and searched for average frequencies of each SNP within *CYP1B1* haploblock 2 among African populations; we inferred the best reconstructed haplotypes and estimated their frequencies (Additional file 2: Table S1). Interestingly, haplotype ACTTGATC was found to be the most common (ranked #1)—among the sum of all African genomes sequenced to date—with a frequency of 0.356.

Regarding the majority of our HNSCC cohort who were of self-identified Caucasian descent ($N = 196$, i.e., 392 chromosomes), we failed to find an association ($P = 0.249$) between haplotype ACTTGATC and HNSCC risk (Table 10). Using data extracted from the 1000 Genomes Project for European populations (Additional file 2: Table S1) and analyzing it similarly to what had been done for African populations, we found that haplotype ACTTGATC has a frequency of 0.003—ranking #9 among inferred haplotypes for Europeans.

Taking into account the significant difference ($P < 0.0001$) in frequency of haplotype ACTTGATC between European and African populations represented in the 1000 Genomes Project (Table 11), and the apparent lack of significant association between this haplotype and HNSCC risk when African-Americans (Table 9) and Caucasian-Americans (Table 10) were analyzed separately, we are able confidently to conclude that the association found in our initial analysis was spurious and caused by *population stratification*—a common artifact that can lead many statisticians to reporting a false positive (type I error) as a true positive.

Discussion

The present study represents a candidate-gene analysis, using the extreme discordant phenotype (EDP) method, attempting to determine a genotype-phenotype association—in which the trait being studied is HNSCC among cigarette smokers. The four candidate genes included *AHR*, *CYP1A1*, *CYP1A2*, and *CYP1B1*. Based on innumerable laboratory animal studies, these four genes were selected because they encode three CYP1 enzymes, plus AHR which regulates the levels of these enzymes. This gene battery [27] reflects one of the earliest responses to cigarette smoke as an environmental signal. Moreover, HNSCC is most likely to develop as the result of "direct-contact" of head-and-neck epithelial cells with chronic irritants in cigarette smoke. For example, if any nonsmoker were to smoke even one cigarette—these three *CYP1* genes would become up-regulated by AHR (to genetically varying degrees) within a few hours; without further "stimulation" (i.e., no more cigarettes), AHR-regulated *CYP1* genes would quickly return to baseline. In laboratory animals, there is ample evidence that these four genes play an "up-front" role in PAH-induced cancer [5, 27, 38].

Statistical power of EDP analysis

An accurate power calculation of EDP design—based on exposure level [EDP design 2 of the [46] article [to a known risk factor (e.g., cigarette smoking)] —would require quantification of several features: (a) phenotypic variance attributable to the environmental exposure, (b) truncation selection areas (e.g., 1–40 cigarette pack-years for HS cases, >80 cigarette pack-years for HR controls), and (c) disease prevalence. Given that these values are not well known, it would be very difficult to provide an accurate power calculation.

However, if we can roughly assume that cigarette smoking explains >20% variance in liability of HNSCC that the selection areas (1–40 vs. >80 cigarette pack-years) represent the bottom and the top 10% exposure groups, respectively, and that the prevalence of HNSCC in all smokers is ~1%, then our EDP design would have similar power to a normal case/control study comprising 500~600 samples (Fig. 9 of the [46] article).

Previous genotype-phenotype association studies: HNSCC and our four candidate genes

We found no studies in the literature concerning *AHR* polymorphisms associated with HNSCC. With regard to

Table 6 Comparison of haplotype frequencies (additive model), as tested by chi-square analysis and permutations test, HS vs. HR Caucasian-only sample ($N = 196$)

Haploblock and inferred haplotypes	Haplotype total frequency	Cases (HS) frequencies	Controls (HR) frequencies	*P* value (chi-square analysis)	*P* value, following 100 K permutations
AHR Block 1					
ACTTTC	0.477	0.485	0.468	0.7405	1.00
AATTCC	0.173	0.144	0.205	0.1069	0.6743
AACCTG	0.14	0.178	0.100	0.0259 *	0.142
AACTTG	0.097	0.089	0.105	0.589	1.00
GCTTTC	0.094	0.074	0.116	0.1598	0.8367
AATTTC	0.01	0.015	0.005	0.3454	1.00
AHR Block 2					
GACTCAGGG	0.231	0.232	0.231	0.9865	1.00
AGCTCAGGG	0.198	0.193	0.205	0.7632	1.00
GATTTAAGG	0.159	0.184	0.132	0.16	0.8367
GGCTCAGGG	0.128	0.119	0.137	0.5925	1.00
GATCTAAGG	0.081	0.073	0.089	0.5716	1.00
GATTCTGTA	0.069	0.059	0.079	0.4451	1.00
GGCTCTGGG	0.066	0.054	0.079	0.3323	1.00
GATCCAAGG	0.033	0.030	0.037	0.6932	1.00
GATTCAGGA	0.01	0.020	0.000	0.0512	0.3157
CYP1A1					
AACT	0.656	0.649	0.663	0.7604	1.00
ACCT	0.189	0.193	0.184	0.8228	1.00
GCCC	0.094	0.104	0.084	0.5039	0.9829
ACTC	0.054	0.045	0.063	0.4136	0.9422
CYP1A2					
GAT	0.64	0.614	0.668	0.2606	0.8454
GCC	0.286	0.292	0.279	0.7736	1.00
GAC	0.048	0.059	0.037	0.2985	0.888
AAC	0.026	0.035	0.016	0.2365	0.8278
CYP1B1 Block 1					
TTGG	0.377	0.390	0.362	0.5704	1.00
GTAC	0.249	0.276	0.220	0.1999	0.8872
TTAC	0.207	0.179	0.238	0.1535	0.8046
TCGG	0.161	0.144	0.179	0.3404	1.00
CYP1B1 Block 2					
ACTCGATT	0.268	0.277	0.258	0.6658	1.00
CGTCGGTC	0.258	0.248	0.268	0.6364	1.00
CGTTAGTC	0.194	0.188	0.200	0.7667	1.00
CGCTGACC	0.186	0.163	0.210	0.2309	0.9115
CGTCAATC	0.036	0.054	0.016	0.0392 *	0.1666
CGTTAATC	0.015	0.015	0.016	0.9386	1.00
CGCTAGCC	0.013	0.015	0.011	0.7029	1.00

AHR Block 1 SNPs 1, 2, 3, 4, 5, 7, *AHR Block 2* SNPs 8, 9, 10, 11, 12, 13, 14, 15, 16, *CYP1A1* SNPs 1, 2, 3, 4, *CYP1A2* SNPs 1, 2, 3, *CYP1B1 Block 1* SNPs 1, 2, 3, 4, *CYP1B1 Block 2* SNPs 5, 6, 7, 8, 9, 10, 11, 12

*Statistically significant ($P < 0.05$)

Table 7 Detailed studies of chromosomal frequencies of *CYP1B1* haplotype ACTTGATC in haploblock 2

Caucasian-Americans and African-Americans in present cohort, combined (Latino-American excluded)			
	ACTTGATC*	Other haplotypes	Total chromosomes
HS	11	213	224
HR	0	196	196
Total	11	409	420

* Statistically significant ($P = 0.0011$) by Fisher's exact test, comparing HS with HR

CYP1 polymorphisms and HNSCC risk, six of the seemingly most relevant studies are detailed below.

(**a**) The Leu-432 allele of *CYP1B1* has been associated with a greater frequency [odds-ratio (O.R.) >4] of *TP53* gene mutations and HNSCC in cigarette smokers [16]; however, this "greater risk" is difficult to reconcile experimentally—given the fact that the Leu-432 mutant was shown in an *E. coli* expression assay in vitro to metabolize PAHs only 1.2- to 1.5-fold better than the Val-432 variant [42]. (**b**) Four nonsynonymous coding SNPs (p.R48G, p.A119S, p.L432V, and p.N453S) were studied in 150 cases of HNSCC and 150 controls [43]; when the four SNPs were analyzed as a haplotype, amino acid changes p.A48G and p.A119S exhibited complete LD in all cases and controls, and significant differences ($P < 0.05$) were reported in the two haplotypes—GTCA and GTGA—as being associated with increased risk of HNSCC. (**c**) Another study, comparing 312 HNSCC cases and 300 noncancer controls, looked at the impact of 22 sequence variations in *CYP1A1, CYP1B1, CYP2E1, GSTM1*, and six other genes encoding either xenobiotic-metabolizing or DNA repair enzymes [13]; using logit regression and a Bayesian version of logit regression, authors found significant ($P < 0.05$) associations of *CYP1B1* p.L432V and *CYP2E1* −70G > T (O.R. 10.84; 95% CI, 1.64–71.53), as well as *CYP1B1* p.L432V and *GSTM1 (0/0)* null/null (OR 11.79; 95% CI, 2.18–63.77) with HNSCC risk. (**d**) In another study of 153 HNSCC cases and 145 controls with no current or previous diagnosis of cancer [37], authors examined *CYP1A1, CYP1A2, CYP2E1, GSTM1*, and *GSTT1* SNPs as risk factors in HNSCC; a significant difference ($P < 0.001$) was detected for tobacco and alcohol consumption between cases and controls; moreover, the *CYP1A2**1D* (OR 16.24) variant and *GSTM1(0/0)* null alleles (OR 0.02) conferred increased risk of HNSCC, and alcohol consumption in HNSCC patients was associated with the *CYP2E1**5B* variant allele ($P < 0.0001$, OR 190.6). (**e**) Performing meta-analysis of six published case-control studies of HNSCC—which included some studies that had concluded no significant risk—authors [41] found that the *CYP1B1* p.L432V polymorphism was significantly related with HNSCC risk (OR = 1.13, 95% CI = 1.03–1.25, $P = 0.014$), whereas no significant association between *CYP1B1* p.N453S polymorphism and HNSCC risk was found. (**f**) Finally, a case-control study of 750 HNSCC patients and equal number of healthy controls investigated the association of polymorphisms in *CYP1A1, CYP1B1, CYP2E1*, and *GSTM1* with HNSCC [24]; cases having variant haplotypes of both *CYP1A1**2A* and *CYP1A1**2C* or *CYP1B1**2* and *CYP1B1**3* or *CYP2E1**5B* and *CYP2E1**6* were at significant ($P < 0.05$) risk of developing HNSCC; statistical analysis revealed a more than multiplicative interaction between combinations of these variant *CYP* genotypes plus the *GSTM1(0/0)* null genotype and between variant genotypes and tobacco smoking or chewing or alcohol consumption—indicating these genes contribute a "modest risk factor" for developing HNSCC via gene-gene and gene-environment interactions. To learn more about each *CYP* allele described above, please refer to http://www.cypalleles.ki.se/

What sets our study apart from all previous studies?

None of the studies mentioned above, however, had distinguished between HNSCC in smokers and nonsmokers. Also, in contrast to the present study that used the EDP method of sample selection, all the studies mentioned above used, as their control population—nonsmokers, or any smoker not having cancer. Furthermore, none of these studies employed any form of multi-locus statistical testing; this is a very serious recurring problem in the medical literature with regard to studies attempting to discover associations between one or more SNPs and a multifactorial trait [4, 36], which is a type I (false-positive) error variously called "$P < 0.05$ studies" [36] and the "incidentalome" [18].

Table 8 Detailed studies of chromosomal frequencies of *CYP1B1* haplotype ACTTGATC in haploblock 2

Ethnicity (self-identified)	ACTTGATC*		Other haplotypes		Total chromosomes
	HS	HR	HS	HR	
Caucasian-American (CAU)	3	0	199	190	392
African-American (AFR)	8	0	14	6	28
Latino-American	0	0	0	2	2
Total	11	0	213	198	422

*Statistically significant ($P < 0.0001$) by Fisher's exact test (two-tailed), comparing AFR with CAU (HS patients)

Table 9 Detailed studies of chromosomal frequencies of *CYP1B1* haplotype ACTTGATC in haploblock 2

African-Americans in present cohort			
	ACTTGATC*	Other haplotypes	Total chromosomes
HS	8	14	22
HR	0	6	6
Total	8	20	28

*Not significantly different ($P = 0.141$) by Fisher's exact test, comparing HS with HR

A type I error (false-positive) likely resulting from population stratification

The present study, at least based on the size of our cohort, rejected all four *AHR*, *CYP1A1*, *CYP1A2*, and *CYP1B1* genes as having any statistically significantly detectable contribution to HNSCC in smokers; further, we first found a false-positive observation—that the *CYP1B1* haplotype ACTTGATC might be associated with HNSCC risk in our cohort (Table 7). Instead, this finding was realized to be extremely unlikely, due to haplotype frequency differences of more than 100-fold between African and Caucasian populations in the 1000 Genomes Project (Additional file 2: Table S1); hence, our study is an excellent teaching example of the consequences of a population stratification artifact.

Increasing appreciation of CYP1-mediated detoxication, as well as metabolic activation

One additional consideration is that we now know from transgenic mouse studies that all three CYP1 enzymes can participate in both metabolic activation and detoxication—depending on dose, length of exposure, route of administration, and target organ specificity [5, 35, 38]. Therefore, it seems clear that virtually all previous epidemiological studies are ambiguous because such studies have considered only the "metabolic activation" component [29]. Yet, when cancer initiation occurs via "direct contact" of a carcinogen—such as cigarette smoke exposure causing HNSCC—one would presume that CYP1-mediated metabolic activation [27] would be a much more likely determinant than detoxication [35].

However, in the population studied herein, we failed to demonstrate a correlation between the genotype of *AHR* and all three *CYP1* genes and HNSCC in smokers. This does not mean that—if a much larger cohort were collected—an association between HNSCC and small-effect contribution by one or another of these four genes might not emerge. We found a lack of association because our Caucasian and African cohorts were apparently too small, especially the African population (both HS and HR).

Table 10 Detailed studies of chromosomal frequencies of *CYP1B1* haplotype ACTTGATC in haploblock 2

Caucasian-only sample			
	ACTTGATC*	Other haplotypes	Total chromosomes
HS	3	199	202
HR	0	190	190
Total	3	389	392

*Not significantly different ($P = 0.249$) by Fisher's exact test, comparing HS with HR

Table 11 Detailed studies of chromosomal frequencies of *CYP1B1* haplotype ACTTGATC in haploblock 2

Caucasians (EUR) and Africans (AFR) in The 1000 Genomes Project				
	ACTTGATC*	Other haplotypes	Total chromosomes	Haplotype frequency
EUR	3	1003	1006	0.003
AFR	471	851	1332	0.356
Total	474	1854	2328	

*Statistically significant ($P < 0.0001$) by Fisher's exact test, comparing EUR with AFR

Environmental risk factors for HNSCC

Tobacco (smoking cigarettes, cigars or pipe, second-hand smoke, chewing tobacco, and using snuff) probably represents the single greatest risk factor for developing HNSCC. The vast majority (~85%) of HNSCC is associated with tobacco usage. Heavy and frequent alcohol consumption is the second highest risk factor. Often intertwined with tobacco and heavy alcohol usage, poor oral and dental hygiene, malnutrition, and cannabis usage may enhance HNSCC risk. Men are two- to three-fold more likely than women to develop HNSCC, and patients above age 40 are at higher risk. Exposure to occupational inhalants (wood dust, paint fumes, and certain chemicals) may increase a person's HNSCC risk. Finally, gastroesophageal-reflux disease (GERD), laryngopharyngeal-reflux disease (LPRD), and a weakened immune system can raise the risk of HNSCC. HPV is commonly seen in some types of HNSCC, as detailed below.

Past and present concepts of the role of HPV in HNSCC

Over the past two decades, HPV-positive squamous cell carcinoma has emerged as a distinct subset of HNSCC, based on multiple lines of evidence. Such evidence for an etiologic role of HPV includes (**a**) observations of higher rate of anti-HPV antibody sero-positivity—even after adjustment for smoking—among patients with HNSCC, compared with that among cancer-free individuals [25]; (**b**) preferential occurrence of HPV-positive tumors in the oropharynx [12]; (**c**) unique demographic features, including an increased number of lifetime sexual partners and less exposure to tobacco and alcohol [11]; (**d**) lower rates of *TP53* gene mutations, in HPV-positive than HPV-negative oropharyngeal tumors [12]; and (**e**) improved survival among HPV-positive oropharynx cancers, compared to HPV-negative oropharynx carcinomas in both retrospective [2] and prospective [9] analyses of clinical trial data.

Given the clinical impact of HPV infection in oropharyngeal SCCs, clinical testing for the presence of virus in newly diagnosed tumors at this site is considered mandatory and guides therapy—including enrollment in clinical trials investigating de-intensified radiotherapy regimens [23]. Such testing may include immunohistochemical staining for p16, the overexpression of which occurs when HPV proteins E6 and E7 inactivate TP53 and RB1 proteins, thereby serving as a surrogate marker of viral infection. Alternatively, HPV-specific methods such as in situ hybridization staining of viral DNA or polymerase chain reaction (PCR)-based detection of viral DNA or RNA may be used. Many argue that p16 immunohistochemistry is sufficient for oropharyngeal carcinomas [22] and that PCR-based tests may be overly sensitive and detect latent HPV unrelated to carcinogenesis in a given tumor [21].

Despite the well-established role of HPV in oropharyngeal SCC and its attendant clinical implications, classic carcinogenic effects of tobacco and alcohol appear to abrogate the improved survival of these tumors. Specifically, HPV infection may occur after significant exposure to tobacco and/or alcohol and therefore may not be associated with a statistically different outcome than HPV-negative tumors [45]. On the basis of these mitigating effects of tobacco exposure in HPV-positive tumors, risk stratification schemes that incorporate both smoking history and tumor HPV status have been developed [2]. It should also be noted that the etiologic role and clinical impact of HPV in nonoropharyngeal SCC has not been clearly established; for example, the prevalence of virus in a study of oral cavity SCCs was low, and clinical features of patients with, vs. without, tumor-containing HPV were similar [19].

Whereas our cohort did include 15-20% oropharyngeal SCCs, we believe our results are still applicable to this subset, although we did not analyze the tumors for HPV infection. Specifically, it may be argued that—because the carcinogenic effects of tobacco appear to supersede the effects associated with HPV-driven oropharyngeal SCC, and not all oropharyngeal tumors are etiologically due to HPV—the lack of genotype-phenotype correlation that we have clearly demonstrated in this study is highly likely to hold true in the oropharynx, as well as in other head-and-neck subsites.

Conclusions

With regard to our EDP method of rigorously selecting cohorts of $N = 112$ highly-sensitive (HS) light smokers with HNSCC and $N = 99$ highly resistant (HR) heavy smokers with no cancer, we conclude that the carefully chosen single-nucleotide variants located in and near four genes—*AHR*, *CYP1A1*, *CYP1A2*, *CYP1B1*, alone, or in combination—are not statistically significantly associated with risk of cigarette-smoking-induced HNSCC. One haploblock, ACTTGATC in the 5′ portion of *CYP1B1*, retained statistical significance after 100,000 permutations, but this was discovered to be a type I error (false positive) finding, due to spurious association by population stratification.

Abbreviations

AHR: Aryl hydrocarbon receptor protein; *AHR*: Gene encoding aryl hydrocarbon receptor; Cig-Pk-Yr: Cigarette-pack-year (20 cigarettes smoked per day for 1 year); *CYP1A1*: Gene encoding cytochrome P450 family 1 subfamily A member 1; CYP1A1: Name of enzyme; *CYP1A2*: Gene encoding cytochrome P450 family 1 subfamily A member 2; CYP1A2: Name of enzyme; *CYP1B1*: Gene encoding cytochrome P450 family 1 subfamily B member 1; CYP1B1: Name of enzyme; EDP: Extreme discordant phenotype method of population analysis; gDNA: Genomic DNA; HNSCC: Head-and-neck squamous cell carcinoma; HPV: Human papilloma virus; LD: Linkage disequilibrium; SNPs: Single nucleotide polymorphisms, variant alleles

Acknowledgements

We thank Drs. Michael Wolfe, Lyon Gleich, Reena Dhanda-Patil, and Li Jin for their help early in this study. We are grateful to nurses Cathy Bailey, Nancy Emrath, JaMyllah Payne, and especially Victoria Straughn for interfacing tirelessly between the Department of Otolaryngology-Head and Neck Surgery Outpatient Clinics (Barrett Cancer Center) and Cincinnati VA Hospital and our laboratories. We thank colleagues for valuable discussions and careful reading of this manuscript.

Funding

Grant support included National Institutes of Health Grants R01 DE016325 (to D.W.N.), P30 ES006096 (to D.W.N. and S.M.P.), and a 2003–04 Pilot Program Project sponsored by ES009066 (to D.W.N and J.L.G.).

Authors' contributions

LFJ-N carried out the DNA isolation and molecular genetic studies. LFJ-N, GZ, and ZJ performed statistical analyses. JLG and DWN originally designed the study. LFJ-N, GZ, ZJ, RB, and DWN drafted the manuscript. JLG, KMW, and DWN determined the clinical status for participation in the study. SMP interfaced with nurses drawing bloods, coding of data, and assessing questionnaires. All authors read and approved the final manuscript.

Competing interests

All authors declare that they have no competing interests.

Consent for publication

Not applicable.

Author details

[1]Department of Environmental Health and Center for Environmental Genetics, University of Cincinnati Medical Center, Cincinnati, OH 45267-0056, USA. [2]Division of Human Genetics, Department of Pediatrics & Molecular

Developmental Biology, Cincinnati Children's Hospital, Cincinnati, Ohio 45229-2899, USA. [3]Department of Otolaryngology-Head and Neck Surgery, University of Cincinnati College of Medicine, Cincinnati, OH 45267-0528, USA. [4]Department of Pathology and Laboratory Medicine, University of Cincinnati College of Medicine, Cincinnati, OH 45267-0533, USA. [5]Present address: Genesky Diagnostics, Suzhou, China.

References

1. Abel J, Haarmann-Stemmann T. Introduction to the molecular basics of aryl hydrocarbon receptor biology. Biol Chem. 2010;391(11):1235–48.
2. Ang KK, Harris J, Wheeler R, Weber R, Rosenthal DI, Nguyen-Tan PF, Westra WH, Chung CH, Jordan RC, Lu C, Kim H, Axelrod R, Silverman CC, Redmond KP, Gillison ML. Human papillomavirus and survival of patients with oropharyngeal cancer. N Engl J Med. 2010;363(1):24–35.
3. Barrett JC, Fri B, Maller J, Daly MJ. HaploView: analysis and visualization of LD and haplotype maps. Bioinformatics. 2005; 21(2):263–5.
4. Bernstein JA, Zhang G, Jin L, Abbott C, Nebert DW. Olfactory receptor gene polymorphisms and nonallergic vasomotor rhinitis. J Asthma. 2008;45(4):287–92.
5. Conney AH, Chang RL, Cui XX, Schiltz M, Yagi H, Jerina DM, Wei SJ. Dose-dependent differences in the profile of mutations induced by carcinogenic (R, S, S, R)bay- and fjord-region diol-epoxides of polycyclic aromatic hydrocarbons. Adv Exp Med Biol. 2001;500:697–707.
6. Corchero J, Pimprale S, Kimura S, Gonzalez FJ. Organization of the *CYP1A* cluster on human chromosome 15: implications for gene regulation. Pharmacogenetics. 2001;11(2):1–6.
7. Divanovic S, Dalli J, Jorge-Nebert LF, Flick LM, Galvez-Peralta M, Boespflug ND, Stankiewicz TE, Fitzgerald JM, Somarathna M, Karp CL, Serhan CN, Nebert DW. Contributions of the three CYP1 monooxygenases to pro-inflammatory and inflammation-resolution lipid mediator pathways. J Immunol. 2013;191(6):3347–57.
8. Dragin N, Shi Z, Madan R, Karp CL, Sartor MA, Chen C, Gonzalez FJ, Nebert DW. Phenotype of the *Cyp1a1/1a2/1b1(–/–)* triple-knockout mouse. Mol Pharmacol. 2008;73(6):1844–56.
9. Fakhry C, Westra WH, Li S, Cmelak A, Ridge JA, Pinto H, Forastiere A, Gillison ML. Improved survival of patients with human papillomavirus-positive head-and-neck squamous cell carcinoma in a prospective clinical trial. J Natl Cancer Inst. 2008;100(4):261–9.
10. Fernandez-Salguero P, Pineau T, Hilbert DM, McPhail T, Lee SS, Kimura S, Nebert DW, Rudikoff S, Ward JM, Gonzalez FJ. Immune system impairment and hepatic fibrosis in mice lacking the dioxin-binding AH receptor. Science. 1995;268(5211):722–6.
11. Gillison ML, D'Souza G, Westra W, Sugar E, Xiao W, Begum S, Viscidi R. Distinct risk factor profiles for human papillomavirus type 16-positive and human papillomavirus type 16-negative head-and-neck cancers. J Natl Cancer Inst. 2008;100(6):407–20.
12. Gillison ML, Koch WM, Capone RB, Spafford M, Westra WH, Wu L, Zahurak ML, Daniel RW, Viglione M, Symer DE, Shah KV, Sidransky D. Evidence for a causal association between human papillomavirus and a subset of head-and-neck cancers. J Natl Cancer Inst. 2000;92(9):709–20.
13. Harth V, Schafer M, Abel J, Maintz L, Neuhaus T, Besuden M, Primke R, Wilkesmann A, Thier R, Vetter H, Ko YD, Bruning T, Bolt HM, Ickstadt K. Head-and-neck squamous cell cancer and its association with polymorphic enzymes of xenobiotic metabolism and DNA repair. J Toxicol Environ Health A. 2008;71(13–14):887–97.
14. Jiang Z, Dalton TP, Jin L, Wang B, Tsuneoka Y, Shertzer HG, Deka R, Nebert DW. Toward the evaluation of function in genetic variability: characterizing human SNP frequencies and establishing BAC-transgenic mice carrying the human *CYP1A1_CYP1A2* locus. Hum Mutat. 2005;25(2):196–206.
15. Jorge-Nebert LF, Jiang Z, Chakraborty R, Watson J, Jin L, McGarvey ST, Deka R, Nebert DW. Analysis of human *CYP1A1* and *CYP1A2* genes and their shared bidirectional promoter in eight world populations. Hum Mutat. 2010; 31(1):27–40.
16. Ko Y, Abel J, Harth V, Brode P, Antony C, Donat S, Fischer HP, Ortiz-Pallardo ME, Thier R, Sachinidis A, Vetter H, Bolt HM, Herberhold C, Bruning T. Association of *CYP1B1* codon 432 mutant allele in head-and-neck squamous cell cancer is reflected by somatic mutations of *TP53* gene in tumor tissue. Cancer Res. 2001;61(11):4398–404.
17. Koch WM, Lango M, Sewell D, Zahurak M, Sidransky D. Head-and-neck cancer in nonsmokers: a distinct clinical and molecular entity. Laryngoscope. 1999;109(10):1544–51.
18. Kohane IS. (Mis)treating the pharmacogenetic incidentalome. Nat Rev Drug Discov. 2012;11(2):89–90.
19. Krüger M, Pabst AM, Walter C, Sagheb K, Günther C, Blatt S, Weise K, Al-Nawas B, Ziebart T. The prevalence of human papilloma virus (HPV) infections in oral squamous cell carcinomas: a retrospective analysis of 88 patients and literature overview. J Craniomaxillofac Surg. 2014;42(7):1506–14.
20. Lahvis GP, Bradfield CA. *Ahr* null alleles: distinctive or different? Biochem Pharmacol. 1998;56(7):781–7.
21. Leemans CR, Braakhuis BJM, Brakenhoff RH. The molecular biology of head-and-neck cancer. Nat Rev Cancer. 2011;11:9–22.
22. Lewis Jr JS. p16 immunohistochemistry as a stand-alone test for risk stratification in oropharyngeal squamous cell carcinoma. Head Neck Pathol. 2012;6 Suppl 1:S75–82.
23. Marur S, Forastiere AA. Head-and-neck squamous cell carcinoma: update on epidemiology, diagnosis, and treatment. Mayo Clin Proc. 2016;91(3):386–96.
24. Maurya SS, Anand G, Dhawan A, Khan AJ, Jain SK, Pant MC, Parmar D. Polymorphisms in drug-metabolizing enzymes and risk of head-and-neck cancer: evidence for gene-gene and gene-environment interaction. Environ Mol Mutagen. 2014;55(2):134–44.
25. Mork J, Lie AK, Glattre E, Hallmans G, Jellum E, Koskela P, Moller B, Pukkala E, Schiller JT, Youngman L, Lehtinen M, Dillner J. Human papillomavirus infection as a risk factor for squamous cell carcinoma of the head and neck. N Engl J Med. 2001;344(15):1125–31.
26. Murray IA, Patterson AD, Perdew GH. Aryl hydrocarbon receptor ligands in cancer: friend and foe. Nat Rev Cancer. 2014;14(12):801–14.
27. Nebert DW. The *Ah* locus: genetic differences in toxicity, cancer, mutation, and birth defects. Crit Rev Toxicol. 1989;20(3):153–74.
28. Nebert DW. Extreme discordant phenotype methodology: an intuitive approach to clinical pharmacogenetics. Eur J Pharmacol. 2000;410(2–3):107–20.
29. Nebert DW, Dalton TP. The role of cytochrome P450 enzymes in endogenous signalling pathways and environmental carcinogenesis. Nat Rev Cancer. 2006;6(12):947–60.
30. Nebert DW, Dalton TP, Okey AB, Gonzalez FJ. Role of aryl hydrocarbon receptor-mediated induction of the CYP1 enzymes in environmental toxicity and cancer. J Biol Chem. 2004;279(23):23847–50.
31. Nebert DW, Galvez-Peralta M, Shi Z, Dragin N. Inbreeding and epigenetics: beneficial as well as deleterious effects. Nat Rev Genet. 2010;11(9):662.
32. Nebert DW, Karp CL. Endogenous functions of aryl hydrocarbon receptor (AHR): intersection of cytochrome P450 1 (CYP1)-metabolized eicosanoids and AHR biology. J Biol Chem. 2008;283(52):36061–5.
33. Nebert DW, Roe AL, Dieter MZ, Solis WA, Yang Y, Dalton TP. Role of the aromatic hydrocarbon receptor and [*Ah*] gene battery in the oxidative stress response, cell cycle control, and apoptosis. Biochem Pharmacol. 2000;59(1):65–85.
34. Nebert DW, Shi Z, Galvez-Peralta M, Uno S, Dragin N. Oral benzo[*a*]pyrene: understanding pharmacokinetics, detoxication, and consequences—*Cyp1* knockout mouse lines as a paradigm. Mol Pharmacol. 2013;84(3):304–13.
35. Nebert DW, Wikvall K, Miller WL. Human cytochromes P450 in health and disease. Philos Trans R Soc Lond B Biol Sci. 2013;368(1612): 20120431.
36. Nebert DW, Zhang G, Vesell ES. From human genetics and genomics to pharmacogenetics and pharmacogenomics: past lessons, future directions. Drug Metab Rev. 2008;40(2):187–224.
37. Olivieri EH, da Silva SD, Mendonca FF, Urata YN, Vidal DO, Faria MA, Nishimoto IN, Rainho CA, Kowalski LP, Rogatto SR. *CYP1A2*1C*, *CYP2E1*5B*, and *GSTM1* polymorphisms are predictors of risk and poor outcome in head-and-neck squamous cell carcinoma patients. Oral Oncol. 2009;45(9):e73–9.
38. Pelkonen O, Nebert DW. Metabolism of polycyclic aromatic hydrocarbons: etiologic role in carcinogenesis. Pharmacol Rev. 1982;34(2):189–222.
39. Pinney SM, Freyberg RW, Levine GE, Brannen DE, Mark LS, Nasuta JM, Tebbe CD, Buckholz JM, Wones R. Health effects in community residents near a uranium plant at Fernald, Ohio, USA. Int J Occup Med Environ Health. 2003; 16(2):139–53.
40. Sartor MA, Schnekenburger M, Marlowe JL, Reichard JF, Wang Y, Fan Y, Ma C, Karyala S, Halbleib D, Liu X, Medvedovic M, Puga A. Genome-wide analysis of aryl hydrocarbon receptor-binding targets reveals an extensive array of gene clusters that control morphogenetic and developmental programs. Environ Health Perspect. 2009;117(7):1139–46.

41. Shen M, Hu YY, Hu YK, Xie LC, Xu XM, Wu MY, Niu YM. Quantitative assessment of the influence of *CYP1B1* polymorphisms and head-and-neck squamous cell carcinoma risk. Tumour Biol. 2014;35(4):3891–7.
42. Shimada T, Watanabe J, Kawajiri K, Sutter TR, Guengerich FP, Gillam EM, Inoue K. Catalytic properties of polymorphic human cytochrome P450 1B1 variants. Carcinogenesis. 1999;20(8):1607–13.
43. Singh AP, Shah PP, Mathur N, Buters JT, Pant MC, Parmar D. Genetic polymorphisms in cytochrome P450 1B1 and susceptibility to head-and-neck cancer. Mutat Res. 2008;639(1–2):11–9.
44. Vasiliou V, Gonzalez FJ. Role of CYP1B1 in glaucoma. Annu Rev Pharmacol Toxicol. 2008;48:333–58.
45. Weinberger PM, Yu Z, Haffty BG, Kowalski D, Harigopal M, Brandsma J, Sasaki C, Joe J, Camp RL, Rimm DL, Psyrri A. Molecular classification identifies a subset of human papillomavirus-associated oropharyngeal cancers with favorable prognosis. J Clin Oncol. 2006;24(5):736–47.
46. Zhang G, Nebert DW, Chakraborty R, Jin L. Statistical power of association using the extreme discordant phenotype design. Pharmacogenet Genomics. 2006;16(6):401–13.

7

Genetic determinants of clinical heterogeneity of the coronary artery disease in the population of Hyderabad, India

Rayabarapu Pranavchand[1], Arramraju Sreenivas Kumar[2] and Battini Mohan Reddy[1*]

Abstract

Background: Genetic predisposition to the clinical categories of coronary artery disease (anatomical viz., insignificant, single, double, and triple vessel diseases and phenotypic severity categories viz., angina, acute coronary syndrome, and myocardial infarction) is poorly understood. Particularly, the apolipoprotein genes clustered at 11q23.3 chromosomal region play a vital role in cholesterol homeostasis, and a large number of SNPs identified in this region need to be explored for their association with the clinical categories of CAD.

Methods: Using fluidigm SNP genotyping platform, a prioritized set of 96 SNPs of 11q23.3 chromosomal region were genotyped on 508 CAD cases and 516 ethnicity matched controls, enrolled from Hyderabad, India, and its vicinity.

Results: The association analysis suggests 19 and 15 SNPs to be significantly associated ($p \leq 0.05$) with at least one of the anatomical and/or phenotypic severity categories, respectively. Overall, the six SNPs rs17440396:G>A, rs6589566:A>G, rs2849165:G>A, rs10488699:G>A, rs1263163:G>A, and rs1263171:G>A were significant even after correction for multiple testing. Three of these (rs17440396:G>A, rs6589566:A>G, and rs2849165:G>A) that belong to *BUD13*, *ZPR1*, and *APOA5-APOA4* intergenic regions, respectively, were found to be associated across the anatomical categories of CAD. However, no particular trend in the genotypic odds ratios with the increasing severity was apparent. The association analysis of the variants with phenotypic severity categories suggests that a high degree of phenotypic severity could be a result of more number of risk alleles. While the risk score analysis suggests high discriminative power of the variants towards the individual clinical categories of CAD, the complex network of interactions seen between the intronic variants of *BUD13* and *ZPR1* regulatory genes and intergenic variants of *APOA5-APOA4* suggests pleiotropic effects of regulatory genes in the manifestation of these CAD categories.

Conclusion: The complex network of interactions observed in the present study between the regulatory and protein-coding genes suggests their role in the manifestation of distinct clinical categories of CAD, which needs to be functionally validated.

Keywords: Coronary artery disease, Clinical heterogeneity, Genetic association, Pleiotropy, Epistasis

* Correspondence: bmrisi@gmail.com
[1]Molecular Anthropology Group, Biological Anthropology Unit, Indian Statistical Institute, Hyderabad, India
Full list of author information is available at the end of the article

Background

Coronary artery disease (CAD) results from a progressive damage of blood vessels that supply blood to the heart muscle. It is caused by a process of hardening of arteries called atherosclerosis, which involves both genetic and environmental factors and interaction between them. A large number of genes that belong to lipoprotein metabolism are found associated with monogenic and polygenic forms of CAD [1]. Particularly, *APOA1*, *APOC3*, *APOA4*, and *APOA5* genes clustered in 11q23.3 human chromosomal region are predominantly expressed in the liver and intestine and crucial in regulating lipoprotein metabolism and cholesterol homeostasis [2]. However, several variants that belong to this apolipoprotein gene cluster region were found to be associated with increased risk towards overall CAD phenotype and/or elevated lipid traits among the Indians [3–6]. Our recent study based on the 96 SNPs from this region suggested distinct patterns of association of these variants with CAD and dyslipidemia. This pleiotropic nature of association was explained by the observed SNP-SNP interactions between regulatory and apolipoprotein coding genes [6].

CAD is identified as a broad phenotype that manifests as less severe stable angina (SA) or unstable angina (UA) and more severe forms like acute coronary syndrome (ACS) and myocardial infarction (MI). Further, the advanced coronary angiography modalities classify CAD cases into distinct anatomical categories depending upon the number of vessels and length and diameter of the atherosclerotic lesions in the coronary blood vessels. However, defining CAD cases is difficult due to the process of multi-decadal atherosclerotic phenomenon that can manifest any of the CAD conditions. With the characteristic clinical heterogeneity and the difficulty in defining CAD cases, it has become a challenging trait to study [7]. There was substantial phenotypic heterogeneity of the samples hitherto considered for genetic association studies and that might be one of the reasons for the lack of consistency in the findings of those studies [8]. The couple of studies on the genetic predisposition to anatomical categories were restricted to 9p21.3 locus among the Caucasians [9, 10], except for an increasing genotypic odds ratio observed with increasing number of diseased blood vessels in case of a single polymorphism of the *APOA1* gene (–75G>A (*MspI*)) in two independent studies, among the north Indian [11] and Australian populations [12]. Despite the vital role in the process of atherosclerosis envisaged for the apolipoprotein genes clustered at 11q23.3 chromosomal region, no further attempts were made to understand their associations with anatomical or phenotypic severity categories of CAD. Given the characteristic dyslipidemic feature of Indian populations in general and particularly of the southern Indians, it is pertinent to explore the pattern of association of the variants of this genomic region with each of the anatomical and phenotypic severity categories of CAD, which might throw light on the possible genetic mechanisms responsible for progression of the disease. We present here the results of our association analysis of a prioritized set of 96 SNPs at 11q23.3 region with sub phenotypes of CAD classified according to anatomical and phenotypic severity.

Methods

The study design and population

The population of Hyderabad is a conglomeration of people from different parts of the undivided state of Andhra Pradesh, and the mother tongue of most of its populations is Telugu, one of the four Dravidian languages. It would be also pertinent to note that despite the subdivision of Telugu population into a number of traditionally endogamous castes and subcastes, Reddy et al. [13] observed genetic differentiation among the populations of Andhra Pradesh to be very low and insignificant; the Markov chain Monte Carlo analysis of population structure, which implements model-based clustering method for grouping individuals into populations, did not reveal any unique population clusters, suggesting high degree of genetic homogeneity.

A total of 1024 individuals, including 508 CAD cases and 516 controls, representing the population of Hyderabad, participated in our case-control study. Patients with characteristic symptoms of stable/unstable angina pectoris along with varying degrees (generally >40%) of stenosis in at least one of the major coronary arteries as determined through angiogram were included in the study. Cases with monogenic diseases, valvular heart disease, cardiomyopathy, renal disease, acute and chronic viral or bacterial infections, asthma, tumors or connective tissue diseases, and other vascular diseases were excluded from the study. All the cases were recruited at the CARE Hospitals, Hyderabad, and evaluated by interventional cardiologists for the above mentioned criteria. Control samples were recruited by conducting free health camps in and around Hyderabad, mostly representing subjects aged above 45 years and with similar ethnic backgrounds as that of the cases. The individuals with characteristic features of any of the above mentioned disease conditions were not included as part of the controls. However, certain proportions of both the case and control subjects were found with T2DM, dyslipidemia, and hypertension.

Data and sample collection

Data pertaining to present age, sex, and age at diagnosis for cases and other background information such as history and current status of smoking, alcoholism, and food habits were obtained through a detailed questionnaire. Information regarding the current status of the subjects on diabetes, dyslipidemia, and hypertension were drawn from hospital records for the cases and through personal interviews for the controls. About 5–6 ml of fasting blood sample was collected peripherally by certified medical lab technicians. Clinical investigations were done for lipid profile and blood sugar for all the samples at Tapadia Diagnostic Centre, Hyderabad, using Auto Analyzer. Blood pressure along with height, weight, waist circumference, and hip circumference were also measured in the field for all the controls, and for the cases, these data were obtained from hospital records.

DNA isolation, SNP selection, and genotyping

DNAs were isolated from all the samples using phenol chloroform method [14] and quantified with the help of Thermo Scientific Varioskan™ Flash Multimode Reader using Quant-iT™ PicoGreen® dsDNA Assay Kit. In order to comprehensively genotype the variants at 11q23.3 chromosomal region, we gathered information on SNPs pertaining to this region from earlier candidate gene and sequencing studies and from databases particularly EBI-NHGRI GWAS database, HAPMAP and dbSNP. Given the key role of *BUD13* in splicing mechanism and *ZPR1* as essential protein for normal cell proliferation and signal transduction, in addition to the SNPs of *APOAI-CIII-AIV-AV* genes clustered at 11q23.3 chromosomal region, we also included SNPs related to these regulatory protein-coding genes. A total of 130 SNPs, studied through candidate gene and GWAS approaches, were subjected to Fluidigm D3 Assay design software [15], and a panel of 96 SNPs with high efficiency for genotyping was chosen. Genotyping was performed using fluidigm nanofluidic SNP genotyping system. Eleven 96.96 IFC chips were utilized for genotyping wherein the selected 96 SNPs were analyzed against 96 samples in each chip. These chips were thermal cycled, and the endpoint fluorescent values were measured on Biomark™ system. Final sample wise genotype calls were obtained using Fluidigm SNP Genotyping Analysis software. A subset of 240 samples was genotyped prior to genotyping of the total 1024 samples. The observed call concordance was 100%.

Prior to genetic association analysis, data quality control was achieved by limiting the sample wise call rate to ≥90%. This resulted in a genotype call rate of 99% in 386 cases and 462 controls, which were considered for further analysis. Further, after excluding SNPs that showed either minor allele frequency <1% and/or deviated from hardy Weinberg equilibrium ($p < 0.001$), only 75 of the 96 SNPs were qualified for final analysis.

Clinical categories of the CAD cases

The CAD cases were categorized into the following four anatomical sub types : (i) cases with 40–70% stenosis and symptomatic for CAD with characteristic atherosclerotic lesions as "insignificant disease," (ii) with >70% stenosis in any one of the major coronary blood vessel as "single vessel disease (SVD)," (iii) with >70% stenosis in two major coronary blood vessels as "double vessel disease (DVD)," and (iv) with >70% stenosis in three major coronary blood vessels are categorized as "triple vessel disease (TVD)." After the genotype pruning, 93 cases with insignificant stenosis, 121 with SVD, 75 with DVD, and 70 with TVD remained with approximately 99% genotype call rate. We also categorized the cases based on the phenotypic severity into three broad classes, (i) those with characteristic symptoms of stable or unstable angina (SA/US), (ii) with symptoms of acute coronary syndrome (ACS), and (iii) with reported myocardial infarction (MI). After the genotype pruning, 73 cases with SA/UA, 165 with ACS, and 76 with MI remained with approximately 99% genotype call rate. The cases that could not be assigned to any of the anatomical/phenotypic severity categories were excluded from the association analysis.

Statistical methods

The descriptive statistical analysis of the background data on quantitative variables was done using MINITAB (version 17). Genotyping quality check and association analysis of alleles were done using PLINK [16]. Genotype-phenotype association analysis using logistic regression assuming different genetic models-dominant, co-dominant, recessive, over dominant, and log-additive were performed using "SNPassoc" package of R PROGRAM [17].

Results and discussion

The means and standard deviations of quantitative parameters along with t values for the difference in means of controls and each of the anatomical and phenotypic severity categories are provided in Additional files 1 and 2, respectively. The average age of individuals of the TVD category is significantly higher than that of controls as well as other less severe anatomical categories. Analysis of variance suggests significant heterogeneity in the mean age of subjects among the anatomical categories, which is because of the difference between the average age of TVD and all other anatomical categories. When compared to controls, a significant increase in the means of fasting blood sugar (FBS) and decrease in mean high density lipoprotein cholesterol (HDLC) and the low density lipoprotein cholesterol (LDLC) levels were observed in each of the anatomical and phenotypic severity categories. The elevated levels of triglycerides (TG) and very low density lipoproteins (VLDL) were observed for the SVD group in comparison to the other anatomical categories and for the ACS in comparison to other phenotypic severity categories.

Association of variants at 11q23.3 chromosomal region with anatomical categories of CAD

The test of association of variants at the 11q23.3 chromosomal region yielded 19 SNPs to be significantly associated ($p \leq 0.05$) with at least one of the four anatomical categories of CAD which clearly implicates the potential role of these variants in the manifestation of these categories. The minor allele frequencies of the associated SNPs with respective odds ratios are presented in Table 1. Seven, ten, nine, and 11 of these 19 SNPs are observed to be associated with insignificant, SVD, DVD, and TVD categories, respectively. However, while the risk-reducing effect of the intronic *BUD13* variant rs17440396:G>A and two *APOA5-APOA4* intergenic variants rs633389:C>T and rs2849165:G>A is consistent across the anatomical categories, rs6589566:A>G of *ZPR1* gene is consistently associated with increased risk. Of the remaining three SNPs associated with insignificant category, rs10488699:G>A and rs625524:G>A confer risk and rs3741298:A>G is protective. Of the remaining six of the nine SNPs associated with SVD, rs664059:C>T, rs2075294:G>T, rs633867:C>T, rs5081:A>T, and rs6321 53:G>T are associated with increased risk while only rs5072:C>T is associated with decreased risk. Among the five remaining SNPs associated with DVD, while rs1263163:G>A and rs7396835:C>T are associated with decreased risk rs1263171:G>A, rs2727793:G>A, and rs25420 63:G>A are with increased risk towards the condition. Similarly, of the remaining seven SNPs associated with TVD, while rs10488699:G>A, rs664059:C>T, rs20752 94:G>T, and rs1263171:G>A confer risk rs11600380:T>C, rs1263163:G>A, and rs625524:G>A are protective in nature. We applied a powerful yet less stringent Benjamin-Hochberg (BH) correction for multiple testing and found rs17440396:G>A to be highly significantly associated across the anatomical categories, while the two SNPs, rs6589566:A>G and rs28491 65:G>A, remained significantly associated with insignificant, SVD, and DVD categories. The results of genotypic association analysis of the above SNPs with anatomical categories under the log-additive model (Additional file 3) suggest a similar pattern of association of the SNPs as in the case of allelic association, except for rs3741298:A>G in case of insignificant category and rs7396835:C>T in case of DVD and both rs664059:C>T and rs2075294:G>T in case of TVD. Among the variants that belong to *BUD13*, *ZPR1*, and *APOA5-APOA4* intergenic regions (rs17440396:G>A, rs6589566:A>G, and rs2849165: G>A), respectively, and found to be associated across the anatomical categories of CAD, no particular trend with increasing severity of the disease was evident in the genotypic odds ratios. The observed pattern of association may thus suggest the implicit pathophysiology of the disease where the abnormal lipid metabolism only triggers the process of atherosclerosis and may not play a major role in further progression of the disease [18]. Further, an increasing genotypic odds ratio of the -75G>A (MspI) polymorphism in the promoter region of *APOA1* gene observed with increasing anatomical severity among the north Indians [11] and in an Australian population [12] could not be observed in our population. Concurrently, the significant association of two variants, rs1225006 (Chromosome 3; *CPNE4* gene) and rs6745588 (Chromosome 2; *STK39* gene), with triple vessel disease discovered in a recent GWAS on Koreans were not associated in a replication cohort [19]. Except for this single GWAS study and a few validating studies on 9p21.3 variants and Msp1 *APOA1* conventional polymorphism, genetic determinants of the clinical heterogeneity of CAD are hitherto poorly understood.

Association of variants at 11q23.3 chromosomal region with phenotypic severity categories

The results of allelic association with each of the phenotypic severity categories are presented in Table 2. Overall, 15 SNPs were significantly associated ($p \leq 0.05$) with at least one of the three phenotypic severity categories. However, only four of the 15 SNPs in case of angina, eight in case of ACS and 11 in case of MI phenotypes were individually associated. Among these 15, only two SNPs, rs17440396:G>A and rs633389:C>T, were commonly associated with decreased risk towards all the three phenotypic severity categories. The remaining two SNPs rs6589566:A>G and rs1263167:A>G associated with angina are risk conferring in nature. Of the six remaining SNPs associated with ACS, while rs2187126:A>G and rs2849165:G>A were associated with decreased risk, rs664059:C>T, rs6589566:A>G, rs2075294:G>T, and rs633867:C>T were associated with increased risk towards the condition. In case of MI, while the seven of the nine remaining SNPs (rs10488699:G>A, rs664059:C>T, rs2075294:G>T, rs1263171:G>A, rs284917 6:G>C, rs5081:A>T ,and rs632153:G>T) were found to be risk conferring, rs1263163:G>A and rs2849165:G>A are risk reducing in nature. The observed trend of increasing number of risk conferring SNPs associated with increasing phenotypic severity (angina to MI) imply that the variants at 11q23.3 chromosomal region might regulate the phenotypic severity. However, when non MI-angina cases were used as controls in the analysis, none of these additional risk variants turned out to be significant except for the risk reducing effect of rs1263163:G>A on MI, reflecting its specificity towards MI. After correction for multiple testing, while only rs17440396:G>A exhibited significant association with all the three phenotypic categories, rs658956 6:A>G and rs2849165:G>A show significant association with angina and ACS and ACS and MI categories, respectively.

Table 1 Association of variants at 11q23.3 chromosomal region with anatomical categories of CAD

SNP (major/ minor allele)	Nearby/ associated gene	MAF in Controls $N = 462$	Insignificant ($n = 93$)			Single vessel disease ($n = 121$)			Double vessel disease ($n = 75$)			Triple vessel disease ($n = 70$)		
			MAF	*p* value	OR (95% CI)	MAF	*p* value	OR (95% CI)	MAF	*p* value	OR (95% CI)	MAF	*p* value	OR (95% CI)
rs17440396:G>A	BUD13	0.21	0.02	3.5×10^{-10}	0.06 (0.02–0.20)	0.03	1.3×10^{-10}	0.13 (0.06–0.27)	0.03	2.3×10^{-07}	0.13 (0.05–0.32)	0.04	2.1×10^{-06}	0.15 (0.06–0.37)
rs10488699:G>A		0.20	0.26	0.040	1.47 (1.02–2.13)							0.27	0.042	1.55 (1.01–2.35)
rs664059:C>T		0.30				0.38	0.017	1.43 (1.06–1.92)				0.39	0.042	1.47 (1.01–2.13)
rs6589566:A>G	ZPR1	0.25	0.43	1.5×10^{-06}	2.25 (1.61–3.14)	0.37	0.0005	1.71 (1.26–2.33)	0.42	1.9×10^{-05}	2.18 (1.52–3.13)	0.39	0.002	1.86 (1.25–2.76)
rs3741298:A>G		0.35	0.27	0.038	0.68 (0.48–0.98)									
rs2075294:G>T		0.05				0.10	0.012	1.93 (1.15–3.24)				0.09	0.048	1.89 (1.00–3.59)
rs633389:C>T	APOA5-APOA4	0.16	0.08	0.004	0.44 (0.25–0.78)	0.10	0.033	0.62 (0.39–0.97)	0.07	0.003	0.38 (0.19–0.73)	0.07	0.005	0.38 (0.19–0.76)
rs633867:C>T		0.06				0.10	0.020	1.81 (1.09–2.99)						
rs11600380:T>C		0.31										0.22	0.030	0.62 (0.41–0.96)
rs1263163:G>A		0.21							0.11	0.009	0.50 (0.30–0.85)	0.10	0.004	0.44 (0.25–0.78)
rs625524:G>A		0.05	0.10	0.006	2.21 (1.24–3.92)							0.01	0.028	0.15 (0.02–1.07)
rs1263171:G>A		0.43							0.53	0.029	1.47 (1.04–2.08)	0.55	0.008	1.62 (1.13–2.32)
rs2727793:G>A		0.37							0.45	0.044	1.43 (1.01–2.02)			
rs7396835:C>T		0.44							0.35	0.042	0.69 (0.48–0.99)			
rs2542063:G>A		0.37							0.45	0.047	1.43 (1.00–2.02)			
rs2849165:G>A		0.36	0.20	3.1×10^{-05}	0.45 (0.31–0.66)	0.22	2.9×10^{-05}	0.49 (0.35–0.69)	0.21	0.0002	0.47 (0.31–0.71)	0.23	0.004	0.54 (0.35–0.83)
rs5081:A>T	APOA1	0.03				0.07	0.008	2.25 (1.22–4.16)						
rs5072:C>T		0.37				0.30	0.037	0.72 (0.53–0.98)						
rs632153:G>T		0.03				0.06	0.020	2.11 (1.11–4.02)						

Italic font—significant after Benjamin Hochberg Correction, blank cell—not significant
MAF minor allele frequency, *OR* odds ratio obtained from logistic regression analysis

Table 2 Association of variants at 11q23.3 chromosomal region with stable/unstable angina, ACS, and MI categories

SNP	Control ($n = 462$)	Angina ($n = 73$)			Acute coronary syndrome ($n = 165$)			Myocardial infarction ($n = 76$)		
	MAF	MAF	p value	OR (95% CI)	MAF	p value	OR (95% CI)	MAF	p value	OR (95% CI)
rs17440396:G>A	0.21	0.01	1.36×10^{-08}	0.05 (0.01–0.21)	0.03	5.15×10^{-14}	0.12 (0.06–0.23)	0.03	6.89×10^{-08}	0.10 (0.04–0.28)
rs10488699:G>A	0.20							0.34	0.0001	2.08 (1.41–3.07)
rs2187126:A>G	0.13				0.09	0.025	0.61 (0.39–0.94)			
rs664059:C>T	0.30				0.37	0.029	1.34 (1.03–1.75)	0.41	0.0080	1.61 (1.13–2.30)
rs6589566:A>G	0.25	0.46	5.63×10^{-07}	2.56 (1.75–3.73)	0.43	3.36×10^{-09}	2.23 (1.70–2.92)			
rs2075294:G>T	0.05				0.08	0.049	1.63 (1.00–2.66)	0.10	0.026	1.97 (1.07–3.62)
rs633389:C>T	0.16	0.06	0.0027	0.36 (0.18–0.72)	0.09	0.0054	0.56 (0.37–0.85)	0.07	0.0053	0.41 (0.22–0.78)
rs633867:C>T	0.06				0.09	0.015	1.77 (1.11–2.82)			
rs1263163:G>A	0.20							0.03	1.64×10^{-08}	0.11 (0.04–0.29)
rs1263167:A>G	0.09	0.15	0.037	1.70 (1.03–2.82)						
rs1263171:G>A	0.43							0.52	0.040	1.44 (1.02–2.03)
rs2849165:G>A	0.36				0.17	1.18×10^{-10}	0.35 (0.26–0.49)	0.23	0.0013	0.51 (0.34–0.77)
rs2849176:G>C	0.49							0.59	0.032	1.46 (1.03–2.08)
rs5081:A>T	0.03							0.07	0.0203	2.28 (1.12–4.65)
rs632153:G>T	0.03							0.06	0.035	2.18 (1.04–4.59)

Blank cell—not significant
OR odds ratio obtained from logistic regression analysis

Except for the three of the above 15 SNPs, rs1263167:A>G, rs2075294:G>T, and rs632153:G>T that failed to show significant association with angina, ACS, and MI categories, respectively, the results of genotypic association with the phenotypic severity categories under the log-additive model suggest similar pattern as that of the allelic association (Additional file 4). Nevertheless, those three SNPs which lost significance were the ones with marginal significance ($p = \sim 0.05$) in case of allelic association, and these departures can be attributed to chance fluctuations because of small sample sizes.

Pattern of association of genetic variants with anatomical and phenotypic severity categories of the CAD patients using subsets of controls

The control cohort of the present study constitutes two kinds of subjects: (1) those presenting any one or more of the conditions like dyslipidemia, diabetes, and hypertension and (2) those having none of these conditions. In order to determine the confounding effects of these clinical risk factors, particularly dyslipidemia, we repeated the allelic association analysis by using only nondyslipidemic cohort of controls (cohort 1; $n = 270$) and those devoid of any of the three conditions—dyslipidemia, diabetes and hypertension (cohort 2; $n = 129$) and present here only the salient features of the findings of this exploratory analysis.

Overall, a consistent pattern of association of the SNPs rs17440396:G>A, rs6589566:A>G, rs2849165:G>A, rs10488699:G>A, rs1263163:G>A, and rs1263171:G>A is observed with both anatomical and/or phenotypic severity categories, which needs to be explored further for their functional role. The respective p values for these SNPs after BH correction for multiple testing in the three sets of controls (total controls, cohort 1, and cohort 2) are presented in Table 3. Certain inconsistencies observed with reference to SNPs that showed significant association at $p \leq 0.05$ could be due to sampling errors possibly emanated from reduced sample size for the subsets of control. The salient features of the results are the following:

1. The rs17440396:G>A is significantly associated across the anatomical and phenotypic categories, irrespective of control cohorts used in the analysis.
2. The rs6589566:A>G is significantly associated with insignificant, SVD, and DVD categories in the analysis using total controls and insignificant and DVD anatomical categories in cohort 1 analysis. Further, this variant is associated with angina and ACS phenotypic severity categories irrespective of control cohorts used in the analysis.
3. The rs2849165:G>A is significantly associated with insignificant, SVD, and ACS categories irrespective of the control cohort used. However, it is associated with DVD in the analysis using total controls or cohort 1 and MI using total controls.
4. The rs10488699:G>A and rs1263163:G>A variants are found associated with MI irrespective of control cohorts used in the analysis.

While rs633389:C>T is significantly associated with DVD when total controls were used, rs5081:A>T, rs633867:C>T,

Table 3 *P* values for variants that were significantly associated after Benjamin Hochberg correction for multiple testing

SNP	Control choice	Anatomical category (*p* values)				Phenotypic severity categories (*p* values)		
		Insignificant	SVD	DVD	TVD	ANGINA	ACS	MI
rs17440396:G>A	Total Controls	4.92×10^{-08}	1.49×10^{-08}	2.02×10^{-05}	0.0001	1.43×10^{-06}	1.04×10^{-11}	6.79×10^{-06}
	Cohort 1	2.36×10^{-06}	2.13×10^{-06}	0.0005	0.0027	2.54×10^{-05}	4.78×10^{-09}	0.0001
	Cohort 2	1.77×10^{-06}	2.47×10^{-06}	0.0004	0.0021	2.18×10^{-05}	1.55×10^{-08}	0.0001
rs6589566:A>G	Total Controls	0.0001	0.012	0.0008		2.61×10^{-05}	1.26×10^{-07}	
	Cohort 1	0.017		0.044		0.004	0.0005	
	Cohort 2					0.036	4.70×10^{-06}	
rs2849165:G>A	Total Controls	0.001	0.001	0.0093			7.65×10^{-09}	0.026
	Cohort 1	0.011	0.010	0.044			4.30×10^{-07}	
	Cohort 2	0.015	0.016				0.020	
rs10488699:G>A	Total Controls							0.004
	Cohort 1							0.0007
	Cohort 2							0.018
rs1263163:G>A	Total Controls							7.83×10^{-06}
	Cohort 1							0.0033
	Cohort 2							0.011
rs633389:C>T	Total Controls			0.014				
rs5081:A>T	Cohort 1		0.047					
rs633867:C>T	Cohort 1		0.047					
rs632153:G>T	Cohort 1		0.047					
rs1263171:G>A	Cohort 1				0.046			

Blank cell—not significant, total control samples ($n = 462$), cohort 1—nondyslipidemic control cohort ($n = 270$), cohort 2—control cohort devoid of dyslipidemia, diabetes, and hypertension ($n = 129$)

and rs632153:G>T showed significant association with SVD and rs1263171:G>A with TVD in case of control cohort 1.

Results of SNP-SNP interaction analysis

While the rs17440396:G>A of *BUD13* gene showed risk reducing effects, rs10488699:G>A variant that belongs to the same gene is associated with increased risk particularly for severe forms of anatomical and phenotypic conditions. Similarly, rs3741298:A>G that belong to ZPR1 gene is associated with risk-reducing effect specific to the insignificant category despite the high risk conferring nature towards CAD due to rs6589566:A>G variant. Therefore, we performed pair wise logistic regression analysis to understand the interaction effects of these SNPs on the anatomical and phenotypic categories of CAD. From the pairs of SNPs that showed significant interactions, the following types of interactions can be inferred:

1. Interactions between intronic SNPs of *BUD13* regulatory gene and intergenic SNPs of *APOA5-APOA4* genes.
2. Intronic SNPs of *ZPR1* and *BUD13* genes.
3. Intronic SNPs of *ZPR1* gene and Intergenic SNPs of *APOA5-APOA4*.
4. Interactions among the variants of *APOA5-APOA4* intergenic region.

The pair wise SNP-SNP interactions associated with anatomical categories of CAD are presented in Table 4. The interaction odds ratio of SNPs suggests epistatic effects of rs3741298:A>G over rs6589566:A>G resulting in reducing risk for SVD. While rs2187126:A>G–rs1263171:G>A (OR = 6.07, p value = 3.81×10^{-06}) that is characteristic of type 1 interaction is associated with increased risk towards SVD, rs6589566:A>G–rs1263163:G>A (OR = 5.39, $p = 1.55 \times 10^{-05}$) of type 3 interaction showed increased risk towards insignificant category. Most of the interactions of variants of *BUD13* with variants of *APOA4-APOA5* intergenic region exhibit risk reducing effects to SVD. Further, a type 4 SNP-SNP interaction, rs1263163:G>A–rs2849165:G>A is observed to exhibit risk reducing effects across the anatomical categories while the type 2 interaction, namely, rs6589566:A>G–rs3741298:A>G, showed similar effect only towards SVD.

The interaction analysis with phenotypic severity categories suggests significant SNP-SNP interactions towards angina and ACS categories but not with MI category. Pertaining to angina, only two SNP pairs, namely, rs2187126:A>G–rs1263163:G>A (odds ratio = 0.01; *p* value

Table 4 Significant SNP-SNP interaction odds ratios from pair wise logistic regression with anatomical categories

Type of interaction	SNP pair	Associated category	Odds ratio	*p* value
BUD13–intergenic variants of APOA5-APOA4 genes	rs10488699:G>A–rs1263163:G>A	Insignificant	0.13	4.43×10^{-06}
	rs2187126:A>G–rs1263163:G>A	Insignificant	0.04	3.11×10^{-06}
	rs10488699:G>A–rs1263163:G>A	SVD	0.06	2.26×10^{-07}
	rs2187126:A>G–rs633389:C>T	SVD	0.06	8.28×10^{-07}
	rs2187126:A>G–rs1263163:G>A	SVD	0.02	3.00×10^{-07}
	rs2187126:A>G–rs1263171:G>A	SVD	6.07	3.81×10^{-06}
Intronic variants of ZPR1–BUD13 genes	rs6589566:A>G–rs3741298:A>G	SVD	0.29	1.36×10^{-05}
ZPR1–intergenic variants of APOA5-APOA4 genes	rs6589566:A>G–rs1263163:G>A	Insignificant	5.39	1.55×10^{-05}
Within the intergenic variants of APOA5-APOA4 genes	rs1263163:G>A–rs2849165:G>A	Insignificant	0.03	1.12×10^{-10}
		SVD	0.06	7.70×10^{-09}
		DVD	0.03	1.90×10^{-07}
		TVD	0.06	9.06×10^{-06}

$= 6.4 \times 10^{-05}$) and rs1263163:G>A–rs2849165:G>A (odds ratio = 0.03; $p = 1.5 \times 10^{-09}$) characteristic of type 1 and type 2 interactions, respectively, showed significant risk-reducing effects towards the condition. The interactions associated with ACS (Table 5) are of type 1, and type 4, and confer significant risk-reducing effects towards the condition. Further, eight SNPs of *BUD13* gene individually showed significant interaction with rs6589566:A>G of ZPR1 gene with six of these interactions being risk conferring towards the ACS (Table 5). Besides this, the interaction of rs6589566:A>G with an APOA5-APOA4 intergenic variant rs1263163:G>A is also risk conferring towards ACS.

The observed trait specific associations might be due to the pleiotropic nature of *BUD13* and *ZPR1* regulatory genes and their interaction with variants at *APOA5-APOA4* intergenic region. Being the component of retention and spliceosome (RES) complex, *BUD13* protein acts as splicing factor for a number of genes and is involved in controlling pre-mRNA retention in the nucleus [20, 21]. *ZPR1* is another prominent pleiotropic regulatory gene whose deficiency causes defects in DNA replication, transcription, and cell cycle regulation [22]. However, the putative pleiotropic effects of these genes in atherosclerosis need to be validated through in vitro functional experiments. Although the present study

Table 5 Significant SNP-SNP interaction odds ratios from pair wise logistic regression with acute coronary syndrome[a]

Type of interaction	SNP pair		Odds ratio	*p* value
BUD13–intergenic variants of APOA5-APOA4 genes	rs10488699:G>A	rs633389:C>T	0.17	1.48×10^{-05}
	rs10488699:G>A	rs1263163:G>A	0.10	9.82×10^{-09}
	rs2187126:A>G	rs633389:C>T	0.04	1.95×10^{-07}
	rs2187126:A>G	rs1263163:G>A	0.06	5.17×10^{-08}
Intronic variants of BUD13–ZPR1–genes	rs11216126:A>C	rs6589566:A>G	2.62	2.11×10^{-05}
	rs11216129:C>A		2.48	8.22×10^{-05}
	rs180326:A>G		0.36	5.61×10^{-05}
	rs2075295:T>C		2.76	2.46×10^{-06}
	rs17119975:T>C		2.66	1.71×10^{-05}
	rs1263149:A>G		0.34	1.78×10^{-06}
	rs623908:A>G		2.39	4.17×10^{-05}
	rs2041967:A>G		2.69	4.67×10^{-06}
ZPR1–intergenic variants of APOA5-APOA4 genes	rs6589566:A>G	rs1263163:G>A	4.31	4.59×10^{-06}
Within the intergenic variants of APOA5-APOA4 genes	rs1263163:G>A	rs1263171:G>A	0.35	2.87×10^{-05}
	rs1263163:G>A	rs2849165:G>A	0.04	2.85×10^{-12}

[a]No SNP-SNP interaction was significant in case of angina and MI

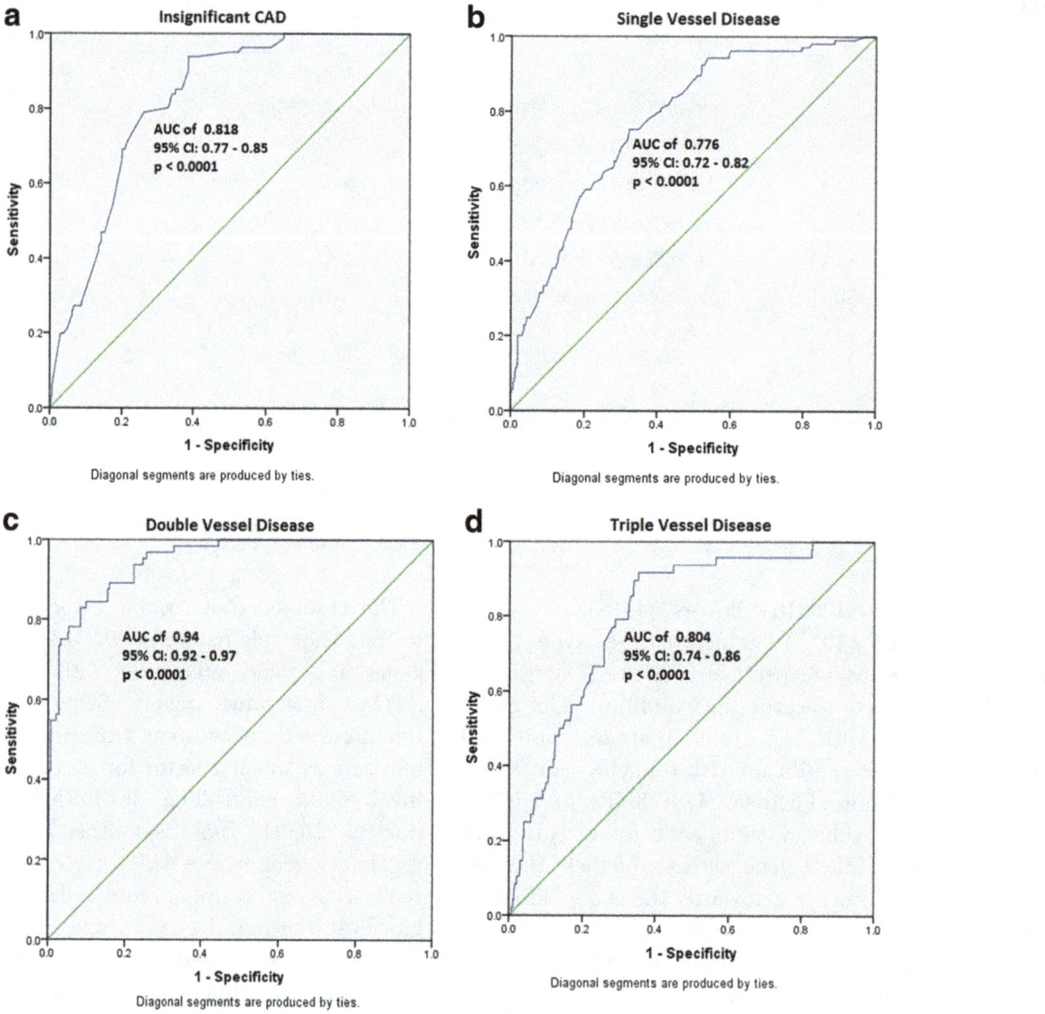

Fig. 1 ROC analysis indicating discriminative power of the variants for anatomical categories of CAD. **a** Insignificant CAD. **b** Single vessel disease. **c** Double vessel disease. **d** Triple vessel disease

included (i) missense and nonsense variants and (ii) variants in the noncoding regions such as promoters, 5′ or 3′ untranslated regions (UTR), upstream and downstream elements, and intergenic regions, we observed the intronic SNPs of BUD13 and ZPR1 genes and intergenic SNPs of APOA5-APOA4 genes to be mostly associated with CAD and/or its categories. In fact, most of the SNPs that belong to this region and associated with abnormal lipid traits through GWAS approach are localized to noncoding domains, which has also been the characteristic pattern of association of the genomic regions with other complex genetic diseases. Further, the candidate gene association studies for risk factors of CAD observed several conventional polymorphisms such as 3238 C>G Sacl (3′UTR SNP of APOC3 gene), −1131 C>T (promoter SNP of APOA5 gene), −75G>A, and +83C>T (promoter SNPs of APOA1 gene) that belong to the noncoding domains of this genomic region [3, 4, 23, 24]. The outcome of ENCODE project also suggests that the noncoding variants may be involved in regulation of gene expression depending on the cell type, developmental stage, and environmental factors [25, 26] and thus have multiple targets. Prioritization of intergenic SNPs through computational tools and subsequent validation through expression GWAS presents a promising strategy for understanding the underlying mechanisms of CAD [27]. For example, Musunuru et al. [28], identified a single noncoding DNA variant rs12740374 at 1p13.3 chromosomal region to influence LDLC and MI risk via liver specific transcriptional regulation of the SORT1 gene by C/EBP transcription factors, providing insights into mechanisms by which the noncoding genetic variants can lead to clinical phenotypes.

Risk score analysis

In order to determine the combined risk effect of the associated SNPs towards each of the anatomical and

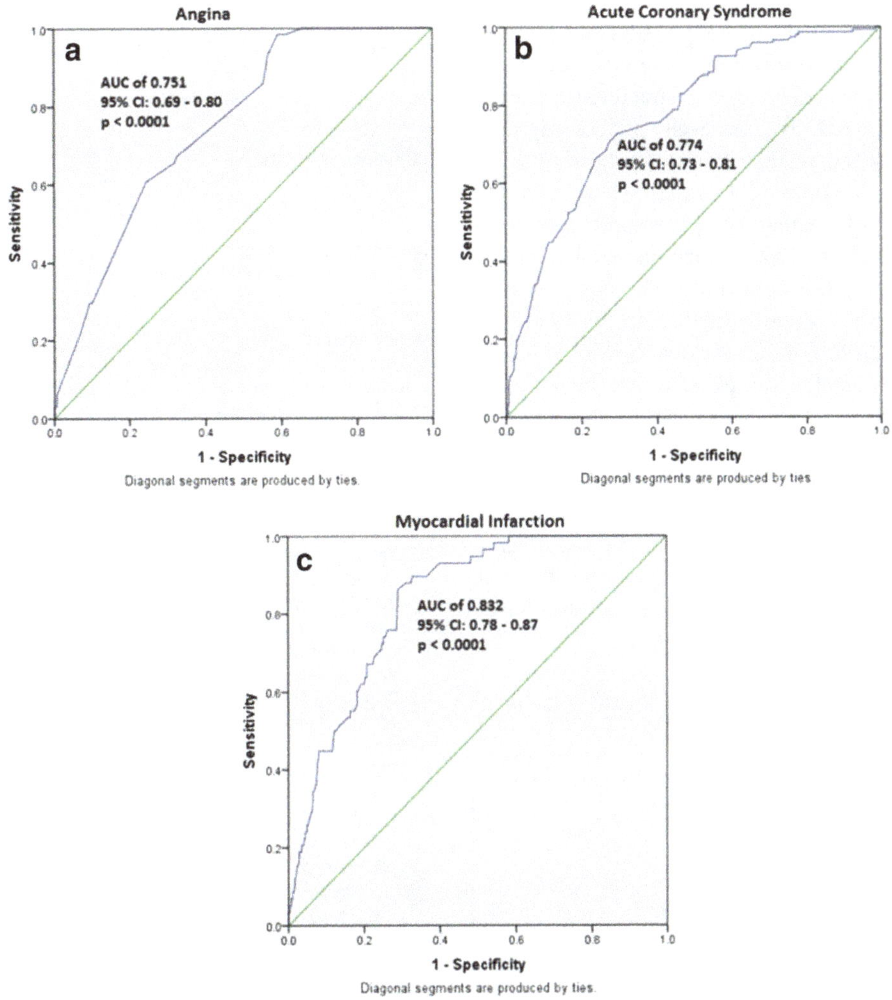

Fig. 2 ROC analysis indicating discriminative power of the variants for phenotypic severity categories of CAD. **a** Angina. **b** Acute coronary syndrome. **c** Myocardial infarction

phenotypic categories, we computed the weighted mean proportion of the risk alleles of the SNPs by taking 2 for two risk alleles, 1 for one risk allele, and 0 for no risk alleles with weights as relative log odds ratios of different SNPs. The cumulative risk allele score for each individual is obtained by multiplying with number of SNPs associated with the clinical category. The risk scores for individuals ranged from 1.5 to 12.5, 0.5 to 15.5, 0.5 to 18.5, and 5.5 to 17.5 in case of insignificant, SVD, DVD, and TVD anatomical conditions of CAD, respectively. Similarly, the ranges are 0 to 8.5, 0.5 to 14.5, and 3.5 to 18.5 in case of angina, ACS, and MI, respectively. We grouped the risk scores with very low frequencies into the adjacent category. The details of risk categories and the risk scores constituting each of those categories are furnished in (Additional file 5: Table S5) and (Additional file 6: Table S6), according to each of the anatomical and phenotypic severity conditions. The results suggest a clear trend of increase in the proportion of cases with increasing risk score. With reference to baseline category with lowest risk scores, we computed odds ratios for each of the remaining risk categories. An increasing trend of OR values with increasing risk score is apparent with all the clinical conditions of CAD (Additional files 7 and 8). Except for the risk category 1 of ACS, the odds ratios suggest significant associations with clinical categories of CAD. To gauge discriminative power of the risk scores, we constructed the ROC (receiver operating curve) plot (Figs. 1 and 2) for the risk scores and status of the clinical categories of CAD. The observed area under curve (AUC) indicates that this study has substantial and significant power to confer these genetic variants as predictors for clinical conditions of CAD.

Conclusions

The intrinsic biological processes and methodological aspects have always been confounding factors for studying complex diseases, which led to inconsistent results in genetic studies based on candidate gene and genome wide approaches [29, 30]. Although several attempts were made to minimize the confounding effects of these factors, the genetic background of characteristic clinical heterogeneity of CAD is not well addressed until recently [7]. From the results of our study, it may be hypothesized that a complex interaction between the intronic domains of regulatory genes and intergenic protein coding genes at 11q23.3 chromosomal region is involved in regulating the lipid traits. The high discriminative power of the variants further implicates their significant role in developing various forms of CAD. Given the relatively smaller sample sizes for the sub phenotypes, the statistical power for our study is somewhat limited and hence this study can be considered only exploratory in nature. Nevertheless, because of the paucity of such studies, we still believe that our results would provide useful insights into the possible genetic mechanisms behind the phenotypic heterogeneity of CAD, which might help in aptly designing future studies based on much larger samples.

Additional files

Additional file 1: Table S1. Baseline characteristics of the controls and anatomical categories of CAD and the *p* values (*t* test) for mean difference between controls and each of the anatomical categories.

Additional file 2:Table S2. Baseline characteristics of the controls and phenotypic severity categories of CAD and the *p* values (*t* test) for mean difference between controls and each of the phenotypic severity categories.

Additional file 3: Table S3. Genotypic odds ratios for the SNPs significantly associated under log-additive model with anatomical categories.

Additional file 4: Table S4. Genotypic odds ratios for the SNPs significantly associated under log-additive model with phenotypic severity.

Additional file 5: Table S5. Groups of risk scores for anatomical categories.

Additional file 6: Table S6. Groups of risk scores for phenotypic severity categories.

Additional file 7: Figure S1. Plot of odds ratios for risk categories of different anatomical categories of CAD.

Additional file 8: Figure S2. Plot of odds ratios for risk categories of different phenotypic severity categories of CAD.

Acknowledgements

We are thankful to the Director, Indian Statistical Institute for the logistics support and the participants in the study for their cooperation. We also thank ALN Murthy and GSR Murthy, Indian Statistical Institute for their suggestions during statistical analysis.

Funding

This study is funded by the Indian Statistical Institute.

Authors' contributions

BMR and PR conceived and designed the study. ASK characterized the patients. BMR supervised the study as principle investigator of this project, and PR carried out the genotyping, statistical analysis, and drafting of the manuscript. All authors read and approved the final manuscript.

Competing interests

The authors declare that they have no competing interests.

Consent for publication

All the authors consented for publication of this manuscript.

Author details

[1]Molecular Anthropology Group, Biological Anthropology Unit, Indian Statistical Institute, Hyderabad, India. [2]Department of Cardiology, Care Hospitals, Banjara Hills, Hyderabad, India.

References

1. Pranavchand R, Reddy BM. Current status of understanding of the genetic etiology of coronary heart disease. J Postgrad Med. 2013;59:30–41.
2. Karathanasis SK. Apolipoprotein multigene family: tandem organization of human apolipoprotein AI, CIII, and AIV genes. Proc Natl Acad Sci U S A. 1985;82(19):6374–8.
3. Shanker J, Perumal G, Rao VS, Khadrinarasimhiah NB, John S, Hebbagodi S, et al. Genetic studies on the APOA1-C3-A5 gene cluster in Asian Indians with premature coronary artery disease. Lipids Health Dis. 2008;7:33.
4. Bhanushali AA, Das BR. Influence of genetic variants in the apolipoprotein A5 and C3 gene on lipids, lipoproteins, and its association with coronary artery disease in Indians. J Community Genet. 2010;1:139–48.
5. Walia GK, Gupta V, Aggarwal A, Asghar M, Dudbridge F, Timpson N, et al. Association of common genetic variants with lipid traits in the Indian population. PLoS ONE. 2014;9:7. e101688.
6. Pranavchand R, Sreenivas KA, Anuj K, Vishnupriya S, Reddy BM. Distinct patterns of association of variants at 11q23.3 chromosomal region with coronary artery disease and dyslipidemia in the population of Andhra Pradesh, India. PLoS ONE. 2016;11:6. e0153720.
7. Luo AK, Jefferson BK, Garcia MJ, Ginsburg GS, Topol EJ. Challenges in the phenotypic characterisation of patients in genetic studies of coronary artery disease. J Med Genet. 2007;44:161–5.
8. Lieb W, Vasan RS. Brief review: genetics of coronary artery disease. Circulation. 2013;128:10. doi:10.1161/CIRCULATIONAHA.113.005350.
9. Dandona S, Stewart AF, Chen L, Williams K, So D, O'Brien E, et al. Gene dosage of the common variant 9p21 predicts severity of coronary artery disease. J Am Coll Cardiol. 2010;56:479–86.
10. Patel RS, Su S, Neeland IJ, Ahuja A, Veledar E, Zhao J, et al. The chromosome 9p21 risk locus is associated with angiographic severity and progression of coronary artery disease. Eur Heart J. 2010;31:3017–23.
11. Chhabra S, Narang R, Lakshmy R, Das N. APOA1-75 G to A substitution associated with severe forms of CAD, lower levels of HDL and apoA-I among northern Indians. Dis Markers. 2005;21:169–74.
12. Wang XL, Liu SX, McCredie RM, Wilcken DE. Polymorphisms at the 5'-end of the apolipoprotein AI gene and severity of coronary artery disease. J Clin Investig. 1996;98:372–7.
13. Reddy BM, Naidu VM, Madhavi VK, Thangaraj LK, Kumar V, Langstieh BT, et al. Microsatellite diversity in Andhra Pradesh, India: genetic stratification versus social stratification. Hum Biol. 2005;77:803–23.
14. Sambrook J, Fritschi EF, Maniatis T. Molecular cloning: a laboratory manual. Newyork: Cold Spring Harbor Laboratory Press; 1989.
15. Wang J, Lin M, Crenshaw A, Hutchinson A, Hicks B, Yeager M, et al. High-throughput single nucleotide polymorphism genotyping using nanofluidic Dynamic Arrays. BMC Genomics. 2009;10:561.
16. Purcell S, Neale B, Todd-Brown K, Thomas L, Ferreira MAR, Bender D, et al. PLINK: a toolset for whole-genome association and population-based linkage analysis. Am J Human Genet. 2007;81:559–75.
17. González JR, Armengol L, Solé X, Guinó E, Mercader JM, Estivill X, et al. SNPassoc: an R package to perform whole genome association studies. Bioinformatics. 2007;23:644–5.

18. Lusis AJ, Fogelman AM, Fonarow GC. Genetic basis of atherosclerosis: part I: new genes and pathways. Circulation. 2004;110:1868–73.
19. Lee JY, Kim G, Park S, Kang SM, Jang Y, Lee SH. Associations between genetic variants and angiographic characteristics in patients with coronary artery disease. J Atheroscler Thromb. 2015;22:363–71.
20. Dziembowski A, Ventura AP, Rutz B, Caspary F, Faux C, Halgand F, et al. Proteomic analysis identifies a new complex required for nuclear pre-mRNA retention and splicing. EMBO J. 2004;23:4847–56. http://doi:10.1038/sj.emboj.7600482.
21. Wysoczanski P, Zweckstetter M. Retention and splicing complex (RES)—the importance of cooperativity. RNA Biol. 2016;13:128–33.
22. Gangwani L. Deficiency of the zinc finger protein ZPR1 causes defects in transcription and cell cycle progression. J Biol Chem. 2006;281:40330–40.
23. Eichenbaum-Voline S, Olivier M, Jones EL, Naoumova RP, Jones B, Gau B, et al. Linkage and association between distinct variants of the APOA1/C3/A4/A5 gene cluster and familial combined hyperlipidemia. Arterioscler Thromb Vasc Biol. 2004;24:167–74.
24. Delgado-Lista J, Perez-Jimenez F, Ruano J, Perez-Martinez P, Fuentes F, Criado-Garcia J, et al. Effects of variations in the *APOA1/C3/A4/A5* gene cluster on different parameters of postprandial lipid metabolism in healthy young men. J Lipid Res. 2010;51:63–73.
25. Dimas AS, Deutsch S, Stranger BE, Montgomery SB, Borel C, Attar-Cohen H, et al. Common regulatory variation impacts gene expression in a cell type dependent manner. Science. 2009;325:1246–50.
26. The ENCODE Project Consortium. An integrated encyclopedia of DNA elements in the human genome. Nature. 2012;489:57–74.
27. Hsu J, Smith JD. Genome wide studies of gene expression relevant to coronary artery disease. Curr Opin Cardiol. 2012;27:210–3.
28. Musunuru K, Strong A, Frank-Kamenetsky M, Lee NE, Ahfeldt T, Sachs KV, et al. From noncoding variant to phenotype via *SORT1* at the 1p13 cholesterol locus. Nature. 2010;466:714–9.
29. McCarthy MI, Abecasis GR, Cardon LR, Goldstein DB, Little J, Ioannidis JP, et al. Genome-wide association studies for complex traits: consensus, uncertainty and challenges. Nat Rev Genet. 2008;9:356–69.
30. Manolio TA, Collins FS, Cox NJ, Goldstein DB, Hindorff LA, Hunter DJ, et al. Finding the missing heritability of complex diseases. Nature. 2009;461:747–53.

8

Identification of novel genes associated with HIV-1 latency by analysis of histone modifications

Kyung-Chang Kim*, Sunyoung Lee, Junseock Son, Younghyun Shin, Cheol-Hee Yoon, Chun Kang and Byeong-Sun Choi*

Abstract

Background: A reservoir of HIV-1 is a major obstacle in eliminating HIV-1 in patients because it can reactivate in stopping antiretroviral therapy (ART). Histone modifications, such as acetylation and methylation, play a critical role in the organization of chromatin domains and the up- or downregulation of gene expression. Although many studies have reported that an epigenetic mechanism is strongly involved in the maintenance of HIV-1 transcriptional latency, neither the epigenetic control of viral replication nor how HIV-1 latency is maintained is not fully understood.

Results: We re-analyzed a high throughput parallel DNA sequencing (ChIP-seq) data from previous work to investigate the effect of histone modifications, H3K4me3 and H3K9ac, on HIV-1 latency in terms of chromosome distribution. The outputs of ChIP-seq from uninfected CD4+ T cell lines and HIV-1 latently infected cells were aligned to hg18 using bowtie and then analyzed using various software packages. Certain chromosomes (16, 17, 19, and 22) were significantly enriched for histone modifications in both decreased and increased islands. In the same chromosomes in HIV-1 latently infected cells, 38 decreased and 41 increased islands from common islands of H3K4me3 and H3K9ac were selected for functional annotation. In Gene Ontology analysis, the 38 genes associated with decreased islands were involved in the regulation of biological process, regulation of cellular process, biological regulation, and purinergic receptor signaling pathway, while the 41 genes associated with increased islands were involved in nucleic acid binding, calcium-activated cation channel activity, DNA binding, and zinc ion binding. In Pathway Commons analysis, the 38 genes were strongly involved in the p63 transcription factor network, while the 41 genes were involved in the RNA polymerase III transcription termination pathway. Several genes such as Nuclear factor I X (*NFIX*) and TNF receptor association factor 4 (*TRAF4*) were selected as candidate genes for HIV latency. Especially, *NFIX* was highly expressed in HIV-1 latently infected cell lines and showed a dramatic reduction in expression after phorbol-13-myristate-12-acetate (PMA) treatment.

Conclusions: These results show that the unique enrichment of histone modifications and its linked genes in specific chromosomes might play a critical role in the establishment and maintenance of HIV-1 latency.

Keywords: HIV latency, Histone modification, Chromosome, H3K4me3, H3K9ac

* Correspondence: joytodeath@korea.kr; byeongsun@korea.kr
Division of AIDS, Center for Immunology and Pathology, Korea National Institute of Health, Chung-buk 28160, Republic of Korea

Background

HIV-1 can establish a state of latency in the early stages of infection. The integration of viral cDNA into host chromosomes is a fundamental step, after which, viral genomes can persist for the lifespan of infected cell [1]. Despite combination antiretroviral therapy (cART), which can repress HIV-1 replication and delay the progression of AIDS, HIV-1 reemerges rapidly after an interruption in treatment [2, 3]. Due to the long half-life of viral reservoirs, it has been reported that the eradication of HIV-1 with current drugs would require over 60 years [4]. The transcription of HIV-1 genome depends on both viral and cellular factors [5]. The activity of HIV promoter is closely connected to the activation of host cells [6]. Additionally, gene expression of HIV-1 latency can be dynamically regulated through epigenetic changes in chromatin structure surrounding and within integrated HIV-1 provirus [7, 8]. In particular, acetylation and methylation have considered as critical elements of chromatin activity and play a critical role in gene regulation [9]. Recently, Park et al. [10] generated the genome-wide maps of histone modifications for histone 3 lysine 4 trimethylated (H3K4me3) and histone 3 lysine 9 acetylated (H3K9ac) using HIV-1 latently infected cells and uninfected cells. They found that histone modification near HIV-1 integrated regions did not show significantly different patterns compared with control and that the enrichment of histone modification was generally no difference before and after HIV-1 integration. In addition, HIV-1 is able to establish interchromosomal interactions to control viral transcription [11]. Although many studies have reported that an epigenetic mechanism is strongly involved in the maintenance of HIV-1 transcriptional latency [12–16], neither the epigenetic control of viral replication nor how HIV-1 latency is maintained is not fully understood.

Here, we re-analyzed chromatin immunoprecipitation-high throughput parallel DNA sequencing (ChIP-seq) data from Park et al. in AIDS [10] to examine the effect of histone modifications on HIV-1 latency in terms of chromosomal distribution and to find novel genes related to HIV-1 latency.

Results

Chromosomal distribution of enriched histone binding sites in HIV-1 latently infected cells

To identify binding sites of histone modifications such as H3K4me3 and H3K9ac, ChIP-seq data was obtained from the Gene Expression Omnibus (GEO) data repositories (GSE58246). Spatial clustering for identification of ChIP-enriched regions (SICER) 1.1 [17] was used to find differentially enriched histone islands. The number of decreased islands at H3K4me3 and H3K9ac histone modifications ranged from 1587 to 13,173 in HIV-1 latently infected cells (ACH2, NCHA1, and J1.1 cells). Increased islands ranged from 2078 to 9875 (Table 1). These results suggest that the binding sites of histone modifications may differ according to cell lines. Next, to determine the binding distribution pattern of H3K4me3 and H3K9ac across individual chromosomes, the cis-regulatory element annotation system (CEAS) software [18] was used with the decreased and increased island data (Fig. 1). Significant chromosomes were determined as those with a p value $<10^4$, those with a ratio of histone modification binding percentage to genome background binding percentage on each chromosome >1.3 and those showing both decreased and increased islands together in all HIV-1 latently infected cells. Chromosomes 16, 17, 19, and 22 were selected by these criteria for comparison of decreased and increased islands on the same chromosome (Additional files 1, 2, 3, and 4). The results suggested that histone modifications binding to host genomes are non-random and preferentially bind to specific chromosomes.

Identification of common decreased and increased islands in chromosomes 16, 17, 19, and 22

We investigated common decreased and increased islands in selected chromosomes of HIV-1 latently infected cells. As a result, we identified 126 and 302 decreased islands in H3K4me3 and H3K9ac, respectively (Fig. 2a). In particular, 38 islands were identified as common decreased islands in both H3K4me3 and H3K9ac (Fig. 2b). Meanwhile, 130 and 164 increased islands were identified in H3K4me3 and H3K9ac, respectively, and 41 islands were identified as common increased islands in both H3K4me3 and H3K9ac. To find out where islands were distributed on each chromosome, the location of 38 decreased and 41 increased islands were visualized on chromosomes 16, 17, 19, and 22 using the idiographica webtool (Fig. 2c). The results suggested that HIV-1 latency may be dominated by common factors.

Table 1 Numbers of islands identified by SICER 1.1 in HIV-1 latently infected cells according to histone modifications

Cell line	Dec or Inc	Histone modifications	
		H3K4me3	H3K9ac
ACH2	Dec	1587	3344
	Inc	2078	5563
J1.1	Dec	8634	13,173
	Inc	6941	9875
NCHA1	Dec	5275	7124
	Inc	6622	8084

Dec decreased islands identified by SICER 1.1, *Inc* increased islands identified by SICER 1.1, *ACH2 and J1.1* HIV-1 latently infected cells, *NCHA1* novel chronic HIV-1 infected cells derived from A3.01

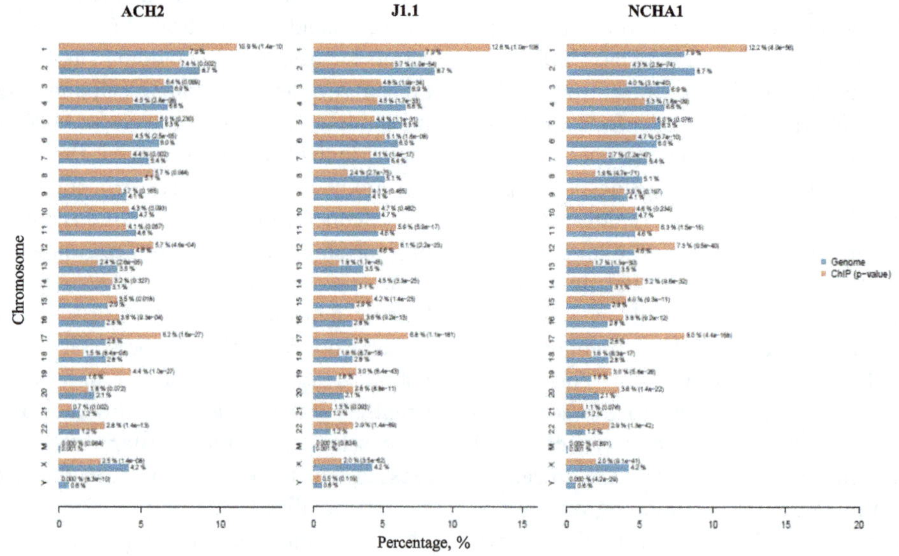

Fig. 1 Chromosomal distribution of H3K4me3 binding sites in HIV-1 latently infected cells. Enrichment pattern of decreased H3K4me3 islands in individual chromosomes is shown as a bar chart. Percentage of total H3K4me3 islands (*red bar*) and expected results by random chance (*blue bars*) for each chromosome is shown. The *p* value is shown in *parentheses*

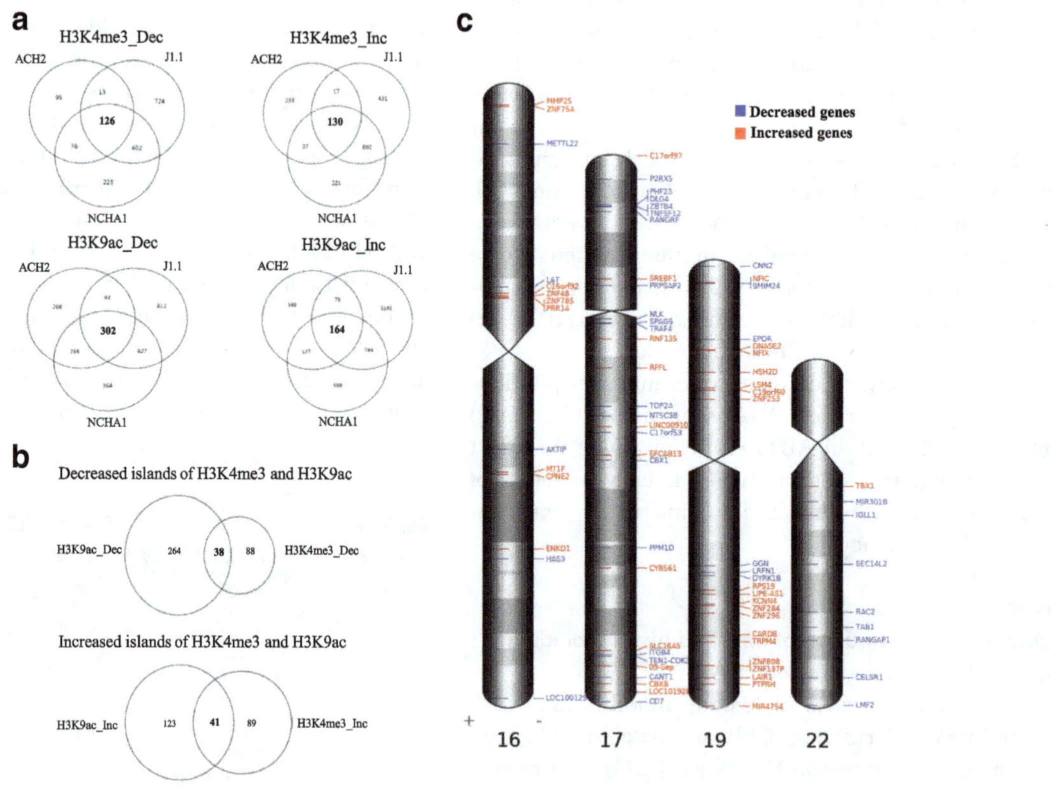

Fig. 2 Identification of H3K4me3 and H3K9ac islands in chromosomes 16, 17, 19, and 22 of HIV-1 latently infected cells. **a** Venn diagram of differentially enriched histone modification islands in chromosomes 16, 17, 19, and 22 of HIV-1 latently infected cells. **b** The differential histone H3K4me3 and H3K9ac islands were compared and the co-occurrence of two histone modification islands was examined. **c** Human chromosome map showing the localization of 38 decreased and 41 increased genes of islands in chromosomes 16, 17, 19, and 22 of HIV-1 latently infected cells. Genes associated with decreased islands are highlighted in *blue*, and genes associated with increased islands are highlighted in *red*

Functional annotation of 38 decreased and 41 increased islands in both H3K4me3 and H3K9ac

First, 38 decreased and 41 increased islands were transformed to genes corresponding to each island (Additional file 5). To investigate the functional annotation of the genes associated with the 38 decreased and 41 increased islands, gene ontology enrichment analysis was performed using the WEB-based GEne SeT AnaLysis Toolkit (WebGestalt) [19]. The 38 genes associated with decreased islands were involved in the regulation of biological process, regulation of cellular process, biological regulation, and purinergic receptor signaling pathway. The 41 genes associated with increased islands were involved in nucleic acid binding, calcium-activated cation channel activity, DNA binding, and zinc ion binding (Additional file 6). Subsequently, Pathway Commons analysis was performed to further investigate the biological pathways of genes associated with decreased and increased islands. The p63 transcription factor network was strongly linked to the 38 genes with decreased islands (Fig. 3a), and the RNA polymerase transcription termination pathway was linked to the 41 genes with increased islands (Fig. 3b). Especially, DNA topoisomerase 2 alpha (*TOP2A*), integrin beta 4 subunit (*ITGB2*), TNF receptor association factor 4 (*TRAF4*), and SEC14-like lipid binding 2 (*SEC14L2*) genes associated with decreased islands were related to the p63 transcription factor network. While, nuclear factor IC (*NFIC*) and nuclear factor IX (*NFIX*) with increased islands were linked to the RNA polymerase transcription termination pathway. To examine interactions between *TOP2A*, *ITGB2*, *TRAF4*, *SEC14L2*, *NFIC*, and *NFIX*, the Pathway Commons Network Visualizer was used (Fig. 4a). *NFIX* was found to interact with *TRAF4*, which suggests that genes related to decreased and increased islands can mutually interact in the maintenance of HIV-1 latency.

Investigation of the interaction between NFIX and TRAF4 for HIV-1 latency

Lastly, the expression of *NFIX* and *TRAF4* was examined using real-time PCR between HIV-1 latently infected cell lines, ACH2 and NCHA1, and an uninfected cell line, A3.01 (Fig. 4b). *NFIX* was highly expressed in HIV-1 latently infected cell lines compared with the uninfected cell line. The expression of *TRAF4* was similar in all cell lines. After treatment by phorbol-13-myristate-12-acetate (PMA), *NFIX* showed a dramatic reduction in expression (nearly fourfold) in HIV-1 latently infected cell lines compared with untreated cells. *TRAF4* was increased threefold in only A3.01 cells but was decreased in NCHA1 cells. These results suggest that *NFIX* might affect the reactivation and maintenance of HIV-1 latency.

Discussion

In this study, we re-analyzed ChIP-seq data obtained from GEO data repositories to investigate whether specific chromosomal distribution of histone modifications effects on HIV-1 latency and to find novel latency-related genes. As whole distribution of histone modification was unchanged before and after HIV-1 integration [10], chromatin domains which show enriched epigenetic features partially might be potential factors for the establishment or maintenance of HIV-1 latency. To elucidate the connection between chromosomal distribution of histone modification and HIV-1 latency, H3K4me3 and H3K9ac

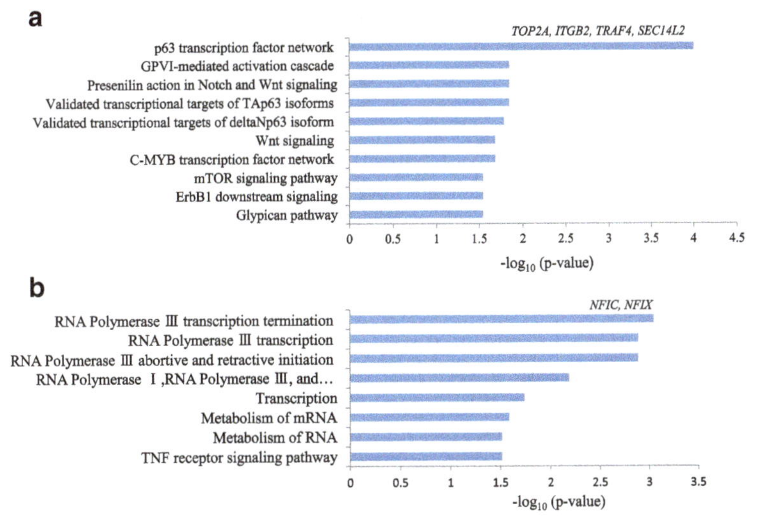

Fig. 3 Functional annotation of 38 genes associated with decreased (**a**) and 41 genes associated with increased (**b**) islands on chromosomes 16, 17, 19, and 22 of HIV-1 latently infected cells using Pathway Commons analysis. *TOP2, ITGB2, TRAF4*, and *SEC14L2* in the 38 genes associated with decreased islands were strongly linked to the p63 transcription factor network. Meanwhile, *NFIC* and *NFIX* in the 41 genes associated with increased islands were related to RNA polymerase III transcription termination. The x-axis values are $-\log_{10}$ of raw *p* values

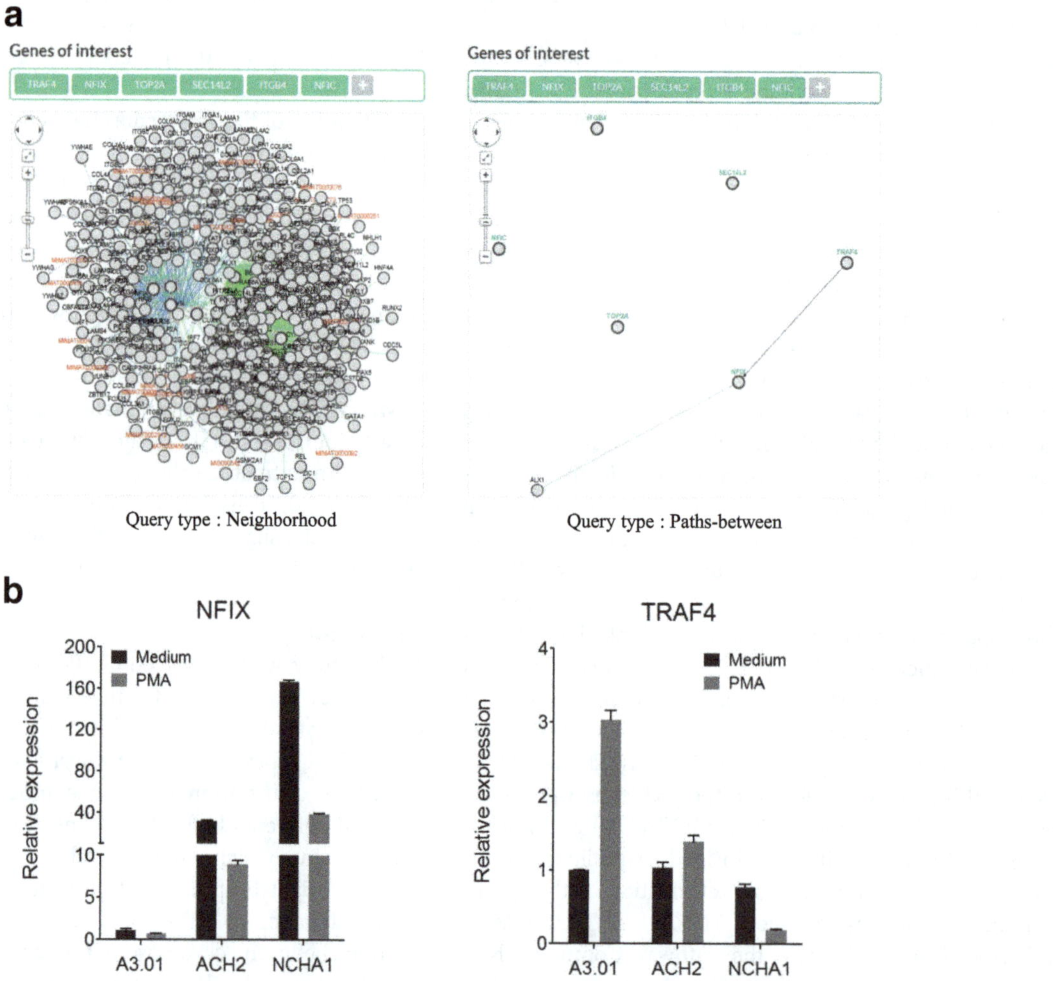

Fig. 4 Identification of the interaction of *NFIX* with *TRAF4* in HIV-1 latently infected cells. **a** A visualization of the interactions of *TOP2A, ITGB2, TRAF4*, and *SEC14L2* with *NFIC* and *NFIX* by PCViz (Pathway Commons Network Visualizer). The query type can control the size of the network or interactions by filtering genes. The neighborhood in query type shows the overall frequency of alteration for each gene in the network and the paths between in query type shows the direct interaction of each gene in the network. **b** The result of real-time PCR in treatment with/without PMA in HIV-1 latently infected cells. Cells were treated with PMA at a final concentration of 10 ng/ml for 24 h. All experiments were performed in triplicate. *Error bars* indicate the standard deviation

binding distribution patterns were determined using CEAS software. Chromosomes 16, 17, 19, and 22 were significantly enriched in both histone modifications of decreased and increased binding sites.

Accordingly, the alternation of chromatin conditions in host cells induced by HIV-1 infection might play an important role in viral latency. These results indicate that there may be common factors governing HIV-1 latency. Thought genome-wide analysis of histone modifications in HIV-1 latently infected cell lines, cyclin-dependent kinase inhibitor 1A (*CDKN1A*) and *cyclin D2* were given into latency-related genes [10].

Based on Pathway Commons analysis, *NFIX* from RNA polymerase transcription termination pathway and *TRAF4* from p63 transcription factor network were strongly linked to each other.

Nuclear factor one (*NFI*) is a family of transcription factors which consist of four related members, *NFIA, NFIB, NFIC*, and *NFIX* [20]. *NFIX* is known to bind to the palindromic sequence in viral and cellular promoters [21]. These binding sites may act as activators or repressors of transcription [22].

HIV-1 long terminal repeat (LTR) is comprised of U5, R, and U3 regions. The U3 region can be divided into the core promoter, enhancer, and modulatory regions [23]. The modulatory region includes the negative regulatory element (*NRE*) [22]. Schwartz et al. [24] has shown that core sequence in the *NRE*, TGATTGGC, was the binding site for *NFI* family, and the *NFI* binding on this site has negative effect on the control of HIV-1 transcription in Jurkat cells. In this study, *NFIX* showed the increased acetylation (H3K9ac) and methylation

(H3K4me3) in HIV-1 latently infected cell lines. These will allow transcription factors to easily access to the promoter of *NFIX* and increase the mRNA expression level of *NFIX* in HIV-1 latently infected cells. Consequently, increased *NFIX* may bind to core sequence in the *NRE* and then repress the transcription of HIV-1. In addition, CCAAT/enhancer binding protein (*C/EBP*) was known to be necessary for HIV-1 replication [25] and HIV-1 LTR contains three *C/EBP* binding sites. *NFIX* was previously named as CCAAT-binding transcription factor. This will provide a possibility that *C/EBP* and *NFIX* may compete for binding sites of HIV-1 LTR each other. Therefore, *C/EBP* or *NFIX* binding on HIV-1 LTR will induce or repress HIV-1 transcription, respectively. Lastly, Biressi et al. [26] reported that *NFIX* is strongly expressed in the fetus and makes a complex with *PKC* theta and then activates *MEF2A*. These results suggest that *NFIX* may negatively control HIV-1 transcription via *PKC* pathway.

TRAF4 is known to encode a member of the TNF receptor association factor (*TRAF*) family and interact with neurotrophin receptor p75 (*NTR*). Also, it negatively controls *NTR*-induced cell death and *NF-kappa B* activation [27, 28]. Meanwhile, Xu et al. [29] showed that co-expression of p47 (*phox*) and *TRAF4* increase oxidant production and c-Jun N-terminal kinase (*JNK*) activation and that HIV-1 Tat activates *JNK* via signaling pathway at the level of *TRAF4*. In contrast to *NFIX*, *TRAF4* showed the decreased (H3K9ac) and methylation (H3K4me3) on HIV-1 latency. As a result, the decreased expression of *TRAF4* may reduce activity of *JNK* to maintain HIV-1 latency. However, *TRAF* was not seen to directly control the expression of *NFIX* in our study.

Conclusions

In conclusion, these results suggest that the unique enrichment of histone modifications and its linked genes in specific chromosomes might play a critical role in the establishment and maintenance of HIV-1 latency.

Methods

Cell culture

A3.01, ACH2, and NCHA1 cells were maintained in RPMI 1640 medium supplemented with 10% FBS, 5% penicillin-streptomycin, and 2 mM glutamine. For comparison of *NFIX* expression between the normal and reactivation state, PMA was used at a final concentration of 10 ng/ml for 24 h.

ChIP-seq data analysis

To analysis of histone modifications for H3K4me3 and H3K9ac in HIV-1 latently infected cells, ChIP-seq data was obtained from GEO data repositories (GSE58246). Enriched histone islands of the genome were identified by comparing the chromatin immunoprecipitated (ChIPed) samples to the input samples using the SICER 1.1 program with the following parameters: window size = 200, gap size = 400, *E* value = 0.01, and FDR = 0.05. The overlapped histone islands were identified by hypergeometric optimization of motif enrichment (HOMER) programs [30] with maximum distance to merge ("given"). For genomic distribution of differential binding sites, CEAS program was used. Genomic position was plotted onto the chromosome map using the idiographica webtool (http://www.ncrna.org/idiographica). Gene ontology enrichment analysis was performed using WebGestalt, and interaction of genes was visualized using the Pathway Commons Network Visualizer (PCViz, http://www.pathwaycommons.org/pcviz).

Quantitative real-time PCR

Total RNA was extracted from cells using TRIzol reagent, and 1 μg of RNA was reverse-transcribed using Superscript III cDNA synthesis kit (Invitrogen). *NFIX* and *TRAF4* quantitative reverse transcription PCR was performed on a 7500 real-time PCR system (Applied Biosystems) using SYBR Green PCR Master Mix (Applied Biosystems). The relative expression was calculated as follows using the ΔΔCt method: fold change of enrichment = 2 delta ((Ct-ChIPed)-(Ct-Input)). ΔCt values were determined using a the 7500 real-time PCR system software (Applied Biosystems) with *GAPDH* as an endogenous control. The following primers were used for real-time PCR:

> *NFIX* forward primer (5′ -AGGAGATGCGGACATCAAA-3′), *NFIX* reverse primer (5′ -TACTCTCACCAGCTCCGTCA-3′), *TRAF4* forward primer (5′ -AGGAGTTCGTCTT TGACACCATC-3′), *TRAF4* reverse primer (5′ -CT TTGAATGGGCAGAGCACC-3′), *GAPDH* forward primer (5′ -GAAGGTGAAGGTCGGAGTC-3′), and *GAPDH* reverse primer (5′ -GAAGATGGTGATGG GATTTC-3′).

Additional files

Additional file 1: Chromosomal distribution of H3K4me3 binding sites in HIV-1 latently infected cells. Enrichment pattern of increased islands of H3K4me3 among individual chromosomes is shown as bar chart. Percent of total H3K4me3 islands (red bar) and what would be expected by random chance (blue bars) for each chromosome is shown. The value in parenthesis means *p* value.

Additional file 2: Chromosomal distribution of H3K9ac binding sites in HIV-1 latently infected cells. Enrichment pattern of decreased islands of H3K9ac among individual chromosomes is shown as bar chart. Percent of total H3K9ac islands (red bar) and what would be expected by random chance (blue bars) for each chromosome is shown. The value in parenthesis means *p* value.

Additional file 3: Chromosomal distribution of H3K9ac binding sites in HIV-1 latently infected cells. Enrichment pattern of increased islands of H3K9ac among individual chromosomes is shown as bar chart. Percent of total H3K9ac islands (red bar) and what would be expected by random chance (blue bars) for each chromosome is shown. The value in parenthesis means p value.

Additional file 4: The ratio of percentage of total H3K4me3 or H3K9ac islands to percentage expected by random chance for each chromosome in three different HIV-1 latently infected cell lines.

Additional file 5: Lists of 38 decreased genes and 41 increased genes identified in the co-occurrence of two histone modifications, H3K4me3 and H3K9ac.

Additional file 6: Functional annotation of 38 decreased (A) and 41 increased (B) genes in chromosomes 16, 17, 19, and 22 of HIV-1 latently infected cells using WebGestalt. The x-axis values are $-\log_{10}$ of raw p values.

Abbreviations

ART: Antiretroviral therapy; cART: Combination antiretroviral therapy; *CDKN1A*: Cyclin-dependent kinase inhibitor 1A; CEAS: Cis-regulatory element annotation system; ChIPed: Chromatin immunoprecipitated; ChIP-seq: Chromatin immunoprecipitation-high throughput parallel DNA sequencing; H3K4me3: Histone 3 lysine 4 trimethylated; H3K9ac: Histone 3 lysine 9 acetylated; HOMER: Hypergeometric optimization of motif enrichment; *ITGB2*: Integrin beta 4 subunit; *NFIX*: Nuclear factor IX; PMA: Phorbol-13-myristate-12-acetate; *SEC14L2*: SEC14-like lipid binding 2; SICER: Spatial clustering for identification of ChIP-enriched regions; *TOP2A*: DNA topoisomerase 2 alpha; *TRAF4*: TNF receptor association factor 4; WebGestalt: WEB-based GEne SeT AnaLysis Toolkit

Acknowledgements

We thank Tae-Young Roh (Pohang University of Science and Technology) and Jihwan Park (Pohang University of Science and Technology) for access to epigenetics data.

Funding

This research was supported by a grant from Research of Korea Centers for Disease Control and Prevention (Grant No. 2016-NG51003-00).

Authors' contributions

KCK and BSC conceived and designed the experiments. JSS, YHS, and CHY performed the experiments. SYL and CK analyzed the data. KCK and BSC wrote the manuscript. All authors read and approved the final manuscript.

Competing interests

The authors declare that they have no competing interests.

Consent for publication

Not applicable

References

1. Siliciano RF, Greene WC. HIV latency. Cold Spring Harb Perspect Med. 2011; 1(1):a007096.
2. Finzi D, et al. Identification of a reservoir for HIV-1 in patients on highly active antiretroviral therapy. Science. 1997;278(5341):1295–300.
3. Wong JK, et al. Recovery of replication-competent HIV despite prolonged suppression of plasma viremia. Science. 1997;278(5341):1291–5.
4. Finzi D, et al. Latent infection of CD4+ T cells provides a mechanism for lifelong persistence of HIV-1, even in patients on effective combination therapy. Nat Med. 1999;5(5):512–7.
5. Kingsman SM, Kingsman AJ. The regulation of human immunodeficiency virus type-1 gene expression. Eur J Biochem. 1996;240(3):491–507.
6. Jordan A, Bisgrove D, Verdin E. HIV reproducibly establishes a latent infection after acute infection of T cells in vitro. EMBO J. 2003;22(8):1868–77.
7. Felsenfeld G, Groudine M. Controlling the double helix. Nature. 2003; 421(6921):448–53.
8. Tamaru H. Confining euchromatin/heterochromatin territory: jumonji crosses the line. Genes Dev. 2010;24(14):1465–78.
9. Jenuwein T, Allis CD. Translating the histone code. Science. 2001;293(5532): 1074–80.
10. Park J, et al. Genome-wide analysis of histone modifications in latently HIV-1 infected T cells. Aids. 2014;28(12):1719–28.
11. Lusic M, Giacca M. Regulation of HIV-1 latency by chromatin structure and nuclear architecture. J Mol Biol. 2015;427(3):688–94.
12. Ay E, et al. Epigenetics of HIV infection: promising research areas and implications for therapy. AIDS Rev. 2013;15(3):181–8.
13. Blazkova J, et al. CpG methylation controls reactivation of HIV from latency. PLoS Pathog. 2009;5(8):e1000554.
14. Blazkova J, et al. Paucity of HIV DNA methylation in latently infected, resting CD4+ T cells from infected individuals receiving antiretroviral therapy. J Virol. 2012;86(9):5390–2.
15. Verdin E, Paras Jr P, Van Lint C. Chromatin disruption in the promoter of human immunodeficiency virus type 1 during transcriptional activation. Embo j. 1993;12(8):3249–59.
16. Williams SA, et al. NF-kappaB p50 promotes HIV latency through HDAC recruitment and repression of transcriptional initiation. Embo j. 2006;25(1):139–49.
17. Zang C, et al. A clustering approach for identification of enriched domains from histone modification ChIP-Seq data. Bioinformatics. 2009;25(15):1952–8.
18. Shin H, et al. CEAS: cis-regulatory element annotation system. Bioinformatics. 2009;25(19):2605–6.
19. Wang J, et al. WEB-based GEne SeT AnaLysis Toolkit (WebGestalt): update 2013. Nucleic Acids Res. 2013;41(Web Server issue):W77–83.
20. Gronostajski RM. Roles of the NFI/CTF gene family in transcription and development. Gene. 2000;249(1-2):31–45.
21. Apt D, Liu Y, Bernard HU. Cloning and functional analysis of spliced isoforms of human nuclear factor I-X: interference with transcriptional activation by NFI/CTF in a cell-type specific manner. Nucleic Acids Res. 1994; 22(19):3825–33.
22. Pereira LA, et al. A compilation of cellular transcription factor interactions with the HIV-1 LTR promoter. Nucleic Acids Res. 2000;28(3):663–8.
23. Gaynor R. Cellular transcription factors involved in the regulation of HIV-1 gene expression. Aids. 1992;6(4):347–63.
24. Schwartz C, et al. Characterization of nuclear proteins that bind to the regulatory TGATTGGC motif in the human immunodeficiency virus type 1 long terminal repeat. Nucleic Acids Res. 1997;25(6):1177–84.
25. Henderson AJ, Connor RI, Calame KL. C/EBP activators are required for HIV-1 replication and proviral induction in monocytic cell lines. Immunity. 1996; 5(1):91–101.
26. Biressi S, et al. Intrinsic phenotypic diversity of embryonic and fetal myoblasts is revealed by genome-wide gene expression analysis on purified cells. Dev Biol. 2007;304(2):633–51.
27. Ye X, et al. TRAF family proteins interact with the common neurotrophin receptor and modulate apoptosis induction. J Biol Chem. 1999;274(42):30202–8.
28. Krajewska M, et al. TRAF-4 expression in epithelial progenitor cells. Analysis in normal adult, fetal, and tumor tissues. Am J Pathol. 1998;152(6):1549–61.
29. Xu YC, et al. Involvement of TRAF4 in oxidative activation of c-Jun N-terminal kinase. J Biol Chem. 2002;277(31):28051–7.
30. Heinz S, et al. Simple combinations of lineage-determining transcription factors prime cis-regulatory elements required for macrophage and B cell identities. Mol Cell. 2010;38(4):576–89.

9

Partial uniparental isodisomy of chromosome 16 unmasks a deleterious biallelic mutation in *IFT140* that causes Mainzer-Saldino syndrome

Benjamin M. Helm[1,2], Jason R. Willer[3], Azita Sadeghpour[3], Christelle Golzio[3,4], Eric Crouch[5], Samantha Schrier Vergano[1,6], Nicholas Katsanis[3*] and Erica E. Davis[3*]

Abstract

Background: The ciliopathies represent an umbrella group of >50 clinical entities that share both clinical features and molecular etiology underscored by structural and functional defects of the primary cilium. Despite the advances in gene discovery, this group of entities continues to pose a diagnostic challenge, in part due to significant genetic and phenotypic heterogeneity and variability. We consulted a pediatric case from asymptomatic, non-consanguineous parents who presented as a suspected ciliopathy due to a constellation of retinal, renal, and skeletal findings.

Results: Although clinical panel sequencing of genes implicated in nephrotic syndromes yielded no likely causal mutation, an oligo-SNP microarray identified a ~20-Mb region of homozygosity, with no altered gene dosage, on chromosome 16p13. Intersection of the proband's phenotypes with known disease genes within the homozygous region yielded a single candidate, *IFT140*, encoding a retrograde intraflagellar transport protein implicated previously in several ciliopathies, including the phenotypically overlapping Mainzer-Saldino syndrome (MZSDS). Sanger sequencing yielded a maternally inherited homozygous c.634G>A; p.Gly212Arg mutation altering the exon 6 splice donor site. Functional studies in cells from the proband showed that the locus produced two transcripts: a majority message containing a mis-splicing event that caused a premature termination codon and a minority message homozygous for the p.Gly212Arg allele. Zebrafish in vivo complementation studies of the latter transcript demonstrated a loss of function effect. Finally, we conducted post-hoc trio-based whole exome sequencing studies to (a) test the possibility of other causal loci in the proband and (b) explain the Mendelian error of segregation for the *IFT140* mutation. We show that the proband harbors a chromosome 16 maternal heterodisomy, with segmental isodisomy at 16p13, likely due to a meiosis I error in the maternal gamete.

Conclusions: Using clinical phenotyping combined with research-based genetic and functional studies, we have characterized a recurrent *IFT140* mutation in the proband; together, these data are consistent with MZSDS. Additionally, we report a rare instance of a uniparental isodisomy unmasking a deleterious mutation to cause a ciliary disorder.

Keywords: Skeletal ciliopathy, Conorenal dysplasia, Intraflagellar transport, Zebrafish, Whole exome sequencing, Heterodisomy

* Correspondence: nicholas.katsanis@duke.edu; erica.davis@duke.edu
[3]Center for Human Disease Modeling, Duke University Medical Center, Durham, NC 27701, USA
Full list of author information is available at the end of the article

Background

Genetic medicine in the post-genome era has benefited from increasingly sophisticated methodologies to detect both chromosomal lesions, as well as point mutations across subsets of functionally related genes, or indeed the entire genome [1]. Such approaches are particularly useful for human genetic conditions that are unified by a common cellular basis, but for which extensive phenotypic variability and genetic heterogeneity pose clinical diagnostic challenges. The ciliopathies—multi-systemic disorders caused by dysfunction, altered assembly, and/or altered maintenance of cilia—exemplify a class of inherited disorders which have achieved accelerated molecular discovery in the past ~15 years in part from the advent of whole exome sequencing (WES) and chromosomal array technologies [2–5].

Given the near ubiquitous presence of cilia on different cell types of the vertebrate body plan, it is not surprising that mutations in genes critical for ciliary function can give rise to a broad range of phenotypic features, often characterized by developmental defects in single or multiple systems [6–8]. Some estimates suggest that ~1000 proteins may localize to or be required for ciliary processes [9], thus representing a substantial mutational target. Within the ciliary organelle, macromolecular complexes that perform discrete functions have been characterized biochemically; these include the intraflagellar transport (IFT) complex [10], BBSome [11], and transition zone complex [12], and some phenotype correlations with these complexes have emerged [13]. For example, the skeletal ciliopathies, including Jeune asphyxiating thoracic dystrophy (JATD; MIM 208500), short-rib polydactyly (SRP; MIM 208500), Sensenbrenner syndrome (MIM 218330), and Mainzer-Saldino syndrome (MZSDS; MIM 266920) are enriched for mutations in genes involved in IFT, especially retrograde transport (IFT-complex A). For example, Sensenbrenner syndrome (also known as cranioectodermal dysplasia [14]), a ciliopathy characterized by sagittal craniosynostosis with accompanying facial, skeletal, and ectodermal anomalies, is caused by mutations in the IFT-A genes *IFT43* [15], *WDR35* [16], *IFT122* [17], *WDR19* [18], and *IFT140* [19]. Similarly, the neonatal lethal JATD [20] is hallmarked by a constricted thoracic cage, shortening of the long bones, polydactyly, renal cystic disease, hepatic insufficiency, and retinal degeneration in the individuals who survive past birth. Like Sensenbrenner, JATD can also be caused by mutations in retrograde IFT components, including the motor *DYNC2H1* [21, 22] and IFT-A genes *TTC21B* [23], *WDR19* [18], and *IFT140* [24]. However, neither of these two ciliopathies is caused exclusively by IFT-A dysfunction; for example, JATD can be caused by mutations in the anterograde IFT genes (complex B) *IFT80* [25] or *IFT172* [26]. Furthermore, recessive mutations in a causal gene might not necessarily give rise to a skeletal phenotype, adding further complexity to genotype-phenotype correlations within the skeletal ciliopathies. For example, mutations in *IFT172* can also cause syndromic or non-syndromic retinal degeneration [27].

Here, we report a pediatric case with hallmark retinal, renal, and skeletal phenotypes whose molecular diagnosis was initially unclear. Multi-tiered in vitro and in vivo functional studies suggested that loss of function of *IFT140* is the likely driver, as caused by a combination of mis-splicing and reduction to homozygosity of a missense variant through a uniparental isodisomy mechanism. Together, these results support a diagnosis of MZSDS, represent a rare example of a ciliopathy caused by a non-Mendelian mutational mechanism, and reinforce the need to examine all possible mutation inheritance paradigms when analyzing diagnostic exomes.

Results

Case report

We consulted a 5-year-old male proband who had a history of complex medical issues, including the following triad of symptoms: retinal dystrophy, acute-onset renal failure, and joint pain (Fig. 1; Additional file 1: Figure S1; Table 1). He was the product of a normal pregnancy with no known teratogenic exposures and was born at term via induced vaginal delivery from his 38-year-old G5 P3-P4 mother. He had a normal birth weight of 3543 g, was measured 48 cm in length, and was the fourth child born to a non-consanguineous couple of northern European ancestry (Fig. 2a). Both parents and all older siblings (two males and one female) were reported healthy. The extended family history was also unremarkable, with the exception of a paternal family history of type 1 diabetes, a paternal grandmother with alpha-1-antitrypsin deficiency, and a maternal cousin who was reported to have hearing loss and retinitis pigmentosa of unknown etiology. The proband reached his developmental milestones on time; although, there were some mild difficulties with visual tracking and speech articulation.

At 3 years of age, the proband was referred to an ophthalmologist for concerns of nyctalopia, high myopia, and continued poor visual tracking. Fundus exams and electroretinograms (ERG) revealed a rod-cone dystrophy in both eyes with mottling of the retinal pigmentary epithelium and dystrophy of the internal limiting membrane (Additional file 1: Figure S1). The fundus photos show non-specific subretinal hypopigmentation consistent with early onset retinal degeneration typical of rod-cone dystrophies (Additional file 1: Figure S1a). The retinal vasculature appeared attenuated in both eyes. There was no relative afferent pupillary defect and no nystagmus. His high myopia was progressive. At 5 years of age, his refraction was −4.5 and −6.75 diopters in the right

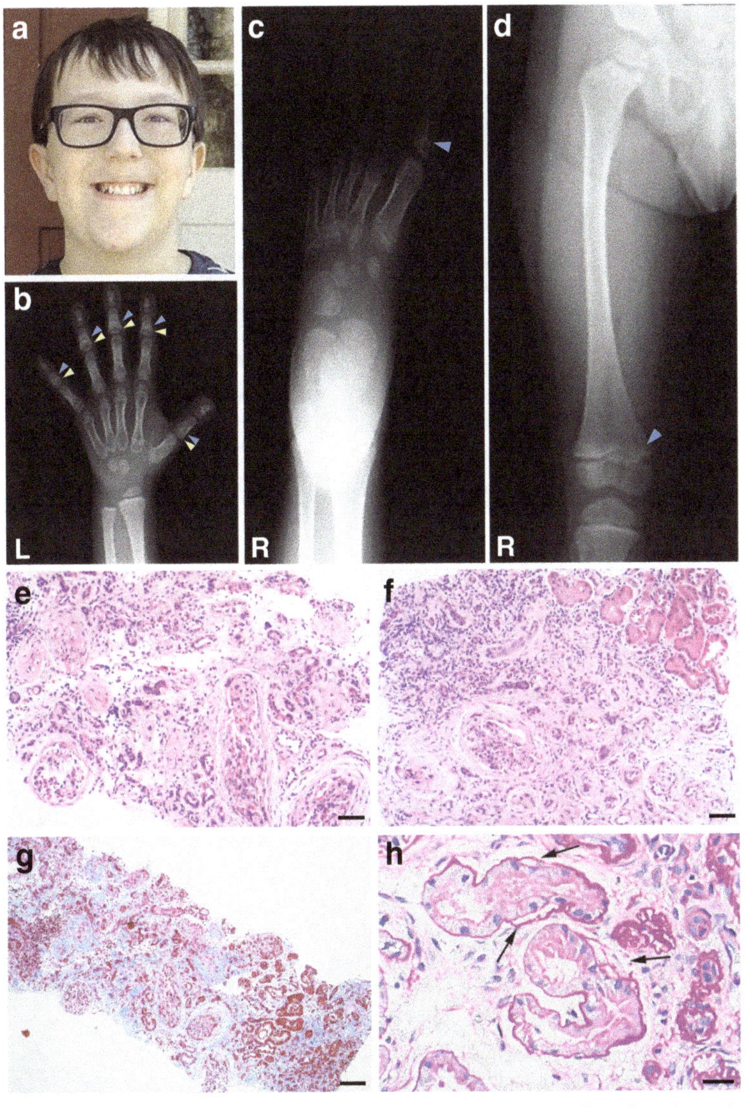

Fig. 1 Radiographic and renal features of the proband. **a** Photograph of the index case at 10 years of age. **b** Radiograph of the left (*L*) hand at 5 years and 1 month of age shows cup-shaped metaphyses (*blue arrowheads*), cone-shaped epiphyses (*yellow arrowheads*), and short metacarpals. Bone age is consistent with chronological age. **c** Radiograph of right (R) foot at 5 years and 1 month of age shows shortened proximal phalanges with cup-shaped metaphyses (*blue arrowhead*) and short metatarsals with irregular metaphyses. **d** Radiograph of the right (R) femur showed bilateral coxa valga, broad femoral necks, abnormal femoral metaphyses, and mild bowing of the femoral diaphysis. **e** Renal glomeruli with prominent, periglomerular fibrosis; hematoxylin and eosin (H&E) stain; scale bar, 50 μm. **f** Tubulointerstitial fibrosis, chronic inflammation, tubular atrophy, and focal non-atrophic tubules, (*upper right of panel*); H&E; scale bar 50 μm. **g** Diffuse, severe tubulointerstitial fibrosis and tubular atrophy; Masson's trichrome stain; scale bar 100 μm. **h** Focal non-atrophic proximal tubules with segmentally thin and duplicated (*arrows*) basement membranes; periodic acid-Schiff stain; scale bar 20 μm

and left eyes, respectively. By the age of 8 years old, his best-corrected visual acuity was 20/50 OD and 20/50 OS with glasses and his refraction had increased to −10.25 +2.75 ×106 OD and −11.00 +1.50 ×61 OS. The ERG testing demonstrated reduced cone function under photopic conditions with reduced a-wave implicit time and amplitudes and slowed implicit times for the b-wave (Additional file 1: Figure S1b). Under scotopic testing, both a-waves and b-waves showed slower implicit time and diminished amplitudes (Additional file 1: Figure S1c). The a-wave was substantially delayed, and there were reduced incremental increases in a-wave and b-wave. Audiometric testing was normal bilaterally.

In addition to retinal dystrophy, the proband developed acute-onset renal failure. Initial symptoms included pulmonary edema, severe hypertension, and fluid overload requiring hemodialysis. At 5 years of age, he was transitioned to peritoneal dialysis and remained on dialysis for 5 months, until he received a living related donor kidney transplant. A renal biopsy of the diseased kidney confirmed the cause of renal failure: histological analyses showed diffuse severe tubulointerstitial fibrosis with

Table 1 Phenotype summary

	Case reported here (DM165-0001)	Family V-1 (Perrault et al. 2012) [24]	Family V-2 (Perrault et al. 2012) [24]	Family VII-1 (Perrault et al. 2012) [24]
Sex	M	M	F	M
Age at examination	5 years	4 years; 10 years	2 years; 4 years	Birth; 8 months; 12 months; 18 months
IFT140 allele 1	c.634G>A (p.Gly212Arg)	c.634G>A (p.Gly212Arg)	c.634G>A (p.Gly212Arg)	c. 2399+1G>T (splice)
IFT140 allele 2	c.634G>A (p.Gly212Arg)	c.3916dup (p.Ala1306Glyfs*56)	c.3916dup (p.Ala1306Glyfs*56)	c. 634G>A (p.Gly212Arg)
Diagnosis	Mainzer-Saldino syndrome (MIM 266920)	Mainzer-Saldino syndrome (MIM 266920)	Mainzer-Saldino syndrome (MIM 266920)	Jeune syndrome (MIM 208500)
Psychomotor development	Dysgraphia (HP:0010526), Reading difficulties (HP:0010522), Adjustment disorder with mixed disturbance of emotions and conduct (HP:0100851)	Normal	Normal	Hypotonia (HP:0001290), Poor feeding at birth (HP:0008872), Developmental delay (HP:0001263)
Cerebral MRI	NA	Normal	Normal	Normal
Retina				
Nystagmus	No	No	No	Yes (HP:0000639)
Refraction	High myopia (HP:0011003)	Myopia (HP:0000545)	NA	NA
Light behavior	Nyctalopia (HP:0000662)	NA	NA	NA
Visual field	NA	Tubular (HP:0030588)	NA	NA
Visual acuity	Mild difficulty with visual tracking (HP:0030532)	60/200 RE; 120/200 LE with color vision	120/200 RE;100/200 LE	Able to fix and follow
Fundus	Retinitis pigmentosa (HP:0000510)	Retinitis pigmentosa (HP:0000510)	Retinitis pigmentosa (HP:0000510)	NA
ERG	Rod-cone dystrophy	Severely altered	Severely altered	NA
Craniofacial features	No obvious facial dysmorphisms	Craniosynostosis (HP:0001363), Scaphocephaly (HP:0030799), Facial dysmorphism (HP:0001999)	Craniosynostosis (HP:0001363), Scaphocephaly (HP:0030799), Facial dysmorphism (HP:0001999), Microcephaly (HP:0000252)	Normal
Stature	120.2 cm (50th centile) at 6 years and 11 months	Short (HP:0004322)	Short (HP:0004322)	Normal

Table 1 Phenotype summary *(Continued)*

Limbs				
Hands	Cone-shaped epiphyses of phalanges (HP:0010230), Brachydactyly (HP:0011927), Normal bone age	Cone-shaped epiphyses of phalanges (HP:0010230),	Cone-shaped epiphyses of phalanges (HP:0010230)	Cone-shaped epiphyses of phalanges (HP:0010230)
Feet	Shortened metacarpals (HP:0010049), Shortened proximal phalanges (HP:0001831), Shortened metatarsals (HP:0010743), Irregular proximal metatarsal metaphyses (HP:0010630), Brachydactyly (HP:0001831)	NA	NA	NA
Long bones	Abnormal proximal femoral metaphyses (HP:0003411), Sclerotic changes of the of proximal femoral growth plate (HP:0008797), Broad femoral necks (HP:0006429), Mild bowing of the femoral diaphysis (HP:0002980)	NA	NA	Trident acetabulum (HP:0003170)
Other skeletal	Small left joint effusion of the hip (HP:0001384), Bilateral coxa vara (HP:0002812)	NA	NA	Short thorax (HP:0010306), Short ribs (HP:0000773)
Renal features				
Function	Acute-onset renal failure (5 years) (HP:0001919);	Chronic renal failure (4 years) (HP:0000083), End-stage renal disease (15 years) (HP:0003774)	Chronic renal failure (9 months) (HP:0000083); End-stage renal disease (3 years) (HP:0003774)	Elevated urea (HP:0000093 Proteinuria (18 months) (HP:0000093)
Ultrasonography	Pre-transplant renal ultrasound: No evidence of hydronephrosis, Renal cortical medullary parenchymal disease (HP:0025327)	Cysts, Small kidneys (-1DS) (HP:0000107) Hyperechogenicity (HP:0004719), Cortico-medullary differentiation loss	1 cyst (cortico-medullar, right kidney) (HP:0000107) Hyperechogenicity (HP:0004719),	Increased echogenicity (HP:0004719)

Table 1 Phenotype summary *(Continued)*

			Cortico-medullary differentiation	
	Pre-transplant abdominal ultrasound: Echogenic kidneys bilaterally	(HP:0005565)	loss	
		Growth retardation (-2DS)	(HP:0005565)	
	(HP:0004719)	(HP:0000089)	Growth retardation	
	Post-transplant renal ultrasound: Normal transplanted kidney, normal bladder, atrophic appearance of native kidney		(HP:0000089)	
Biopsy	Diffuse severe tubulointerstitial fibrosis			Non-specific tubulointerstitial nephritis
	(HP:0005576)			(HP:0001970)
	Tubular atrophy			
	(HP:0000092			
	Only focal non-atrophic tubules			
Liver	Pre-transplant abdominal ultrasound: Hepatomegaly	Moderate cholestasis	Moderate cholestasis	NA
		(HP:0001396)	(HP:0001396)	
	(HP:0002240)	Hepatic cytolisis with hepatomegaly		
	Possible fatty infiltration			
	(HP:0001397)	(HP:0002240),		
		Portal fibrosis		
		(HP:0006580)		
Other	Pulmonary edema	Bilateral hypoacousia		
	(HP:0100598),	(HP:0000407)		
	Severe hypertension			
	(HP:0000822),			
	Fluid overload			
	(HP:0011105);			
	Mild sclerosis of arterioles			
	(HP:0002634)			

NA not available

tubular atrophy and only focal non-atrophic tubules (Fig. 1e, f). Half of the 14 glomeruli examined were obsolescent. The remaining glomeruli appeared partially collapsed and some had segmental capsular adhesion and scarring. Non-obsolescent glomeruli had periglomerular fibrosis with an onion skin-like appearance (Fig. 1g), as evidenced by Masson's trichrome staining. The tubulointerstitium contained a patchy infiltrate of lymphocytes, and eosinophils were present focally. Periodic acid-Schiff stain (PAS) showed that non-atrophic tubules had segmentally thick and thin tubular basement membranes and segmental duplication (Fig. 1h). Together, these data were suggestive of nephronophthisis [28]; although, we did not detect tubular cysts in the sampled tissue; this may have reflected an absence of the cortico-medullary junction in the biopsy. Electron microscopy also showed tubular basement membranes that were focally thick or thin. Arteries were normal and arterioles had mild sclerosis. Except for some segmental entrapment, glomeruli were negative for IgG, IgA, IgM, C3, and C1q by direct immunofluorescence (not shown), ruling out the possibility of C1q steroid-resistant nephrotic syndrome [29].

At the time of the initial evaluation, the proband also reported joint pain with popping of the knees and hips bilaterally. An X-ray bone age study of the hand showed cone-shaped epiphyses of all phalanges with shortened metacarpals and otherwise normal bone age (Fig. 1b). The lower proximal phalanges were short with cup-shaped metaphyses, and the metatarsals were also short (Fig. 1c). Imaging of the knees was normal; however, radiographs of the lower extremities showed bilateral

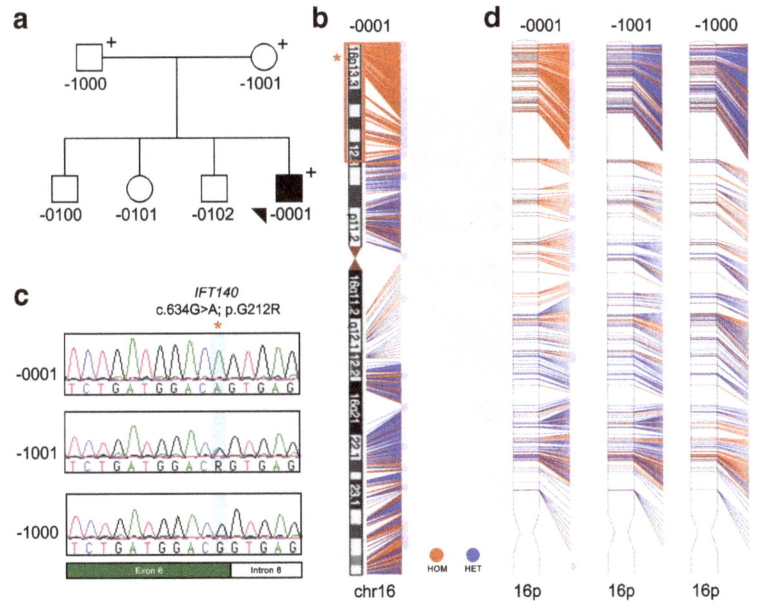

Fig. 2 Clinical and research-based genetic analysis. **a** Pedigree of family DM165. **b** Schematic of human chromosome 16. The ~20 Mb region of homozygosity identified in the proband is outlined with a *red rectangle*. The genomic location of *IFT140* is indicated with a *red star*. The 823 variants extracted from the exome data are indicated as homozygous (HOM; *red*) or heterozygous (HET; *blue*); *pale blue circles* indicate homozygous variants of unambiguous maternal origin ($n = 111$). **c** Sequence chromatograms identified a homozygous *IFT140* c.634G>A; p.Gly212Arg mutation that does not segregate in a Mendelian fashion. The mutation location at the exon 6 canonical splice donor site is indicated with a *red star*. **d** Expanded representations of chr16p in each of the proband, maternal, and paternal samples. *Colors* are indicated as described for panel **b**

coxa vara, abnormal femoral metaphyses, and sclerotic changes along the metaphyseal side of the femoral growth plate (Fig. 1d). There was broadening of the femoral necks and mild bowing of the femoral diaphysis. Although we could not rule out that some of the joint pain was caused by renal osteodystrophy, the radiographic studies suggested metaphyseal dysplasia. At the time of the most recent examination (6 years and 11 months), the proband weighed 28.8 kg (90th centile), had a height of 120.2 cm (50th centile), and had a fronto-occipital circumference of 52.7 cm (50th centile). A physical exam showed well-healed scars over his abdomen from prior surgeries, with a G-tube in place. His hands showed brachydactyly with shortened distal phalanges, and his facial appearance was not overtly dysmorphic (Fig. 1a). The remainder of the physical examination was otherwise unremarkable.

Candidate gene analysis identifies *IFT140* as the likely causal gene

Prompted by the acute renal failure of the proband, we conducted clinical panel sequencing of genes implicated in renal conditions (Additional file 1: Table S1). First, we screened six genes that cause nephrotic syndrome or glomerulosclerosis [30–37]. No sequence alterations were identified in *NPHS2*, *LAMB2*, *ACTN4*, or *TRPC6*. In *WT1*, a gene known to be associated with autosomal dominant nephrotic syndrome [32, 36], we identified a heterozygous, intronic variant (c.1099-9T>C); although rare, its frequency in presumably healthy controls in the Exome Aggregation Consortium (ExAC) database indicated that it is likely a benign variant (0.004 minor allele frequency (MAF); 4 homozygotes in 120,852 individuals). Additionally, we detected a rare heterozygous missense change in *NPHS1* (c.1175T>C; p.Leu392Pro) deemed similarly as likely benign due to the presence of healthy homozygous individuals in ExAC (18 of 121,144 samples). Next, we carried out clinical genetic testing of another five genes known to cause the isolated or syndromic renal ciliopathies, nephronophthisis or Senior-Løken syndrome, respectively [38–42]. We identified a missense change that was common in healthy controls (>1% MAF; *SDCCAG8*), three synonymous variants (*GLIS2, IQCB1, SDCCAG8*), and two deep intronic variants (*IQCB1* and *INVS*), all of which were considered likely benign (Additional file 1: Table S1).

We next considered the possibility that the proband might harbor chromosomal aberrations. We therefore conducted an oligonucleotide (oligo)-single nucleotide polymorphism (SNP) microarray (Affymetrix 6.0). We identified a ~20-Mb region of homozygosity on 16p13.3p12.3 (hg19, chr16:94,807-20,250,946; Fig. 2b). This region encompasses 411 genes, 44 of which are associated currently with disease. When we intersected the clinical synopsis of each disease-related gene with the phenotypes of the proband, we

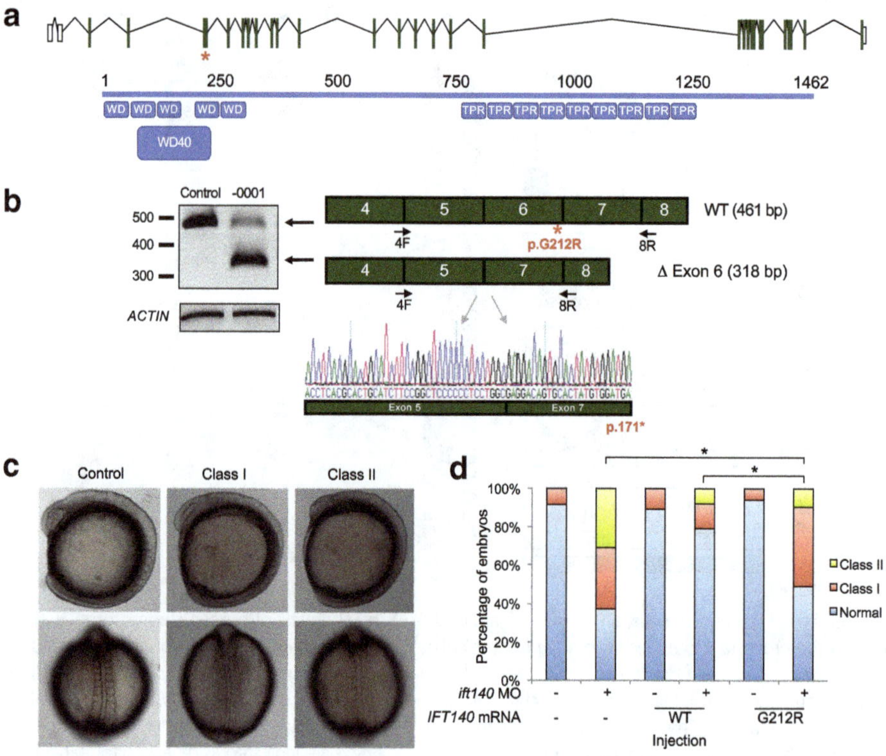

Fig. 3 In vitro and in vivo functional analysis of *IFT140* c.634G>A; p.Gly212Arg. **a** Schematic of the human *IFT140* locus at chr16:1,510,427-1,612,110 (hg19; *top*) and translated protein (*bottom*). Exons are depicted as *green boxes*; untranslated regions are shown as *white boxes* (NM_014714.3). Protein schematic (*blue*; NP_055529.2) indicates predicted WD40 repeat (WD40) and tetratricopeptide-like helical domains (TPR). IFT140 c.634G > A; p.Gly212Arg location is indicated with a *red star*. **b** RT-PCR from proband lymphocyte cell lines indicating aberrant mRNA splicing. *IFT140* cDNA was amplified with primers flanking the exon 6 canonical splice donor site (*black arrows*). Sequence chromatograms from purified PCR product indicates that the majority of *IFT140* message is missing exon 6 (*bottom*). **c** Lateral (*top*) and dorsal (*bottom*) views of *ift140* morphants exhibiting gastrulation defects. **d** In vivo complementation studies. Live embryo batches were scored to assay variant pathogenicity. Class I, modest shortening of the body axis and reduction in size of anterior structures; class II, severe shortening of the body axis and decreased anterior structures accompanied by broadening and/or kinking of the notochord and thinning of the somites. To compare pairs of embryo batches, χ^2 tests were used; $p < 0.0001$ indicated with (*asterisk*)

identified a single likely candidate, *IFT140*. This locus encodes a component of the retrograde intraflagellar transport complex [10]. Crucially, mutations in *IFT140* cause a spectrum of ciliopathies, including MZSDS [24], JATD [24], syndromic and non-syndromic retinal dystrophy [43–45], Opitz trigonocephaly C syndrome [46], and Sensenbrenner syndrome [19]. To investigate the possibility that mutations in *IFT140* could contribute to the proband's clinical presentation, we conducted research-based Sanger sequencing of the 29 coding exons and intron-exon boundaries. We identified a single rare homozygous variant at the exon 6 donor splice site (NM_014714.3: c.634G>A; p.Gly212Arg) in the proband (Fig. 2c). This change is ultra-rare in publicly available control datasets (5/121,320 alleles in ExAC, all heterozygous), and, notably, has been reported in three individuals from two unrelated families with skeletal ciliopathies [24] (Table 1). Segregation of the c.634G>A variant in both parents revealed a Mendelian violation: his mother was a heterozygous carrier, but his father was wild type at the same genomic position (Fig. 2c).

Functional studies of the IFT140 c.634G>A variant indicate loss of function

We considered two mechanisms by which the c.634G>A variant could ablate the function of *IFT140*: (1) disruption of mRNA splicing by abolishing the canonical donor spice site and (2) diminished protein function due to an amino acid alteration of a conserved residue predicted to reside within a WD40 domain (Fig. 3a). We tested both hypotheses.

First, to investigate whether c.634G>A results in aberrant *IFT140* mRNA processing, we established lymphocyte cell lines (LCL) from a whole-blood sample from the proband and we monitored mRNA splicing. We harvested total RNA from proband LCLs and matched control LCLs, reverse transcribed to generate cDNA, and PCR-amplified a 461-bp region spanning exons 4–8 of the canonical *IFT140* transcript. In comparison to the expected sized amplicon in the control sample, we observed two different-sized PCR products in the proband (Fig. 3b). Sequencing revealed that the smaller splice variant was missing the 143-bp *IFT140* exon 6, resulting in a transcript that shifts the reading frame and results in a stop codon at residue 171. The larger mRNA species amplified from proband cDNA was equivalent in size to that of the control, but harbored the c.634G>A mutation (Fig. 3b). Although we do not know whether this correctly spliced mRNA would give rise to a stable or functional protein, the p.Gly212 position is conserved (present in all vertebrates to *Danio rerio*) and p.Gly212Arg is predicted to be probably damaging (PolyPhen-2; score of 1). Together, our in vitro studies revealed the production of two different *IFT140* mRNA species, one resulting in an unambiguous loss of function and another harboring a candidate pathogenic missense variant.

We have shown previously that suppression of both anterograde and retrograde IFT proteins in zebrafish results in gastrulation defects in mid-somite stage embryos [4, 23, 47, 48] and that an assay of missense variants through an in vivo complementation approach [49] is highly concordant with in vitro cilia length complementation studies [23, 47]. We therefore used this approach to test the pathogenic potential of *IFT140* c.634G>A; p.Gly212Arg. Reciprocal BLAST identified a single zebrafish *IFT140* ortholog (59% identity; 76% similarity versus human IFT140), which we suppressed with a previously validated morpholino (MO) [50] that targets the exon 2 splice donor site of the *D. rerio* transcript (GRCz10; ENSDART00000129889.4; Additional file 1: Figure S2a-c). We injected progressively increasing doses of splice-blocking MO (sb-MO) into embryos at the one-to-four-cell stage (3, 6, and 9 ng; $n = 59–70$ embryos/injection batch, repeated twice; Additional file 1: Figure S2d). First, we scored live embryos for gastrulation defects at the nine somite stages according to established scoring criteria (class I is defined as having a shortened anterior-posterior body axis with small anterior structures and mild somite defects; class II is defined as having a severely shortened body axis, severely affected anterior structures, broadening and kinking of the notochord, thinning and widening of the somites, and tail extension defects [4, 23, 47]; Fig. 3c). We observed dose-dependent defects (Additional file 1: Figure S2d). Co-injection of 100 pg WT human *IFT140* mRNA with 8 ng sb-MO resulted in a significant rescue of gastrulation defects (21 versus 62% affected for WT rescue compared to MO alone; $p < 0.0001$; Fig. 3d). Next, we compared the ability of equivalent doses of *IFT140* WT or p.G212R-encoding mRNA to rescue the MO-induced gastrulation defects observed at the mid-somite stage. Embryos injected with the *IFT140* p.G212R-encoding change were worse than that of WT rescue, but improved from the MO-injected batches alone, indicating that it is a likely partial loss of function variant ($p < 0.0001$ for all pairwise tests; $n = 51–78$ embryos/injection; repeated twice masked to injection cocktail; Fig. 3d). We saw no significant difference between WT and p.Gly212Arg-encoding mRNA alone. In sum, these data suggest that full-length mRNA encoding IFT140 p.Gly212Arg has impaired function.

Whole exome sequencing informs the origin of the Mendelian error

The stretch of homozygosity on 16p13 was suggestive of a partial uniparental isodisomy, and identification of a maternal, but not paternal, carrier of the *IFT140* c.634G > A; p.Gly212Arg allele was consistent with this hypothesis. To explore further the basis of this chromosomal event, we conducted trio-based whole sequencing. This experiment offered us the opportunity not only to determine the parental origin of the proband chromosome 16 but also to test whether other variants elsewhere in the exome might contribute to proband phenotype.

We generated 100-bp paired-end reads from exome capture libraries corresponding to each of the proband, mother, and father and conducted variant calling and alignment as described [51]. All three exomes exceeded our quality control thresholds with an average coverage of 117x, and greater than 95% of target bases with ≥20x coverage (Additional file 1: Table S2). We filtered variant data from each individual to retain functional variants (predicted to alter mRNA splicing or protein sequence) with a MAF <1% in two public databases, the National Heart, Lung and Blood Institute (NHLBI) Exome Sequencing project ($n = 6500$ exomes) [52], and the Atherosclerosis Risk in Communities cohort (ARIC; $n = 2300$ exomes) [53]. Next, we compared trio exomes to identify a final candidate gene list harboring rare homozygous, compound heterozygous, X-linked, or de novo variants. Reassuringly, we identified the same homozygous c.634G>A variant in *IFT140* discovered previously by Sanger sequencing. We detected two additional rare variants within the ~20-Mb stretch of homozygosity on 16p13: a missense variant in *ABCA3* (c.4420C>T, p.Arg1474Trp) and another change in *ABCC1* (c.569G>A, p.Cys190Tyr; Additional file 1: Table S3). We also identified compound heterozygous missense variants in *ARHGEF10* (c.1601_1603delTGA, p.Met535del het; c.3967C>T, p.Arg1323Trp het). Although we cannot rule out the possibility that these three genes might contribute to the proband phenotype, they harbor variants of unclear significance that were either not associated previously with any human condition (*ABCC1* and *ARHGEF10*) or have reported roles in human genetic disorders that do not overlap with the proband phenotype (neonatal respiratory distress, *ABCA3* [54, 55]). Thus, *IFT140* remained the most plausible genetic driver of disease.

Finally, we compared chromosome (chr) 16 variant calls across the trio (Fig. 2d). To eliminate sequencing artifacts, we set a stringent single nucleotide variant (SNV) call threshold at ≥10 exome sequencing reads and a quality score >30; 823 of such variants were present in the proband with sufficient coverage to also call the parental genotypes. The majority of variants were uninformative because (1) the genotype was identical across all three individuals, (2) proband genotype was homozygous and both parental genotypes were heterozygous, or (3) proband genotype was heterozygous and one of the two parents was heterozygous. Next, we asked how many SNVs were consistent with the Mendelian violation paradigm observed for *IFT140* c.634G>A. Not only did we identify 76 of such variants within the 16p13 region reported by the oligo-SNP array, but we also observed another 15 variants on the remainder of 16p and 20 variants on 16q that were homozygous in the proband, heterozygous in the maternal sample, and homozygous reference in the father (Fig. 2d). Finally, we tested whether the proband harbored variants that could only have been inherited paternally (not present in the maternal sample); no variants fulfilled these criteria. Importantly, we found an equivalent contribution of rare variation (<1% MAF) from each parent on all other autosomes, excluding the possibility of non-paternity. In sum, our exome data indicate that the proband has a chr16 maternal heterodisomy, with a segmental isodisomy at 16p13 and suggest that an early error in meiosis occurred in the maternal gamete.

Discussion

Here, we report a case with phenotypic features and a molecular lesion consistent with Mainzer-Saldino syndrome (MZSDS). This disorder, known historically as conorenal dysplasia, is a rare disease characterized by phalangeal cone-shaped epiphyses, chronic renal disease, retinal dystrophy, and mild abnormalities of the proximal femur [56]. Other features may include short stature, hepatic fibrosis, and cerebellar ataxia. MZSDS was first classified as a ciliopathy when recessive mutations in *IFT140* were reported to be causative [24], and it is now known also to be caused by dysfunction of *IFT172* [26]. Three individuals from two unrelated skeletal ciliopathy pedigrees also harbor c.634G>A; p.Gly212Arg, but this allele is in trans with likely null mutations (frameshifting (MZSDS) or canonical splice mutation (JATD)) [24]. Based on this observation, we predict that p.Gly212Arg retains at least partial function. This is supported by reports from IFT gene mutations in humans in which two loss of function mutations are likely incompatible with life past early embryogenesis; recessive null mutations are typically found in *trans* with hypomorphic changes [15–19, 23–27, 43–47, 57, 58]. However, further studies on the protein level will be required to determine whether p.Gly212Arg produces a stable and functional peptide. Although the phalangeal cone-shaped epiphyses, renal phenotypes, and retinal degeneration are consistent between our case and the two reported siblings in family V with MZSDS [3], we note the absence of any overt craniofacial dysmorphisms or craniosynostosis in the case reported here (Table 1). Based on the genetic complexity of the ciliopathies, stochastic, or mutational burden in additional ciliary genes might be contributing to these differences [13].

Our report also represents a rare case of a ciliopathy caused by a segmental uniparental isodisomy. The incidence of uniparental disomy (UPD), in which both homologs of a chromosomal region are inherited from one parent, is not exceedingly rare (estimated ~1:3500 live births [59]) and may not necessarily result in pathology. However, uniparental isodisomy, i.e., the inheritance of a single copy of a parental chromosome or chromosomal segment with concomitant homozygosity, can potentially unmask recessive deleterious alleles inherited from a carrier parent. This phenomenon was first reported in cystic fibrosis in 1988 and is now a recognized, albeit rare, mechanistic basis for inherited disease [60]. There are multiple mechanisms proposed to underlie UPD [61]. Based on the observation of maternal chr16 heterodisomy with isodisomy at 16p13, we consider two possibilities originating from meiosis I error to be the most likely (1) gamete complementation or (2) trisomy rescue. In gamete complementation, fertilization of a disomic chr16 maternal gamete and a nullisomic chr16 paternal gamete would have occurred. In the latter possibility, a disomic chr16 maternal gamete would have been fertilized by a normal male gamete, and a mitotic rescue event would have rebalanced the chromosome number. To determine the likelihood of a trisomy rescue, we evaluated the exome reads covering the 76 homozygous variants on 16p13 in the proband (Fig. 2b, d) for evidence of mosaicism. The majority of variants harbored exclusively alternate reads (49/76; 64%); however, the remainder (27/76; 36%) had 1–3 reference reads at positions annotated as homozygous (Additional file 1: Figure S3). Mosaicism, if any, is likely to be low in the proband, and deeper sequencing would be required to confirm the possibility of a trisomy rescue.

Conclusions

In summary, we report the successful combination of clinical evaluation and research-based genetic and functional testing to render a molecular diagnosis of Mainzer-Saldino syndrome.

Additionally, we report a rare instance of a uniparental isodisomy unmasking a deleterious mutation to cause a ciliary disorder, highlighting the importance of assessing Mendelian errors in both chromosomal array and trio-based whole exome sequencing studies.

Methods

Clinical genetic testing

Panel sequencing was conducted in compliance with the Clinical Laboratory Improvement Amendments (CLIA) regulations as part of standard of care.

Bidirectional Sanger sequencing of coding regions and intron-exon junctions was performed using standard methodology as part of a nephrotic syndrome panel (*NPHS2, NPHS1, LAMB2, WT1, ACTN4,* and *TRPC6*; Athena Diagnostics, Worcester, MA) or nephronophthisis panel (*GLIS2, IQCB1, NPHP3, SDCCAG8,* and *INVS*; Prevention Genetics, Marshfield, WI). All test results were reviewed, interpreted, and reported by ACMG-certified Clinical Molecular Geneticists. Chromosome microarray analysis was also performed in a CLIA environment as part of standard of care (Quest Diagnostics, Chantilly, VA). The oligo-SNP array (Affymetrix 6.0) contained over 1.8 million probes, including 900,000 copy-number probes and 900,000 SNP probes with an average inter-probe distance of 700bp; calls were made against the GRCh37/hg19 human genome assembly. Thresholds for genome-wide screening were set at >200 kb for gains and >50 kb for losses, and the threshold used for calling regions of homozygosity was set at >3–5 Mb.

Bidirectional sequencing of IFT140

Subsequent to informed consent for participation in research, we obtained a blood sample from each member of the trio and extracted DNA from lymphocytes (Gentra Puregene Blood Kit, Qiagen). We PCR-amplified coding regions and intron-exon boundaries and conducted bidirectional cycle sequencing with BigDye v3.1 chemistry on an ABI3730 capillary sequencer. Sequencing reads were aligned to reference and analyzed with Lasergene (DNASTAR) software.

Lymphocyte cell culture and mRNA splicing studies

B cell lymphocytes were separated from whole-blood samples and were transformed with Epstein-Barr virus (EBV). We cultured cells in RPMI-1640 media supplemented with 10% fetal bovine serum. Wild type cell lines originated from the NHGRI Sample Repository for Human Genetic Research (GM19153; Coriell Institute). We extracted total RNA from LCLs using Trizol reagent (ThermoFisher), and 1 μg was DNAse treated and reverse transcribed using the QuantiTect Reverse Transcription kit (Qiagen). *IFT140* exons 4–8 were amplified and subjected to electrophoresis on a 2% gel, and resulting fragments were gel purified (QIAquick Gel Extraction kit; Qiagen) and sequenced.

Whole exome sequencing

We generated capture libraries and 100-bp paired-end reads essentially as described [51]. Briefly, we fragmented genomic DNA by sonication, ligated it to Illumina multiplexing PE adapters, and PCR-amplified samples using primers with sequencing indexes. The pre-capture library was then hybridized to biotin-labeled VCRome 2.1 exome probes. Post-capture library DNA was then subjected to sequence analysis on an Illumina Hiseq platform to generate ~9–12 Gb of data. Output data were converted from bcl files to FastQ files by Illumina CASAVA 1.8 software and mapped to the hg19 reference genome by the Burrows Wheeler Alignment (BWA) program. Variant calling was performed using Atlas-SNP and Atlas-Indel [62]. For chr16 variant comparisons, we used Enlis Genome Research software to set thresholds (≥10 exome sequencing reads and a quality score >30) and extract genotype lists for each of the proband, maternal, and paternal exome data. Variant read depth and variant calls were confirmed visually with the Integrated Genomics Viewer (IGV). Data were visualized with PhenoGram (Center for System Genomics, Pennsylvania State University; http://visualization.ritchielab.psu.edu/).

Zebrafish manipulation and injections

A splice-blocking morpholino (Gene Tools) targeting *ift140* (5′-AGTGATCATGTCTTACCTGCTGCAG-′3) [50] was injected at the indicated concentration into wild-type zebrafish embryos at the one-to-four-cell stage (1 nl/embryo; all injections repeated at least twice, with masked scoring). To validate MO efficiency, we harvested embryos in Trizol reagent at the eight-to-ten-somite stage, extracted total RNA, and conducted RT-PCR studies as described above for human lymphocyte cell lines. To generate human wild-type and mutant mRNA, we first cloned the full-length open reading frame (ORF) (IOH56213; Ultimate ORF, ThermoFisher) into a Gateway-compatible pCS2+ plasmid using LR clonase II (ThermoFisher), linearized with Not I, and performed in vitro transcription with the mMessage mMachine SP6 transcription kit (ThermoFisher). For dose curve or rescue experiments, embryos were scored live at the eight-to-ten-somite stage using previously defined objective scoring criteria [4, 23, 47]. Images were acquired on a Nikon AZ100 microscope facilitated by NIS Elements software. χ^2 tests were used to perform statistical comparisons of embryo batches in a pairwise manner.

Additional file

Additional file 1: Table S1. Clinical genetic testing summary. **Table S2.** Whole exome sequencing capture statistics. **Table S3.** Final variant list following trio-based exome analysis (<1% minor allele frequency (MAF), homozygous, compound heterozygous, de novo, or X-linked). **Figure S1.** Fundus imaging and full field electroretinograms indicate retinal dystrophy. a. Fundus imaging of the proband at 5 years 6 months in the right (top) and left (bottom) eyes show mottling of the retinal pigmentary epithelium. b. Photopic (light adapted) testing of cone function demonstrated reduced a-wave implicit time and amplitudes with slowed implicit times for the b-wave. c. Scotopic (dark adapted) testing of rod function demonstrated slower implicit time and diminished amplitudes for both a- and b-waves. The a-wave was substantially delayed with reduced incremental increases in a-wave and b-wave. **Figure S2.** Validation of the *ift140* splice-blocking (sb) morpholino (MO). a. Schematic of the D. rerio *ift140* locus at chr24:37,937,737-38,013,596 (GRCz10; top) and translated protein (bottom). Exons are depicted as green boxes; untranslated regions are shown as white boxes (ENSDART00000129889.4). The sb-MO targets the splice donor of exon (Ex) 2 (red box). Protein schematic (blue) indicates predicted WD40 repeat (WD40) and tetratricopeptide-like helical domains (TPR). b. *ift140* sb-MO induces aberrant splicing of endogenous transcript. *ift140* transcript was evaluated by RT-PCR; embryos were injected with 9 ng MO at the one-to-four-cell stage and harvested for RNA extraction at the mid-somitic stage. Resulting cDNA was amplified with primers flanking the MO target site (arrows in panel a), and migrated on a 2.5% agarose gel. β–actin was used as a control to ensure RNA integrity. c. Chromatograms indicate splicing defects in *ift140* mRNA in morphants. Sequence analysis of purified PCR product indicates that the majority of *ift140* message is missing exon 2 (bottom), which contains the translation initiation codon. Modest amounts of wild type (wt; center) and mRNA containing a 33 nucleotide in-frame intronic insertion (top) are also detectable. d. Representative dose curve of the *ift140* sb-MO. Embryo batches were injected at the one-to-four-cell stage with increasing amounts of *ift140* sb-MO and scored for gastrulation defects at the mid-somitic stage (eight to ten somites). Embryos were scored live according to previously established phenotypic criteria: class I, modest shortening of the body axis and reduction in size of anterior structures; class II, severe shortening of the body axis and decreased anterior structures accompanied by broadening and/or kinking of the notochord and thinning of the somites (See Figure 3; $n = 59–70$ embryos/injection; repeated twice). **Figure S3.** Read depth and variant calls of the homozygous variants called on 16p13. Plot of the 76 maternally inherited homozygous variants identified in the 16p13 region of the proband. Red indicates alternate (alt) reads; blue indicates reference (ref) reads; 49/76 (64%) are comprised only of alternate reads; 27/76 (36%) had one to three reference reads at positions annotated as homozygous.

Abbreviations
bp: Base pair; chr: Chromosome; IFT: Intraflagellar transport; LCL: Lymphocyte cell line; MO: Morpholino; ORF: Open reading frame; SNP: Single nucleotide polymorphism; SNV: Single nucleotide variant; WT: Wild-type

Acknowledgements
We thank the family for their active engagement and enthusiasm for our research. The authors would like to honor Dr. Virginia K. Proud for her input on this work and commitment to patient care. We are grateful to Caroline B. Drucker for her assistance with the manuscript preparation and to our colleagues in the Center for Human Disease Modeling for the thoughtful discussions.

Funding
This work was funded by NIH grants DK075972, HD042601 (N.K.); DK072301 (N.K. and E.E.D.); and EY021872 (E.E.D.).

Authors' contributions
BMH, NK, and EED designed the study. BMH, EC, and SSV conducted patient phenotyping and clinical genetic testing. AS enrolled the family in genetics research and obtained biospecimens from the family. JRW, NK, and EED performed research-based genetic studies. CG and EED performed zebrafish studies. BMH, SSV, NK, and EED wrote the manuscript, which was revised and approved by all authors.

Consent for publication
Written informed consent for publication of clinical details and clinical images was obtained from the parent of the proband.

Competing interests
SSV is on the medical advisory board for Ambry Genetics. NK is a paid consultant for and holds significant stock of Rescindo Therapeutics, Inc. All other authors confirmed that they have no competing interests.

Author details
[1]Division of Medical Genetics and Metabolism, Children's Hospital of The King's Daughters, Norfolk, VA 23507, USA. [2]Department of Medical & Molecular Genetics, Riley Hospital for Children at IU Health, Indiana University School of Medicine, Indianapolis, IN 46202, USA. [3]Center for Human Disease Modeling, Duke University Medical Center, Durham, NC 27701, USA. [4]Institute of Genetics and Molecular and Cellular Biology, 67404 Illkirch, France. [5]Department of Ophthalmology, Children's Hospital of the King's Daughters, Norfolk, VA 23507, USA. [6]Department of Pediatrics, Eastern Virginia Medical School, Norfolk, VA 23507, USA.

References
1. Katsanis SH, Katsanis N. Molecular genetic testing and the future of clinical genomics. Nat Rev Genet. 2013;14(6):415–26.
2. Bruel AL, Franco B, Duffourd Y, Thevenon J, Jego L, Lopez E, Deleuze JF, Doummar D, Giles RH, Johnson CA, et al. Fifteen years of research on oral-facial-digital syndromes: from 1 to 16 causal genes. J Med Genet. 2017;54(6):371–80.
3. Huber C, Cormier-Daire V. Ciliary disorder of the skeleton. Am J Med Genet C Semin Med Genet. 2012;160C(3):165–74.
4. Lindstrand A, Frangakis S, Carvalho CM, Richardson EB, McFadden KA, Willer JR, Pehlivan D, Liu P, Pediaditakis IL, Sabo A, et al. Copy-number variation contributes to the mutational load of Bardet-Biedl syndrome. Am J Hum Genet. 2016;99(2):318–36.
5. Szymanska K, Hartill VL, Johnson CA. Unraveling the genetics of Joubert and Meckel-Gruber syndromes. J Pediatr Genet. 2014;3(2):65–78.
6. Badano JL, Mitsuma N, Beales PL, Katsanis N. The ciliopathies: an emerging class of human genetic disorders. Annu Rev Genomics Hum Genet. 2006;7:125–48.
7. Hildebrandt F, Benzing T, Katsanis N. Ciliopathies. N Engl J Med. 2011; 364(16):1533–43.
8. Waters AM, Beales PL. Ciliopathies: an expanding disease spectrum. Pediatr Nephrol. 2011;26(7):1039–56.
9. Gherman A, Davis EE, Katsanis N. The ciliary proteome database: an integrated community resource for the genetic and functional dissection of cilia. Nat Genet. 2006;38(9):961–2.
10. Rosenbaum JL, Witman GB. Intraflagellar transport. Nat Rev Mol Cell Biol. 2002;3(11):813–25.
11. Nachury MV, Seeley ES, Jin H. Trafficking to the ciliary membrane: how to get across the periciliary diffusion barrier? Annu Rev Cell Dev Biol. 2010;26:59–87.
12. Garcia-Gonzalo FR, Reiter JF. Open Sesame: How Transition Fibers and the Transition Zone Control Ciliary Composition. Cold Spring Harb Perspect Biol. 2017;9(2).
13. Davis EE, Katsanis N. The ciliopathies: a transitional model into systems biology of human genetic disease. Curr Opin Genet Dev. 2012;22(3):290–303.
14. Levin LS, Perrin JC, Ose L, Dorst JP, Miller JD, McKusick VA. A heritable syndrome of craniosynostosis, short thin hair, dental abnormalities, and short limbs: cranioectodermal dysplasia. J Pediatr. 1977;90(1):55–61.
15. Arts HH, Bongers EM, Mans DA, van Beersum SE, Oud MM, Bolat E, Spruijt L, Cornelissen EA, Schuurs-Hoeijmakers JH, de Leeuw N, et al. C14ORF179 encoding IFT43 is mutated in Sensenbrenner syndrome. J Med Genet. 2011;48(6):390–5.
16. Gilissen C, Arts HH, Hoischen A, Spruijt L, Mans DA, Arts P, van Lier B,

Steehouwer M, van Reeuwijk J, Kant SG, et al. Exome sequencing identifies WDR35 variants involved in Sensenbrenner syndrome. Am J Hum Genet. 2010;87(3):418–23.

17. Walczak-Sztulpa J, Eggenschwiler J, Osborn D, Brown DA, Emma F, Klingenberg C, Hennekam RC, Torre G, Garshasbi M, Tzschach A, et al. Cranioectodermal dysplasia, Sensenbrenner syndrome, is a ciliopathy caused by mutations in the IFT122 gene. Am J Hum Genet. 2010;86(6):949–56.
18. Bredrup C, Saunier S, Oud MM, Fiskerstrand T, Hoischen A, Brackman D, Leh SM, Midtbo M, Filhol E, Bole-Feysot C, et al. Ciliopathies with skeletal anomalies and renal insufficiency due to mutations in the IFT-A gene WDR19. Am J Hum Genet. 2011;89(5):634–43.
19. Bayat A, Kerr B, Douzgou S, Study DDD. The evolving craniofacial phenotype of a patient with Sensenbrenner syndrome caused by IFT140 compound heterozygous mutations. Clin Dysmorphol. 2017. doi:10.1097/MCD.0000000000000169. [Epub ahead of print].
20. Jeune M, Beraud C, Carron R. Asphyxiating thoracic dystrophy with familial characteristics. Arch Fr Pediatr. 1955;12(8):886–91.
21. Dagoneau N, Goulet M, Genevieve D, Sznajer Y, Martinovic J, Smithson S, Huber C, Baujat G, Flori E, Tecco L, et al. DYNC2H1 mutations cause asphyxiating thoracic dystrophy and short rib-polydactyly syndrome, type III. Am J Hum Genet. 2009;84(5):706–11.
22. Merrill AE, Merriman B, Farrington-Rock C, Camacho N, Sebald ET, Funari VA, Schibler MJ, Firestein MH, Cohn ZA, Priore MA, et al. Ciliary abnormalities due to defects in the retrograde transport protein DYNC2H1 in short-rib polydactyly syndrome. Am J Hum Genet. 2009;84(4):542–9.
23. Davis EE, Zhang Q, Liu Q, Diplas BH, Davey LM, Hartley J, Stoetzel C, Szymanska K, Ramaswami G, Logan CV, et al. TTC21B contributes both causal and modifying alleles across the ciliopathy spectrum. Nat Genet. 2011;43(3):189–96.
24. Perrault I, Saunier S, Hanein S, Filhol E, Bizet AA, Collins F, Salih MA, Gerber S, Delphin N, Bigot K, et al. Mainzer-Saldino syndrome is a ciliopathy caused by IFT140 mutations. Am J Hum Genet. 2012;90(5):864–70.
25. Beales PL, Bland E, Tobin JL, Bacchelli C, Tuysuz B, Hill J, Rix S, Pearson CG, Kai M, Hartley J, et al. IFT80, which encodes a conserved intraflagellar transport protein, is mutated in Jeune asphyxiating thoracic dystrophy. Nat Genet. 2007;39(6):727–9.
26. Halbritter J, Bizet AA, Schmidts M, Porath JD, Braun DA, Gee HY, McInerney-Leo AM, Krug P, Filhol E, Davis EE, et al. Defects in the IFT-B component IFT172 cause Jeune and Mainzer-Saldino syndromes in humans. Am J Hum Genet. 2013;93(5):915–25.
27. Bujakowska KM, Zhang Q, Siemiatkowska AM, Liu Q, Place E, Falk MJ, Consugar M, Lancelot ME, Antonio A, Lonjou C, et al. Mutations in IFT172 cause isolated retinal degeneration and Bardet-Biedl syndrome. Hum Mol Genet. 2015;24(1):230–42.
28. Hildebrandt F, Attanasio M, Otto E. Nephronophthisis: disease mechanisms of a ciliopathy. J Am Soc Nephrol. 2009;20(1):23–35.
29. Jennette JC, Hipp CG. C1q nephropathy: a distinct pathologic entity usually causing nephrotic syndrome. Am J Kidney Dis. 1985;6(2):103–10.
30. Boute N, Gribouval O, Roselli S, Benessy F, Lee H, Fuchshuber A, Dahan K, Gubler MC, Niaudet P, Antignac C. NPHS2, encoding the glomerular protein podocin, is mutated in autosomal recessive steroid-resistant nephrotic syndrome. Nat Genet. 2000;24(4):349–54.
31. Hasselbacher K, Wiggins RC, Matejas V, Hinkes BG, Mucha B, Hoskins BE, Ozaltin F, Nurnberg G, Becker C, Hangan D, et al. Recessive missense mutations in LAMB2 expand the clinical spectrum of LAMB2-associated disorders. Kidney Int. 2006;70(6):1008–12.
32. Jeanpierre C, Denamur E, Henry I, Cabanis MO, Luce S, Cecille A, Elion J, Peuchmaur M, Loirat C, Niaudet P, et al. Identification of constitutional WT1 mutations, in patients with isolated diffuse mesangial sclerosis, and analysis of genotype/phenotype correlations by use of a computerized mutation database. Am J Hum Genet. 1998;62(4):824–33.
33. Kaplan JM, Kim SH, North KN, Rennke H, Correia LA, Tong HQ, Mathis BJ, Rodriguez-Perez JC, Allen PG, Beggs AH, et al. Mutations in ACTN4, encoding alpha-actinin-4, cause familial focal segmental glomerulosclerosis. Nat Genet. 2000;24(3):251–6.
34. Kestila M, Lenkkeri U, Mannikko M, Lamerdin J, McCready P, Putaala H, Ruotsalainen V, Morita T, Nissinen M, Herva R, et al. Positionally cloned gene for a novel glomerular protein—nephrin—is mutated in congenital nephrotic syndrome. Mol Cell. 1998;1(4):575–82.
35. Reiser J, Polu KR, Moller CC, Kenlan P, Altintas MM, Wei C, Faul C, Herbert S, Villegas I, Avila-Casado C, et al. TRPC6 is a glomerular slit diaphragm-associated channel required for normal renal function. Nat Genet. 2005;37(7):739–44.
36. Schumacher V, Scharer K, Wuhl E, Altrogge H, Bonzel KE, Guschmann M, Neuhaus TJ, Pollastro RM, Kuwertz-Broking E, Bulla M, et al. Spectrum of early onset nephrotic syndrome associated with WT1 missense mutations. Kidney Int. 1998;53(6):1594–600.
37. Winn MP, Conlon PJ, Lynn KL, Farrington MK, Creazzo T, Hawkins AF, Daskalakis N, Kwan SY, Ebersviller S, Burchette JL, et al. A mutation in the TRPC6 cation channel causes familial focal segmental glomerulosclerosis. Science. 2005;308(5729):1801–4.
38. Attanasio M, Uhlenhaut NH, Sousa VH, O'Toole JF, Otto E, Anlag K, Klugmann C, Treier AC, Helou J, Sayer JA, et al. Loss of GLIS2 causes nephronophthisis in humans and mice by increased apoptosis and fibrosis. Nat Genet. 2007;39(8):1018–24.
39. Olbrich H, Fliegauf M, Hoefele J, Kispert A, Otto E, Volz A, Wolf MT, Sasmaz G, Trauer U, Reinhardt R, et al. Mutations in a novel gene, NPHP3, cause adolescent nephronophthisis, tapeto-retinal degeneration and hepatic fibrosis. Nat Genet. 2003;34(4):455–9.
40. Otto EA, Hurd TW, Airik R, Chaki M, Zhou W, Stoetzel C, Patil SB, Levy S, Ghosh AK, Murga-Zamalloa CA, et al. Candidate exome capture identifies mutation of SDCCAG8 as the cause of a retinal-renal ciliopathy. Nat Genet. 2010;42(10):840–50.
41. Otto EA, Loeys B, Khanna H, Hellemans J, Sudbrak R, Fan S, Muerb U, O'Toole JF, Helou J, Attanasio M, et al. Nephrocystin-5, a ciliary IQ domain protein, is mutated in Senior-Loken syndrome and interacts with RPGR and calmodulin. Nat Genet. 2005;37(3):282–8.
42. Otto EA, Schermer B, Obara T, O'Toole JF, Hiller KS, Mueller AM, Ruf RG, Hoefele J, Beekmann F, Landau D, et al. Mutations in INVS encoding inversin cause nephronophthisis type 2, linking renal cystic disease to the function of primary cilia and left-right axis determination. Nat Genet. 2003;34(4):413–20.
43. Bifari IN, Elkhamary SM, Bolz HJ, Khan AO. The ophthalmic phenotype of IFT140-related ciliopathy ranges from isolated to syndromic congenital retinal dystrophy. Br J Ophthalmol. 2016;100(6):829–33.
44. Hull S, Owen N, Islam F, Tracey-White D, Plagnol V, Holder GE, Michaelides M, Carss K, Raymond FL, Rozet JM, et al. Nonsyndromic retinal dystrophy due to bi-allelic mutations in the ciliary transport gene IFT140. Invest Ophthalmol Vis Sci. 2016;57(3):1053–62.
45. Xu M, Yang L, Wang F, Li H, Wang X, Wang W, Ge Z, Wang K, Zhao L, Li H, et al. Mutations in human IFT140 cause non-syndromic retinal degeneration. Hum Genet. 2015;134(10):1069–78.
46. Pena-Padilla C, Marshall CR, Walker S, Scherer SW, Tavares-Macias G, Razo-Jimenez G, Bobadilla-Morales L, Acosta-Fernandez E, Corona-Rivera A, Mendoza-Londono R et al. Compound heterozygous mutations in the IFT140 gene cause Opitz trigonocephaly C syndrome in a patient with typical features of a ciliopathy. Clin Genet. 2016.
47. McIntyre JC, Davis EE, Joiner A, Williams CL, Tsai IC, Jenkins PM, McEwen DP, Zhang L, Escobado J, Thomas S, et al. Gene therapy rescues cilia defects and restores olfactory function in a mammalian ciliopathy model. Nat Med. 2012;18(9):1423–8.
48. Li C, Inglis PN, Leitch CC, Efimenko E, Zaghloul NA, Mok CA, Davis EE, Bialas NJ, Healey MP, Heon E, et al. An essential role for DYF-11/MIP-T3 in assembling functional intraflagellar transport complexes. PLoS Genet. 2008;4(3):e1000044.
49. Niederriter AR, Davis EE, Golzio C, Oh EC, Tsai I-C, Katsanis N. In Vivo Modeling of the Morbid Human Genome using Danio rerio. JoVE. 2013;(78):e50338.
50. Tsujikawa M, Malicki J. Intraflagellar transport genes are essential for differentiation and survival of vertebrate sensory neurons. Neuron. 2004;42(5):703–16.
51. Bainbridge MN, Wang M, Wu Y, Newsham I, Muzny DM, Jefferies JL, Albert TJ, Burgess DL, Gibbs RA. Targeted enrichment beyond the consensus coding DNA sequence exome reveals exons with higher variant densities. Genome Biol. 2011;12(7):R68.
52. Tennessen JA, Bigham AW, O'Connor TD, Fu W, Kenny EE, Gravel S, McGee

S, Do R, Liu X, Jun G, et al. Evolution and functional impact of rare coding variation from deep sequencing of human exomes. Science. 2012;337(6090):64–9.
53. Morrison AC, Bare LA, Chambless LE, Ellis SG, Malloy M, Kane JP, Pankow JS, Devlin JJ, Willerson JT, Boerwinkle E. Prediction of coronary heart disease risk using a genetic risk score: the Atherosclerosis Risk in Communities Study. Am J Epidemiol. 2007;166(1):28–35.
54. Shulenin S, Nogee LM, Annilo T, Wert SE, Whitsett JA, Dean M. ABCA3 gene mutations in newborns with fatal surfactant deficiency. N Engl J Med. 2004; 350(13):1296–303.
55. Verhoeven K, De Jonghe P, Van de Putte T, Nelis E, Zwijsen A, Verpoorten N, De Vriendt E, Jacobs A, Van Gerwen V, Francis A, et al. Slowed conduction and thin myelination of peripheral nerves associated with mutant rho guanine-nucleotide exchange factor 10. Am J Hum Genet. 2003;73(4):926–32.
56. Saldino RM, Mainzer F. Cone-shaped epiphyses (CSE) in siblings with hereditary renal disease and retinitis pigmentosa. Radiology. 1971;98(1):39–45.
57. Cavalcanti DP, Huber C, Sang KH, Baujat G, Collins F, Delezoide AL, Dagoneau N, Le Merrer M, Martinovic J, Mello MF, et al. Mutation in IFT80 in a fetus with the phenotype of Verma-Naumoff provides molecular evidence for Jeune-Verma-Naumoff dysplasia spectrum. J Med Genet. 2011;48(2):88–92.
58. Mill P, Lockhart PJ, Fitzpatrick E, Mountford HS, Hall EA, Reijns MA, Keighren M, Bahlo M, Bromhead CJ, Budd P, et al. Human and mouse mutations in WDR35 cause short-rib polydactyly syndromes due to abnormal ciliogenesis. Am J Hum Genet. 2011;88(4):508–15.
59. Robinson WP. Mechanisms leading to uniparental disomy and their clinical consequences. Bioessays. 2000;22(5):452–9.
60. Spence JE, Perciaccante RG, Greig GM, Willard HF, Ledbetter DH, Hejtmancik JF, Pollack MS, O'Brien WE, Beaudet AL. Uniparental disomy as a mechanism for human genetic disease. Am J Hum Genet. 1988;42(2):217–26.
61. Yamazawa K, Ogata T, Ferguson-Smith AC. Uniparental disomy and human disease: an overview. Am J Med Genet C Semin Med Genet. 2010;154C(3):329–34.
62. Shen Y, Wan Z, Coarfa C, Drabek R, Chen L, Ostrowski EA, Liu Y, Weinstock GM, Wheeler DA, Gibbs RA, et al. A SNP discovery method to assess variant allele probability from next-generation resequencing data. Genome Res. 2010;20(2):273–80.

10

Inferring clonal structure in HTLV-1-infected individuals: towards bridging the gap between analysis and visualization

Amir Farmanbar[1,2†], Sanaz Firouzi[1*†], Wojciech Makałowski[3], Masako Iwanaga[4], Kaoru Uchimaru[1], Atae Utsunomiya[5], Toshiki Watanabe[1,6] and Kenta Nakai[1,2*]

Abstract

Background: Human T cell leukemia virus type 1 (HTLV-1) causes adult T cell leukemia (ATL) in a proportion of infected individuals after a long latency period. Development of ATL is a multistep clonal process that can be investigated by monitoring the clonal expansion of HTLV-1-infected cells by isolation of provirus integration sites. The clonal composition (size, number, and combinations of clones) during the latency period in a given infected individual has not been clearly elucidated.

Methods: We used high-throughput sequencing technology coupled with a tag system for isolating integration sites and measuring clone sizes from 60 clinical samples. We assessed the role of clonality and clone size dynamics in ATL onset by modeling data from high-throughput monitoring of HTLV-1 integration sites using single- and multiple-time-point samples.

Results: From four size categories analyzed, we found that big clones (B; 513–2048 infected cells) and very big clones (VB; >2048 infected cells) had prognostic value. No sample harbored two or more VB clones or three or more B clones. We examined the role of clone size, clone combination, and the number of integration sites in the prognosis of infected individuals. We found a moderate reverse correlation between the total number of clones and the size of the largest clone. We devised a data-driven model that allows intuitive representation of clonal composition.

Conclusions: This integration site-based clonality tree model represents the complexity of clonality and provides a global view of clonality data that facilitates the analysis, interpretation, understanding, and visualization of the behavior of clones on inter- and intra-individual scales. It is fully data-driven, intuitively depicts the clonality patterns of HTLV-1-infected individuals and can assist in early risk assessment of ATL onset by reflecting the prognosis of infected individuals. This model should assist in assimilating information on clonal composition and understanding clonal expansion in HTLV-1-infected individuals.

Keywords: Data-driven modeling, Adult T cell leukemia, Human T cell leukemia virus type 1, Integration site, Clonal expansion, High-throughput sequencing, Prognostic indicator

* Correspondence: firouzisanaz1@gmail.com; knakai@ims.u-tokyo.ac.jp
[†]Equal contributors
[1]Department of Computational Biology and Medical Sciences, Graduate School of Frontier Sciences, The University of Tokyo, Kashiwanoha, Kashiwa-shi, Chiba, Japan
Full list of author information is available at the end of the article

Introduction

Clonal expansion in neoplasms is accepted as a general feature of a broad range of tumors and has been addressed from different perspectives [1–5]. Detecting clones, especially in the tumor's early appearance, and determining the factors that affect those clones should soon have a substantial role in early diagnosis and early therapeutic intervention [3]. Monitoring the clonal expansion of most malignancies is feasible based on the analysis of genetic abnormalities within progressing and/or already developed stages; however, tracking early events has proven challenging [6]. Similar to other types of cancer, clonal expansion of abnormal cells is a hallmark of adult T cell leukemia-lymphoma (ATL), an aggressive T cell malignancy [7]. ATL is a unique neoplasm that is directly caused by infection with human T cell leukemia virus type 1 (HTLV-1) and manifests after a long latency period [8–11]. The course of disease can be monitored from early asymptomatic stages of infection to the final stages of fully developed malignancy [12]. The HTLV-1 integration site, i.e., the position at which the provirus inserted into the host genome, defines individual virus-infected cells and can be used as a marker to characterize ATL clones, as well as to monitor the clonal composition of infected individuals [13–17]. Each host cell has a unique and single integration site [18], and a population of cells with the same integration site originated from the same ancestor cell and is considered a clone [14, 15]. The clonal composition of HTLV-1-infected individuals has been investigated by a broad range of methods, including Southern blotting [7, 13], inverse PCR [16, 19–21] and linker-mediated PCR [22, 23]. Recently, by taking advantage of next-generation sequencing (NGS) technology, researchers have devised two main methods, shear sites [14] and tag systems [15], for high-throughput analysis of clonality based on provirus integration sites. These methods enable an improved understanding of genome-wide integration sites and open up new avenues to quantify clone sizes [14, 15, 20, 24–27]. In this study, we used the data derived from the tag system, which measures clone sizes without using statistical estimations [15].

In the clinical setting, the vast amounts of data rapidly generated by NGS technologies require efficient and accurate processing, analysis, management, interpretation, visualization, and conversion to useful knowledge for rendering more accurate patient outcomes [28]. In response to this demand, interdisciplinary approaches, such as the development of mathematical models in medical research, have been successfully applied over recent years. Mathematical and computational models have invaluable potential to increase the understanding of tumor development and hold promise to suggest new ways to improve the efficacy of therapeutic interventions [4, 29, 30]. To extract reliable fundamental meaning from a mathematical model and strengthen its predictive power, real experimental data must be introduced [31, 32]. Unlike other types of malignancies for which broad ranges of mathematical models are available [33], to our knowledge, there are few mathematical models for ATL [34–41], and only one visual model illustrating clonal expansion [27]. However, access to NGS data and the ability to retrieve clone size and integration site information have provided new opportunities to model clonal expansion in ATL. We recently quantified ATL clones, defined clonality patterns and categorized observed clones using threshold criteria based on clone size [27]. The conventionally described polyclonal, oligoclonal and monoclonal patterns [13, 19] can be intuitively reflected by combinations of very small, small, big, and very big (VS, S, B, and VB) clone sizes using these categorizing criteria [27]. Moreover, we showed that clone size can be a risk indicator for development and progression of ATL. Among asymptomatic carriers (ACs), individuals with polyclonal patterns remain as ACs over time, whereas those with atypical patterns of monoclonal or oligoclonal (VB or B) clones exhibited disease progression [42].

Tree theory belongs to discrete mathematics [43, 44] and can handle the simple representation of biological data and can be used for effective data organization. Tree theory provides informative and graphical representation of concepts to readily illustrate similarities and relationships and to improve the comparison of results across samples [45]. In the present study, we developed a hierarchical tree model using both cross-sectional and longitudinal data of HTLV-1 clonal expansion. Our main goals were to display the relationship between the size and combination of clones and to visualize biological information hidden in clonality data derived from HTLV-1-infected individuals. Figure 1 illustrates a general overview of our study for representing the clonal status of each patient via hierarchical tree structures. The development of a simple model based on observed clonality data should lead to a deeper understanding of ATL clonal expansion and should enable prediction of the prognosis of infected individuals.

Methods

In total, 60 clinical samples from 29 individuals were used, including 9 cross-sectional samples [27], 4 longitudinal samples, and 47 longitudinal samples described in [20] and [42], respectively. All clinical materials were collected by the Joint Study on Predisposing Factors of ATL Development (JSPFAD) [46, 47]. HTLV-1-infected individuals were diagnosed based on the Shimoyama criteria [48]. All sample information is provided in Additional file 1: Table S1.

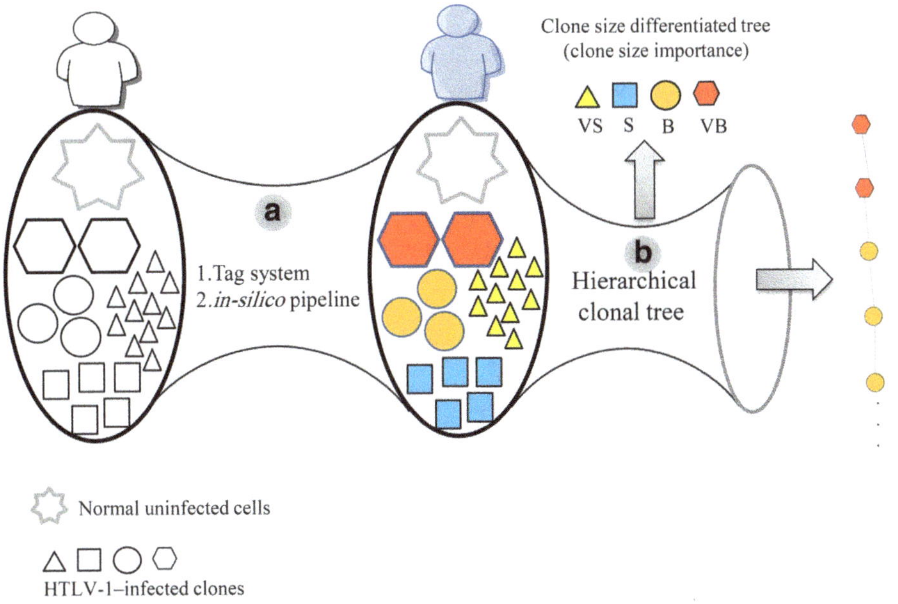

Fig. 1 Overview of clonality analysis and visualization of results. The workflow consists of two main bottlenecks. In bottleneck **a**, our original tag system was used to isolate HTLV-1 integration sites in infected individuals and measure the size of clones. Subsequently, our in silico pipeline was used to analyze the high-throughput sequencing data. In bottleneck **b**, which is the main focus of this manuscript, the observed clone sizes were classified and visualized using different shapes and colors using the concept of hierarchical tree structure

Across all samples, a tag system was used to isolate integration sites and measure clone sizes. All details for experimental design and protocols are provided in [15]. In brief, BLAST searches against the HTLV-1 long terminal repeat were used to detect viral integration sites [49]. Human DNA regions flanking the integration sites were identified by alignment against hg19 using Bowtie [50]. The quality of sequencing outputs was checked with the FastQC tool [51]. Information regarding sequencing outputs and read quality is provided in Additional file 2: Figure S1. Two million mapped reads were used to measure clone sizes. We categorized observed clone sizes into four distinct groups: very small (VS, 1–128 cells), small (S, 129–512 cells), big (B, 513–2048 cells), and very big (VB, >2048 cells) [27]. We used the ordered m-ary tree to visualize the observed data. For a detailed explanation, see Additional file 2: Figure S2.

Results

Overview of analyzed samples and results

We used clonality data from 60 samples derived from cross-sectional and longitudinal analyses of HTLV-1-infected individuals. Longitudinal monitoring of 21 cases (51 samples) was performed over a period of 15–90 months. Complete patient and sample information is provided in Additional file 1: Table S1. We followed each individual over time to capture the dynamics of clonality based on clone size and disease stage. Additional file 3: Table S2 summarizes the status of analyzed samples.

In total, nine ACs (individuals 1–9; in Additional file 1: Table S1) were analyzed at two or more time points. Five retained their AC status (individuals 1–5), whereas four progressed to ATL (individuals 6–9). All individuals who remained ACs had combinations of S and VS clones (polyclonal pattern; for clone sizes, see Additional file 1: Table S1). The size of the largest clone in individuals 1–5 at time point 1 (samples H-3, H-9, H-17, H-2, H-8) was 46, 357, 105, 77, and 388 cells, respectively, and at time point 2 (samples H-1, H-7, H-29, H-4, H-5) was 112, 433, 153, 187, and 244 cells, respectively. All samples from progressed ACs harbored B or VB clones at each analyzed time point.

In addition, 11 individuals with indolent types of ATL (smoldering (SM) or chronic) at time point 1 were analyzed, along with follow-up data at time point 2. Seven patients (individuals 10–16) were SM at time point 1. Of these, five (individuals 11–15) progressed to the chronic subtype at time point 2, one (individual 16) progressed to the acute subtype, and one (individual 10) remained SM. Four patients (individuals 17–20) were in the chronic state at time point 1. Of these, three (individuals 17–19) remained chronic and one (individual 20) progressed to the acute state at time point 2. Individual 21 was an acute patient who was monitored at 4 time points before, during, and after therapy. Finally, clonality data of some cross-sectional samples (individual 22–29) were also included.

Is there a correlation between clone size and number of integration sites?

The maximum and minimum numbers of isolated integration sites across samples were 1832 and 18, respectively. The maximum and minimum sizes of the largest clones across samples were 5377 and 46, respectively. Figure 2 illustrates the number of integration sites belonging to different clone size categories (VB, B, S, VS) in three representative individuals monitored over time. In Fig. 3, we plotted the largest clone size and number of integration sites isolated from all samples. The Pearson correlation coefficient ($R = -0.55$) indicates a moderate negative correlation. Box plots in Fig. 4 illustrate the relationship between the number of integration sites and the size of the largest clone in an individual; the samples were divided into three classes, VS-S, B, and VB. Most samples with larger clone sizes had fewer integration sites. The differences in the number of sites between the VS-S and B categories and between the VS-S and VB categories were significant ($P = 0.05$ and $P = 0.0001$, respectively; Fig. 4). The P value of the difference between B and VB categories was 0.08 (Fig. 4). Samples with larger clones had significantly fewer integration sites (Fig. 4). An exception to this trend was seen in individual 13 (samples H-40 and H-33 in Additional file 1: Table S1 and Additional file 2: Figure S3). This patient was SM at time point 1, with one VB, one S, and 1472 VS clones (H-40) and progressed to chronic at time point 2, with one VB, one S, and 1169 VS clones (H-33). Although the biological reason for the atypically large number of VS clones in this patient is unknown, this observation suggests that there is a low probability of a technical limitation in detecting background S and VS clones in the presence of B and VB clones. Thus, the presence of large clones and a small numbers of S and VS clones might be of prognostic value.

ACs who remained ACs over time had only combinations of VS and S clones (Figs. 5 and 2a). In contrast, ACs who progressed to different subtypes of ATL had B or VB clones in addition to VS or S clones. The B and VB clones in ACs at early time points were also detected during the malignant stage of the disease (Fig. 6 and Additional file 2: Figure S3). All observed clonality patterns for cross-sectional samples are shown in Additional file 2: Figure S4. Only the five largest clones in each sample are shown because significant changes in clone size occur only at this level. Below this level, only S and VS clones are observed.

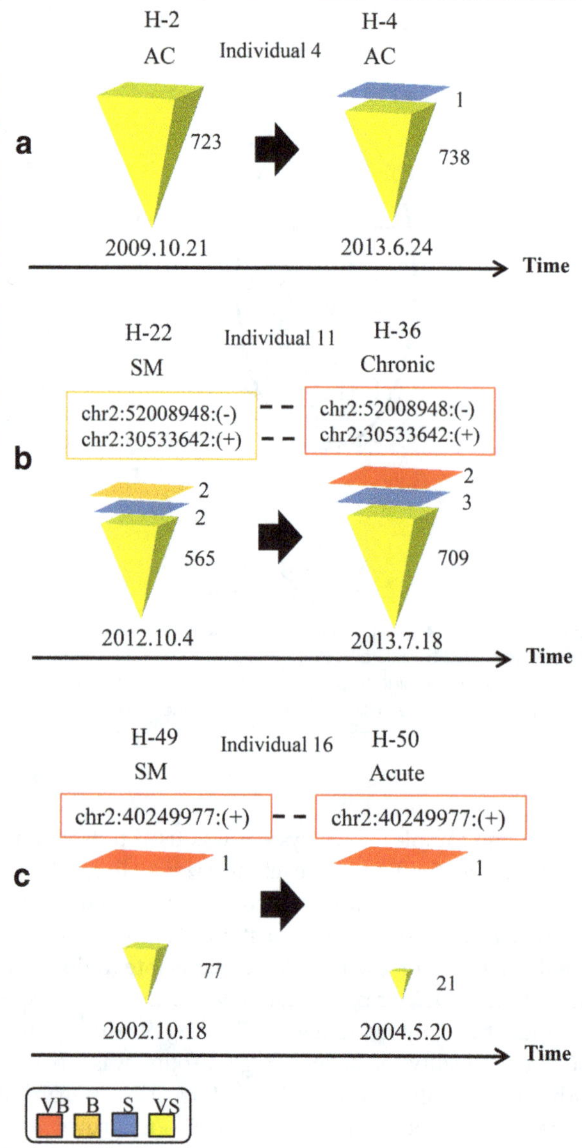

Fig. 2 Clone size and integration sites in three representative individuals. Individuals are indicated at the *top*, followed by sample numbers and disease states at times 1 and 2. The number of integration sites is indicated beside color-coded representations of clone size. Clones with identical integration sites are shown in *boxes* color-coded for clone size and connected by *dashed lines*. **a** Individual 4 remained an asymptomatic carrier (AC) over time and had a large number of integration sites that fluctuated over time. **b** Individual 11 progressed from smoldering (SM) to chronic ATL. The integration sites of the two largest clones were identical over time. **c** Individual 16 progressed from SM to acute ATL. The integration site of the largest clone was identical over time. Results for other samples are shown in Additional file 2: Figure S3

An informative way to interpret and visualize data

After processing high-throughput clonality data and measuring clone sizes for each sample, we ordered clones based on size in a descending manner and categorized them into VB, B, S, or VS groups. Subsequently, we represented the clonal structure of ACs and different subtypes of ATL using different shapes and colors.

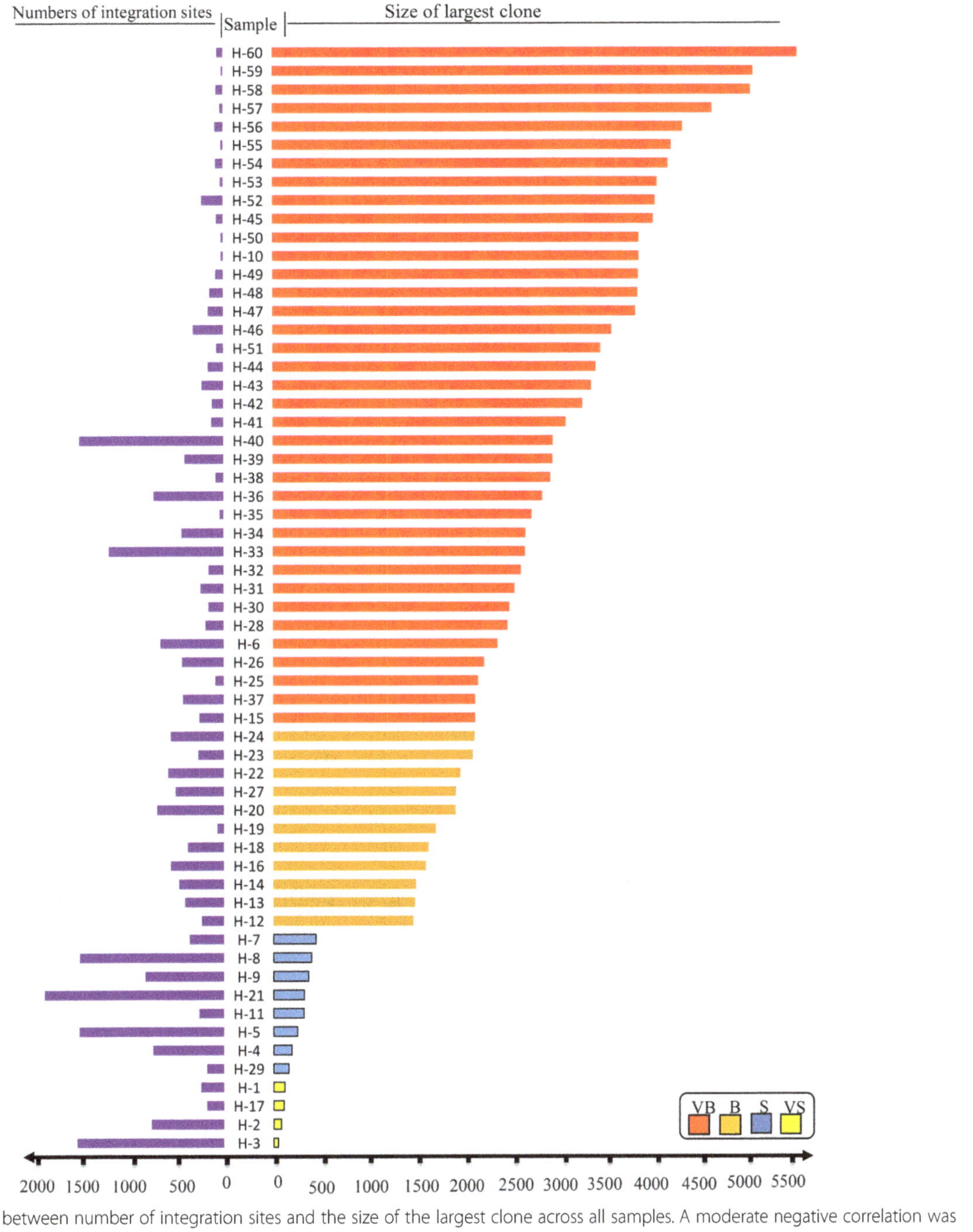

Fig. 3 Correlation between number of integration sites and the size of the largest clone across all samples. A moderate negative correlation was observed ($R = -0.55$ based on Pearson correlation coefficient)

To reflect clonality behaviors and prognosis status of HTLV-1–infected individuals, we modeled the observed data using the concept of a rooted ordered max-tree and m-ary ($1 \leq m \leq 4$) tree (Additional file 2: Figure S2). The tree vertices (V) are defined by the four clone size categories: $V = \{VB, B, S, VS\}$. The tree edges (E) have ten possible members $E = \{VB\text{-}VB, VB\text{-}B, VB\text{-}S, VB\text{-}VS, B\text{-}B, B\text{-}S, B\text{-}VS, S\text{-}S, S\text{-}VS, VS\text{-}VS\}$. In the proposed tree structures, the key values indicate clone size. To discriminate and visualize clone sizes, different shapes and colors are used. We first provided four basic ordered max-trees based on clone size as primary rules (Additional file 2: Figure S2B). We then constructed tree structures for the clonality pattern of each sample using these four basic trees that are ordered based on clone size (Additional file 2: Figure S2C). These trees represent the relationship between the clone sizes in each analyzed sample. We defined a path in the proposed trees as a set

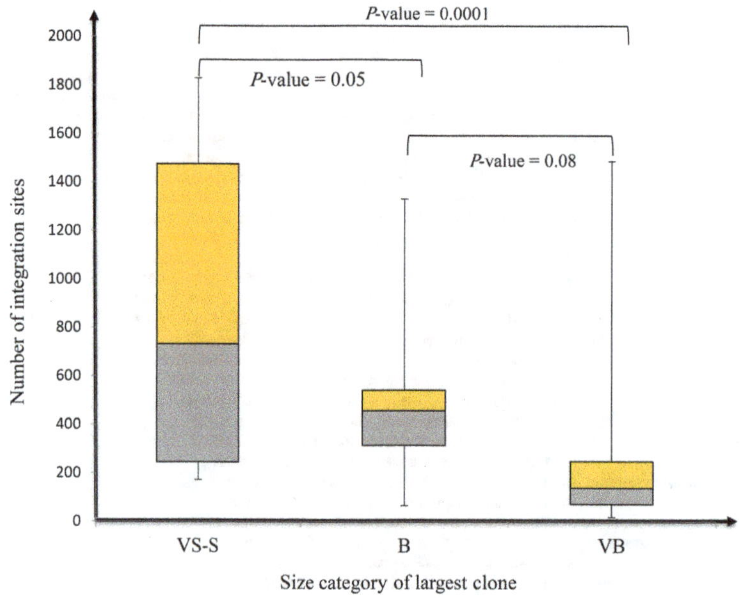

Fig. 4 Number of clones belonging to size categories across analyzed samples. Samples were divided into three groups (VS-S, B, and VB) based on the category of the largest clone. The number of integration sites isolated from each group was then plotted. *P* values were calculated by Student's *t* test

of edges from an ancestor node to a specific leaf of that tree (i.e., to a terminal node). In the proposed hierarchical tree model, each node of a tree corresponds to a unique observed clone. The edges between vertices and the order of vertices are designed based on the clone size. Each path represents the clonality status of a sample, and clone size is seen to decrease along the path.

In Fig. 7, we augmented this sample-by-sample view by including disease progression, which increases from left to right as the disease approaches full malignancy and the prognosis worsens. This hierarchical structure presents clones from each sample based on the size, order, and disease progression stage, allowing comparisons with other samples (Fig. 7). This illustration can be used as a guide for predicting prognosis based on clonality among infected individuals. Importantly, to construct these final trees and determine the order of paths, we used real clonality data obtained from our cross-sectional and longitudinal analysis. It is noteworthy that no sample had more than two VB clones and three B clones. Also, the number of S clones was limited as compared with that of VS clones, which were much more numerous. VB clones should be located at level 0 and/or 1, and B clones can be observed up to level 3 (4th level of tree). In principle, nodes can be shown for any number of levels (as indicated by dots in Fig. 7). Here, we used only five levels because our data indicate that diversity in clone sizes is observable only up to the 4th level, after which only S and/or VS clones are detectable. The number of vertices in m-ary trees theoretically is calculated from Eqs. 1 and 2 (Additional file 2: Figure S2). The actual number of observed vertices based on observed clonality patterns is much less than full m-ary tree because of the pruning of some edges. The proposed tree was validated by the clonality of all 60 analyzed samples.

Representative examples to intuitively depict how the model works

Trees of all samples without progression from AC (individuals 1–5) are represented in Fig. 7a, b. All other samples (patients with progression to different subtypes of ATL) are represented in Fig. 7c, d. The order of paths in each panel is determined based on disease progression. For example, trees of individual 15 belong to Fig. 7c. This patient was SM at time point 1 (H-14), and the tree was located to the left; at time point 2 (H-23), following progression to the chronic state, the tree moved to the right (Additional file 2: Figure S3). Trees of all acute individuals belong to the right side of Fig. 7d. Samples with a single B or VB clone had worse prognoses than those with two or more, and thus, their paths were located toward the right.

As an example, we present the case of individual 7, who was monitored at four time points over 6 years (Fig. 8). Unlike typical ACs, who manifest a combination of S and/or VS clones (polyclonal pattern), this individual showed largely expanded clones, with one B clone at time point 1 (H-19) and one VB clone at time point 2 (H-25). Therefore, the path for the clonality of this

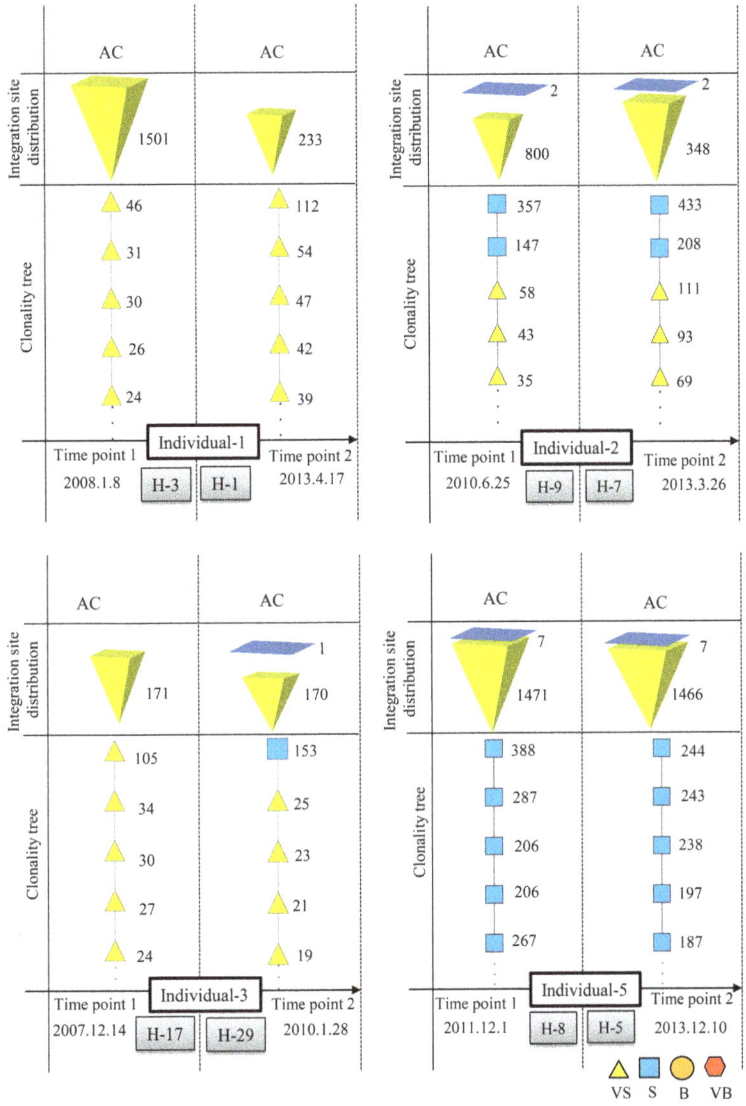

Fig. 5 Clonality data of ACs who remained ACs. High numbers of integration sites were detected from each sample. Only S and VS clones were observed

sample was located on the right side of the global tree (Fig. 7c), which indicates a higher risk of ATL development in this sample. Consistent with our prediction from the model, this individual subsequently progressed to the chronic state at time point 3 and the acute state at time point 4. Clonality trees for the remaining samples are presented in Figs. 5 and 6 and Additional file 2: Figures S3 and S4. As noted above, individuals 1–5 harbored only S and VS clones and remained as ACs over time; their clonality trees were located on the left side of the global tree, which is considered very low risk (Fig. 5).

Individual 6 was an invaluable AC, with two atypically large B clones at time point 1. Therefore, the clonality tree was located in a higher risk area. Consistent with the model, the patient developed SM ATL at time point 2 (H-16) and remained SM at time point 3 (H-18). At time point 4 (H-15), this patient showed a VB clone, a B clone, and an increased proviral load (PVL) (Fig. 6). Individual 8 was an atypical AC with a VB clone at time point 1 (H-34). This individual progressed to SM at time point 2 (H-31), remained SM at time point 3 (H-52) and finally progressed to chronic state at time point 4 (H-42). Individual 9 was a rare AC with an atypical VB clone at time point 1 (H-28), and thus, the clonality tree was located in a higher risk area. This individual progressed into SM ATL at time point 2 (H-32) and remained SM at time point 3 (H-41). Ultimately, this patient developed acute ATL and died of the disease, but clonality data

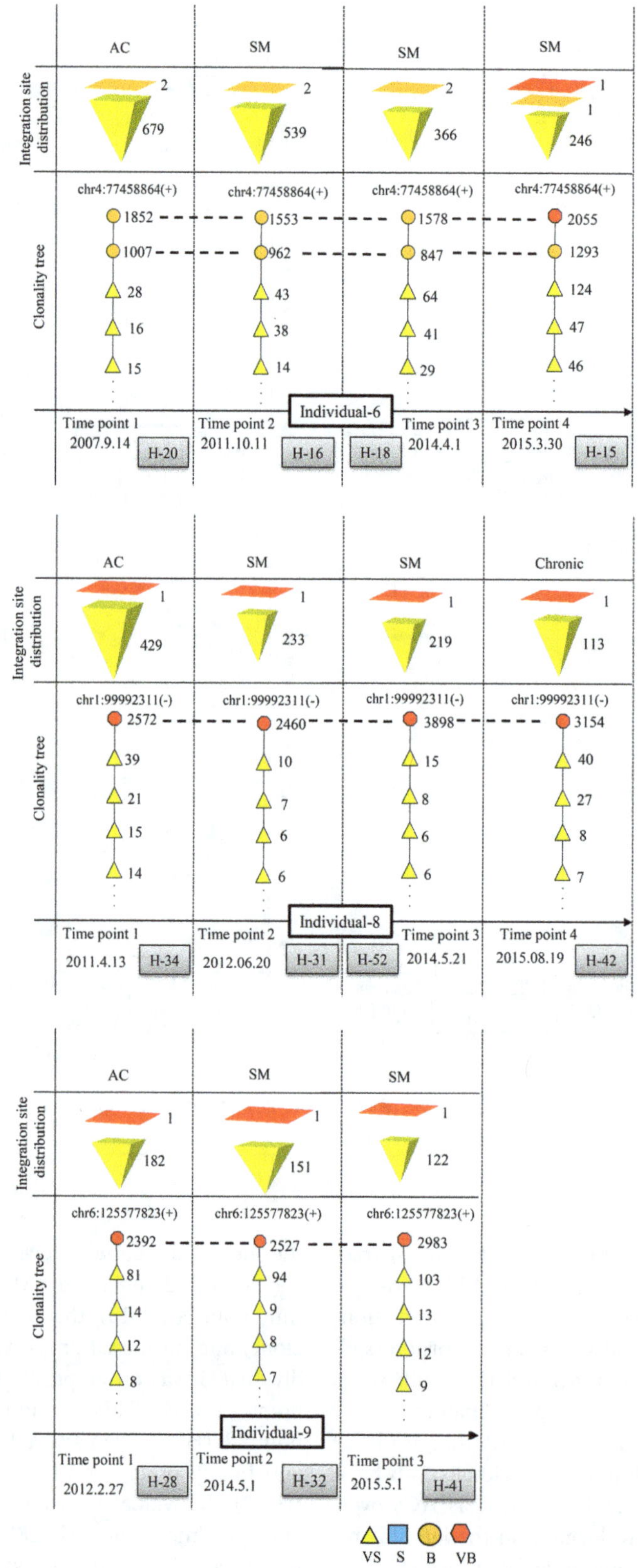

Fig. 6 Clonality data of ACs who progressed to ATL. Clones with identical integration sites are connected by *horizontal dashed lines*. B or VB clones were detected in addition to VS and/or S clones, and their integration sites were constant over time

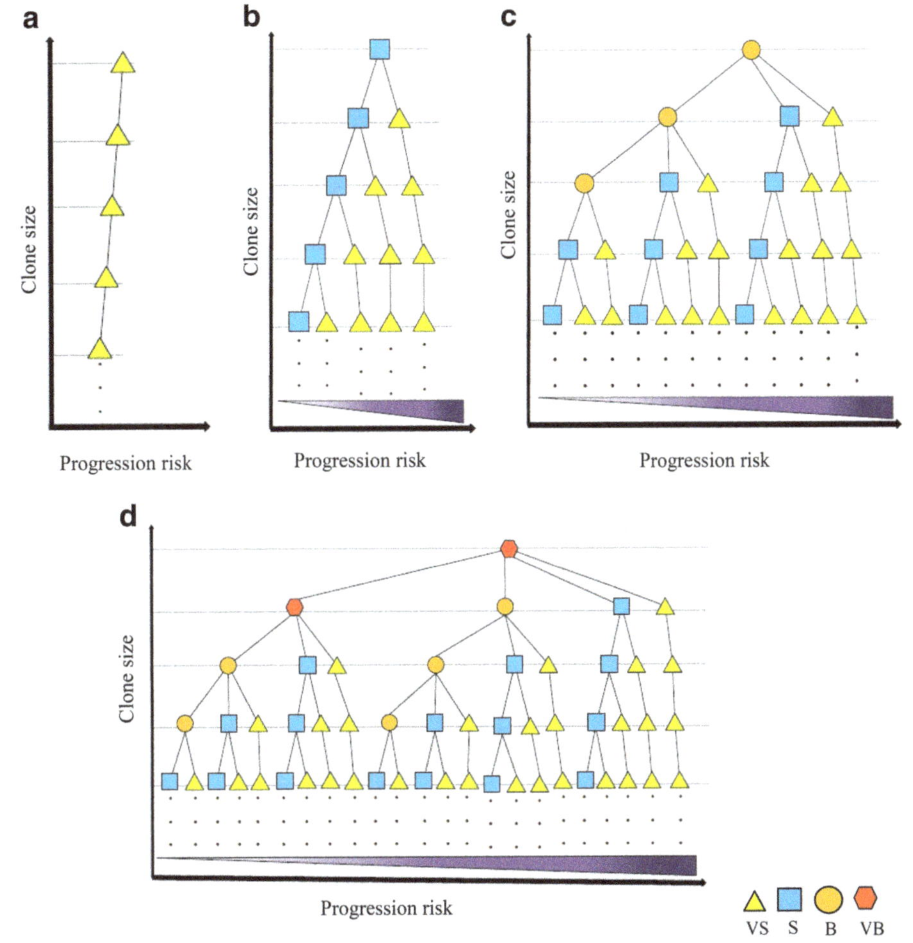

Fig. 7 a–d Global scheme of clonality dynamics among analyzed samples represented by tree structures. Each tree begins with a root of a different clone size (VS, S, B, and VB). The clonality status of each sample can be presented by an individual path. Within each tree, clone size decreases from the root to the leaves. In addition, samples on the right side of a tree have a greater risk of disease progression compared with those on the left. Thus, clone size, order, and disease progression of each sample can be compared to the other samples

at this stage were not available (Fig. 6). The clonality paths for all acute samples were located at the right side of the global tree (Additional file 2: Figure S3).

Another interesting case was an acute patient (individual 21) who was monitored at four time points before, during, and after treatment. The absolute sizes of the largest clone at each time point were 2848, 2411, 2147, and 4066 cells, and the integration site was located at chr9:22270886 (–). This clone belonged to the VB group; the patient remained resistant to therapy and ultimately relapsed. The number of background VS and S clones was H-30: 152 VS; H-39: 394 VS, 2 S; H-26: 420 VS, 4 S, and H-55: 21 VS. The number of integration sites fluctuated during therapy but was greatly reduced after relapse (Additional file 2: Figure S3).

Discussion

Cancer is a complex disease characterized by many different traits, making it an ideal target for various modeling approaches [28]. Well-designed visualization models can transform raw data into logically structured and visually informative representations capable of revealing patterns and structures that otherwise remain hidden [45]. Motivated by the increased availability of NGS data on clonality, we focused on the problem of constructing a progression model of ATL. Development of such progression models is very important for predicting prognosis and, hence, for timely therapeutic decision-making. We focused on clonal expansion as a hallmark of ATL and provide an overall picture of clonality using intra- and inter-sample clinical data.

HTLV-1-infected individuals have an about 3–5% lifetime risk of developing ATL [11]. Currently, there is no established method to clearly predict the risk of progression to ATL. The current risk indicator is PVL; however, individual patient prognosis based on PVL alone is difficult because the range of PVLs among patients with disease overlaps that of patients who remain lifelong

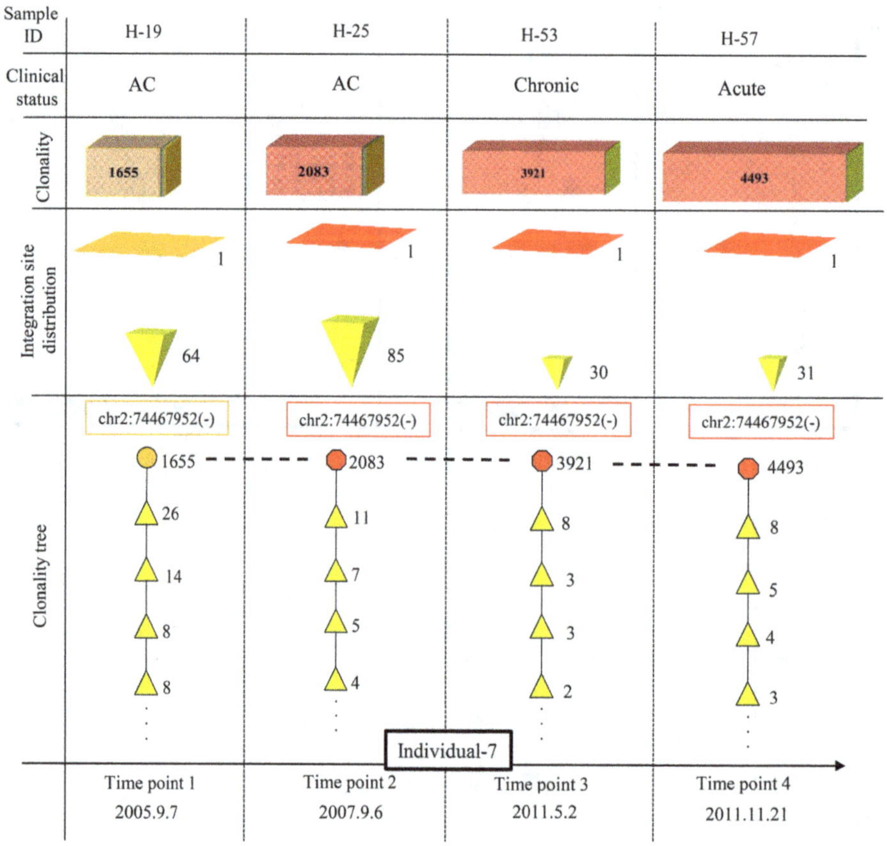

Fig. 8 Clonality tree for a patient (individual 7) monitored over 6 years who progressed from AC status to ATL onset. The patient had a B clone at time point 1 and a VB clone at time points 2, 3, and 4. The integration site of the largest clone was identical across all time points, as indicated by the *dashed horizontal lines*. Clonality of this high-risk sample was notably different from that of a typical AC

ACs [52–54]. Therefore, additional prognosis indicators are required. It is known that clonal proliferation of infected cells is primarily responsible for maintaining the PVL of HTLV-1 [55]. Thus, investigating the clonal composition of HTLV-1-infected individuals as a potential risk indicator of ATL would be of great clinical and biological significance. Understanding how clonal expansion occurs could be used as a key factor in finding effective ways for early detection of ATL, enabling more timely and precise treatment. To address this problem, we converted complex clonal expansion data into a manageable model of relationships among a set of clones that incorporates the characteristics of clone sizes, clone order, and disease progression over time. Using this approach, we were able to provide answers to several important questions [12, 26, 56, 57]:

1. What is the role of clone size and combinations in ATL progression?

Only ACs with B or VB clones progressed to ATL [42]. Samples with a single B or VB clone had worse prognoses than those with two or more, and thus, their paths were located toward the right of the proposed model. All individuals who progressed to higher states of ATL harbored B and/or VB clones. The presence of VB and B clones, as well as small numbers of S and VS clones might be of prognostic value. (Figs. 5 and 6, Additional file 2: Table S2, Fig. 7).

2. How many distinct HTLV-1$^+$ clones of various sizes are detectable from a single host?

We detected a maximum of two VB clones, three B clones, a small numbers of S clones, and a large number of VS clones.

3. What is the relationship between clonal composition and the risk of ATL onset?

Based on the observed clonality data and clinical status of each individual, we ordered the observed trees of each sample and proposed a global tree model. Samples for which trees were located at the left side of the global tree had less risk of disease development, whereas those located at the right side had higher risks (Fig. 7).

Our longitudinal data provided further direct evidence that HTLV-1 can be integrated and persist in humans for years. We also confirmed that each HTLV-1-infected individual carries a large number of infected clones of different sizes. S and VS clones were much more numerous than B and VB clones. We showed that selective clonal expansion or survival of certain transduced or infected cell clones is a limited occurrence, as no more than two VB clones and three B clones was detected in any individual. Moreover, only certain clones that were already dominant (B/VB) at a non-malignant state gave rise to dominant clones during the malignant disease. None of the samples with only S and VS clones belonged to patients who developed ATL over time. Thus, consistent with what we have reported previously [42], the clonality of HTLV-1-infected cells could be a useful predictive marker of ATL onset and progression.

Combining biological experimental data with computational analysis and mathematical modeling facilitates quantification and accurate description of biological phenomena. Using simpler mathematics is recommended to improve communication with experimental biologists [58]. In general, understanding analytical deviations, complicated mathematical equations and formalisms is difficult for the average biologist. It has been suggested that a large number of equations in a paper is associated with a reduced impact of the study among biologists [58, 59]. Therefore, strong and complicated technical presentation of mathematical models may reduce their impact and comprehension among biologists. Moreover, avoiding an overdependence on theory and integrating experimental data are important and appropriate strategies that enhance the effectiveness of models [32]. Bridging analysis and visualization of data—particularly omics data—not only facilitates the presentation of information but also may uncover unknown concepts, conclusions, and insights about the results.

The model of ATL clonal expansion presented here permits investigation of the impact of clonal expansion and related parameters on the risk of ATL development by offering a new understanding of how clonality patterns contribute to disease progression. This easy-to-understand model is an appropriate data-driven model for ATL clonality because it uses real biological data without theoretical assumptions and relies on a small number of variables and a simple set of relationships to explain the clonality status of samples.

Conclusions

Here, we used tree-based structures to show clonality patterns and to display the importance of clone size and clonal composition in ATL development. This integration site-based clonality tree model (1) has both descriptive and prognostic capacities, (2) provides clear visualization of the results, (3) uses simple mathematics, and (4) is fully data-driven and is not dependent on theoretical assumptions. The proposed hierarchical tree model will be useful for organizing and summarizing a mechanistic understanding of observed results and for providing significant insights into clonal history based on the integration sites of HTLV-1-infected cells. To our knowledge, this is the first model that enables the prediction of prognosis among infected individuals based on clonality information (clone size and dynamics of clones). Designing a scoring system for representing the clonal expansion process and implementing a user-friendly software tool for clinical applications are of high interest for future studies.

Abbreviations

AC: Asymptomatic carrier; ATL: Adult T cell leukemia; B: Big; HTLV-1: Human T cell leukemia virus type 1; JSPFAD: Joint Study on Predisposing Factors of ATL Development; NGS: Next-generation sequencing; PVL: Proviral load; S: Small; SM: Smoldering; VB: Very big; VS: Very small

Acknowledgements

AF expresses deep respect and gratitude to the Otsuka Toshimi Scholarship Foundation for supporting his graduate studies. We gratefully appreciate Y. Suzuki for his supports in sequencing of samples, K. Abe, K. Imamura, Y. Ishikawa, T. Horiuchi, and Y. Kuze for technical support; H. Farmanbar for mathematics advice; and U. Firouzi for help with the design of figures. Computational analyses were provided by the Super Computer System at the Human Genome Center, Institute of Medical Science, University of Tokyo. We appreciate the technical assistance of H. Nishijima.

Funding

This work was supported by Kakenhi grants of the Japan Society for the Promotion of Science (JSPS) (17K15044, 16H06277, and 16H05248) and grants by Japan Agency for Medical Research and Development (AMED) (17fk0108112h0001)).

Authors' contributions

The first two authors (AF and SF) contributed equally. AF and SF conceived the project, designed and carried out the experiments, performed in silico data analysis, and interpreted the data. AF performed modeling of observed data. AF and SF designed the figures and wrote the initial manuscript. WM contributed in interpretation of data and writing the manuscript. MI, KU, and AU contributed to the medical data interpretation and sample collection. TW contributed to interpretation of medical and biological data. KN supervised the study and assisted writing the manuscript. All authors assisted in critically revising the manuscript and have read and approved the final manuscript.

Competing interests

The authors declare that they have no competing interests.

Author details
[1]Department of Computational Biology and Medical Sciences, Graduate School of Frontier Sciences, The University of Tokyo, Kashiwanoha, Kashiwa-shi, Chiba, Japan. [2]Laboratory of Functional Analysis in silico, Human Genome Center, The Institute of Medical Science, The University of Tokyo, Tokyo, Japan. [3]Institute of Bioinformatics, Faculty of Medicine, University of Muenster, Muenster, Germany. [4]Department of Frontier Life Science, Graduate School of Biomedical Sciences, Nagasaki University, Nagasaki, Japan. [5]Department of Hematology, Imamura General Hospital, Kagoshima, Japan. [6]Department of Advanced Medical Innovation, St. Marianna University Graduate School of Medicine, Kanagawa, Japan.

References

1. Nowell PC. The clonal evolution of tumor cell populations. Science. 1976;194:23–8.
2. Crespi B, Summers K. Evolutionary biology of cancer. Trends Ecol Evol. 2005;20:545–52.
3. Merlo LMF, Pepper JW, Reid BJ, Maley CC. Cancer as an evolutionary and ecological process. Nat Rev Cancer. 2006;6:924–35.
4. Greaves M. Evolutionary determinants of cancer. Cancer Discov. 2015;5:806–20.
5. Sun YJ, Yao J, Nowak NJ, Goodison S. Cancer progression modeling using static sample data. Genome Biol. 2014;15:440. DOI:10.1186/s13059-014-0440-0.
6. Sprouffske K, Merlo LM, Gerrish PJ, Maley CC, Sniegowski PD. Cancer in light of experimental evolution. Curr Biol. 2012;22:R762–771.
7. Yoshida M, Seiki M, Yamaguchi K, Takatsuki K. Monoclonal integration of human T-cell leukemia provirus in all primary tumors of adult T-cell leukemia suggests causative role of human T-cell leukemia virus in the disease. Proc Natl Acad Sci U S A. 1984;81:2534–7.
8. Fujino T, Nagata Y. HTLV-I transmission from mother to child. J Reprod Immunol. 2000;47:197–206.
9. Yamaguchi K, Watanabe T. Human T lymphotropic virus type-I and adult T-cell leukemia in Japan. Int J Hematol. 2002;76:240–5.
10. Gallo RC. The discovery of the first human retrovirus: HTLV-1 and HTLV-2. Retrovirology. 2005;2:17.
11. Ishitsuka K, Tamura K. Human T-cell leukaemia virus type I and adult T-cell leukaemia-lymphoma. Lancet Oncol. 2014;15:e517–526.
12. Watanabe T. Adult T-cell leukemia (ATL): molecular basis for clonal expansion and transformation of HTLV-1-infected T cells. Blood. 2017;129(9):1071–81.
13. Tsukasaki K, Tsushima H, Yamamura M, Hata T, Murata K, Maeda T, Atogami S, Sohda H, Momita S, Ideda S, et al. Integration patterns of HTLV-I provirus in relation to the clinical course of ATL: frequent clonal change at crisis from indolent disease. Blood. 1997;89:948–56.
14. Gillet NA, Malani N, Melamed A, Gormley N, Carter R, Bentley D, Berry C, Bushman FD, Taylor GP, Bangham CR. The host genomic environment of the provirus determines the abundance of HTLV-1-infected T-cell clones. Blood. 2011;117:3113–22.
15. Firouzi S, Lopez Y, Suzuki Y, Nakai K, Sugano S, Yamochi T, Watanabe T. Development and validation of a new high-throughput method to investigate the clonality of HTLV-1-infected cells based on provirus integration sites. Genome Med. 2014;6:46.
16. Etoh K-I, Tamiya S, Yamaguchi K, Okayama A, Tsubouchi H, Ideta T, Mueller N, Takatsuki K, Matsuoka M. Persistent clonal proliferation of human T-lymphotropic virus type I-infected cells in vivo. Cancer Res. 1997;57:4862–7.
17. Okayama A, Stuver S, Matsuoka M, Ishizaki J, Tanaka G, Kubuki Y, Mueller N, Hsieh CC, Tachibana N, Tsubouchi H. Role of HTLV-1 proviral DNA load and clonality in the development of adult T-cell leukemia/lymphoma in asymptomatic carriers. Int J Cancer. 2004;110:621–5.
18. Cook LB, Rowan AG, Melamed A, Taylor GP, Bangham CR. HTLV-1-infected T cells contain a single integrated provirus in natural infection. Blood. 2012;120:3488–90.
19. Takemoto S, Matsuoka M, Yamaguchi K, Takatsuki K. A novel diagnostic method of adult T-cell leukemia: monoclonal integration of human T-cell lymphotropic virus type I provirus DNA detected by inverse polymerase chain reaction. Blood. 1994;84:3080–5.
20. Aoki S, Firouzi S, Lopez Y, Yamochi T, Nakano K, Uchimaru K, Utusnomiya A, Iwanaga M, Watanabe T. Transition of adult T-cell leukemia/lymphoma clones during clinical progression. Int J Hematol. 2016;104:330–7.
21. Ohshima K, Mukai Y, Shiraki H, Suzumiya J, Tashiro K, Kikuchi M. Clonal integration and expression of human T-cell lymphotropic virus type I in carriers detected by polymerase chain reaction and inverse PCR. Am J Hematol. 1997;54:306–12.
22. Cavrois M, Wain-Hobson S, Wattel E. Stochastic events in the amplification of HTLV-I integration sites by linker-mediated PCR. Res Virol. 1995;146:179–84.
23. Cavrois M, Gessain A, Wain-Hobson S, Wattel E. Proliferation of HTLV-1 infected circulating cells in vivo in all asymptomatic carriers and patients with TSP/HAM. Oncogene. 1996;12:2419–23.
24. Melamed A, Laydon DJ, Gillet NA, Tanaka Y, Taylor GP, Bangham CR. Genome-wide determinants of proviral targeting, clonal abundance and expression in natural HTLV-1 infection. PLoS Pathog. 2013;9:e1003271.
25. Cook LB, Melamed A, Niederer H, Valganon M, Laydon D, Foroni L, Taylor GP, Matsuoka M, Bangham CR. The role of HTLV-1 clonality, proviral structure, and genomic integration site in adult T-cell leukemia/lymphoma. Blood. 2014;123:3925–31.
26. Bangham CR, Cook LB, Melamed A. HTLV-1 clonality in adult T-cell leukaemia and non-malignant HTLV-1 infection. Semin Cancer Biol. 2014;26:89–98.
27. Farmanbar A, Firouzi S, Park SJ, Nakai K, Uchimaru K, Watanabe T. Multidisciplinary insight into clonal expansion of HTLV-1-infected cells in adult T-cell leukemia via modeling by deterministic finite automata coupled with high-throughput sequencing. BMC Med Genomics. 2017;10:4.
28. Masoudi-Nejad A, Wang E. Cancer modeling and network biology: accelerating toward personalized medicine. Semin Cancer Biol. 2015;30:1–3.
29. Wolkenhauer O. Why model? Front Physiol. 2014;5:21.
30. Michor F, Beal K. Improving cancer treatment via mathematical modeling: surmounting the challenges is worth the effort. Cell. 2015;163:1059–63.
31. Chambers RB. The role of mathematical modeling in medical research: "research without patients?". Ochsner J. 2000;2:218–23.
32. Gunawardena J. Models in biology: 'accurate descriptions of our pathetic thinking'. BMC Biol. 2014;12:29.
33. Beerenwinkel N, Schwarz RF, Gerstung M, Markowetz F. Cancer evolution: mathematical models and computational inference. Syst Biol. 2015;64:e1–25.
34. Laydon DJ, Melamed A, Sim A, Gillet NA, Sim K, Darko S, Kroll JS, Douek DC, Price DA, Bangham CR, Asquith B. Quantification of HTLV-1 clonality and TCR diversity. PLoS Comput Biol. 2014;10:e1003646.
35. Nowak MA, Bangham CR. Population dynamics of immune responses to persistent viruses. Science. 1996;272:74–9.
36. Wodarz D, Bangham CR. Evolutionary dynamics of HTLV-I. J Mol Evol. 2000;50:448–55.
37. Stilianakis NI, Seydel J. Modeling the T-cell dynamics and pathogenesis of HTLV-I infection. Bull Math Biol. 1999;61:935–47.
38. Wang L, Li MY, Kirschner D. Mathematical analysis of the global dynamics of a model for HTLV-I infection and ATL progression. Math Biosci. 2002;179:207–17.
39. Katri P, Ruan S. Dynamics of human T-cell lymphotropic virus I (HTLV-I) infection of CD4+ T-cells. C R Biol. 2004;327:1009–16.
40. Cai L, Li X, Ghosh M. Global dynamics of a mathematical model for HTLV-I infection of CD4+ T-cells. App Math Model. 2011;35:3587–95.
41. Lim AG, Maini PK. HTLV-I infection: a dynamic struggle between viral persistence and host immunity. J Theor Biol. 2014;352:92–108.
42. Firouzi S, Farmanbar A, Nakai K, Iwanaga M, Uchimaru K, Utsunomiya A, Suzuki Y, Watanabe T. Clonality of HTLV-1-infected T-cells as a risk indicator for development and progression of adult T-cell leukemia. Blood advances; 2017;1:1195-205.
43. Grimaldi RP. Discrete and combinatorial mathematics: an applied introduction. Addison-Wesley; 2004:980.
44. Epp SS. Discrete mathematics with applications. Brooks/Cole; 2011:820.
45. Pavlopoulos GA, Soldatos TG, Barbosa-Silva A, Schneider R. A reference guide for tree analysis and visualization. Bio Data Min. 2010;3:1.
46. Yamaguchi K, Uozumi K, Taguchi H, Kikuchi H, Okayama A, Kamihira S, Hino S, Nosaka K, Watanabe T. Nationwide cohort study of HTLV-1 carriers in Japan: joint study on predisposing factors of ATL development (JSPFAD). Aids Research Human Retroviruses. 2007;23:582.
47. Biomaterial resource bank of HTLV-1 carriers JSPFAD: http://www.htlv1.org/old/bank-en.html . Accessed 10 June 2017.
48. Shimoyama M. Diagnostic criteria and classification of clinical subtypes of adult T-cell leukaemia-lymphoma. A report from the Lymphoma Study Group (1984-87). Br J Haematol. 1991;79:428–37.

49. Altschul SF, Gish W, Miller W, Myers EW, Lipman DJ. Basic local alignment search tool. J Mol Biol. 1990;215:403–10.
50. Langmead B, Trapnell C, Pop M, Salzberg SL. Ultrafast and memory-efficient alignment of short DNA sequences to the human genome. Genome Biol. 2009;10:R25.
51. FastQC: a quality control tool for high throughput sequence data. https://www.bioinformatics.babraham.ac.uk/projects/fastqc/ . Accessed 10 June 2017.
52. Iwanaga M, Watanabe T, Utsunomiya A, Okayama A, Uchimaru K, Koh KR, Ogata M, Kikuchi H, Sagara Y, Uozumi K, et al. Human T-cell leukemia virus type I (HTLV-1) proviral load and disease progression in asymptomatic HTLV-1 carriers: a nationwide prospective study in Japan. Blood. 2010;116:1211–9.
53. Iwanaga M, Watanabe T, Yamaguchi K. Adult T-cell leukemia: a review of epidemiological evidence. Front Microbiol. 2012;3:322.
54. Kamihira S, Dateki N, Sugahara K, Hayashi T, Harasawa H, Minami S, Hirakata Y, Yamada Y. Significance of HTLV-1 proviral load quantification by real-time PCR as a surrogate marker for HTLV-1-infected cell count. Clin Lab Haematol. 2003;25:111–7.
55. Wattel E, Vartanian JP, Pannetier C, Wain-Hobson S. Clonal expansion of human T-cell leukemia virus type I-infected cells in asymptomatic and symptomatic carriers without malignancy. J Virol. 1995;69:2863–8.
56. Taylor GP, Matsuoka M. Natural history of adult T-cell leukemia/lymphoma and approaches to therapy. Oncogene. 2005;24:6047–57.
57. Cavrois M, Wain-Hobson S, Gessain A, Plumelle Y, Wattel E. Adult T-cell leukemia/lymphoma on a background of clonally expanding human T-cell leukemia virus type-1-positive cells. Blood. 1996;88:4646–50.
58. Fawcett TW, Higginson AD. Heavy use of equations impedes communication among biologists. Proceedings of the National Academy of Sciences of the United States of America 2012, 109:11735-11739.
59. Fusco G. For a new dialogue between theoretical and empirical studies in evo-devo. Front Ecol Evol. 2015;3. doi:10.3389/fevo.2015.00097.

11

Distinct transcriptional and metabolic profiles associated with empathy in Buddhist priests

Junji Ohnishi[1,2*], Satoshi Ayuzawa[3,4], Seiji Nakamura[5], Shigeko Sakamoto[2], Miyo Hori[2], Tomoko Sasaoka[4], Eriko Takimoto-Ohnishi[2,6], Masakazu Tanatsugu[7] and Kazuo Murakami[2,8]

Abstract

Background: Growing evidence suggests that spiritual/religious involvement may have beneficial effects on both psychological and physical functions. However, the biological basis for this relationship remains unclear. This study explored the role of spiritual/religious involvement across a wide range of biological markers, including transcripts and metabolites, associated with the psychological aspects of empathy in Buddhist priests.

Methods: Ten professional Buddhist priests and 10 age-matched non-priest controls were recruited. The participants provided peripheral blood samples for the analysis of gene expression and metabolic profiles. The participants also completed validated questionnaires measuring empathy, the Health-Promoting Lifestyle Profile-II (HPLP-II), and a brief-type self-administered diet history questionnaire (BDHQ).

Results: The microarray analyses revealed that the distinct transcripts in the Buddhist priests included up-regulated genes related to type I interferon (IFN) innate anti-viral responses (i.e., MX1, RSAD2, IFIT1, IFIT3, IFI27, IFI44L, and HERC5), and the genes C17orf97 (ligand of arginyltranseferase 1; ATE1), hemoglobin γA (HBG1), keratin-associated protein (KRTAP10-12), and sialic acid Ig-like lectin 14 (SIGLEC14) were down-regulated at baseline. The metabolomics analysis revealed that the metabolites, including 3-aminoisobutylic acid (BAIBA), choline, several essential amino acids (e.g., methionine, phenylalanine), and amino acid derivatives (e.g., 2-aminoadipic acid, asymmetric dimethyl-arginine (ADMA), symmetric dimethyl-arginine (SMDA)), were elevated in the Buddhist priests. By contrast, there was no significant difference of healthy lifestyle behaviors and daily nutrient intakes between the priests and the controls in this study. With regard to the psychological aspects, the Buddhist priests showed significantly higher empathy compared with the control. Spearman's rank correlation analysis showed that empathy aspects in the priests were significantly correlated with the certain transcripts and metabolites.

Conclusions: We performed in vivo phenotyping using transcriptomics, metabolomics, and psychological analyses and found an association between empathy and the phenotype of Buddhist priests in this pilot study. The up-regulation of the anti-viral type I IFN responsive genes and distinct metabolites in the plasma may represent systemic biological adaptations with a unique signature underlying spiritual/religious practices for Buddhists.

* Correspondence: johnishi@tokyo-kasei.ac.jp
[1]Department of Food and Nutrition, Tokyo Kasei University, Itabashi, Tokyo, Japan
[2]Foundation for Advancement of International Science, Kasuga, Tsukuba, Japan
Full list of author information is available at the end of the article

Introduction

Spirituality/religiosity is one of several unique aspects of human social environments and generally consists of psychological aspects accompanied by one's behavior and social relationships. According to psycho-neuro-immune models of health regulation, spirituality/religiosity can be positively associated with psychological and physical health [1, 2]. Spirituality/religiosity provide coping resources that may improve mental health by increasing the frequency of positive psychological aspects and positive emotions [3, 4]. These positive psychological aspects include empathy and altruism, and the beneficial positive emotions include general well-being, happiness, and self-esteem [1, 5]. Coping resources also reduce negative stress that could result in emotional disorders, such as depression or anxiety. Because greater physical activity is associated with better mental health [6, 7], spirituality/religiosity should have a favorable impact on the physical state by balancing mental health. Furthermore, longitudinal studies by Miller et al. revealed that a high personal importance of spirituality/religiosity was associated with thicker cortices in certain brain regions (i.e., the left and right parietal and occipital regions, the mesial frontal lobe of the right hemisphere, and the cuneus and precuneus in the left hemisphere) and may confer protective benefits against the depressive symptoms in individuals with a high familial risk of major depression [8].

The surrounding social environments can change the basal expression profiles of certain genes (i.e., basal transcriptome) that are critical for the internal biological processes in our body [9, 10]. Notably, these effects are reliably associated with psychological states and are often induced by individuals' subjective perceptions of their experiences in the surrounding social environments [11]. Stressors can activate the autonomic nervous system and/or the hypothalamic-pituitary-adrenal (HPA) axis to allow for physiological adaptation [12]. Individuals can manage these perceived stressors if the stressors fall within their coping abilities. However, once events or environmental demands exceed one's coping ability, psychological stress ensues [11, 12]. Perceived adverse environmental events, such as psychological stress (e.g., perceived social isolation, bereavement, or long-term caregiving), could elicit negatively affected states, such as feelings of anxiety or depression. These negative psychological states, in turn, directly influence neural-endocrine processes with an immune dysfunction to enhance susceptibility to disease (e.g., viral infection, cardiovascular disease, and type II diabetes) and shape complex behavioral phenotypes (e.g., poorer sleep or appetite and addiction) [11, 12]. Meanwhile, these adversity-associated processes can influence the genome function. Consequently, these stressors can change the basal gene expression activity in peripheral blood leukocytes by up-regulating the expression of pro-inflammatory genes (e.g., interleukin-1β (IL1B), interleukin-8 (IL8), and tumor necrosis factor (TNF)) and down-regulating the expression of genes involved in type I interferon (IFN) innate anti-viral responses (e.g., IFIT-family genes and MX-family genes) [9, 10]. Due to the sensitivity of leukocytes to social adversity, these adverse experiences may shift the leukocyte's basal transcriptional resources from a default anti-viral defense mode to an inflammation-promoting mode. In contrast, Fredrickson et al. have recently shown that a positive psychological state and social conditions (e.g., well-being) could oppose the adversity-associated transcriptome in peripheral blood [13, 14]. In another experimental design, we presented evidence that mirthful laughter with positive emotion can up-regulate distinct genes in the peripheral leukocytes of type II diabetes patients and adjust the activity of natural killer (NK) cells [15].

Notably, most studies reporting a positive connection between spirituality/religiosity and health have focused on "Christian" populations. These studies have typically been conducted in Western and Middle Eastern countries that are generally religious and economically well developed [2]. In contrast, a limited number of studies have been performed in non-Christian unique religious cultures in China [16] and Japan.

In this pilot study, we investigated the impacts of spirituality/religiosity on the functional activity of the human genome and metabolic reactions required for social regulation in qualified Buddhist priests. The objectives of this study were to (1) characterize the basal gene expression and metabolite profiles of qualified Japanese Buddhist priests; (2) evaluate the empathy psychological aspects in the priests, which are core elements of the fundamental concept of Buddhism [17, 18]; and (3) examine the association between empathy and the identified molecular markers.

Methods

Recruitment of participants

The data in this study were collected as a part of a research project that was conducted over 3 years to explore the role of spiritual/religious involvement across a wide range of biological markers. Ten male Japanese Buddhist priests and 10 male healthy age-matched Japanese non-priest controls were recruited from six different urban or rural areas (e.g., Tokyo, Kanagawa, Ibaraki, Kyoto, Gunma, and Wakayama in Japan) via leaflets and electronic media.

All Buddhist participants in this study belonged to the "Shingon" sect of Japanese Buddhism [19] and have performed the spiritual/religious main practices and duties for more than 7 years (median length, 15.50 years; range 7–35 years). These participants were qualified priests and commuted daily from their own homes to temples in towns and cities. All of the non-priest controls had steady full-time jobs and commuted daily from their own residences to their workplaces in towns and cities. Their occupational titles included manager, professional, technician, office worker, and service worker. Additionally, the non-priest controls had not received any training related to meditation or religious practices. The commuting methods of both the priests and the non-priest controls were similar and included travel on foot, using one's own car or via public transportations, with commutes completed within approximately 50 min.

Both the priests and the non-priest controls met the following criteria: (i) decent health status with no subjective symptoms in the previous month and (ii) no history of receiving any dietary counseling or therapy from a doctor or dietitian. All participants were given a complete explanation of the study, which was approved on September 11, 2013, by the ethics committee of Tsukuba University of Technology (approval number TUT 20130911) and provided written informed consent. All participants were paid for their participation. During the study, the participants were asked a series of questions about their age, physical health, and medication use. The participants included a heterogeneous group of middle-aged adults (29–52 years old) as shown in Additional files 1, 2, 3 and 4. One priest had a previous history of venous thrombosis and had been administered an anticoagulant agent (Warfarin). He was also taking an anti-hypertensive drug (candesartan). Two priests suffered from diabetes; one of which was prescribed oral hypoglycemic agents (glimepiride and sitagliptin phosphate hydrate) and rosuvastatin to treat the accompanying hypercholesterolemia, and the other priest declared his diabetic nephropathy. In this pilot study, we did not exclude these three priests due to the limited number of recruited priests. A general serum chemical analysis and hematological tests were performed to examine their general conditions. These results are presented in Additional file 1. All participants had relative proportions of the peripheral blood cell types within normal ranges (27–90% neutrophils, 20–51% lymphocytes, 2–12% monocytes, 0–3% basophils, and 0–10% eosinophils).

Measures

Genome-wide expression profiling

For RNA preparation, 2 × 2.5 mL samples of peripheral blood were drawn into PaxGene Blood RNA tubes (PreAnalytiX/QIAGEN Inc., Valencia, CA) from each subject. After blood collection, PaxGene tubes were left stand at room temperature (RT) for 2 h to ensure complete lyses of all blood cells, followed by stored at 4 °C for overnight. All tubes were stored at − 80 °C until RNA isolation. Based on the convenience of the participants, we performed blood sample collection on three different days (Dec 8, 2013, Apr 28, 2014, and Jun 8, 2014) within 6-month period. Total RNA was isolated within 6 months after storage. Total RNA was isolated using PaxGene Blood RNA Kit (PreAnalytiX/QIAGEN) according to the manufacturers' instructions. Quantity and purity of the RNA was tested using the NanoDrop ND-1000 spectrophotometer (Thermo Fisher Scientific, Waltham, MA) and the Agilent 2100 Bioanalyzer (Agilent Technologies, Santa Clara, CA). In any event, only high-quality RNA samples containing intact 18S and 28S RNA were used for subsequent microarray and quantitative RT-PCR analyses.

Microarray analysis

One hundred nanograms of the total RNA was converted to complementary DNA (cDNA), amplified, and labeled with Cy3-labeled CTP using the Low Input Quick Amp Labeling Kit (Agilent Technologies, Santa Clara, CA) according to the protocol supplied by the manufacturer. Following labeling and clean up, the amplified RNA and dye incorporation were quantified using the NanoDrop ND-1000 spectrophotometer (Thermo Fisher Scientific, Waltham, MA). We used Agilent SurePrint G3 Human GE v2 8x60k Microarrays (Agilent Technologies, Santa Clara, CA) containing 50,599 unique genes. All samples were assayed in a single batch. After hybridization at 65 °C for 17 h, the arrays were washed consequently using the Gene Expression Wash Pack (Agilent Technologies, Santa Clara, CA). The microarrays were scanned using the Agilent Scanner, and the fluorescence intensities of the scanned images were quantified using Feature Extraction software ver. 10.7.3.1 (Agilent Technologies, Santa Clara, CA). Normalization was performed using Agilent GeneSpring GX version 13.1.1. (per chip normalization, 75 percentile shift; per gene normalization, none) with the range of expression intensities for inter-microarray. Only those genes whose expression data were available in more than 50% of hybridizations were included for further analysis [20]. Microarray raw data were deposited in the National Center for Biotechnology Information Gene Expression Omnibus (accession number GSE77676).

Differential expression analysis: the rank products method

Due to small sample sizes (n = 10 for each group) in this study, we could not evaluate whether transcriptional data showed the normal distribution. Therefore, we chose to employ the Rank Products Method, a "non-parametric" test as a cautious approach to determine significantly differential expressed genes. This method is reported to be relatively powerful especially for small sample sizes and when the data is non-homogeneous [21]. A gene considered as significantly differential expressed if the adjusted P value (false discovery rate, FDR) was equal to or less than 5% (0.05).

Quantitative RT-PCR analysis

We quantified the expression of 12 selected genes identified in the priests, including anti-viral signaling molecules (MX1, RSAD2, IFIT1, IFIT3, IFI27, IFI44L, HERC5), DEFA4, FOLR3, HBG1, C17orf97, and S100P through quantitative real-time reverse transcription polymerase chain reaction (qRT-PCR). cDNA was synthesized from 1000 ng total RNA using a High-Capacity cDNA Reverse Transcription Kit (Thermo Fisher Scientific). PCR reactions were carried out using ABI7500 Real Time PCR System (Thermo Fisher Scientific). Specific sets of primers and TaqMan probes were obtained from Thermo Fisher Scientific. Gene expression level of the target transcript was normalized with the values of an endogenous control gene (GAPDH). The data were normalized using the ΔC_T method (ΔC_T = ΔC_T target-- ΔC_T control) by measuring cycle threshold ratios between candidate genes and an internal control gene, GAPDH. Expression level was described inversely as "$-\Delta C_T$"; thus, higher relative quantities indicate greater expression of target genes [20]. Differences of each transcript between the priests and the controls were compared using a Mann–Whitney *U* test, genes were considered as significantly differential expressed if the *P* value was equal to or less than 5% (0.05).

Metabolomics analysis

Peripheral blood samples were collected in Vacutainer tubes voted with ethylendiaminetetraacetic acid (VP-NA070K; Terumo Corporation, Tokyo, Japan) and immediately centrifuged at × 1200*g* for 10 min to separated plasma. The plasma samples were then frozen in dry ice and stored at − 80 °C until use. The Human Metabolome Technologies, Inc., (Tsuruoka, Japan) performed all metabolome analysis [22].

For "hydrophilic" metabolites, capillary electrophoresis (CE)-coupled a time-of-flight mass spectroscopy (TOF/MS) was performed. Briefly, 50 μL of plasma samples were mixed with 450 μL methanol containing internal standards (solution ID H3304-1002, Human Metabolome Technologies, Inc., Tsuruoka, Japan) at 0 °C in order to inactivate enzymes. Then, 500 μL chloroform and 200 μL Milli-Q water were added, and the mixed solution was centrifuged at × 2300*g* for 5 min at 4 °C. The 350 μL of upper aqueous phase was filtered through a Millipore 5-kDa cutoff filter to remove proteins. The filtrate was centrifugally lyophilized and dissolved in 50 μL of Milli-Q water for following CE-TOF/MS analysis. CE-TOF/MAS was carried out using an Agilent CE Capillary Electrophoresis System equipped with an Agilent 6210 Time of Flight mass spectrometer, Agilent 1100 isocratic HPLC pump, Agilent G1603A CE-MS adapter kit, and Agilent G1607A CE-ESI-MS sprayer kit (Agilent Technologies, Waldbronn, Germany). The systems were controlled by Agilent G2201AA ChemStation software version B.03.01 for CE (Agilent Technologies, Waldbronn, Germany). The metabolites were analyzed by using a fused silica capillary (50 μm *i.d.* × 80 cm total length), with commercial electrophoresis buffer (solution ID H3301-1001 for cation analysis and H3302-1021 for anion analysis, Human Metabolome Technologies) as the electrolyte. The sample was injected at a pressure of 50 mbar for 10 s (approximately 10 nL) in cation analysis and 25 s (approximately 25 nL) in anion analysis. The spectrometer was scanned from *m/z* 50 to 1000.

For "hydrophobic" metabolites, liquid chromatography (LC) time-of-flight mass spectrometry (TOFMS) was performed. Briefly, 500 μL of plasma samples were mixed with 1500 μL 1% formic acid/acetonitrile containing internal standard solution (solution ID H3304-1002, Human Metabolome Technologies, Inc., Tsuruoka, Japan) at 0 °C in order to inactivate enzymes. Then, the mixture was centrifuged at × 2300*g* for 5 min at 4 °C. The supernatant was filtered through Hybrid SPE phospholipid cartridge (55261-U; Supelco, Bellefonte, PA, USA) to remove phospholipids. The filtrate 400 μL was lyophilized and dissolved in 100 μL of 50% isopropanol/Milli-Q water solution for analysis. LC-TOFMS was performed using an Agilent LC System (Agilent 1200 series RRLC system SL) equipped with an Agilent 6230 time-of-light mass spectrometer (Agilent Technologies, Waldbronn, Germany). The systems were controlled by Agilent G2201AA ChemStation software version B.03.01 for CE (Agilent Technologies, Waldbronn, Germany). The cationic and anionic compounds were measured by using ODS column (2 × 50 mm, 2 μm).

Peaks were extracted using the MasterHands automatic integration software ver.2.16.0.15 (Keio University, Tsuruoka, Japan) to obtain peak information including the *m*/z ratio, migration time for (MT) for CE-TOF/MS measurement, retention time (RT) for LC-TOF/MS measurement and peak area. Signal peaks corresponding to "isotopomers," "adduct ions," and other "ions" of known metabolites were excluded. The remaining peaks were annotated with putative metabolites from the metabolite database and Known-Unknown library database established in the HMT, based on MT/RT and *m*/z ratio values as determined by TOF/MS. The tolerance range for the peak annotation was configured at ± 0.5 min for MT/RT, and ± 10 ppm for *m*/z ratio. In addition, the peak areas were normalized against those of the internal standards. Finally, the resultant relative area values were further normalized by sample amount. For metabolomics analysis, the relative area was defined as the relative concentration of each metabolite. The Human Metabolome Technologies, Inc., performed hierarchical cluster analysis (HCA) and principal component analysis (PCA) by our proprietary software, PeakStat and SampleStat, respectively. Detected metabolites were plotted on metabolic pathway maps using VANTED (Visualization and Analysis of Networks containing Experimental Data) software.

Health-promoting lifestyle profile (HPLP)

The HPLP was originally developed by Walker, Sechrist, and Pender in 1987 and revised as HPLP-II in 1995 [23]. Wei et al. developed the Japanese version of the HPLP-II and established its validity and credibility with a Cronbach's α internal consistency coefficient of 0.90 [24]. The HPLP-II is a 52-item questionnaire composed of two main categories and six sub-dimension scales. The health-promoting behaviors category includes the health responsibility (9 items), physical activity (8 items), and nutrition (9 items) subscales. The psychosocial well-being category includes the spiritual growth (9 items), interpersonal relationship (9 items), and stress management (8 items) subscales. Each subscale can be used independently. Higher scores indicate a healthier lifestyle. All items on the HPLP-II are affirmative with no reverse questions. The answers were provided on a four-point Likert-type scale, and ratings of "never," "sometimes," "frequently," and "regularly" were scored as 1, 2, 3, and 4 points, respectively. Therefore, the total scores on the HPLP-II ranged between 52 and 208. The scores on the "physical activity" and "stress management" subscales range between 8 and 32, and the scores on the other four subscales range between 9 and 36. In this study, we subsequently divided the total scores and the sum of each subscale score of the HPLP-II by the number of items; therefore, the mean item score ranged from 1 to 4. The internal consistency of the HPLP-II scale used in this study was demonstrated with a Cronbach's α of 0.902. We divided the HPLP-II ratings into three groups of "good," "moderate," and "poor." For the total HPLP-II scale, a score of 3.01–4 was considered "good," a score of 2.01–3 was considered "moderate," and a score of 1–2 was considered "poor." All measurements obtained using the HPLP-II are summarized in Additional file 2.

Brief-type self-administered diet history questionnaire (BDHQ)

The BDHQ is a 58-item short version of the self-administered diet history questionnaire (DHQ) that assesses Japanese dietary habits. The BDHQ was reported to have a satisfactory ability to rank the energy-adjusted intakes of many nutrients in healthy Japanese adults [25]. The BDHQ assesses the dietary habits during the preceding month and consists of the following five sections: (i) intake frequency of 46 food and non-alcoholic beverage items; (ii) daily intake of rice, including the type of rice (refined or unrefined, etc.) and miso soup; (iii) frequency of drinking alcoholic beverages and the amount per drink of five different types of alcoholic beverages; (iv) usual cooking methods; and (v) general dietary behaviors. Most food and beverage items listed on the DHQ are very commonly consumed in Japan, with some modifications using a food list provided by the National Health and Nutrition Survey of Japan as additional information. Standard portion sizes and adult sizes of bowls for rice and cups for miso soup were derived from several recipe books for Japanese dishes. All measurements obtained using the BDHQ are summarized in Additional file 3.

Empathy process scale

Empathy was evaluated using the Empathetic Process Scale, which was recently developed by Hayama et al. based on the Interpersonal Reactivity Index proposed by Davis [26]. This Japanese 30-item questionnaire was designed with six sub-dimension scores to assess both the cognitive and emotional aspects of empathy [27]. This scale focuses on more detailed emotional aspects than that developed by Davis. The emotional aspects of empathy include "Sharing positive emotions with others," "Good feeling for others' positive emotions," "Sharing negative emotions with others," and "Sympathy for others' negative emotions." The cognitive aspects of empathy include "Perspective taking" and "Sensibility about others' emotions." The self-reported responses were provide on a 5-point Likert-type scale with the anchors of 1 = Strongly disagree and 5 = Strongly agree. All six subscales consisted of five items with scores ranging from 5 to 25. The internal consistency of the Empathy Scale used in this study was demonstrated with a Cronbach's α of 0.821. Two professional native-speaking English translators at Tokyo Kasei University translated the Japanese questionnaire into English (Additional file 5).

Statistical analysis

All data are presented as medians with interquartile ranges (25–75th percentile). All statistical analyses were performed using SPSS version 19 (IBM Inc., Armonk, NY, USA) and GraphPad Prism 6.0 (GraphPad Software Inc., San Diego, CA, USA). Because the sample sizes in this study were relatively small (n = 10 in each group), we could not determine whether the transcripts, metabolites, blood, and psychological indices followed a normal distribution. Therefore, we chose the non-parametric Mann–Whitney U test to compare the detected values between the priest group and the control group. A P value of 0.05 or less was considered significant. The effect sizes (r) for the Mann–Whitney U test were calculated based on the Z value as described below. N represents the total sample size. Cohen's guidelines indicate that a large effect size is $r = 0.5$, a medium effect size is $r = 0.3$, and a small effect size is $r = 0.1$ [28].

$$r = \frac{Z}{\sqrt{N}}$$

To explore the significance of the deviated transcripts and metabolites in the "spiritual/religious" trained priests, Spearman's rank correlation coefficients were calculated to assess the correlation between the psychological indices (Empathy Scales) and the significant transcriptional or metabolic candidates. Statistical significance was defined as $P < 0.05$.

Results

Descriptive characteristics of the participants' lifestyle behaviors and blood chemistries

Lifestyle behaviors are considered essential components that influence well-being and health. Dietary habits, regular physical exercise, and religiosity/spirituality are major factors associated with these behaviors [29]. Therefore, we firstly compared the healthy lifestyle behaviors and dietary habits between the priests (n = 10) and the age-matched non-priest controls (n = 10). The healthy lifestyle behaviors and dietary habits were evaluated using the HPLP-II questionnaires and BDHQ questionnaires, respectively. The outcomes of the six HPLP-II sub-dimensions are dot-plotted in Additional file 4 and summarized in Additional file 2. Measurements from the BDHQ were summarized in Additional file 3. Additional files 2, 3 and 4 show that no significant differences were observed in the healthy lifestyle behaviors and daily nutrient intakes between the priests and the non-priest controls in this study.

Additional file 1 presents the serum chemistries of the participants in this study. The priests and the non-priest controls were matched with respect to age, gender (all males), ethnic group (all Japanese), and BMI, and there were no significant differences between the groups regarding these parameters (P > 0.05). In most of the blood indices, no significantly differences were observed at the basal state between the priests (n = 10) and the non-priest controls (n = 10). Only the sodium (Na) and potassium (K) concentrations were significantly different between the groups (P = 0.03 and 0.04, respectively); however, these values were within the normal healthy ranges. Two priests with diabetes had high blood sugar (133 and 164 mg/dL) and HbA1c (6.2 and 7.6%) levels; one priest also showed a high level of plasma creatinine (1.87 mg/dL) due to diabetic nephropathy. Notably, no significant difference (P = 0.47) in their levels of the systemic inflammatory marker C-reactive protein (CRP) was observed comparing to other healthy priests or all non-priest controls. The median CRP level in the priest group was 0.05 (0.03–0.08, 25–75%) (mg/dL) and that in the non-priest control group was 0.04 (0.03–0.05, 25–75%) (mg/dL). Notably, the results of the comparison of the serum nutritional variables between the groups shown in Additional file 1 (protein, cholesterol, triglyceride, glucose, Na, and K) were consistent with the findings obtained using the BDHQ (protein, cholesterol, fat, carbohydrate, sodium, and potassium) shown in Additional file 3.

Microarray results and selection of candidate genes

To determine the gene expression profiles of the qualified priests at the basal state, we performed one-color-microarray experiments using Agilent platforms. We tested 50,599 probe sets to investigate the expression of 23,284 unique human genes.

To obtain an initial list of candidate genes, we performed a Ranked Product analysis to compare the ranks of the top differentially expressed genes between the priests and the controls. We set the threshold at a P value < 0.05; 162 probe sets representing 111 genes (42 up-regulated genes and 69 down-regulated genes) were differentially expressed between the priests and the non-priest controls after the probe sets passed filtering criteria of 5% false discovery rate (FDR). Additional file 6 provides a list of all 111 differentially expressed candidate genes. The 42 up-regulated genes and 69 down-regulated genes were subjected to a Gene Ontology (GO) analysis using AmiGO (http://amigo2.geneontology.org/amigo). The most significantly enriched functions include "immune effector process (GO:0002252)," "type I interferon signaling pathway (GO:0060337)," "cellular response to type I interferon (GO:0071357)," "innate immune response (GO:0045087)," "response to type I interferon (GO:0034340)," "defense response (GO:0006952)," and "defense response to virus (GO:0051607)."

Tables 1 and 2 show the expression values of representative ranked genes that were differentially expressed by more than 1.5-fold with their corresponding P values, FDRs, and average raw signals (n = 10, priests) in the microarray experiments. Table 1 lists the genes that were up-regulated in the priests, including type I IFN innate anti-viral response genes (e.g., MX1, RSAD2, IFIT1, IFIT3, IFI27, IFI44L, and E3 ubiquitin ligase (HERC5)), defensin4 (DEFA4), and soluble-type folate receptor (FOLR3). Table 2 lists the genes that were down-regulated in the priests, including hemoglobin γA (HBG1), keratin-associated protein (KRTAP10-12), sialic acid Ig-like lectin 14 (SIGLEC14), calcium-binding protein (S100P), and C17orf97.

Result validation by qRT-PCR

To further verify the microarray results using an independent technique, we performed quantitative real-time RT-PCR (qRT-PCR). We compared the basal expression levels of 12 selected representative transcripts in the priests, including MX1, RSAD2, IFIT1, IFIT3, IFI27,

Table 1 Up-expressed genes in the priests

Up genes	Enterz Gene ID	Gene name	Fold change	*P* value	FDR	Average raw
MX1	4599	MX dyn amin-like GTPase 1	2.00	5.7E − 05	0.0352	4503
IFI27	3429	Interferon-α-inducible protein 27	2.60	9.9E − 08	0.0003	1990
MYOM2	9172	Myomesin 2	1.84	6.6E − 08	0.0003	1957
IFIT3	3437	Interferon-induced protein with tetratricopeptide repeats 3	1.77	7.7E − 05	0.0410	1890
IFIT1	3434	Interferon-induced protein with tetratricopeptide repeats 1	2.20	5.8E − 06	0.0073	1682
RSAD2	91,543	Radical S-adenosyl methionine domain containing 2	2.99	< E − 09	0.0000	1347
FOLR3	2352	Folate receptor 3 (γ)	1.82	1.3E − 05	0.0129	1079
BTNL3	10,917	Butyrophilin-like 3	2.23	< E − 09	0.0000	880
DEFA4	1669	Defensin, α4, Corticostatin	2.09	2.1E − 06	0.0034	824
IFI44L	10,964	Interferon-induced protein 44-like	2.58	2.0E − 07	0.0005	726
BATF2	116,071	Basic leucine zipper transcription factor, ATF-like 2	1.54	8.7E − 05	0.0448	540
C4BPA	722	Complement component 4 binding protein, α	2.14	6.6E − 08	0.0003	353
CTSG	1511	Cathepsin G	1.73	7.2E − 05	0.0395	301
HERC5	51,191	HECT and RLD domain containing E3 ubiquitin protein ligase 5	2.05	2.0E − 05	0.0174	266
S100B	6285	S100 calcium binding protein B	1.52	4.5E − 05	0.0305	195
ARPC5	10,092	Actin related protein 2/3 complex, subunit 5, 16 kDa	1.66	2.4E − 05	0.0193	120
LOC 728715	728,715	Ovostatin homolog 2-like	2.75	3.3E − 07	0.0008	72
LOC 100509445	100,509,445	Uncharacterized LOC100509445	4.29	< E − 09	0.0000	67

Representative list of genes up-expressed (> 1.5-fold) in priests compared to the controls. The described data is in descending ordered of the average raw signals (> 50) in the priests' group microarray data. MX1, IFI27, IFIT1, IFIT3, RSAD2, IFI44L, and HERC5 have been validated by quantitative real-time PCR (Fig. 1)

Table 2 Down-expressed genes in the priests

Down genes	Enterz Gene ID	Gene name	Fold change	*P* value	FDR	Average raw
HBG1	3047	Hemoglobin, γA	0.40	6.6E − 08	0.0003	93,019
KRTAP 10-12	386,685	Keratin associated protein 10–12	0.65	1.7E − 05	0.0146	49,880
NPPA	4878	Natriuretic peptide A	0.64	4.0E − 05	0.0254	17,657
BPIFB2	80,341	BPI fold containing family B, member 2	0.60	2.6E − 05	0.0196	14,002
TUBB2A	7280	Tubulin, β2A class IIa	0.58	2.7E − 06	0.0055	8610
S100P	6286	S100 Ca^{2+} binding protein P	0.43	6.6E − 08	0.0003	5677
G3BP1	10,146	GTPase activating protein (SH3 domain) binding protein 1	0.66	2.9E − 05	0.0207	4888
NCR3	259,197	Natural cytotoxicity triggering receptor 3	0.58	9.8E − 05	0.0375	1941
SIGLEC14	100,049,587	Sialic acid Ig-like lectin 14	0.05	< E − 09	0.0000	1625
LAIR2	3904	Leukocyte-associated immunoglobulin-like receptor 2	0.62	7.1E − 05	0.0312	510
IGF2	3481	Insulin-like growth factor 2	0.57	7.4E − 05	0.0314	378
LOC 102467076	102,467,076	Uncharacterized LOC102467076	0.49	6.6E − 08	0.0004	347
RNF182	221,687	Ring finger protein 182	0.40	1.3E − 07	0.0000	140
C17orf97	400,566	Ligand of arginyltransferase 1 (LIAT1)	0.26	< E − 09	0.0000	121
C8orf17	100,507,249	Chromosome 8 open reading frame 17	0.40	1.6E − 05	0.0148	91
TEKT2	27,285	Tektin 2 (testicular)	0.59	5.3E − 05	0.0284	66
KCNJ18	100,134,444	Potassium inwardly rectifying channel subfamily J, member 18	0.65	3.6E − 05	0.0237	61
MDGA1	266,727	MAM domain containing glycosylphosphatidylinositol anchor 1	0.47	2.3E − 07	0.0007	54

Representative list of genes up-expressed (< 0.66-fold) in priests comparing to the controls. The described data is in descending ordered of the average raw signals (> 50) in the priests' group microarray data. C17orf97 has been validated by quantitative real-time PCR (Fig. 1)

IFI44L, HERC5, DEFA4, FOLR3, HBG1, C17orf97, and S100P, between the priests and the non-priest controls. As shown in Fig. 1, the participant priests showed a significantly up-regulated expression of seven genes that are involved in type I IFN anti-viral responses (e.g., MX1, IFI27, IFIT1, IFIT3, RSAD2, IFI44L, and HERC5). C17orf97 was validated by qRT-PCR as a significantly down-regulated gene in the priests. This gene product is currently known as LIAT1 or ligand of arginyltransferase 1 (ATE1), because C17orf97 interacts with ATE1, which is a component of the N-end rule pathway of proteins degradation [30]. Notably, C17orf97 is also down-regulated in the hedonic well-being status [13]. No significant differences were observed in the expression levels of the remaining examined genes between the groups ($P > 0.05$).

Metabolomics analysis

To clarify the underlying metabolic alterations in the spiritually trained Buddhists, we performed global metabolomics profiling to compare the priest participants with the non-priest controls. The complete data sets are shown in Additional file 7. In total, 275 metabolites were detected in the plasma of the participants as follows: 149 hydrophilic metabolites (105 and 44 metabolites in cationic and anionic modes, respectively) and 126 hydrophobic metabolites (62 and 64 metabolites in positive and negative modes, respectively). We detected 20 candidate

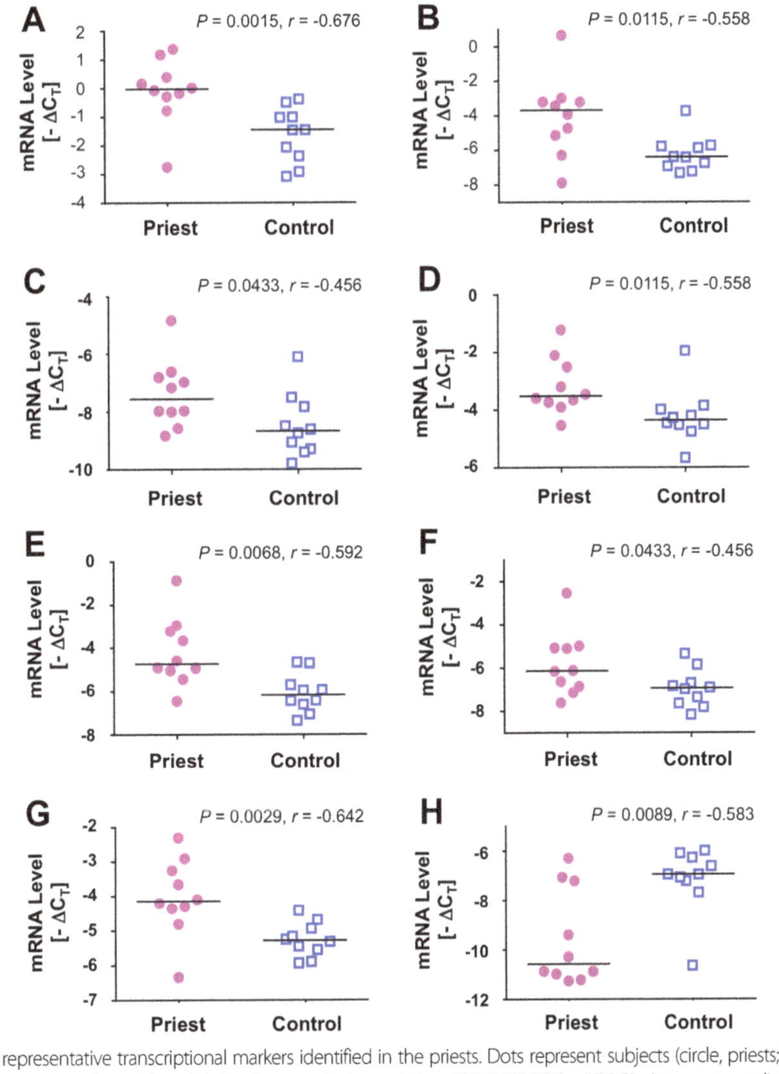

Fig. 1 The dot plots of the representative transcriptional markers identified in the priests. Dots represent subjects (circle, priests; square, controls), and line represents median. Eight genes were selected for validation by quantitative real-time PCR (qRT-PCR). qRT-PCR data are normalized to the housekeeping gene GAPDH. Differences of each transcript were compared using a Mann–Whitney *U* test, indicating *P* value and the effect size *r*. **a** MX1 ($P = 0.0015$, $r = -0.676$), **b** IFI27 ($P = 0.0115$, $r = -0.558$), **c** IFIT1 ($P = 0.0433$, $r = -0.456$), **d** IFIT3 ($P = 0.0115$, $r = -0.558$), **e** RSAD2 ($P = 0.0068$, $r = -0.592$), **f** IFI44L ($P = 0.0433$, $r = -0.456$), **g** HERC5 ($P = 0.0029$, $r = -0.642$), and **h** C17orf97 ($P = 0.0089$, $r = -0.583$). Statistical significance was defined as $P < 0.05$. Cohen's guidelines for the effect sizes (*r*) for Mann–Whitney *U* test are that a large effect is 0.5, a medium effect is 0.3, and small effect is 0.1 [28]

metabolites in the priests, and the relative concentrations of these metabolites were significantly higher in the priests than those in the non-priest controls based on non-parametric Mann–Whitney *U* test ($P < 0.05$, Additional file 7). We chose 14 certain metabolites from the 20 candidate metabolites because these metabolite peaks were detected in all 20 participants. Table 3 showed the chosen metabolites in the priests, including 3-aminoisobutyric acid (BAIBA), choline, amino acids (methionine, phenylalanine, histidine, valine, isoleucine, and leucine), amino acid derivatives (symmetric dimethylarginine (SDMA), asymmetric dimethylarginine (ADMA), 2-aminoadipic acid (AABA)), creatine, and acylcarnitine (13:1). Figure 2 shows the following four representative metabolites that were significantly higher ($P < 0.01$) in the priests than in the non-priest controls: methionine (1.43-fold, $P = 0.0002$, $r = -0.761$); BAIBA (3.06-fold, $P = 0.0015$, $r = -0.676$); phenylalanine (1.36-fold, $P = 0.0029$, $r = -0.642$); and choline (1.38-fold, $P = 0.0052$, $r = -0.608$).

Distribution of empathy sub-dimension scores

We evaluated the psychological trait of empathy in the priests and the non-priest controls. Table 4 presents the empathy sub-dimension scores in the priests and the non-priest controls. The priest group exhibited significantly higher empathy in four sub-dimensions than the non-priest controls as follows: "Sharing positive emotions with others" (priests 22.0 vs. controls 18.0, $P = 0.018$, $r = -0.521$); "Good feeling for others' positive emotions" (priests 22.0 vs. controls 20.0, $P = 0.029$, $r = -0.487$); "Sharing negative emotions with others" (priests 18.5 vs. controls 16.0, $P = 0.039$, $r = -0.462$); and "Sensibility about others' emotions" (priests 22.0 vs. controls 18.5, $P = 0.042$, $r = -0.455$).

Bivariate correlations (Spearman's ρ) between the identified transcripts, metabolites, and the empathy sub-dimension scores

To examine the correlations among the identified transcripts, metabolites, and the four empathy sub-dimensions, we performed a correlation analysis using Spearman's rank correlation coefficients (two-tailed) in the combined samples ($n = 10 + 10$: priests + non-priest controls). We observed a significant correlation ($\rho = -0.636$ or $0.448 \sim 0.700$) between the identified transcripts and metabolites (Table 5), between the identified transcripts and the four empathy sub-dimension scores (Table 6), and between the identified metabolites and the four empathy sub-dimension scores (Table 7).

As shown in Table 5, the BAIBA metabolite was significantly correlated with six distinct transcripts as follows: BAIBA was negatively correlated with C17orf97 ($\rho = -0.636$, $P = 0.026$, Fig. 3) and positively correlated with five anti-viral genes including RSAD2 ($\rho = 0.571$, $P = 0.008$), HERC5 ($\rho = 0.559$, $P = 0.010$), MX1 ($\rho = 0.556$, $P = 0.011$), IFIT3 ($\rho = 0.516$, $P = 0.020$), and IFIT1 ($\rho = 0.501$, $P = 0.025$). Meanwhile, in addition to BAIBA, the MX1 gene was correlated with five different metabolites, including methionine, SDMA, phenylalanine, and glycodeoxycholic acid.

As shown in Table 6, "Sharing positive emotions with others" was associated with HERC5 ($\rho = 0.595$, $P = 0.006$), RSAD2 ($\rho = 0.537$, $P = 0.015$), MX1 ($\rho = 0.511$, $P = 0.019$), IFIT3 ($\rho = 0.483$, $P = 0.031$), IFI44L ($\rho = 0.479$, $P = 0.033$),

Table 3 Fourteen metabolite profiling in the priests

Metabolites	Priests ($n = 10$)	Control ($n = 10$)	Fold change	*P* value	*r*
Methionine	5.2E − 03 (4.7–6.0) E − 03	3.7E − 03 (3.1–4.1) E − 03	1.43	*0.0002*	− 0.761
3-Aminoisobutyric acid (BAIBA)	10.0E − 04 (9.6–13.7) E − 04	2.5E − 04 (2.0–5.4) E − 04	3.06	*0.0015*	− 0.676
Phenylalanine	2.7E − 02 (2.4–3.0) E − 02	2.2E − 02 (2.1–2.3) E − 02	1.36	0.0029	− 0.642
Choline	8.1E − 03 (7.5–10.2) E − 03	6.7E − 03 (5.6–7.3) E − 03	1.38	0.0052	− 0.608
SDMA (symmetric dimethylarginine)	2.2E − 04 (2.0–2.3) E − 04	2.0E − 04 (1.6–2.0) E − 04	1.20	0.010	− 0.563
2-Aminoadipic acid	3.3E − 04 (2.9–3.8) E − 04	2.0E − 04 (1.7–2.4) E − 03	1.45	0.012	− 0.558
2-Aminobutyric acid (AABA)	8.7E − 03 (7.6–10.7) E − 03	6.8E − 03 (5.9–7.3) E − 03	1.38	0.019	− 0.524
ADMA (asymmetric dimethylarginine)	1.8E − 04 (1.6–2.2) E − 04	1.5E − 04 (1.4–1.7) E − 04	1.20	0.025	− 0.507
Valine	10.3E − 02 (9.4–10.9) E − 02	9.0E − 02 (8.6–10.5) E − 02	1.09	0.023	− 0.507
AC(13:1)	5.9E − 05 (4.5–10.3) E − 05	3.2E − 05 (2.2–4.3) E − 05	1.79	0.027	− 0.491
Creatinine	1.3E − 02 (1.0–2.0) E − 02	0.9E − 02 (0.8–1.1) E − 02	1.64	0.030	− 0.483
Leucine	7.8E − 02 (7.5–9.2) E − 02	7.1E − 02 (6.8–7.8) E − 02	1.14	0.036	− 0.473
Histidine	2.3E − 02 (2.1–2.4) E − 02	2.0E − 02 (1.7–2.3) E − 02	1.10	0.043	− 0.456
Isoleucine	4.5E − 02 (4.2–5.2) E − 02	3.8E − 02 (3.4–4.4) E − 02	1.21	0.043	− 0.456

Values are expressed as median and interquartile range (25–75th percentile). A *P* value < 0.05 is statistically significant by Mann–Whitney *U* test. A *P* value < 0.05 is statistically significant by Mann–Whitney *U* test. Cohen's guidelines for the effect sizes (*r*) for Mann–Whitney *U* test are that a large effect is 0.5, a medium effect is 0.3, and small effect is 0.1 [28]

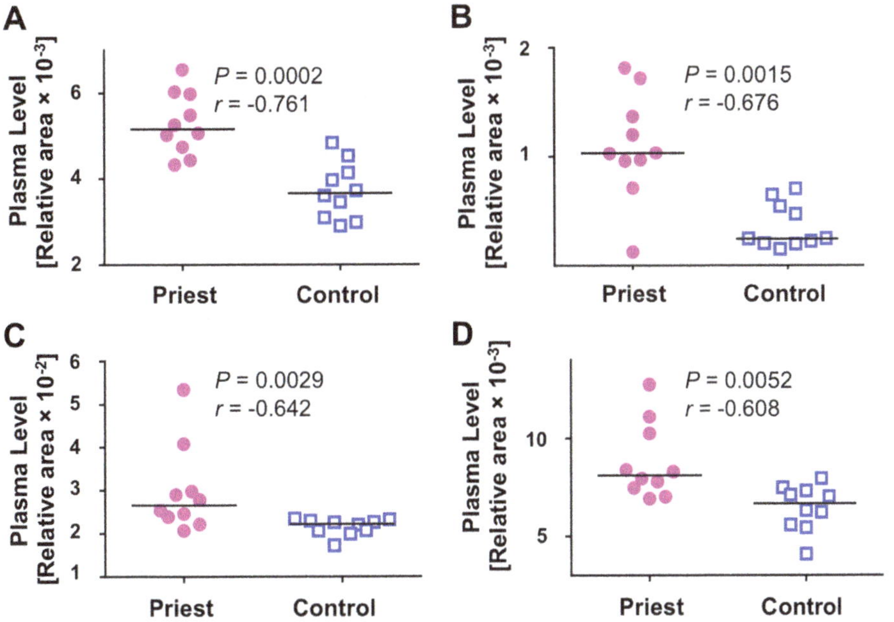

Fig. 2 Representative four plasma metabolite markers identified in the priests. Dots represent subjects (circle, priests; square, controls), and line represents median. **a** Methionine in plasma of priests vs. the controls ($P = 0.0002$, $r = -0.761$). **b** 3-Aminoisobutyric acid (BAIBA) in plasma of priests vs. the controls ($P = 0.0015$, $r = -0.676$). **c** Phenylalanine in plasma of priests vs. the controls ($P = 0.0029$, $r = -0.642$). **d** Choline in plasma of priests vs. the controls ($P = 0.0052$, $r = -0.608$). Plasma level of each metabolite was shown as the relative area value calculated for metabolomics analysis by the Human Metabolome Technologies, Inc. The relative area value was defined as the relative concentration of each metabolite. Differences of each plasma metabolites were compared using a Mann–Whitney U test, indicating P value and the effect size r. Statistical significance was defined as $P < 0.05$. Cohen's guidelines for the effect sizes (r) for Mann–Whitney U test are that a large effect is 0.5, a medium effect is 0.3, and small effect is 0.1 [28]

and C17orf97 ($\rho = -0.459$, $P = 0.042$), while "Sharing negative emotions with others" was associated with IFIT3 ($\rho = 0.603$, $P = 0.005$, Fig. 3), IFIT1 ($\rho = 0.569$, $P = 0.009$), IFI44L ($\rho = 0.536$, $P = 0.015$), HERC5 ($\rho = 0.487$, $P = 0.030$), and RSAD2 ($\rho = 0.448$, $P = 0.047$). "Sensitivity about others' emotions" was correlated with IFIT3 ($\rho = 0.496$, $P = 0.026$) and HERC5 ($\rho = 0.470$, $P = 0.037$). "Good feeling for others' positive emotions" was negatively correlated with C17orf97 ($\rho = -0.528$, $P = 0.017$).

As shown in Table 7, "positive" emotion-related empathy was correlated with five distinct metabolites. "Sharing positive emotions with others" was associated with BAIBA ($\rho = 0.700$, $P = 0.001$, Fig. 3), ADMA ($\rho = 0.602$, $P = 0.005$), SDMA ($\rho = 0.561$, $P = 0.010$), and methionine ($\rho = 0.477$, $P = 0.033$). "Good feeling for others' positive emotions" was associated with SDMA ($\rho = 0.649$, $P = 0.002$), ADMA ($\rho = 0.569$, $P = 0.009$), histidine ($\rho = 0.491$, $P = 0.028$), BAIBA ($\rho = 0.455$, $P = 0.044$). Finally, "Sensibility about others' emotions" was positively related with creatinine ($\rho = 0.573$, $P = 0.008$).

Discussion

Spirituality/religiosity is one of the unique aspects of human social environments and can be positively associated with psychological and physical health according to psycho-neuro-immune models of health regulation [1]. Our current cross-sectional study defined the systemic

Table 4 Empathy scores in priests and the controls

Psychology	Priests ($n = 10$)	Control ($n = 10$)	P value	r
Sharing positive emotions with others	22.0 (18.0–24.0)	18.0 (15.0–20.0)	0.018	− 0.521
Good feeling for others' positive emotions	22.0 (19.8–25.0)	20.0 (17.8–21.0)	0.029	− 0.487
Sharing negative emotions with others	18.5 (17.8–20.3)	16.0 (13.8–19.0)	0.039	− 0.462
Sympathy for others' negative emotions	20.5 (19.5–23.5)	20.0 (18.8–20.0)	0.118	− 0.368
Sensibility about others' emotions	22.0 (20.0–25.0)	18.5 (17.0–23.0)	0.042	0.455
Perspective taking	20.5 (18.5–25.0)	19.0 (17.3–20.0)	0.106	0.368

Values are expressed as median and interquartile range (25–75th percentile). A P value < 0.05 is statistically significant by Mann–Whitney U test. Cohen's guidelines for the effect sizes (r) for Mann–Whitney U test are that a large effect is 0.5, a medium effect is 0.3, and small effect is 0.1 [28]

Table 5 Bivariate correlations (Spearman's ρ) between the transcriptional markers and metabolite markers

Transcriptional marker vs. metabolite marker	ρ	P value
C17orf97 vs. 3-amino isobutyric acid (BAIBA)	− 0.636	0.026
RSAD2 vs. 3-amino isobutyric acid (BAIBA)	0.571	0.008
HERC5 vs. 3-amino isobutyric acid (BAIBA)	0.559	0.010
MX1 vs. 3-amino isobutyric acid (BAIBA)	0.556	0.011
IFIT3 vs. 3-amino isobutyric acid (BAIBA)	0.516	0.020
IFIT1 vs. 3-amino isobutyric acid (BAIBA)	0.501	0.025
MX1 vs. methionine	0.538	0.014
IFI27 vs. methionine	0.492	0.028
MX1 vs. phenylalanine	0.480	0.032

signatures of the gene expression and the metabolic profiles in Buddhist priests compared with those in the non-priest controls. The list of identified transcripts included components of type I IFN responses involved in innate anti-viral protection. Additionally, the identified metabolites indicated that enhanced proteolytic metabolism might occur in the priests. In contrast, no significant differences were observed in the healthy lifestyle behaviors and daily nutrient intake between the two groups. Interestingly, we observed a significant correlation between empathy and the molecular markers identified in the priests.

Among the top ranked list of significantly up-regulated transcripts in the priests, seven gene products (MX1, RSAD2, IFIT1, IFIT3, IFI27, IFI44L, and HERC5) were involved in anti-viral protection mechanisms with type I IFN signaling [31–33]. It is unclear whether the observed basal physiological state in the priest participants controls inflammatory responses or the pathological inflammatory states. However, acute infectious illnesses or chronic inflammation likely did not cause the up-regulation of the type I IFN responsive genes in the priests because no apparent infection signs and symptoms were observed in any of the participants on the day of the experiment, and no significant difference were detected in the plasma CRP levels (P = 0.47), which is a systemic inflammatory marker, between the priests and the non-priest controls (Additional file 1). Notably, the fold change value of these seven transcripts (1.77- to 2.99-fold) seemed to be within the normal range of physiological variations and were one-digit less than those that are observed under infection conditions [34, 35]. Furthermore, no significant differences were detected between the two groups in the expression of inflammation-related genes. One unique feature of the type I IFN system is a weak signal that constitutively produces IFN-α, which is critical to "rev-up" for efficient and robust responses to viral infections in innate immune cells [36]. The up-regulation of the seven genes might possibly provide a foundation for more efficient or sensitive cellular responses for anti-viral protection. "Trained immunity" was recently described as a concept related to the memory-like innate immune function after microbial encounters [37]. Repeated exposure to specific environments or conditions during common religious/spiritual practices might trigger long-lasting changes as "allostatic responses" [38] in the anti-viral response of the type I IFN pathways. Furthermore, the trained spiritual/religious priests may have conditioned themselves to respond more quickly or with a higher sensitivity to environmental changes, particularly viral infections. Notably, six of the seven identified anti-viral genes were inversely down-regulated under socially adverse conditions, such as chronic loneliness [9, 10]. In this pilot study, we did not exclude three priests with known physical ailments from statistical analyses due to the limited number of recruited priests. Nonetheless, the results of analyses performed when these priests were excluded exhibited similar significances (Additional file 8).

Table 6 Bivariate correlations (Spearman's ρ) between the transcriptional markers and empathy sub-dimension scores

Transcriptional marker vs. empathy	ρ	P value
HERC5 vs. sharing positive emotions with others	0.595	0.006
RSAD2 vs. sharing positive emotions with others	0.537	0.015
MX1 vs. sharing positive emotions with others	0.511	0.019
IFIT3 vs. sharing positive emotions with others	0.483	0.031
IFI44L vs. sharing positive emotions with others	0.479	0.033
C17orf97 vs. sharing positive emotions with others	− 0.459	0.042
C17orf97 vs. good feeling for other's positive emotions	− 0.528	0.017
IFIT3 vs. sharing negative emotions with others	0.603	0.005
IFIT1 vs. sharing negative emotions with others	0.569	0.009
IFI44L vs. sharing negative emotions with others	0.536	0.015
HERC5 vs. sharing negative emotions with others	0.487	0.030
RSAD2 vs. sharing negative emotions with others	0.448	0.047
IFIT3 vs. sensibility about other's emotions	0.496	0.026
HERC5 vs. sensibility about other's emotions	0.470	0.037

We uncovered that Buddhists had higher levels of plasma free amino acids and amino acid-derivatives than the non-priest controls. These results indicated that the Buddhists might exhibit higher intracellular protein turnover/metabolism. The two major systems involved in cellular proteolysis are the ubiquitin-proteasome system [39] and the autophagy system [40]. Both systems play crucial roles in adapting and properly sustaining protein turnover (i.e., proteostasis) under a large variety of different environmental conditions. In particular, the autophagy pathway plays a crucial role in the resistance

Table 7 Bivariate correlations (Spearman's ρ) between the metabolite markers and empathy sub-dimension scores

Metabolite marker vs. empathy	ρ	P value
3-Amino isobutyric acid (BAIBA) vs. sharing positive emotions with others	0.700	0.001
Asymmetric dimethylarginine (ADMA) vs. sharing positive emotions with others	0.602	0.005
Symmetric dimethylarginine (SDMA) vs. sharing positive emotions with others	0.561	0.010
Methionine vs. sharing positive emotions with others	0.477	0.033
Symmetric dimethylarginine (SDMA) vs. good feeling for other's positive emotions	0.649	0.002
Asymmetric dimethylarginine (ADMA) vs. good feeling for other's positive emotions	0.569	0.009
Histidine vs. good feeling for other's positive emotions	0.491	0.028
3-Amino isobutyric acid (BAIBA) vs. good feeling for other's positive emotions	0.455	0.044
Creatinine vs. sensitivity about other's emotions	0.573	0.008

against infections and inflammatory conditions [41]. Therefore, autophagy may mediate the cytoprotective processes by maintaining protein and organelle quality control in the system better in the Buddhists than in the non-priest-controls. Among the identified plasma metabolites, BAIBA was correlated with the expression of six transcripts, including C17orf97, RSAD2, HERC5, MX1, IFIT3, and ITIF1 (Table 5). Furthermore, BAIBA can be generated by the catabolism of the branched-chain amino acid valine and functions as a myokine secreted from the skeletal muscle cells to alter the functions of other tissues. For example, BAIBA increases the expression of brown-adipocyte-specific genes in white adipocytes and β-oxidation genes in hepatocytes through a peroxisome proliferator-activated receptor alpha (PPARα)-dependent mechanism [42]. Although we cannot determine a causal relationship responsible for the observed correlations, we speculate that circulating BAIBA may have greater effects on leukocyte gene expression. Future studies are needed to determine the direct action of BAIBA on the up-regulation of anti-viral genes in peripheral leukocytes.

Choline, which is another notable metabolite in the priests, serves as a component in blood and membrane phospholipids and structural lipoproteins and is a precursor for the neurotransmitter acetylcholine. Choline and its oxidative product betaine, which serves as a methyl group donor, are important sources of one-carbon units. In this study, the priests showed significantly higher levels of plasma choline (8.1E − 3 vs. 6.7E − 3: 1.38-fold, P = 0.0052, r = − 0.608, Table 3) than the non-priest controls, and their levels of betaine were higher (2.7E − 2 vs. 2.3E − 2: 1.17-fold, P = 0.052, Additional file 7). Choline promotes homocysteine remethylation to methionine and affects the

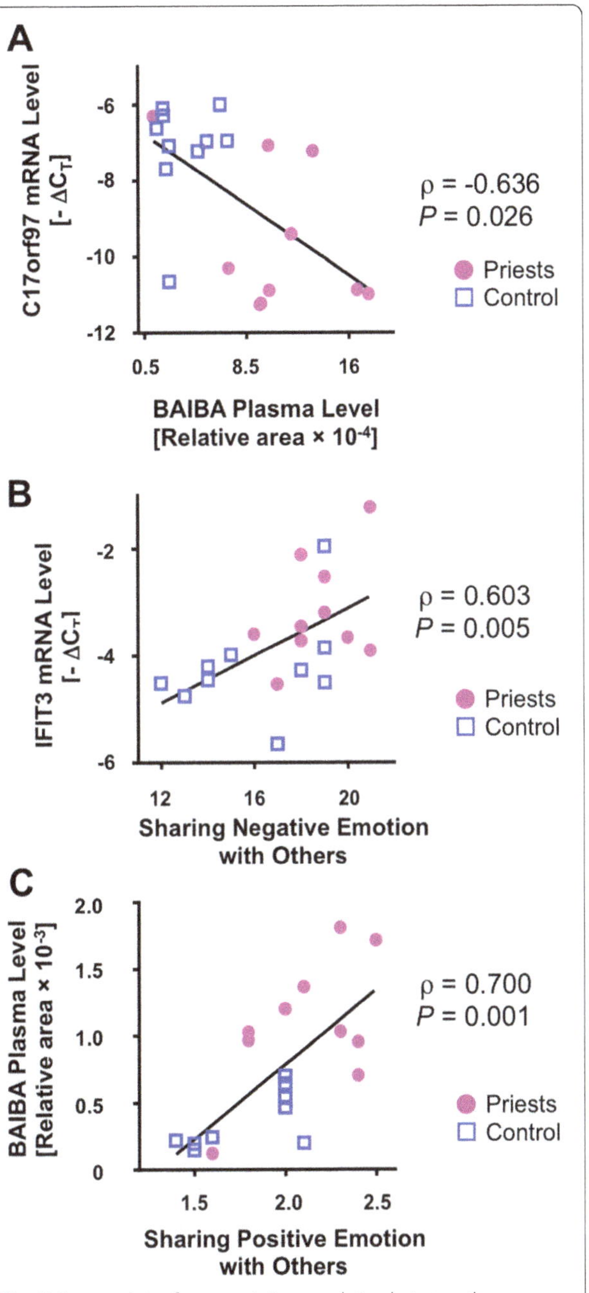

Fig. 3 Scatter plots of representative correlation between the molecular markers identified in the priests and empathy. Spearman's rank correlation coefficients were calculated. Dots represent subjects (circle, priests; square, controls). **a** C17orf97 transcript vs. 3-aminoisobutyric acid (BAIBA) metabolite (ρ = − 0.636). **b** IFIT3 transcript vs. empathy aspect "Sharing negative emotions with others" (ρ = 0.603). **c** 3-Aminoisobutyric acid (BAIBA) metabolite vs. empathy aspect "Sharing positive emotions with others" (ρ = 0.700)

concentration of the universal methyl donor S-adenosylmethionine (SAM). Altered concentrations of SAM may influence DNA methylation at cytosine bases, thereby influencing gene transcription, genomic imprinting, and genomic stability [43]. Plasma choline is delivered across the blood-brain-barrier by a specific transporter [44].

Therefore, in this study, the higher levels of plasma choline likely increased the brain choline levels in the priests compared with that in the non-priest controls. Experimental studies suggest that increased levels of brain choline and phosphatidylcholine improve memory performance and cognitive function as well as enhance neuroprotection and neurorepair activities [45]. Additionally, plasma choline levels are inversely associated with high anxiety levels [46]. Altogether, we hypothesized that the higher levels of plasma choline in the priests may exhibit some beneficial effects on the neuropsychological status, such as improvements in cognitive function and reduced anxiety. Troen et al. showed that cognitive dysfunction in folate-deficient rats was related to depletion of phosphatidylcholine in the brain. Notably, dietary methionine could prevent both cognitive impairment and low phosphatidylcholine [47]. Therefore, the higher levels of plasma methionine in the priests in this study could also present further beneficial contributions to cognitive functions.

The highlight of this study was that we observed a significant correlation between empathy and the molecular markers identified in the priests (Tables 6 and 7). The Buddhist priests showed higher levels of empathy than the non-priest control (Table 4). Previous studies have shown that religious people tend to perceive themselves as pro-social and report higher levels of altruism or charitable deeds compared with non-religious people [5]. Individual Buddhists lead their own daily lives according to Buddhism values and virtues. Compassion (or loving-kindness) is a fundamental Buddhist concept of interpersonal relationships and is defined as the deep wish to relieve others' suffering, coupled with the motivation to alleviate such suffering [17]. Empathy is a core element of both compassion (the wish to relieve others' suffering) and loving-kindness (the wish of happiness for others) and an affective response that arises from the comprehension of another's emotional state [18]. Although compassion is interpersonal, compassion has also been empirically linked to personal benefits, including increased positive emotions [48], improved physical health [49], and a reduced immunological stress response [50]. Emerging literature suggests that positive emotional styles (higher activation/arousal of positive emotions) are associated with improved protective immune responses [51, 52]. Notably, this study also showed that the positive emotional aspect of empathy is significantly correlated with the five transcripts for anti-viral responses (IFIT3, HERC5, RSAD2, IFI44L, and MX1) as shown in Table 6. Meanwhile, C17orf97, which was a down-regulated gene in the priests, and three representative metabolites, i.e., BAIBA, SDMA, and ADMA, were significantly correlated with the positive empathy aspects of "Sharing the positive emotions with others" and "Good feeling for others' positive emotions" (Tables 6 and 7). The current results are the first to demonstrate a link between the molecular signatures of Buddhist priests and the psychological aspects of empathy.

Finally, a large body of research indicates that lifestyle interventions such as physical exercise and a healthy diet are effective for inducing phenotypic alterations in circulating cells and influencing immune function as a non-clinical human model of the body's response to physiological stresses [53]. Additionally, both physical exercise and a healthy diet influence the human gene expression profiles in leukocytes [54, 55]. In the current pilot study, no significant differences in dietary habits and physical activity were observed between the priests and the non-priest controls (Additional files 2, 3 and 4). Despite the small sample sizes (n = 10 in each group), the scores on the overall HPLP-II in both groups were within the ranges obtaining in 512 male residents (mean age 45.5 ± 13.1 years) in a mixed rural-urban representative area in Japan. The actual states of the community residents' (n = 1176: 512 males and 664 females) lifestyles were recently investigated using the same HPLP II questionnaires used in this study [24].

Limitations

The identified transcript and metabolite profiles in the priest participants may potentially serve as markers and mediators for daily spiritual/religious practices. However, this study had several limitations. First, the sizes of the participant groups were too small to generalize the observed results. Second, the "cross-sectional design" made it impossible to determine the cause-and-effect relationship among the identified markers (transcriptomics and metabolomics), and spiritual/religious practices, or the empathy aspects. Third, the psychological aspects of empathy were based on subjective self-reports, leading to a potential recall bias, which could have influenced the accuracy of the reported data. Fourth, no direct biological measures of immune cells were available in this study (e.g., NK cell or dendritic cell activities or effector responses to an immunologic challenge). Finally, the priests selected for this study were from only one religious affiliation, Shingon Buddhism; therefore, we must compare the basal transcriptome/metabolome profiles of other religious priests carefully to generalize these results. Despite the relatively small sample size, we were able to detect the inter-individual consistency of the transcriptional/metabolic alterations in the Buddhist priests. In the future, it will be important to determine whether Buddhists acquire the representative transcriptome and metabolome profiles at baseline or during the course of training needed to become a prospective Buddhist priest [19]. Fancourt et al. recently demonstrated that group-drumming interventions produce psychological benefits associated with a shift toward an anti-inflammatory immune profile [56]. Therefore, we need to examine the correlations between the identified molecular markers and other characteristic factors of daily Buddhist practices (e.g., the number of Buddhist prayers one offers each day or the number of hits of a Buddhist drum each day). The up-regulated set of anti-viral responsive genes may indicate that the trained priests have gained unique immunological traits to protect their bodies from environmental parasitic infections based on the "behavioral-immune response hypothesis."

Conclusions

In this pilot study, we integrated in vivo phenotyping with transcriptomics, metabolomics, and psychological analyses, thereby identifying distinguishing biological characteristics and their association with empathy in Buddhism priests. The identification of the distinct transcripts and metabolites in the priests may be a first step toward understanding the molecular context of systemic biological alterations in the unique signature of spirituality/religiosity in Buddhists.

Additional files

Additional file 1: Physiological characteristics of all participants. Values are expressed as median and interquartile range (25–75th percentile). A *P* value < 0.05 is statistically significant by Mann–Whitney *U* test. Cohen's guidelines for the effect sizes (*r*) for Mann-Whitney *U* test are that a large effect is 0.5, a medium effect is 0.3, and small effect is 0.1 (Fritz et al. 2012). BMI, body mass index; T-protein, total protein; AST, aspartate transaminase; ALT, alanine transaminase; γ-GT, gamma-glutamyl transferase; CRP, C-reactive protein; HDL-Chol, high-density lipoprotein-cholersterol; LDL-Chol, low-density lipoprotein-cholesterol; BUN, blood urea nitrogen; HbA1c, hemoglobin A1c; RBC, red blood cell count; WBC, white blood cell count; MCV, mean corpuscular volume; MCH, mean corpuscular hemoglobin; MCH-C, mean corpuscular hemoglobin concentration; PLT-C, platelet count.

Additional file 2: Comparison of health-promoting behaviors between the priests and the controls estimated by health-promoting lifestyle profile-II (HPLP-II). Values are expressed as median and interquartile range (25–75th percentile). A *P* value < 0.05 is statistically significant by Mann–Whitney *U* test. Cohen's guidelines for the effect sizes (*r*) for Mann–Whitney *U* test are that a large effect is 0.5, a medium effect is 0.3, and small effect is 0.1 [28].

Additional file 3: Comparison of mean daily energy intake and crude and energy-adjusted nutrient intakes estimated by BDHQ between the priests and the controls. Representative mean daily crude nutrient intakes were estimated by BDHQ. Values are expressed as median and interquartile range (25–75th percentile). A *P* value < 0.05 is statistically significant by Mann–Whitney *U* test. Cohen's guidelines for the effect sizes (*r*) for Mann-Whitney *U* test are that a large effect is 0.5, a medium effect is 0.3, and small effect is 0.1 [28].

Additional file 4: Comparisons of health-promoting lifestyle profiles (HPLP-II) between the priests and the controls. HPLP-II is a 52-item questionnaire composed of two main categories and six sub-dimension scales. The health-promoting behaviors category includes health responsibility, physical activity, and nutrition subscales. The psychosocial well-being category includes spiritual growth, interpersonal relationship, and stress management subscales. Dots represent subjects (●, priests $n = 10$; □, controls $n = 10$), and line represents median. Differences of each sub-dimension of HPLP-II were compared using a Mann–Whitney *U* test, indicating *P* value and the effect size *r*. (A) health responsibility ($P = 0.864$, $r = -0.042$), (B) physical activity ($P = 0.837$, $r = -0.051$), (C) nutrition ($P = 0.423$, $r = -0.187$), (D) spiritual growth ($P = 0.678$, $r = -0.103$), (E) interpersonal relationship ($P = 0.615$, $r = -0.119$), (F) stress management ($P = 0.593$, $r = -0.127$). Statistical significance was defined as $P < 0.05$. Cohen's guidelines for the effect sizes (*r*) for Mann–Whitney *U* test are that a large effect is 0.5, a medium effect is 0.3, and small effect is 0.1 [28].

Additional file 5: The empathetic process scale questionnaire.

Additional file 6: The list of distinct transcripts identified in the priests.

Additional file 7: The metabolome analysis.

Additional file 8: The statistical analysis without data of three priest participants with known physical ailments (priest, $n = 7$ vs. non-priest controls; $n = 10$). Values are expressed as median and interquartile range (25–75th percentile). A *P* value < 0.05 is statistically significant by Mann–Whitney *U* test. A *P* value < 0.05 is statistically significant by Mann–Whitney *U* test. Cohen's guidelines for the effect sizes (*r*) for Mann–Whitney *U* test are that a large effect is 0.5, a medium effect is 0.3, and small effect is 0.1 [28].

Abbreviations

AABA: 2-Aminoadipic acid; ADMA: Asymmetric dimethyl-arginine; ATE1: Arginyltranseferase 1; BAIBA: 3-Aminoisobutylic acid; BDHQ: A brief-type self-administered diet history questionnaire; CRP: C-reactive protein; DEFA4: Defensin alpha 4; FOLR3: Folate receptor 3; HBG1: Hemoglobin γA; HERC5: HECT and RLD domain containing E3 ubiquitin protein ligase 5; HPLP-II: The Health-Promoting Lifestyle Profile-II; IFI44L: Interferon induced protein 44 like; IFIT: Interferon induced protein with tetratricopeptide repeats; IFN: Interferon; IL: Interleukin; KRTAP10-12: Keratin-associated protein; LIAT1: Ligand of arginyltransferase 1; MX1: MX dynamin-like GTPase 1; NK: Natural killer; PPARα: Peroxisome proliferator-activated receptor alpha; RSAD2: Radical S-adenosyl methionine domain containing 2; SIGLEC14: Sialic acid Ig-like lectin 14; SMDA: Symmetric dimethyl-arginine; TNF: Tumor necrosis factor

Acknowledgments

The authors are eternally grateful to Dr. Honnen Nakamura for his continuous encouragement during this research. We acknowledge Drs. Yukio Ichitani and Eisho Yoshikawa for critical reading of this manuscript and offering constructive suggestions.

Funding

This study was supported by Koyasan University Fujikin Shuhei Ogawa Memorial Fund for the Promotion of Projects and Research and by Mind and Gene Institute of Foundation for Advancement of International Science. The funders had no role in study design, data collection and analysis, decision to publish, or preparation of the manuscript.

Authors' contributions

JO, SS, MH, SN, TS, ET-O, MT, SA, and KM conceived and designed the experiments. JO, SS, MH, TS and SA performed the experiments. JO, SN, SS, and MH analyzed the data. SN contributed reagents/materials/analysis tools for microarray analysis. JO, SN and SA wrote the manuscript. All authors read and approved the final manuscript.

Consent for publication

Not applicable.

Competing interests

The authors declare that they have no competing interests.

Author details

[1]Department of Food and Nutrition, Tokyo Kasei University, Itabashi, Tokyo, Japan. [2]Foundation for Advancement of International Science, Kasuga, Tsukuba,

Japan. [3]Division of Health Sciences, Tsukuba University of Technology, Kasuga, Tsukuba, Japan. [4]Departmemt of Neurosugery, Center for Integrative Medicine, Kasuga, Tsukuba, Japan. [5]DNA Chip Research Inc., Minato, Tokyo, Japan. [6]National Center for Child Health and Development, Tokyo, Japan. [7]Graduate School of Medicine, Kyoto Prefectural University of Medicine, Kyoto, Japan. [8]The Institute of Esoteric Culture, Koyasan University, Ina, Wakayama, Japan.

References

1. Koenig HG. Religion, spirituality, and health: The research and clinical implications. ISRN Psychiatry. 2012;2012:33.
2. Koenig HG, Zaben FA, Khalifa DA. Religion, spirituality and mental health in the West and the Middle East. Asian J Psychiatr. 2012;5(2):180–2.
3. Loewenthal KM, MacLeod AK, Goldblatt V, Lubitsh G, Valentine JD. Comfort and joy? Religion, cognition, and mood in Protestants and Jews under stress. Cognit Emot. 2000;14(3):355–74.
4. McIntosh DN, Poulin MJ, Silver RC, Holman EA. The distinct roles of spirituality and religiosity in physical and mental health after collective trauma: a national longitudinal study of responses to the 9/11 attacks. J Behav Med. 2011;34(6):497–507.
5. Batson CD. Altruism in Humans. New York: Oxford University Press; 2011. ISBN 978-0-19-534106-5.
6. Hill PC, Pargament KI. Advances in the conceptualization and measurement of religion and spirituality. Implications for physical and mental health research. Am Psychol. 2003;58(1):64–74.
7. Steinmo S, Hagger-Johnson G, Shahab L. Bidirectional association between mental health and physical activity in older adults: Whitehall II prospective cohort study. Prev Med. 2014;66:74–9.
8. Miller L, Bansal R, Wickramaratne P, Hao X, Tenke CE, Weissman MM, Peterson BS. Neuroanatomical correlates of religiosity and spirituality: a study in adults at high and low familial risk for depression. JAMA Psychiatry. 2014;71(2):128–35.
9. Slavich GM, Cole SW. The emerging field of human social genomics. Clin Psychol Sci. 2013;1(3):331–48.
10. Cole SW. Human social genomics. PLoS Genet. 2014;10(8):e1004601.
11. Irwin MR, Cole SW. Reciprocal regulation of the neural and innate immune systems. Nat Rev Immunol. 2011;11(9):625–32.
12. Glaser R, Kiecolt-Glaser JK. Stress-induced immune dysfunction: implications for health. Nat Rev Immunol. 2005;5(3):243–51.
13. Fredrickson BL, Grewen KM, Coffey KA, Algoe SB, Firestine AM, Arevalo JMG, et al. A functional genomic perspective on human well-being. Proc Natl Acad Sci U S A. 2013;110(33):13684–9.
14. Fredrickson BL, Grewen KM, Algoe SB, Firestine AM, Arevalo JMG, Ma J, et al. Psychological well-being and the human conserved transcriptional response to adversity. PLoS One. 2015;10(3):e0121839.
15. Takimoto-Ohnishi E, Ohnishi J, Murakami K. Mind–body medicine: effect of the mind on gene expression. Person Med Univ. 2012;1(1):2–6.
16. Wang Z, Koenig HG, Zhang Y, Ma W, Huang Y. Religious involvement and mental disorders in mainland china. PLoS One. 2015;10(6):e0128800.
17. Wallace BA. Intersubjectivity in indo-Tibetan Buddhism. J Consci Stud. 2001; 8(5–7):209–30.
18. de Waal FBM. Putting the altruism back into altruism: the evolution of empathy. Annu Rev Psychol. 2008;59:279–300.
19. Sharf RH. Thinking through Shingon ritual. J Internat Associ Buddhist Stud. 2003;26(1):51–96.
20. Yamaoka M, Maeda N, Nakamura S, Kashine S, Nakagawa Y, Hiuge-Shimizu A, et al. A pilot investigation of visceral fat adiposity and gene expression profile in peripheral blood cells. PLoS One. 2012;7(10):e47377.
21. Jeffery IB, Higgins DG, Culhane AC. Comparison and evaluation of methods for generating differentially expressed gene lists from microarray data. BMC Bioinformatics. 2006;7:359.
22. Yamashita A, Zhao Y, Matsuura Y, Yamasaki K, Moriguchi-Goto S, Sugita C, et al. Increased metabolite levels of glycolysis and pentose phosphate pathway in rabbit atherosclerotic arteries and hypoxic macrophage. PLoS One. 2014;9(1):e86426.
23. Walker SN, Sechrist KR, Pender NJ. The health-promoting lifestyle profile II. Omaha: University of Nebraska Medical Center, College of Nursing; 1995. http://www.unmc.edu/nursing/faculty/health-promoting-lifestyle-profile-II.html.
24. Zhang S-C, Wie C-N, Harada K, Ueda K, Fukumoto K, Matsuo H, et al. Relationship between lifestyle and lifestyle-related factors in a rural-urban population of Japan. Environ Health Prev Med. 2013;18(4):267–74.
25. Kobayashi S, Honda S, Murakami K, Sasaki S, Okubo H, Hirota N, Notsu A, Fukui M, Date C. Both comprehensive and brief self-administered diet history questionnaires satisfactorily rank nutrient intakes in Japanese adults. J Epidemiol. 2012;22(2):151–9.
26. Davis MH. Measuring individual differences in empathy: evidence for a multidimensional approach. J Pers Soc Psychol. 1983;44(1):113–26.
27. Sawada M, Hayama D. Dispositional vengeance and anger on schadenfreude. Psychol Rep. 2012;111(1):322–34.
28. Fritz CO, Morris PE, Richler JJ. Effect size estimates: current use, calculations, and interpretation. J Exp Psychol. 2012;General 141(1):2–18.
29. Ali NS, Ali OSJ. Stress perception, lifestyle behaviors, and emotional intelligence in undergraduate nursing students. Nurs Edu Pract. 2016;6(10):16–22.
30. Brower CS, Rosen CE, Jones RH, Wadas BC, Piatkov KI, Varshavsky A. Liat1, an arginyltransferase-binding protein whose evolution among primates involved changes in the numbers of its 10-residue repeats. Proc Natl Acad Sci U S A. 2014;111(46):E4936–45.
31. Sadler AJ, Williams BR. Interferon-inducible antiviral effectors. Nat Rev Immunol. 2008;8(7):559–68.
32. Seo JY, Yaneva R, Cresswell P. Viperin: a multifunctional, interferon-inducible protein that regulates virus replication. Cell Host Microbe. 2011;10(6):534–9.
33. Diamond MS, Farzan M. The broad-spectrum antiviral functions of IFIT and IFITM proteins. Nat Rev Immunol. 2013;13(1):46–57.
34. Ioannidis I, McNally B, Willette M, Peeples ME, Chaussabel D, Durbin JE, et al. Plasticity and virus specificity of the airway epithelial cell immune response during respiratory virus infection. J Virol. 2012;86(10):5422–36.
35. Zhai Y, Franco LM, Atmar RL, Quarles JM, Arden N, Bucasas KL, et al. Host transcriptional response to influenza and other acute respiratory viral infections––a prospective cohort study. PLoS Pathog. 2015;11(6):e1004869.
36. Takaoka A, Taniguchi T. New aspects of IFN-alpha/beta signalling in immunity, oncogenesis and bone metabolism. Cancer Sci. 2003;94(5):405–11.
37. Netea MG, Latz E, Mills KH, O'Neill LA. Immune memory: a paradigm shift in understanding host defense. Nat Immunol. 2015;16(7):675–9.
38. Karatsoreos IN, McEwen BS. Resilience and vulnerability: a neurobiological perspective. F1000Prime Rep. 2013;5:13.
39. Schmidt M, Finley D. Regulation of proteasome activity in health and disease. Biochim Biophys Acta. 2014;1843(1):13–25.
40. Martinez-Lopez N, Athonvarangkul D, Singh R. Autophagy and aging. Adv Exp Med Biol. 2015;847:73–87.
41. Levine B, Mizushima N, Virgin HW. Autophagy in immunity and inflammation. Nature. 2011;469(7330):323–35.
42. Roberts LD, Boström P, O'Sullivan JF, Schinzel RT, Lewis GD, Dejam A, et al. β-Aminoisobutyric acid induces browning of white fat and hepatic β-oxidation and is inversely correlated with cardiometabolic risk factors. Cell Metab. 2014;19(1):96–108.
43. Ueland PM. Choline and betaine in health and disease. J Inherit Metab Dis. 2011;34(1):3–15.
44. Geldenhuys WJ, Allen DD. The blood-brain barrier choline transporter. Cent Nerv Syst Agents Med Chem. 2012;12(2):95–9.
45. Alvarez-Sabín J, Román GC. The role of citicoline in neuroprotection and neurorepair in ischemic stroke. Brain Sci. 2013;3(3):1395–414.
46. Bjelland I, Tell GS, Vollset SE, Konstantinova S, Ueland PM. Choline in anxiety and depression: the Hordaland health study. Am J Clin Nutr. 2009;90(4):1056–60.
47. Troen AM, Chao WH, Crivello NA, D'Anci KE, Shukitt-Hale B, Smith DE, Selhub J, Rosenberg IH. Cognitive impairment in folate-deficient rats corresponds to depleted brain phosphatidylcholine and is prevented by dietary methionine without lowering plasma homocysteine. J Nutr. 2008;138(12):2502–9.
48. Dunn EW, Aknin LB, Norton MI. Spending money on others promotes happiness. Science. 2008;319(5870):1687–8.
49. Kok BE, Coffey KA, Cohn MA, Catalino LI, Vacharkulksemsuk T, Algoe SB, et al. How positive emotions build physical health: perceived positive social connections account for the upward spiral between positive emotions and vagal tone. Psychol Sci. 2013;24(7):1123–32.
50. Pace TW, Negi LT, Adame DD, Cole SP, Sivilli TI, Brown TD, et al. Effect of compassion meditation on neuroendocrine, innate immune and behavioral responses to psychosocial stress. Psychoneuroendocrinology. 2009;34(1):87–98.

51. Marsland AL, Cohen S, Rabin BS, Manuck SB. Trait positive affect and antibody response to hepatitis B vaccination. Brain Behav Immun. 2006; 20(3):261–9.
52. Dhabhar FS. Enhancing versus suppressive effects of stress on immune function: implications for immunoprotection versus immunopathology. Allergy Asthma Clin Immunol. 2008;4(1):2–11.
53. Gleeson M, Bishop NC, Stensel DJ, Lindley MR, Mastana SS, Nimmo MA. The anti-inflammatory effects of exercise: mechanisms and implications for the prevention and treatment of disease. Nat Rev Immunol. 2011;11(9):607–15.
54. Olsen KS, Skeie G, Lund E. Whole-blood gene expression profiles in large-scale epidemiological studies: what do they tell? Curr Nutr Rep. 2015;4(4):377–86.
55. Ntanasis-Stathopoulos J, Tzanninis HG, Philippou A, Koutsilieris M. Epigenetic regulation on gene expression induced by physical exercise. J Musculoskelet Neuronal Interact. 2013;13(2):133–46.
56. Fancourt D, Perkins R, Ascenso S, Carvalho LA, Steptoe A, Williamon A. Effects of group drumming interventions on anxiety, depression, social resilience and inflammatory immune response among mental health service users. PLoS One. 2016;11(3):e0151136.

12

Genomic variants in the *FTO* gene are associated with sporadic amyotrophic lateral sclerosis in Greek patients

Konstantinos Mitropoulos[1†], Eleni Merkouri Papadima[2†], Georgia Xiromerisiou[3], Angeliki Balasopoulou[2], Kyriaki Charalampidou[2], Vasiliki Galani[2], Krystallia-Vassiliki Zafeiri[2], Efthymios Dardiotis[3], Styliani Ralli[3], Georgia Deretzi[4], Anne John[5], Kyriaki Kydonopoulou[6], Elpida Papadopoulou[6], Alba di Pardo[7], Fulya Akcimen[8], Annalisa Loizedda[9,10], Valerija Dobričić[11], Ivana Novaković[11,12], Vladimir S. Kostić[11], Clint Mizzi[2], Brock A. Peters[13,14], Nazli Basak[8], Sandro Orrù[9], Evangelos Kiskinis[7], David N. Cooper[15], Spyridon Gerou[6], Radoje Drmanac[13], Marina Bartsakoulia[2], Evangelia-Eirini Tsermpini[2], Georgios M. Hadjigeorgiou[3], Bassam R. Ali[5], Theodora Katsila[2] and George P. Patrinos[2,5*]

Abstract

Background: Amyotrophic lateral sclerosis (ALS) is a devastating disease whose complex pathology has been associated with a strong genetic component in the context of both familial and sporadic disease. Herein, we adopted a next-generation sequencing approach to Greek patients suffering from sporadic ALS (together with their healthy counterparts) in order to explore further the genetic basis of sporadic ALS (sALS).

Results: Whole-genome sequencing analysis of Greek sALS patients revealed a positive association between *FTO* and *TBC1D1* gene variants and sALS. Further, linkage disequilibrium analyses were suggestive of a specific disease-associated haplotype for *FTO* gene variants. Genotyping for these variants was performed in Greek, Sardinian, and Turkish sALS patients. A lack of association between *FTO* and *TBC1D1* variants and sALS in patients of Sardinian and Turkish descent may suggest a founder effect in the Greek population. *FTO* was found to be highly expressed in motor neurons, while in silico analyses predicted an impact on *FTO* and *TBC1D1* mRNA splicing for the genomic variants in question.

Conclusions: To our knowledge, this is the first study to present a possible association between *FTO* gene variants and the genetic etiology of sALS. In addition, the next-generation sequencing-based genomics approach coupled with the two-step validation strategy described herein has the potential to be applied to other types of human complex genetic disorders in order to identify variants of clinical significance.

Keywords: Sporadic amyotrophic lateral sclerosis, *FTO* gene, Genomic variants, Whole-genome sequencing, Founder population

* Correspondence: gpatrinos@upatras.gr
[†]Equal contributors
[2]Department of Pharmacy, University of Patras School of Health Sciences, Campus, Rion, GR-26504 Patras, Greece
[5]Department of Pathology, College of Medicine and Health Sciences, United Arab Emirates University, Al-Ain, UAE
Full list of author information is available at the end of the article

Introduction

Amyotrophic lateral sclerosis (ALS), also known as Lou Gehrig's disease, is a neurodegenerative disorder that affects the upper and lower motor neurons in the motor cortex, brain stem, and spinal cord [20]. ALS was first described in the mid-1800s as a rapidly progressing motor neuron disease, with phenotypes of varying severity. Affected patients present muscle weakness, atrophy, and a general progressive paralysis. Eventually, most develop respiratory failure which leads to death on average 3 to 5 years after the onset of symptoms [45]. As yet, there is no effective treatment or pre-symptomatic diagnostic test for ALS [11].

Approximately 90–95% of ALS patients suffer from a sporadic form of ALS (sALS), which has both an environmental etiology and a likely strong genetic component [23]. The remaining 5–10% of ALS patients exhibit a clear genetic etiology for the disease, also known as familial ALS (fALS). Genomic variants in the *SOD1* gene have been shown to lead to fALS, with an autosomal dominant mode of inheritance. To date, more than 170 different *SOD1* genomic variants have been reported, affecting ~ 13% of fALS and ~ 1% of sALS patients [40]. Furthermore, variants in other genomic loci, such as *FUS* [27], *TARDBP* [25], and more recently *C9orf72* [36], have also been implicated in fALS. In the case of *C9orf72*, a non-coding GGGGCC expansion appears to have a very strong association with both fALS and sALS, with an overall prevalence of 40 and 5–20% in fALS and sALS patients, respectively [39].

The etiology of sALS is quite complex. A variety of different factors, including age, overall lifestyle, occupation, and/or exposure to xenobiotics, appear to influence the risk of developing sALS, but no specific chemical or toxin has been identified as a direct cause of ALS [44]. In addition to environmental factors, sALS also has a substantial genetic component. Genome-wide association studies have so far identified over 100 genomic loci that could yet prove to be linked to ALS disease predisposition with low-to-medium penetrance, supporting the notion of pronounced heritability for ALS [reviewed in [28]]. From these genomic loci, *SQSTM1* [19], *UBQLN2* [14], *VCP* [12], *ANG* [29], *PRPH* [38], *DCTN1*, *TAF15*, and *SETX* [10] appear to be more frequently mutated in sALS than in fALS, with an overall prevalence of between 4 and < 1%; these findings have been confirmed in several studies/cohorts as well as in functional and animal studies. Furthermore, the co-occurrence of genomic variants in addition to the presence of de novo pathogenic variants in sALS brings the estimate of overall heritability for ALS to as high as 60% [3, 4]. The identification of the entire lexicon of underlying genetic defect(s) in both fALS and sALS patients and their families would greatly improve disease diagnosis and potentially disease management.

There is little knowledge of the genetic basis of ALS in fALS and sALS patients of Greek origin. Although the incidence of ALS in Greek patients is comparable to other European populations (~ 1–2 per 100,000), only a single study has so far been performed, highlighting a rather high prevalence of the *C9orf72* GGGGCC repeat expansion in Greek ALS patients [35]. Here, we report a novel genomic locus, *FTO*, that appears to be associated with sALS. Genomic data comparison with sALS patients of Greek, Sardinian, and Turkish descent strongly suggest that this association may represent a founder effect that is specific to the Greek population.

Materials and methods

Case selection

Adopting a three-step strategy to determining data reliability, case selection criteria spanned all three patient cohorts studied herein (Fig. 1): (i) whole-genome sequencing analysis in sALS Greek patients ($n = 6$) and their healthy counterparts ($n = 5$), (ii) SNP genotyping (selected variants of interest) in sALS Greek patients ($n = 28$) and their healthy counterparts ($n = 50$), and (iii) SNP genotyping (selected variants of interest) in sALS patients of Greek ($n = 114$), Sardinian ($n = 114$), and Turkish ($n = 148$) origin and their healthy counterparts ($n = 39$, $n = 87$, $n = 74$, respectively). Patients were independently evaluated by at least two senior neurologists and diagnosed on the basis of revised El Escorial criteria [9]. Patients with definite, probable, probable laboratory supported, or possible ALS were included in the study. Healthy individuals were recruited after excluding the occurrence of common neurological disorders.

Whole-genome sequencing and bioinformatics analyses

Genomic DNA isolation was performed either from peripheral blood using the QIAamp Blood Midi Kit (Qiagen GmbH, Hilden, Germany) or by means of the phenol-chloroform extraction method as previously described with minor modifications or saliva as previously described [1]. Whole-genome sequencing was performed using Complete Genomics' (CA, USA) DNA nanoarray platform [18]. DNA sequencing coverage was ×110. Only high-quality call variants were included in the downstream analysis (> 93%). Genomes were aligned with the hg19 reference genome. Next-generation sequencing data (Complete Genomics Inc., CA) were analyzed using Ingenuity Variant Analysis version 3.1.2 (Ingenuity® Systems, www.ingenuity.com). This is well-established software that identifies associations between phenotypes, defined by the user in terms of the classification of the tested individuals, and variants in the

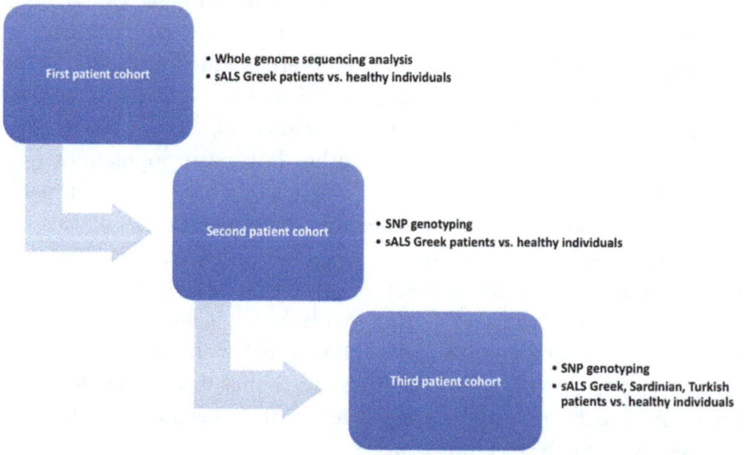

Fig. 1 Our three-step strategy towards data reliability. Our three-step strategy involved three patient cohorts: (i) whole-genome sequencing analysis in sALS Greek patients ($n = 6$) and their healthy counterparts ($n = 5$), (ii) SNP genotyping (selected variants of interest) in sALS Greek patients ($n = 28$) and their healthy counterparts ($n = 50$), and (iii) SNP genotyping (selected variants of interest) in sALS patients of Greek ($n = 114$), Sardinian ($n = 114$), and Turkish ($n = 148$) origin and their healthy counterparts ($n = 39, 87, 74$, respectively). The first patient cohort served as the discovery dataset, while the second and third patient cohorts were the training datasets

sequenced genome. Upon classification by phenotype (sALS vs. healthy individuals), a number of variants were listed; the output was filtered into smaller variant lists of (i) those found only in sALS patients ($n = 174$ single nucleotide polymorphisms (SNPs)), (ii) those not previously annotated or annotated only once, (iii) those with matching Human Gene Mutation Database (HGMD) entries, and (iv) whole-exome findings. All variants were filtered according to the analysis required, using custom scripts and Complete Genomics Analysis Tools (CGA™ Tools).

Linkage disequilibrium and in silico analyses

Pair-wise linkage disequilibrium (LD) calculations were based on phase genotype data (SNAP v2.2) [24], utilizing the HapMap Phase II + III (release 28) [6] and the 1000 Genomes Project (http://www.1000genomes.org) dataset for Caucasians (CEU). Findings were visualized on HaploView 4.2 and by using the LDmatrix module on the LDlink web tool [30]. To further investigate the role of the selected *FTO* and *TBC1D1* variants, we explored their effect on splicing motifs (including the "accept" and "donor" splice sites), the branch point, and auxiliary sequences that enhance (exonic splicing enhancers, ESE) or repress (exonic splicing silencers, ESS) splicing. To this end, in silico prediction was performed, using Human Splicing Finder (http://www.umd.be/HSF3/). This is a system that has rapidly become an international standard, as it combines 12 different algorithms [15]. Intronic variants, if "significant," are often located within gene regions that are characterized by a reduced level of genetic variation. Conserved elements were therefore explored using Variant Effect Predictor [34].

Downstream molecular genetic analysis

The first patient cohort served as our discovery dataset, while the second and third patient cohorts constituted our training datasets (Fig. 1). For this process, selected variants were subsequently validated in sALS Greek patients ($n = 28$) and their healthy counterparts ($n = 50$) (Fig. 1, Tables 1 and 2 and Additional file 1). PCR amplification was carried out according to the KAPA2G Fast HotStart protocol (KAPABIOSYSTEMS, MA, USA); detailed information regarding SNP amplification conditions is available upon request. For *FTO* rs2892469 (C>T) and rs1861869 (G>C), a PCR-based conventional Sanger resequencing approach was employed. Capillary electrophoresis was performed on an ABI Prism 3130xl DNA Analyzer (Applied Biosystems). For *FTO* rs17217144 (C>T) and rs7186521 (A>G) as well as *TBC1D1* rs6850200 (C>A), allele-specific PCR assays were developed (two alternative reverse primers hybridizing exclusively either to the wild-type or the mutant allele). Sanger sequencing was also employed to validate the allele-specific PCR method.

Table 1 Selected *FTO* and *TBC1D1* gene variants of prime interest

Gene	SNPs	Alleles	MAF-all	MAF-EUR	Variant type
FTO	rs2892469	C>T	0.31 (T)	0.49 (T)	
	rs1861869	C>G	0.38 (G)	0.54 (G)	
	rs17217144	C>T	0.31 (T)	0.49 (T)	Intronic
	rs7186521	A>G	0.37 (G)	0.49 (G)	
TBC1D1	rs6850200	C>A	0.41 (A)	0.37 (A)	

Table 2 Genotyping data of sALS patients and their healthy counterparts

Population groups		Genotype frequency (%)																			
		FTO rs2892469				*FTO* rs1861869				*FTO* rs17217144				*FTO* rs7186521				*TBC1D1* rs6850200			
		CC	TT	CT	*p* value	CC	GG	CG	*p* value	CC	TT	CT	*p* value	AA	GG	AG	*p* value	AA	CC	CA	*p* value
Second patient cohort	Greek healthy individuals	26	36	38	0.005	21	41	38	0.003	0	0	100	PA = $1.55e^{-12}$ PB = $7.78e^{-13}$	0	0	100	1.0	18	45	38	0.017
	Greek sALS patients	17	22	61		17	22	61		22	13	65		0	0	100		9	35	57	
Third patient cohort	Greek healthy individuals	26	36	38	0.033	21	41	38	PA = 0.072 PB = 0.071	6	2	92	0.002	0	0	100	$5.13e^{-39}$	14	43	43	0.723
	Greek sALS patients	21	23	56		17	29	54		22	2	76		18	65	18		11	42	48	
	Sardinian healthy individuals	38	25	38	0.63	36	31	33	0.61	24	36	40	0.67	35	22	42	0.64	20	41	39	0.53
	Sardinian sALS patients	31	28	41		29	34	37		27	30	43		29	25	46		15	39	46	
	Turkish healthy individuals	26	26	49	0.62	0	100	0	0.49	26	26	49	0.62	26	26	49	0.09	18	45	38	0.57
	Turkish sALS patients	24	21	55		0	99	1		26	20	54		21	26	53		13	44	43	

Sizing PCR across the *C9orf72* hexanucleotide repeats was performed using previously published primers [13] using repeat-primed PCR [32, 39]. Fragment length analysis was performed with the ABI 3500 genetic analyzer (Applied Biosystems, Foster City, Calif., USA) and analyzed using ABI GeneMapper v4.1 (Applied Biosystems).

qRT-PCR expression analysis for the *FTO* and *TBC1D1* genes

The relative expression levels of *FTO* and *TBC1D1* were determined using human-specific primers in a range of various human cell types; fibroblasts, embryonic stem cells (ESCs), astrocytes, neural PAX6+ progenitors, stem cell-derived purified motor neurons at three different stages of development (*Hb9*::GFP on days 0, 16, and 30), and stem cell-derived cortical neurons (NKX21 on days 45 and 50). Human brain and spinal cord RNA served as controls. Cell pellets were resuspended in Trizol Reagent (Life Technologies), and RNA was isolated following the manufacturer's protocol. 0.5 to 1 μg of RNA was treated with DNase I (Invitrogen) and subsequently used for the generation of cDNA using iSCRIPT Reverse Transcription Supermix (Bio-Rad). RT-PCR was performed using SYBR green on the CFX system (Bio-Rad). All assays were performed in triplicate. The averaged cycle of threshold (Ct) value of the two housekeeping genes (ACTIN/GAPDH) was subtracted from the Ct value of the gene of interest to obtain the ΔCt. Relative gene expression was determined as $^{2-}\Delta Ct$ (ΔΔCt) and expressed relative to undifferentiated human embryonic stem cells (hESCs).

Statistical analysis

Deviations from Hardy-Weinberg equilibrium were tested for using Pearson's goodness-of-fit chi-square (degrees of freedom = 1), log likelihood ratio chi-square (degrees of freedom = 1), and Fisher's Exact tests. De Finetti diagrams were also constructed [41]. Focusing on case-control (healthy individuals) phenotypes, we tested the null hypothesis of no association between rows and columns of the 2 × 3 matrix that contains the counts of the three genotypes (the two homozygotes and the heterozygote) among cases and controls. Genotype and allele frequencies were evaluated using Fisher's exact test. We also performed the Armitage test (Monte Carlo method; it obtains results that are closer to those of an exact test, since the classical Cochran-Armitage trend test is based on approximation) [5]. All tests were performed as two-tailed, and differences were considered to be statistically significant when $P < 0.05$. Further calculations using contingency tables (chi-square test) were performed to explore data stratification, considering (i) the age of onset (below or above 50 years old), (ii) El Escorial criteria, and (iii) spinal or bulbar ALS onset.

Results

Whole-genome sequencing analysis of Greek sALS patients reveals a positive association with *FTO* and *TBC1D1* variants

Our next-generation sequencing-based genomics approach yielded a total of 174 SNPs that were found only in the sALS patients studied; these were variously intergenic ($n = 144$), intronic ($n = 23$), and/or were present in upstream ($n = 2$), downstream ($n = 1$), and 3′UTR ($n = 2$) regions or non-coding RNAs ($n = 2$). Next, genes were clustered by metabolic or disease network and the most prominent ones are shown in Table 3. Our data analysis revealed three genomic variants of prime interest: *FTO* rs2892469 (C>T) and rs1861869 (G>C) and *TBC1D1* rs6850200 (C>A). Findings were further explored in Greek sALS patients and their healthy counterparts (second patient cohort). *FTO* intronic variants, rs2892469 (C>T) and rs1861869 (G>C), reached statistical significance ($p = 0.005$ and $p = 0.003$, respectively). For *TBC1D1* rs6850200 (C>A), a strong association also became evident (p value of 0.017). Data are summarized in Tables 1 and 2 and Additional file 1.

In addition, a *C9orf72* hexanucleotide expansion was identified in 1 of 28 patients with sALS (3.5%; Additional file 1: Table S4). This female patient was diagnosed with unequivocal ALS but was devoid of a positive family history for ALS.

Linkage disequilibrium analyses suggest a haplotype of *FTO* gene variants

Our findings support a disease-associated haplotype constituting *FTO* rs2892469 (C>T), rs1861869 (G>C), rs17217144 (C>T), and rs7186521 (A>G) (Fig. 2). SNAP analysis (HapMap datasets) was in agreement with the HaploView v4.2 analysis (our Greek cohort). The same result was produced for the LDmatrix module on the LDlink web tool. rs2892469 and rs1861869 are in perfect linkage disequilibrium (LD) (D′ = 1, LOD = 23.66, r^2 = 0.884), while rs17217144 and rs7186521 exhibit a strong LD (D′ = 0.882, LOD = 10.1, r^2 = 0.738). Block1 is based on confidence intervals, i.e., it was created due to 95% of informative comparisons being in "strong LD." Such findings serve as a paradigm as to how information technology tools coupled to experimental data may empower discovery.

In silico analyses suggest that *FTO* and *TBC1D1* variants affect splicing

TBC1D1 rs6850200 (C>A) and *FTO* rs2892469 (C>T) were predicted to create a splicing enhancer site which would potentially alter splicing. In the case of *FTO* rs1861869 (G>C), a putative exonic splicing silencer site is disrupted, while *FTO* rs7186521 (A>G) appears to have created a new donor splice site. No impact on

Table 3 Gene clustering per metabolic or disease network using Ingenuity Pathway Analysis (IPA) to assist decision-making

Disease or metabolic network	Genes	*p* value
Neural cell proliferation	ISL1, FTO, ADCYAP1, OTP, SPP1	6.02×10^{-5}
Motor neuron proliferation	ISL1, FTO	9.53×10^{-5}
Motor neuron dysfunction	ISL1, WNT4	1.28×10^{-3}
Development of neuromuscular junction	PDZRN3	1.65×10^{-3}
Branching of sensory neurons	PRKG1	1.65×10^{-3}
ALS	FUS	1.65×10^{-3}
Development of P and M cells of visual nervous system	OTP	1.65×10^{-3}
Growth of cerebellar cortex, regulation of upper cervical ganglia	ADCYAP1	1.65×10^{-3}
Cranial nerve development	ISL1, ERBB4	1.9×10^{-3}
Brain development	ADCYAP1, ERBB4, FTO, OTP, PRKG1	2.56×10^{-3}
Development of specific nerves of the hand, area of dendritic trees	ISL1	3.3×10^{-3}
Movement of the cranial neural crest, participation in the formation of the central nervous system, astrocyte attachment	ERBB4	3.3×10^{-3}
Neural cell number	ADCYAP1, ERBB4, WNT4, SPP1	3.75×10^{-3}
Nervous system morphology	ERBB4, FTO, WNT4, SPP1, YES1, ISL1	6.33×10^{-3}
Oligodendrocyte development, excitatory synapse formation	ERBB4	6.6×10^{-3}
Central nervous system development, dendritic cell development	ERBB4, ADCYAP1	7.4×10^{-3}
Motor neuron sprouting	FTO	9.88×10^{-3}
v3 spinal cord motor neuron sprouting	FTO	9.88×10^{-3}
Midbrain size	FTO	9.88×10^{-3}
Neuronal withdrawal	WNT4	1.32×10^{-2}
Brain morphology	ERBB4, FTO, SPP1, YES1	1.35×10^{-2}
Neuronal branching	ERBB4, ADCYAP1, PRKG1	1.58×10^{-2}
DNA repair	APTX, FTO, FUS	1.65×10^{-2}

splicing was predicted for *FTO* rs17217144 (C>T). Using Variant Effect Predictor, PhyloP, and PhastCons data showed that the selected *FTO* variants are evolutionarily conserved.

Replication analyses

Replication analyses were carried out in sALS patients of Greek origin (to ensure data reliability) and Sardinian and Turkish descent (to account for population differences (third patient cohort)). We found that four *FTO* intronic variants, rs2892469 (C>T) and rs1861869 (G>C), rs17217144 (C>T), and rs7186521 (A>G), reached statistical significance ($p = 0.033$, $p = 0.07$, $p = 0.002$, and $p < 0.001$, respectively), when Greek sALS patients were compared to their healthy counterparts. For *TBC1D1* rs6850200 (C>A), the aforementioned strong association (*p* value of 0.017) no longer held upon replication in Greek sALS patients (*p* value of 0.723). Data stratification based upon (i) the age of onset (below or above 50 years old), (ii) El Escorial criteria, and (iii) spinal or bulbar ALS onset did not alter our findings. It should be noted that all *FTO* and *TBC1D1* variants failed to reach statistical significance, even when stratified for *SOD1*, when Sardinian and Turkish sALS patients were considered. Data are summarized in Tables 1 and 2. The lack of association of these variants with sALS in patients of Sardinian and Turkish descent may suggest a founder effect in the Greek population.

FTO is highly expressed in motor neurons

To begin defining genotype-to-phenotype associations for *FTO*, *TBC1D1*, and ALS, we determined the relative mRNA levels for both genes across several human cell types. We specifically analyzed fibroblasts, undifferentiated embryonic stem cells (ESCs), astrocytes, neural PAX6 + progenitors, and stem cell-derived spinal cord motor neurons (MNs) and cortical neurons (CNs). We used the HUES3-*Hb9*::GFP cell line, which contains a GFP reporter under the control of the MN-specific Hb9 enhancer [17] to differentiate and FACS-purify MNs as described previously [26], on three different stages of development (days 0, 16, and 30). We also used the *NKX2.1*::GFP ESC line to differentiate and FACS-purify cortical neurons as described previously [33] on three different stages of development (days 45 and 50). Our qRT-PCR data show that *FTO* is relatively neural specific and notably is most highly expressed in motor neurons (Fig. 3). *TBC1D1* seems to be rather ubiquitously expressed.

Discussion

sALS is polygenic with a complex interplay with environmental and other factors. Previous studies have proposed genetic associations of a variety of genes with the etiology of sALS. In this article, we adopted a three-step approach to delineate the association of genomic variants with sALS in patients of Greek origin.

Using a next-generation sequencing-based genomics approach, we identified 174 variants that were only present in the initial sALS patients studied. We opted to focus on the *FTO* rs2892469 (C>T) and rs1861869 (G>C) and *TBC1D1* rs6850200 (C>A) variants, based on pathway analysis. Our replication analyses confirmed that the *FTO* but not the *TBC1D1* genomic variants

SNAP analysis (HapMap)

SNP	Proxy	RSquared	DPrime
rs2892469	rs17217144	0.936	1
rs2892469	rs7186521	0.936	1
rs2892469	rs1861869	0.84	0.964
rs1861869	rs17217144	0.846	1
rs1861869	rs7186521	0.846	1
rs1861869	rs2892469	0.84	0.964
rs17217144	rs7186521	1	1
rs17217144	rs2892469	0.936	1
rs17217144	rs1861869	0.846	1
rs7186521	rs17217144	1	1
rs7186521	rs2892469	0.936	1
rs7186521	rs1861869	0.846	1

HaploView v4.2 analysis (Greek Cohort)

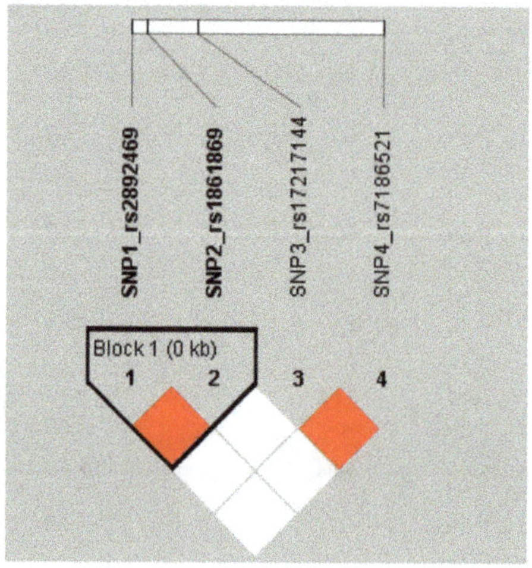

Fig. 2 Linkage disequilibrium and haplotype block analysis for *FTO* rs2892469, rs1861869, rs17217144, and rs7186521. Data suggest a haplotype for the *FTO* genomic variants studied herein. SNAP analysis refers to HapMap datasets, while the HaploView v4.2 analysis corresponds to our Greek cohort. The HaploView 4.2 color scheme denotes the following: white (D′ < 1, LOD < 2) and red (D′ = 1, LOD ≥ 2). rs2892469 and rs1861869 are in perfect LD (D′ = 1, LOD = 23.66, r^2 = 0.884); rs17217144 and rs7186521 exhibit a strong LD (D′ = 0.882, LOD = 10.1, r^2 = 0.738). Block1 is based on confidence intervals, i.e., it was created through 95% of informative comparisons being in "strong LD"

maintained their statistical significance as far as their association with sALS was concerned. Data stratification considering (i) the age of onset (below or above 50 years old), (ii) El Escorial criteria, and (iii) spinal or bulbar ALS onset did not affect our findings (not shown).

We subsequently attempted to explore whether the *FTO* gene variants identified as being associated with sALS in Greek patients might also be associated with sALS in patients of Turkish and Sardinian descent. However, genetic analysis of all *FTO* and *TBC1D1* variants failed to reach statistical significance, even when stratified for *SOD1*, when Sardinian and Turkish sALS patients were considered. This may be due to a founder effect in the Greek population.

In order to explore the functional significance of our findings, we measured the *FTO* and *TBC1D1* gene expression levels in several human cell types, namely fibroblasts, embryonic stem cells (ESCs), astrocytes, neural PAX6+ progenitors, stem cell-derived purified motor neurons at three different stages of development (*Hb9*::GFP on days 0, 16, and 30), and stem cell-derived cortical neurons (NKX21 on days 45 and 50). Our qRT-PCR data demonstrated that *FTO* gene expression is relatively neuron-specific and notably is highly expressed in motor neurons (Fig. 3). By contrast, *TBC1D1* would appear to be rather ubiquitously expressed.

Interestingly, long distance interactions have been reported for the *FTO* intronic region: (i) obesity-associated non-coding sequences within *FTO* interact with the promoter regions of *IRX3* and *FTO* in human, mouse, and zebrafish, and (ii) obesity-associated *FTO* gene variants influence *IRX3* expression in the human brain. The latter gene was associated with body mass, as *IRX3*-deficient mice displayed a decrease in their body weight by 25–30% [42]. Nutritional surveillance of ALS patients has been well established to be of fundamental importance, both in bulbar and spinal onset patients, and nutritional status is a prognostic factor for survival in ALS patients [16]. Jawaid et al. [22] reported that a decrease in body mass index is associated with faster progression of motor symptoms and shorter survival in ALS patients. Moreover, an LD block of 49 kb in the first intron of *FTO* contains numerous highly conserved non-coding elements (HCNEs) and lies within an extended region of conserved synteny involving the *IRX3/5/6* cluster, the closest gene of which is the transcription factor-encoding *IRX3* [37]. *IRX3* is a member of the Iroquois homeobox gene family and plays a role in the early stages of neural development [7].

We also identified an example of *C9orf72* hexanucleotide expansion in 1 of 28 patients with sALS (3.5%; Additional file 1: Table S4). This female patient, harboring this expansion, was diagnosed with unequivocal ALS but was lacked a positive family history of ALS. This massive hexanucleotide repeat expansion mutation in *C9orf72*, which has been linked to fALS, sALS, and frontotemporal dementia, has a single founder who is thought to have originated approximately 6300 years ago in Europe [31, 43]. Our findings differ somewhat from those reported by Mok and coworkers [35], the only study to date in fALS and sALS patients of Greek descent, in which 11 of 136 sALS patients (8.2%) were

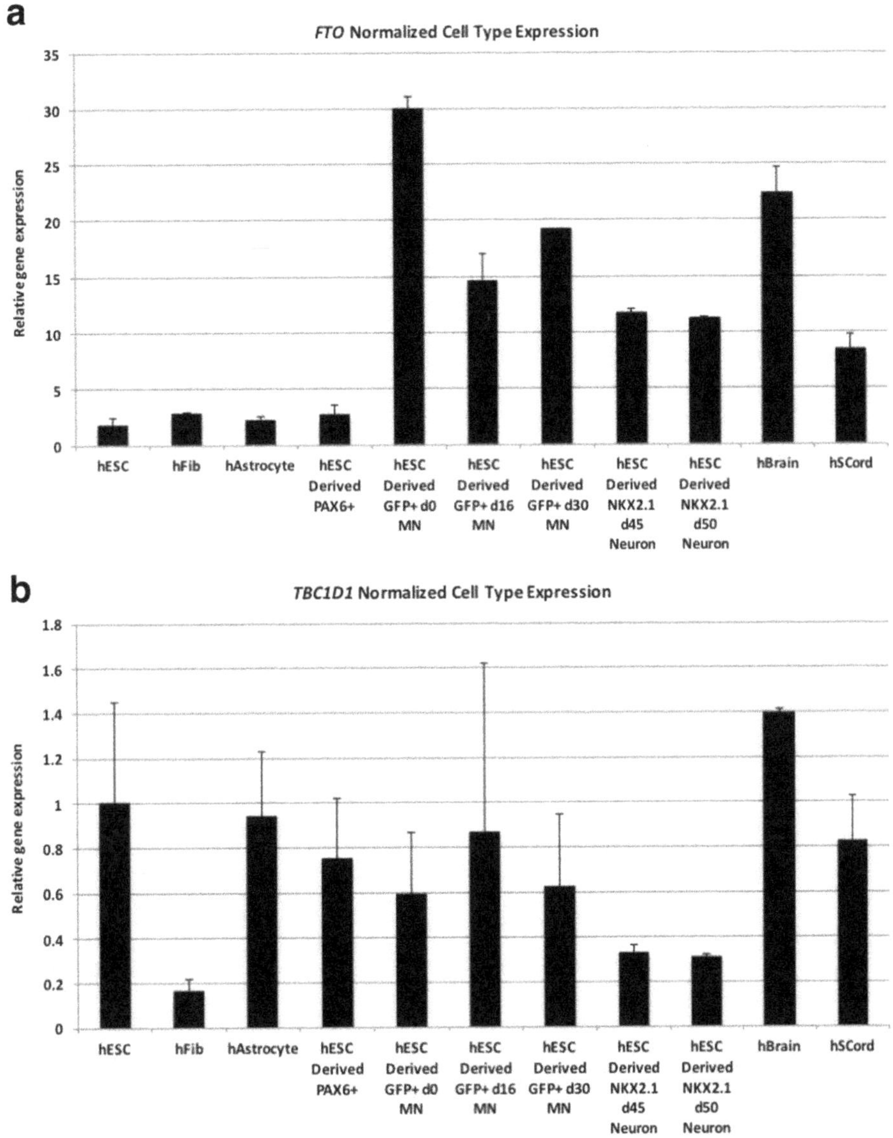

Fig. 3 a, b *FTO* and *TBC1D1* expression levels in a range of various human cell types. *FTO* and *TBC1D1* mRNA levels were determined using human-specific primers in various cell types; fibroblasts, embryonic stem cells (ESCs), astrocytes, neural PAX6+ progenitors, stem cell-derived purified motor neurons at three different stages of development (Hb9:GFP on days 0, 16, and 30), and stem cell-derived cortical neurons (NKX21 on days 45 and 50). Human brain and spinal cord RNA served as controls. Normalized RT-qPCR data are expressed as averaged sample replicates. Error bars represent standard deviations

carriers of the expansion. Such a discrepancy may reflect inter-individual differences not as far as statistical power is concerned, but rather related to the patients' genetic background, as the aforementioned study examined only individuals from Athens, whereas we recruited individuals from across the entire Greek mainland (north, central, and south Greece). Differential exposure to environmental factors are known to contribute to an increased risk of ALS; thus, exposure to lead, mercury, selenium, zinc, and copper is known to increase the risk of ALS [46], while high-risk populations include those who collect their domestic water from surface water [8], possibly reflecting an association between the neurotoxin b-methylamino-l-alanine (BMAA) produced by cyanobacteria and the risk of ALS. Al-Chalabi and Hardiman [4] have proposed a disease model, according to which ALS is the net result of environmental risks and time acting on a pre-existing genetic load, followed by an automatic, self-perpetuating decline until death. Akimoto et al. [2] reported marked differences in genotyping results for GGGGCC-repeat expansions in *C9orf72*. Herein, amplicon length analysis and repeat-primed PCR were combined as a recommended minimum requirement in a research setting [2].

Small sample sizes and data variability, plus our ancestrally diverse patient cohorts, should be taken into account when interpreting our findings. Yet, our datasets may assist future efforts of replicating these findings in independent ALS patient cohorts.

Conclusions

There is no single predominant genetic association underlying sALS. Instead, sALS may be considered to be the net result of the complex interplay of environmental factors and numerous low-risk susceptibility loci. We propose that under this polygenic model, a large number of alleles, each conferring a small genotypic risk, combine additively or multiplicatively to confer a range of susceptibilities in the general population [21, 47].

To our knowledge, this has been the first study to present a possible association between *FTO* variants and the genetic basis of sALS. These findings need to be further validated in a larger sALS patient cohort of Greek origin to overcome the small sample size bias, while a gene ontology analysis to look at links between *FTO* and the other genes known to be associated with ALS could support the validity of these findings.

Abbreviations

ALS: Amyotrophic lateral sclerosis; ESCs: Embryonic stem cells; fALS: Familial amyotrophic lateral sclerosis; HCNEs: Highly conserved non-coding elements; sALS: Sporadic amyotrophic lateral sclerosis; SNPs: Single nucleotide polymorphisms

Acknowledgements

The authors are grateful to the Greek ALS Patients Association for their continuous support. We thank Ms. Aikaterini Dimitropoulou, Vasiliki Delatolla, Maria-Valsamo Founta, Chrysavgi Gritzapi, Venetia Karathanassi, Anna Karra, Dimitra Taroussi, and Paraskevi Tassopoulou for their expert technical assistance.

Funding

Our work was partly funded by grants from the European Commission [RD-Connect (FP7-305444), U-PGx (H2020-668353)], the Greek General Secretariat of Research and Technology [eMoDiA (ΠΔΕ11_0415)], and the Golden Helix Foundation to GPP. VK and IN are funded by the Serbian Ministry of Education, Science and Technology development Grant No. ON175090 and Grant No. ON17509, respectively. BRA is funded by UAEU grants (31M135 and 31M187).

Authors' contributions

GPP conceived the study. KM, EMP, GX, TK and GPP designed the study. GX, ED, SR, GD, SO, NB, and GPP provided the samples. KM, EMP, GX, AB, KC, VG, KVZ, AJ, KK, EP, AP, FA, AL, VD, MB, EET, and TK performed the experiments. AD and EK generated the samples, performed and analyzed the qRT-PCR experiments. AP, FA, AL, VD, CM, BAP, and TK analyzed the data. BRA and GPP provided funding for this work. GX, IN, VSK, NB, SO, EK, DNC, SG, GMH, RD, BRA, TK, and GPP wrote the manuscript. All authors read and approved the manuscript.

Consent for publication

"Not applicable" in this section.

Competing interests

The authors declare that they have no competing interests.

Author details

[1]National and Kapodistrian University of Athens School of Medicine, Athens, Greece. [2]Department of Pharmacy, University of Patras School of Health Sciences, Campus, Rion, GR-26504 Patras, Greece. [3]School of Medicine, University of Thessaly, Larisa, Greece. [4]Papageorgiou hospital, Thessaloniki, Greece. [5]Department of Pathology, College of Medicine and Health Sciences, United Arab Emirates University, Al-Ain, UAE. [6]ANALYSI Diagnostic Laboratories S.A, Thessaloniki, Greece. [7]Departments of Neurology and Physiology, Northwestern University Feinberg School of Medicine, Chicago, IL, USA. [8]Suna and Inan Kirac Foundation, NDAL, Bogazici University, Istanbul, Turkey. [9]Department of Medical Sciences and Public Health, University of Cagliari, Cagliari, Italy. [10]CNR IRGB, Cagliari, Italy. [11]Institute of Neurology CCS, School of Medicine, University of Belgrade, Belgrade, Serbia. [12]Faculty of Medicine, Institute of Human Genetics, University of Belgrade, Belgrade, Serbia. [13]Complete Genomics Inc., Mountain View, CA, USA. [14]BGI Shenzhen, Shenzhen, People's Republic of China. [15]Institute of Medical Genetics, Cardiff University, Cardiff, UK.

References

1. Aidar M, Line SRP. A simple and cost-effective protocol for DNA isolation from buccal epithelial cells. Braz Dent J. 2007;18:148–52.
2. Akimoto, C, Volk, AE, van Blitterswijk, M, Van den Broeck, M, Leblond, CS, Lumbroso, S, Camu, W, Neitzel, B, Onodera, O and van Rheenen, W. (2014). A blinded international study on the reliability of genetic testing for GGGGCC-repeat expansions in C9orf72 reveals marked differences in results among 14 laboratories. *J. Med. Genet.* jmedgenet-2014-102360.
3. Al-Chalabi, A, Fang, F, Hanby, MF, Leigh, PN, Shaw, CE, Ye, W and Rijsdijk, F. (2010). An estimate of amyotrophic lateral sclerosis heritability using twin data. *J Neurol Neurosurg Psychiatry* jnnp 2010.207464.
4. Al-Chalabi A, Hardiman O. The epidemiology of ALS: a conspiracy of genes, environment and time. Nat Rev Neurol. 2013;9:617–28.
5. Balding DJ. A tutorial on statistical methods for population association studies. Nat Rev Genet. 2006;7:781.
6. Barrett JC, Fry B, Maller J, Daly MJ. Haploview: analysis and visualization of LD and haplotype maps. Bioinformatics. 2004;21:263–5.
7. Bellefroid EJ, Kobbe A, Gruss P, Pieler T, Gurdon JB, Papalopulu N. Xiro3 encodes a Xenopus homolog of the Drosophila Iroquois genes and functions in neural specification. EMBO J. 1998;17:191–203.
8. Bradley WG, Borenstein AR, Nelson LM, Codd GA, Rosen BH, Stommel EW, Cox PA. Is exposure to cyanobacteria an environmental risk factor for amyotrophic lateral sclerosis and other neurodegenerative diseases? Amyotrophic Lateral Sclerosis and Frontotemporal Degeneration. 2013;14:325–33.
9. Brooks BR. El Escorial World Federation of Neurology criteria for the diagnosis of amyotrophic lateral sclerosis. J. Neurol. Sci. 1994;124:96–107.
10. Cady J, Allred P, Bali T, Pestronk A, Goate A, Miller TM, Mitra RD, Ravits J, Harms MB, Baloh RH. Amyotrophic lateral sclerosis onset is influenced by the burden of rare variants in known amyotrophic lateral sclerosis genes. Ann Neurol. 2015;77:100–13.
11. Cirulli ET, Lasseigne BN, Petrovski S, Sapp PC, Dion PA, Leblond CS, Couthouis J, Lu Y-F, Wang Q, Krueger BJ. Exome sequencing in amyotrophic lateral sclerosis identifies risk genes and pathways. Science. 2015;347:1436–41.
12. DeJesus-Hernandez M, Desaro P, Johnston A, Ross OA, Wszolek ZK, Ertekin-Taner N, Graff-Radford N, Rademakers R, Boylan K. Novel p. Ile151Val mutation in VCP in a patient of African American descent with sporadic ALS. Neurology. 2011a;77:1102–3.
13. DeJesus-Hernandez M, Mackenzie IR, Boeve BF, Boxer AL, Baker M, Rutherford NJ, Nicholson AM, Finch NA, Flynn H, Adamson J. Expanded GGGGCC hexanucleotide repeat in noncoding region of C9ORF72 causes chromosome 9p-linked FTD and ALS. Neuron. 2011b;72:245–56.
14. Deng H-X, Chen W, Hong S-T, Boycott KM, Gorrie GH, Siddique N, Yang Y, Fecto F, Shi Y, Zhai H. Mutations in UBQLN2 cause dominant X-linked juvenile and adult onset ALS and ALS/dementia. Nature. 2011;477:211.
15. Desmet F-O, Hamroun D, Lalande M, Collod-Béroud G, Claustres M, Béroud C. Human splicing finder: an online bioinformatics tool to predict splicing signals. Nucleic Acids Res. 2009;37:e67.

16. Desport J, Preux P, Truong T, Vallat J, Sautereau D, Couratier P. Nutritional status is a prognostic factor for survival in ALS patients. Neurology. 1999;53:1059.
17. Di Giorgio FP, Boulting GL, Bobrowicz S, Eggan KC. Human embryonic stem cell-derived motor neurons are sensitive to the toxic effect of glial cells carrying an ALS-causing mutation. Cell Stem Cell. 2008;3:637–48.
18. Drmanac R, Sparks AB, Callow MJ, Halpern AL, Burns NL, Kermani BG, Carnevali P, Nazarenko I, Nilsen GB, Yeung G. Human genome sequencing using unchained base reads on self-assembling DNA nanoarrays. Science. 2010;327:78–81.
19. Fecto F, Yan J, Vemula SP, Liu E, Yang Y, Chen W, Zheng JG, Shi Y, Siddique N, Arrat H. SQSTM1 mutations in familial and sporadic amyotrophic lateral sclerosis. Arch Neurol. 2011;68:1440–6.
20. Ferraiuolo L, Kirby J, Grierson AJ, Sendtner M, Shaw PJ. Molecular pathways of motor neuron injury in amyotrophic lateral sclerosis. Nat Rev Neurol. 2011;7:616–30.
21. Houlston RS, Peto J. The search for low-penetrance cancer susceptibility alleles. Oncogene. 2004;23:6471.
22. Jawaid A, Murthy SB, Wilson AM, Qureshi SU, Amro MJ, Wheaton M, Simpson E, Harati Y, Strutt AM, York MK. A decrease in body mass index is associated with faster progression of motor symptoms and shorter survival in ALS. Amyotroph Lateral Scler. 2010;11:542–8.
23. Ji, A-L, Zhang, X, Chen, W-W and Huang, W-J. (2017). Genetics insight into the amyotrophic lateral sclerosis/frontotemporal dementia spectrum. *J Med Genet* jmedgenet-2016-104271.
24. Johnson AD, Handsaker RE, Pulit SL, Nizzari MM, O'donnell CJ, De Bakker PI. SNAP: a web-based tool for identification and annotation of proxy SNPs using HapMap. Bioinformatics. 2008;24:2938–9.
25. Kabashi E, Valdmanis PN, Dion P, Spiegelman D, McConkey BJ, Velde CV, Bouchard J-P, Lacomblez L, Pochigaeva K, Salachas F. TARDBP mutations in individuals with sporadic and familial amyotrophic lateral sclerosis. Nat Genet. 2008;40:572–4.
26. Kiskinis E, Sandoe J, Williams LA, Boulting GL, Moccia R, Wainger BJ, Han S, Peng T, Thams S, Mikkilineni S. Pathways disrupted in human ALS motor neurons identified through genetic correction of mutant SOD1. Cell Stem Cell. 2014;14:781–95.
27. Kwiatkowski TJ, Bosco D, Leclerc A, Tamrazian E, Vanderburg C, Russ C, Davis A, Gilchrist J, Kasarskis E, Munsat T. Mutations in the FUS/TLS gene on chromosome 16 cause familial amyotrophic lateral sclerosis. Science. 2009;323:1205–8.
28. Lattante S, Ciura S, Rouleau GA, Kabashi E. Defining the genetic connection linking amyotrophic lateral sclerosis (ALS) with frontotemporal dementia (FTD). Trends Genet. 2015;31:263–73.
29. Lattante S, Conte A, Zollino M, Luigetti M, Del Grande A, Marangi G, Romano A, Marcaccio A, Meleo E, Bisogni G. Contribution of major amyotrophic lateral sclerosis genes to the etiology of sporadic disease. Neurology. 2012;79:66–72.
30. Machiela MJ, Chanock SJ. LDlink: a web-based application for exploring population-specific haplotype structure and linking correlated alleles of possible functional variants. Bioinformatics. 2015;31:3555–7.
31. Majounie E, Renton AE, Mok K, Dopper EG, Waite A, Rollinson S, Chiò A, Restagno G, Nicolaou N, Simon-Sanchez J. Frequency of the C9orf72 hexanucleotide repeat expansion in patients with amyotrophic lateral sclerosis and frontotemporal dementia: a cross-sectional study. The Lancet Neurology. 2012;11:323–30.
32. Mandic-Stojmenovic G, Stefanova E, Dobricic V, Novakovic I, Stojkovic T, Jesic A, Kostic V. Screening for C9orf72 expansion mutation in Serbian patients with early-onset dementia. Dement Geriatr Cogn Disord. 2015;40:358–65.
33. Maroof AM, Keros S, Tyson JA, Ying S-W, Ganat YM, Merkle FT, Liu B, Goulburn A, Stanley EG, Elefanty AG. Directed differentiation and functional maturation of cortical interneurons from human embryonic stem cells. Cell Stem Cell. 2013;12:559–72.
34. McLaren W, Gil L, Hunt SE, Riat HS, Ritchie GR, Thormann A, Flicek P, Cunningham F. The ensembl variant effect predictor. Genome Biol. 2016;17:122.
35. Mok, KY, Koutsis, G, Schottlaender, LV, Polke, J, Panas, M and Houlden, H. (2012). High frequency of the expanded C9ORF72 hexanucleotide repeat in familial and sporadic Greek ALS patients. Neurobiol Aging 33, e1851-1851.
36. Prudencio, M, Gonzales, PK, Cook, CN, Gendron, TF, Daughrity, LM, Song, Y, Ebbert, MT, van Blitterswijk, M, Zhang, Y-J and Jansen-West, K. (2017). Repetitive element transcripts are elevated in the brain of C9orf72 ALS/FTLD patients. *Hum. Mol. Genet.*
37. Ragvin A, Moro E, Fredman D, Navratilova P, Drivenes Ø, Engström PG, Alonso ME, de la Calle Mustienes E, Skarmeta JLG, Tavares MJ. Long-range gene regulation links genomic type 2 diabetes and obesity risk regions to HHEX, SOX4, and IRX3. Proc Natl Acad Sci. 2010;107:775–80.
38. Ratti A, Corrado L, Castellotti B, Del Bo R, Fogh I, Cereda C, Tiloca C, D'Ascenzo C, Bagarotti A, Pensato V. C9ORF72 repeat expansion in a large Italian ALS cohort: evidence of a founder effect. Neurobiol Aging. 2012;33: e2527–8. e2514
39. Renton AE, Majounie E, Waite A, Simón-Sánchez J, Rollinson S, Gibbs JR, Schymick JC, Laaksovirta H, Van Swieten JC, Myllykangas L. A hexanucleotide repeat expansion in C9ORF72 is the cause of chromosome 9p21-linked ALS-FTD. Neuron. 2011;72:257–68.
40. Rosen DR, Siddiquef T, Patterson D, Figlewicz DA, Sapp P, Hentatif A, O'Regan JP, Dengf H-X, McKenna-Yasek D, Cayabyabi A. Mutations in Cu/Zn superoxide dismutase gene are associated with familial amyotrophic. Nature. 1993;362:4.
41. Sasieni PD. From genotypes to genes: doubling the sample size. Biometrics. 1997;1:1253–61.
42. Smemo S, Tena JJ, Kim K-H, Gamazon ER, Sakabe NJ, Gómez-Marín C, Aneas I, Credidio FL, Sobreira DR, Wasserman NF. Obesity-associated variants within FTO form long-range functional connections with IRX3. Nature. 2014;507:371.
43. Smith BN, Newhouse S, Shatunov A, Vance C, Topp S, Johnson L, Miller J, Lee Y, Troakes C, Scott KM. The C9ORF72 expansion mutation is a common cause of ALS+/− FTD in Europe and has a single founder. Eur J Hum Genet. 2013;21:102.
44. Su F-C, Goutman SA, Chernyak S, Mukherjee B, Callaghan BC, Batterman S, Feldman EL. Association of environmental toxins with amyotrophic lateral sclerosis. JAMA neurology. 2016;73:803–11.
45. Swinnen B, Robberecht W. The phenotypic variability of amyotrophic lateral sclerosis. Nat Rev Neurol. 2014;10:661–70.
46. Trojsi F, Monsurrò MR, Tedeschi G. Exposure to environmental toxicants and pathogenesis of amyotrophic lateral sclerosis: state of the art and research perspectives. Int J Mol Sci. 2013;14:15286–311.
47. Xue Y, Chen Y, Ayub Q, Huang N, Ball EV, Mort M, Phillips AD, Shaw K, Stenson PD, Cooper DN. Deleterious-and disease-allele prevalence in healthy individuals: insights from current predictions, mutation databases, and population-scale resequencing. Am J Hum Genet. 2012;91:1022–32.

13

Identify Down syndrome transcriptome associations using integrative analysis of microarray database and correlation-interaction network

Min Chen[1,2,3,4,5†], Jiayan Wang[1,2,3,4†], Yingjun Luo[7†], Kailing Huang[7], Xiaoshun Shi[8], Yanhui Liu[7], Jin Li[9], Zhengfei Lai[1], Shuya Xue[1,2,3,4,5], Haimei Gao[1,2,3,4,5], Allen Chen[6,7] and Dunjin Chen[1,2,3,4,5*]

Abstract

Background: Long non-coding RNAs (lncRNAs) have previously been emerged as key players in a series of biological processes. Dysregulation of lncRNA is correlated to human diseases including neurological disorders. Here, we developed a multi-step bioinformatics analysis to study the functions of a particular Down syndrome-associated gene DSCR9 including the lncRNAs. The method is named correlation-interaction-network (COIN), based on which a pipeline is implemented. Co-expression gene network analysis and biological network analysis results are presented.

Methods: We identified the regulation function of DSCR9, a lncRNA transcribed from the Down syndrome critical region (DSCR) of chromosome 21, by analyzing its co-expression genes from over 1700 sets and nearly 60,000 public Affymetrix human U133-Plus 2 transcriptional profiling microarrays. After proper evaluations, a threshold is chosen to filter the data and get satisfactory results. Microarray data resource is from EBI database and protein–protein interaction (PPI) network information is incorporated from the most complete network databases. PPI integration strategy guarantees complete information regarding DSCR9. Enrichment analysis is performed to identify significantly correlated pathways.

Results: We found that the most significant pathways associated with the top DSCR9 co-expressed genes were shown to be involved in neuro-active ligand-receptor interaction (GLP1R, HTR4, P2RX2, UCN3, and UTS2R), calcium signaling pathway (CACNA1F, CACNG4, HTR4, P2RX2, and SLC8A3), neuronal system (KCNJ5 and SYN1) by the KEGG, and GO analysis. The A549 and U251 cell lines with stable DSCR9 overexpression were constructed. We validated 10 DSCR9 co-expression genes by qPCR in both cell lines with over 70% accuracy.

Conclusions: DSCR9 was highly correlated with genes that were known as important factors in the developments and functions of nervous system, indicating that DSCR9 may regulate neurological proteins regarding Down syndrome and other neurological-related diseases. The pipeline can be properly adjusted to other applications.

Keywords: lncRNA, DSCR9, Down syndrome, Protein–protein interaction, Correlation-interaction-network, Neurological diseases

* Correspondence: dunjinchen121@163.com
[†]Equal contributors
[1]Department of Fetal Medicine and Prenatal Diagnosis, the Third Affiliated Hospital of Guangzhou Medical University, Guangzhou 510150, China
[2]Obstetrics and Gynecology Institute of Guangzhou, Guangzhou 510150, China
Full list of author information is available at the end of the article

Introduction

Down syndrome (DS) is the most common chromosome disorder occurring in about one per 700 newborns each year [1]. Although it has been well established that an extra copy of chromosome 21 causes DS, the genetic and molecular mechanisms of the disease are yet unclear. Studies on partial trisomy have led to the characterization of a region of the chromosome 21 known as Down syndrome critical region (DSCR), which is located at the distal end of the long arm of chromosome 21 (21q22.1–22.3) and has candidate genes whose imbalance may induce a marked cognitive deficit as well as other pathologies and associated conditions [2]. Although the involvement of DSCR as the sole cause of DS symptoms is still controversial, previous studies have suggested that this region plays a primary role in the genetic interactions related to the pathogenesis of DS. Nevertheless, it has not been completely understood what exact subset of genes that are over-expressed on chromosome 21 generating these DS-related deficiencies. Most studies focus on protein-coding genes in DSCR, whereas little is known about the three long non-coding RNAs (lncRNAs): DSCR8, DSCR9, and DSCR10.

LncRNAs are a large class of non-protein-coding transcripts that are greater than 200 bases in length and are involved in numerous physiological and pathological processes [3]. Only a small number of lncRNAs have been characterized functionally, while most of them were shown to control gene expression by regulating various aspects of gene expression [4]. Many lncRNAs are shown to regulate important cancer hallmarks including proliferation, apoptosis, metastasis, metabolism, senescence, and drug-resistance [5]. In addition, cumulative evidence have demonstrated that lncRNAs contribute to the complex biological system organization and gene regulatory networks of the central nervous system affecting brain patterning, neural stem cell maintenance, neurogenesis and gliogenesis, stress responses, and synaptic and neural plasticity. A number of lncRNAs are linked to neurological diseases such as the dysregulated BACE1-AS and BC200 in Alzheimer's disease [6]. However, evidence of any lncRNA being involved in DS has not yet been fully elucidated. Previous studies showed that NRON (ncRNA repressor of the nuclear factor of activated T cells) was an lncRNA mediating the cytoplasmic to nuclear shuttling of the NFAT transcription factor. In animal models, deregulation of the DSCR1 and DYRK1A acts synergistically to prevent nuclear occupancy of NFATc transcription factors leading to reduced NFATc activity and to a number of features of DS [7]. However, a conclusive link between this lncRNA and DS pathophysiology has not been reported so far.

In current study, we aimed to find lncRNAs that are related to Down syndrome by establishing a systematic bioinformatics analysis as well as the pipeline to predict functions of lncRNAs on human chromosome 21 and by validating their potential regulatory target mRNAs by qPCR. After mining the RNA expression data from Affymetrix transcriptional profiling microarrays, the functions of the DSCR9 lncRNA were found to be enhanced in neurological-related pathways, which might cause Down syndrome and other neurological diseases.

Materials and methods

lncRNA probe localization

The probe sequence of Affymetrix U133 Plus 2 Platform (http://www.ncbi.nlm.nih.gov/geo/query/acc.cgi?acc=GPL570) containing over 4000 sets of data were obtained from NCBI GEO database. They were aligned with the human genome hg19 and GENCODE (version 18) using the BLAT with parameters '-stepSize = 5-repMatch = 1,000,000-minScore = 0-minIdentity = 0' based on an efficient algorithm for microarray probe re-annotations [8]. BLAT results with no more than two mismatches were saved for our study.

Expression data collection and pre-processing

The gene expression data from NCBI GEO database that mentioned above were retrieved from EBI ArrayExpress database [9] by Bioconductor package ArrayExpress [10]. EBI, NCBI, and DDBJ are three high-throughput data exchanging portals, where data will be updated to the newest. In particular, the chip data of EBI is extremely clear, which includes details of the data such as sample information, chip location information, chip signals, and so on. So we chose the EBI database and the data therein for our analysis. Incorporated CEL files were pre-processed using robust multichip average (RMA) normalization method. The standard deviation of the expression levels was calculated. LncRNAs including HOTTIP, HOTAIR, and DSCR9 were analyzed. Datasets with low standard deviation level (< 0.25) were filtered out, and the remaining experimental data were used for proceeding the analysis. According to [11], and using 0.25 as the threshold, we find that the transcriptomic changes of the data are significant enough to determine the lncRNA-related genes. Therefore, we utilize this value (0.25) as the cutoff to guarantee high quality of the data and in the meanwhile capture significantly correlated transcriptome information.

Statistics analysis

Pearson correlation between lncRNA probe (e.g., HOTTIP, HOTAIR, or DSCR9) and other 54,674 probes were calculated. p values were presented as unmodified p values. Multiple testing corrections were performed using the q value package in R [12]. Genes with q values lower than 0.05 were considered as significant to be co-expression genes of lncRNA.

Protein–protein interaction (PPI) network for co-expression genes

The protein–protein interaction (PPI) network, including all the top co-expression genes of DSCR9, was constructed to identify the most important functional relevance of DSCR9. Moreover, all the PPI relationships from the following databases: HPRD, IntAct, MIPS, BIND, DIP, MINT, PDZBase, and Reactome, were combined to gain a more comprehensive understanding of the interactions between DSCR9 and its potential target genes based on a previously described method [13]. We used default options of the abovementioned eight networks and interactomes that were found in at least one of these networks will be integrated into our analysis. In this manner, all relevant interactomes regarding DSCR9 will be considered. Finally, the PPI network was visualized with Cytoscape software [14].

Biological pathway analysis

Genes that showed the highest correlation with lncRNA expression level were used for the biological pathway analysis. Parameters used in our pathway analysis were listed as following: (1) two pathway databases were included: KEGG pathway database [15] and Reactome pathway database [16]; (2) the hypergeometric distribution was employed to calculate the probability of a particular group of genes annotated to the pathway, comparing to all the other human genes in the genome; (3) raw p value was adjusted for multiple testing using the Bonferroni correction method [17]; (4) pathways with adjusted p value < 0.05 were regarded as the significantly enriched pathways.

Collection of DSCR9 transcription data in human brain

DSCR9 expression levels in different human tissues were collected from the Nonhuman Primate Reference Transcriptome Resource project [18]. The DSCR9 RNA levels in various regions of human brain were obtained from Babru Samal's molecular brain project (www.molecular-brain.org/). Data on the transcription factor binding sites were collected from Encyclopedia of DNA Elements (ENCODE) project.

Cell culture

A549 and U231 cells were purchased from American Type Culture Collection. All cells were maintained in DMEM medium (Gibco) supplemented with 10% fetal bovine serum (FBS, Gibco), 100 U/ml penicillin sodium, and 100 mg/ml streptomycin sulfate at 37 °C. All cell lines were passaged for less than 6 months.

Plasmid construction and stable cell line construction

The genomic segment corresponding to DSCR9 was amplified from human genomic DNA and then cloned into the pcDNA3.1 vector. The correct amplified fragments were identified by restriction endonuclease digestion and were confirmed by sequencing.

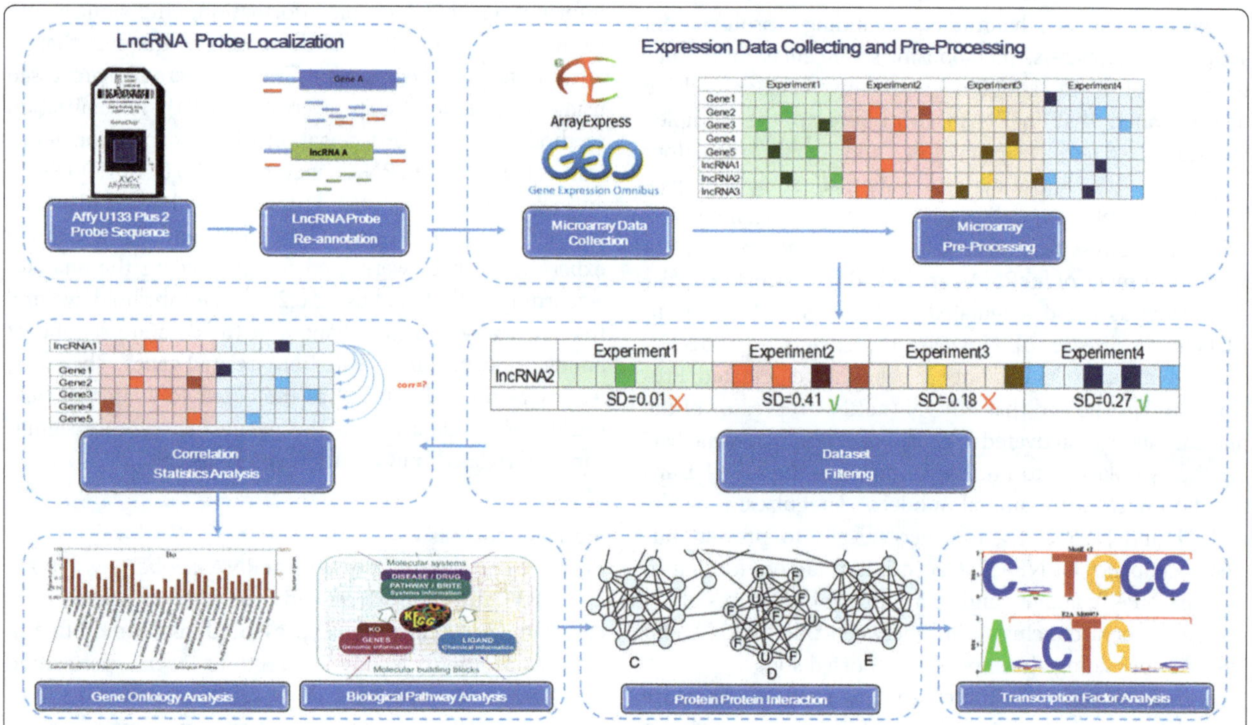

Fig. 1 Schematic diagram of the co-expression-interaction network (COIN) Bioinformatics Analysis

Fig. 2 Prediction of lncRNAs HOTTIP and HOTAIR top co-expression genes and functional enrichment pathways by our COIN analysis pipeline

Glioma cells U251 and lung cancer cells A549 and were transiently transfected with 4 μg empty vector (pcDNA3.1) as a control or recombinant expression plasmid pcDNA3.1-DSCR9 using Lipofectamine 2000 (Invitrogen, Carlsbad, CA, USA) reagent according to manufacturer's instructions. The expression of DSCR9 and other predicted genes were determined by qPCR assay at 48 h post transfection.

Quantitative analysis of DSCR9 and its potential mRNA targets

Total RNAs were extracted from cultured cell lines using the Trizol RNA Reagent (Invitrogen, Carlsbad CA, USA). The RNA concentration was determined by 260/280 nm absorbances using a Nanodrop Spectrophotometer (ND-100, Thermo, USA). QPCR assays were performed using the K1622 RevertAid First Strand cDNA Synthesis Kit (Thermo Scientific) and GoTaq® qPCR Master Mix (Promega) according to the manufacturer's instructions in an Applied Biosystems 7500 Fluorescent Quantitative PCR System (Applied Biosystems, Foster City, CA, USA). The reaction mixtures were incubated at 95 °C for 30 s followed by 45 amplification cycles of 95 °C for 5 s and 60 °C for 30 s. GAPDH and U6 were used as endogenous controls for mRNA and DSCR9 expressions, respectively. Expressions were normalized to endogenous controls, and the fold change of gene expression was calculated as $2^{-\Delta\Delta Ct}$. Three independent experiments were each performed in triplicates. The primer sequences were listed in additional Additional file 1: Table S1.

Results

To identify the potential target genes of lncRNAs, we developed a bioinformatics analysis. Perl and R scripts mainly create the pipeline that implemented. We obtained nearly 6000 lncRNA probes with high confidence in U133 Plus 2.0 Array. All co-expressed probes showing high correlations with the interesting lncRNA were used for the subsequent gene ontology (GO) analysis, KEGG biological pathway analysis, and protein–protein interaction (PPI) analysis. Additionally, the transcription factor binding sites (TFBS) of our target genes were predicted based on the ENCODE

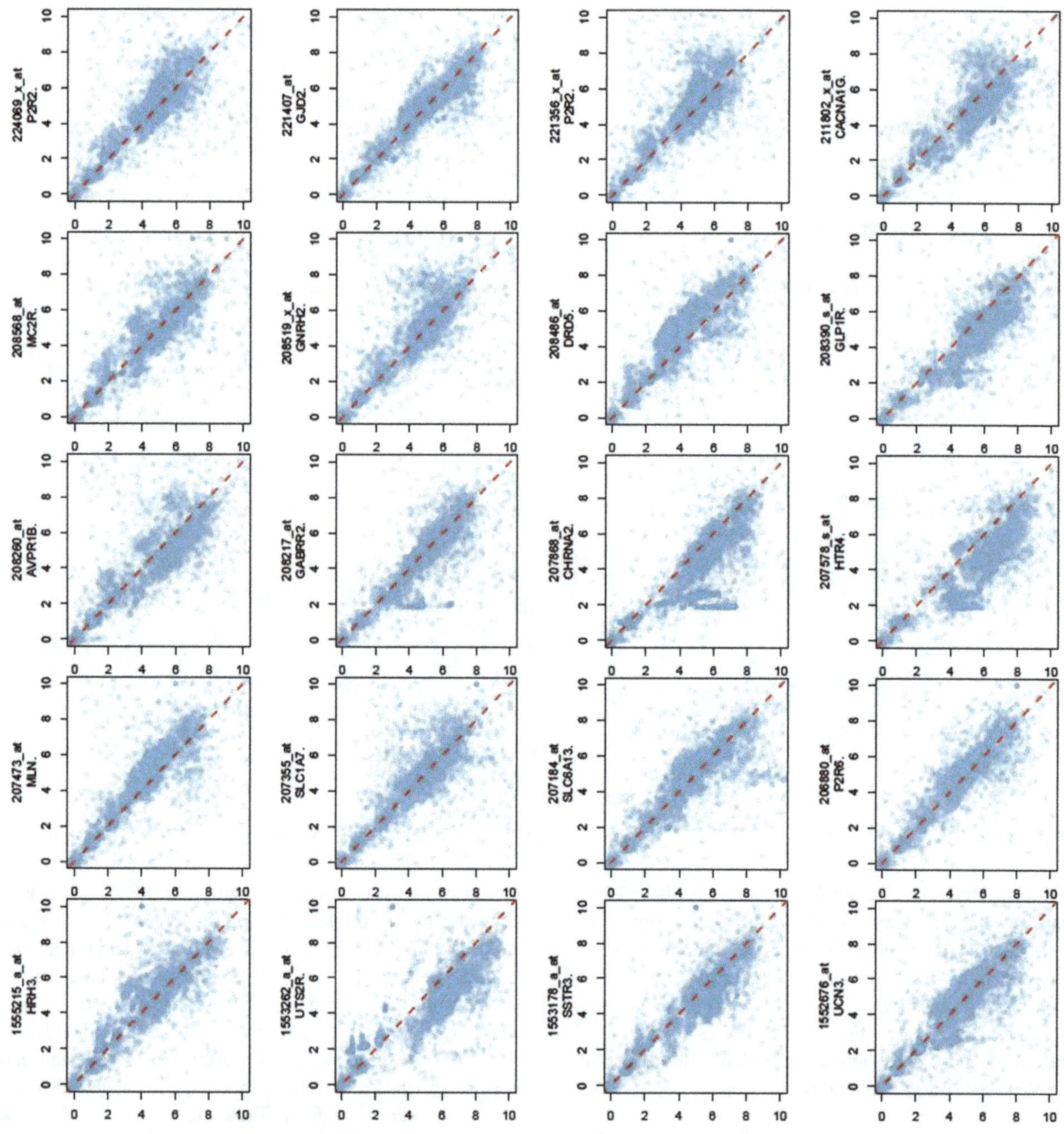

Fig. 3 Correlation analysis between DSCR9 probes and top 20 co-expressed genes in HG U133 Plus 2.0 microarray. *x* axis: DSCR9 expression level in microarray experiments; *y*-axis: expression level of gene (with probeset ID) in the corresponding microarray; red dash line represented correlation of 1.0; blue dot represented the expression level of DSCR9-gene pairs in the same microarray

project datasets. The correlation-interaction-network (COIN) bioinformatics analysis was summarized as Fig. 1.

Validation of COIN prediction with lncRNAs: HOTTIP and HOTAIR

By applying the analysis approach, we found HOTTIP was highly correlated with six HOXA family genes (HOXA13, HOXA11-AS, HOXA10, HOXA11, HOXA9, and HOXA10-AS) that are adjacent to HOTTIP, suggesting the potential regulatory roles of HOTTIP on the HOXA locus genes (Fig. 2). Our HOTTIP target genes and functional analysis was consistent with previous studies, showing that HOTTIP coordinates the activation of several 5' HOXA genes in vivo, and is involved in forelimb morphogenesis as well as proximal/distal pattern formation [19].

Moreover, we were able to show the potential of the COIN method in predicting the regulated targets of

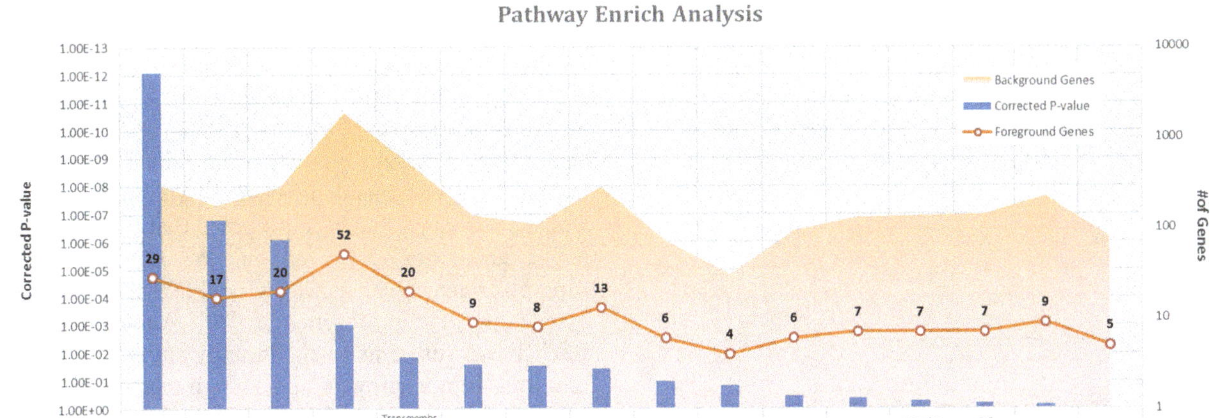

	Neuroactive ligand-receptor interaction	Calcium signaling pathway	Neuronal System	Signal Transduction	Transmembrane transport of small molecules	Cell-Cell communication	GnRH signaling pathway	MAPK signaling pathway	Gastric acid secretion	Phototransduction	Dilated cardiomyopathy	Glutamatergic synapse	Vascular smooth muscle contraction	Cell adhesion molecules (CAMs)	Endocytosis	Cardiac muscle contraction
Background Genes	326	181	283	1918	533	138	110	283	74	31	93	130	135	140	220	81
Corrected P-value	7.92E-13	1.61E-07	8.04E-07	9.34E-04	0.0139	0.02608	0.02851	0.03568	0.10417	0.1509	0.34275	0.42002	0.51857	0.63388	0.70191	0.96557
Foreground Genes	29	17	20	52	20	9	8	13	6	4	6	7	7	7	9	5

Fig. 4 Pathway enrichment analysis of top 1000 co-expression genes. *Blue bars* represented the enrichment significance levels of each pathway. The *orange* empty circle points represented the number of co-expression genes in the corresponding pathway. The area colored in *light orange* represented the number of genes involved

HOTAIR lncRNA (Fig. 2). Our analysis indicated that HOTAIR was highly correlated with six regulated HOXC cluster genes (HOXC10, HOXC9, HOXC6 HOXC11, HOXC13, and HOXC4), and its top pathway functions were embryonic morphogenesis, embryo development, and skeletal system development. Our prediction could be further assessed by the experimental data that was previously reported in Rinn et al. [20].

DSCR9 functions that predicted by COIN procedure

Searching for the most related targets for an lncRNA transcribed from the DSCR on human chromosome 21, and predicting its biological functions with COIN process, we started by calculating the standard deviation of DSCR9 expression level in each dataset. Those with small standard deviation values (< 0.25) were neglected and only 258 sets of experimental data were used in our analysis. Based on Pearson correlation, top 20 DSCR9 co-expression probes were showed in Fig. 3.

From the list of top 1000 co-expression genes of DSCR9, we conducted biological pathway analysis. There are four pathways that were significantly enriched with these co-expression genes (p value < 0.01). The most significant pathways were neuroactive ligand-receptor interaction, calcium signaling pathway, neuronal system, and signal transduction (Fig. 4). Genes that are related to the three most significantly enriched pathways were presented in Table 1.

Moreover, we performed PPI analysis to evaluate the interacting relations of these potential DSCR9 targets with Cytoscape (Fig. 5a). lncRNA DSCR9 exhibited strong interactions with those neuro-related genes (nodes presented in the inner circle). For a further data-mining of this network, we calculated the interaction weight (numbers of neighbors) of each core node (Fig. 5b). Consistently, most of the core genes in the identified PPI network were neuro-related.

DSCR9 expression in human brain tissues

For in-depth examination on DSCR9 expression, we collected data from three international projects and performed integrative analysis of the transcription level of DSCR9. The expression of DSCR9 lncRNA was tissue-specific. Among nine different human tissues, DSCR9 displayed the highest abundance in hearts and brains (Fig. 6a).

The lncRNA levels were also analyzed in various regions of the human brain, and the highest DSCR9 RNA level was found in the hippocampus followed by those in the cerebrum, amygdala, etc. (Fig. 6b). Transcription factor binding site (TFBS) data obtained from ENCODE project was analyzed. Four TFBSs in the third exon of DSCR9 and three transcription factors including USF1, USF2, and Rad21 might be involved in its expression regulation via these binding sites (Fig. 6c).

Genes that relate to DSCR9 co-expression oncology

Using the COIN analysis, we found 14 genes that functioned in more than one neuro-related pathway, including EGFR (Table 1). EGFR was also abundantly detected in the brain by Nieto-Estevez et al. [21]. It was critical in neurological processes including cell death, the survival and differentiation of neurons. EGFR gene has been reported to be dysregulated in the mouse model of Down syndrome (DS), with regulatory impacts on calcium

Table 1 DSCR9 co-expressed genes were related to nervous system by the pathway enrichment analysis

Genes	Genomic Loci	Official Names
1. Neuroactive ligand-receptor interaction (KEGG:04080): 29/326 genes; p-val=7.9E-13		
ADRB3*	[illegible]	adrenoceptor beta 3
AVPR1B*	[illegible]	arginine vasopressin receptor 1B
CHRNA2*	[illegible]	cholinergic receptor, nicotinic, alpha 2 (neuronal)
CHRND*	[illegible]	cholinergic receptor, nicotinic, delta (muscle)
DRD5*	[illegible]	dopamine receptor D5
GABRR3*	[illegible]	gamma-aminobutyric acid (GABA) A receptor, rho 3
GHRHR	[illegible]	growth hormone releasing hormone receptor
GLP1R	[illegible]	glucagon-like peptide 1 receptor
GLRA2	[illegible]	glycine receptor, alpha 2
GNRH2	[illegible]	gonadotropin-releasing hormone 2
GRIN1*	[illegible]	glutamate receptor, ionotropic, N-methyl D-aspartate 1
GRM3	[illegible]	glutamate receptor, metabotropic 3
GRPR*	[illegible]	gastrin-releasing peptide receptor
HCRT	[illegible]	hypocretin (orexin) neuropeptide precursor
HRH3	[illegible]	histamine receptor H3
HTR1B	[illegible]	5-hydroxytryptamine (serotonin) receptor 1B, G protein-coupled
HTR4*	[illegible]	5-hydroxytryptamine (serotonin) receptor 4, G protein-coupled
HTR5A*	[illegible]	5-hydroxytryptamine (serotonin) receptor 5A, G protein-coupled
MC2R	[illegible]	melanocortin 2 receptor (adrenocorticotropic hormone)
MLN	[illegible]	Motilin
NPB	[illegible]	neuropeptide B
P2RX2*	[illegible]	purinergic receptor P2X, ligand gated ion channel, 2
P2RX4*	[illegible]	purinergic receptor P2X, ligand gated ion channel, 4
PTGIR	[illegible]	prostaglandin I2 (prostacyclin) receptor (IP)
SSTR3	[illegible]	somatostatin receptor 3
TACR3*	[illegible]	tachykinin receptor 3
TSPO2	[illegible]	translocator protein 2
UCN3	[illegible]	urocortin 3
UTS2R	[illegible]	urotensin 2 receptor
2. Calcium signaling pathway (KEGG:04020): 17/181 genes; p-val=1.6E-07		
ADCY1*	[illegible]	adenylate cyclase 1 (brain)
ADRB3*	[illegible]	adrenoceptor beta 3
AVPR1B*	[illegible]	arginine vasopressin receptor 1B
CACNA1F	[illegible]	calcium channel, voltage-dependent, L type, alpha 1F subunit
CACNA1G	[illegible]	calcium channel, voltage-dependent, T type, alpha 1G subunit
CAMK2B*	[illegible]	calcium/calmodulin-dependent protein kinase II beta
DRD5*	[illegible]	dopamine receptor D5
EGFR	[illegible]	epidermal growth factor receptor
GRIN1*	[illegible]	glutamate receptor, ionotropic, N-methyl D-aspartate 1
GRPR*	[illegible]	gastrin-releasing peptide receptor
HTR4*	[illegible]	5-hydroxytryptamine (serotonin) receptor 4, G protein-coupled
HTR5A*	[illegible]	5-hydroxytryptamine (serotonin) receptor 5A, G protein-coupled
MYLK	[illegible]	myosin light chain kinase
P2RX2*	[illegible]	purinergic receptor P2X, ligand gated ion channel, 2
P2RX4*	[illegible]	purinergic receptor P2X, ligand gated ion channel, 4
SLC8A3	[illegible]	solute carrier family 8 (sodium/calcium exchanger), member 3
TACR3*	[illegible]	tachykinin receptor 3
3. Neuronal System (REACT:13685): 20/283 genes; p-val=6.5E-07		
ADCY1*	[illegible]	adenylate cyclase 1 (brain)
CACNG4	[illegible]	calcium channel, voltage-dependent, gamma subunit 4
CAMK2B*	[illegible]	calcium/calmodulin-dependent protein kinase II beta
CHRNA2*	[illegible]	cholinergic receptor, nicotinic, alpha 2 (neuronal)
CHRND*	[illegible]	cholinergic receptor, nicotinic, delta (muscle)
GABRR3*	[illegible]	gamma-aminobutyric acid (GABA) A receptor, rho 3
GJD2	[illegible]	gap junction protein, delta 2, 36kDa
GRIN1*	[illegible]	glutamate receptor, ionotropic, N-methyl D-aspartate 1
KCNC3	[illegible]	potassium channel, voltage gated Shaw related subfamily C, member 3
KCNG4	[illegible]	potassium channel, voltage gated modifier subfamily G, member 4
KCNJ5	[illegible]	potassium channel, inwardly rectifying subfamily J, member 5
KCNK16	[illegible]	potassium channel, two pore domain subfamily K, member 16
KCNK9	[illegible]	potassium channel, two pore domain subfamily K, member 9
KCNN1	[illegible]	potassium channel, calcium activated intermediate/small conductance subfamily N alpha, member 1
KCNQ1	[illegible]	potassium channel, voltage gated KQT-like subfamily Q, member 1
KCNQ3	[illegible]	potassium channel, voltage gated KQT-like subfamily Q, member 3
PANX2	[illegible]	pannexin 2
SLC1A7	[illegible]	solute carrier family 1 (glutamate transporter), member 7
SLC6A13	[illegible]	solute carrier family 6 (neurotransmitter transporter), member 13
SYN1	[illegible]	synapsin I

*Genes functioned in more than one neuro-related pathways are bold-faced. Ten genes (P2RX2, SLC8A3, HTR4, CACNG4, UCN3, SYN1, GLP1R, KCNJ5, CACNA1F, and UTS2R) validated by qPCR in two cell lines were highlighted in red

signaling pathway, neuroactive ligand-receptor interaction, and the MAPK signaling pathway [22]. Our analysis results are concordant with the abovementioned reports.

Patients with DS exhibited considerably reduced incidence of most solid tumors [23, 24], and their overall cancer mortality rate was approximately 10% below the normal level of general population. Although the cause was still unclear, it had been proposed that the DS patients might get extra doses of one or multiple cancer-suppressor genes such as DSCR1 and DYRK1A due to the extra copy of chromosome 21 [25]. An extra copy of DSCR1 was sufficient to significantly suppress angiogenesis and tumor growth and its protein, DSCR1, was upregulated in tissues from people with DS [25]. The attenuation of calcineurin activity by DSCR1 together with another chromosome 21 gene DYRK1A, might be sufficient to remarkably diminish angiogenesis. No direct evidence had been reported so far to illustrate that DSCR9 was responsible for cancer progressing. However, it seemed to be a reasonable hypothesis from our analysis.

The proposed COIN approach showed that SRC gene was the most outstanding target candidate in the regulatory network of DSCR9 (Fig. 5b). SRC gene was reported to encode a proto-oncogene tyrosine-protein kinase. Activation of the Src pathway had been observed in about 50% of tumors from colon, liver, lung, breast, and the pancreas [26]. Another significant regulatory target of DSCR9 in the identified network was EGFR, which was a well-studied oncogene. It has been identified as an important drug target and understanding this gene has led to the development of multiple anti-cancer therapeutics (known as 'EGFR inhibitors') such as gefitinib, erlotinib, afatinib, brigatinib and icotinib for lung cancer, and cetuximab for colon cancer. Additionally, gastrin-releasing peptide receptor (GRPR) was also identified as an important target candidate of DSCR9, and diseases that associated with GRPR include lung cancer and prostate adenocarcinoma. In summary, the COIN analysis results indicated that DSCR9 was highly correlated with several oncogenes and therefore it was very likely that its regulatory function could be employed to explain the reduced cancer incidence in patients with DS.

DSCR9 co-expression genes that validated in cell lines

To further examine the predictive capability of the COIN model, a standard gene expression analysis was performed in two DSCR9 over-expressed cell lines. After the construction and validation of DSCR9-overexpression in A549 and U251 cells (Fig. 7a–b), we selected 15 top DSCR9-associated genes in our prediction and determined their expression levels by qPCR. We found that 12 of these 15 predicted co-expressed genes were unregulated upon DSCR9 over-expression in A549 cells, and 11 of them were upregulated in U251 (Fig. 7c–d). In conclusion, our experimental data showed high consistency (> 73%)

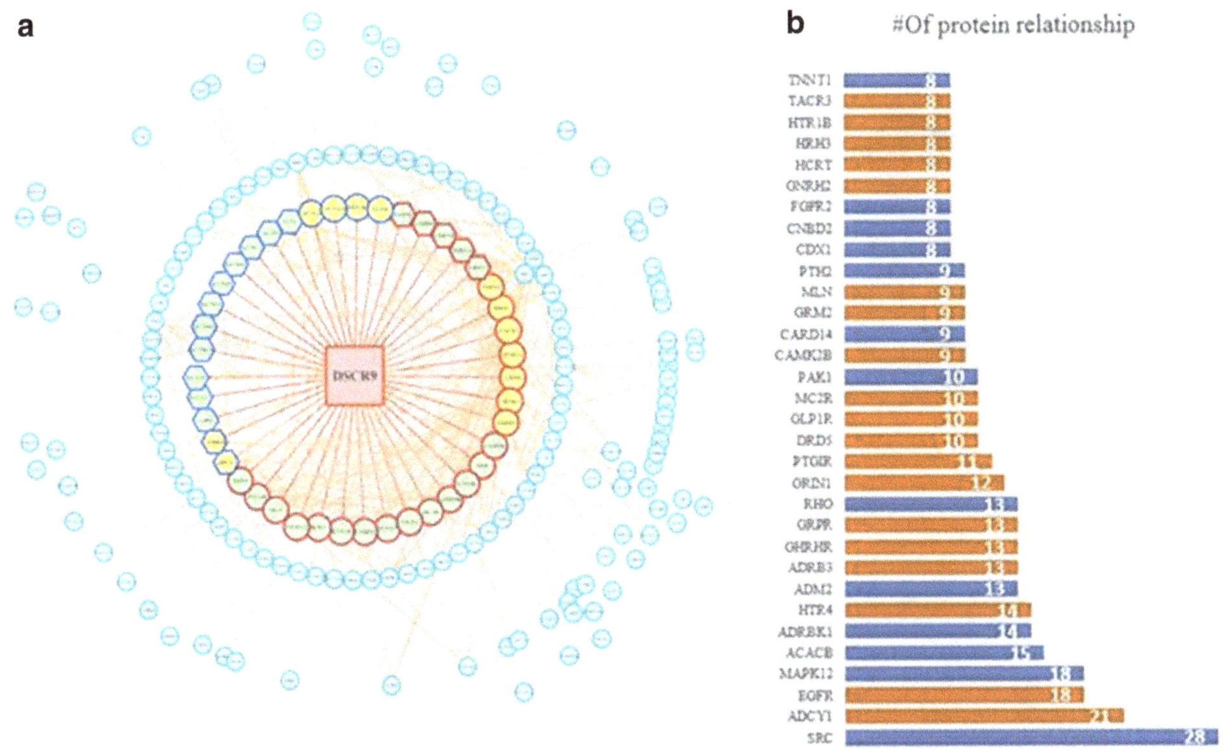

Fig. 5 Protein–protein interaction (PPI) network of DSCR9 co-expression genes. **a** DSCR9 Network structure visualized with Cytoscape. Each node represented one gene. Nodes with red border represented co-expressed genes involved in neuroactive ligand-receptor interaction pathway. Nodes filled with light yellow color represented co-expressed genes functioned in calcium signaling pathway, while nodes shaped in hexagonal represent co-expressed genes related to neuronal system. Orange lines show PPI between those highly correlated co-expressed genes of DSCR9. Red lines represented potential relationships between DSCR9 and its targets. **b** The core DSCR targeted genes in the PPI network were listed with their gene symbols and weights. Numbers in the bars showed the interaction weight of the corresponding genes in DSCR9 network (Fig. 4a). Orange-colored bars indicated that the corresponding genes were members of the neuro-related pathways

with the bioinformatics analysis predictions, supporting the reliability of the proposed COIN method.

Discussion

Although the dysregulation of some lncRNAs had been linked to human diseases, the underlying molecular mechanisms, including the regulatory function of many lncRNAs have not been fully elucidated. An effective approach is in need to provide a feasible guidance for functional studies of these non-coding RNAs. COIN method aimed for the prediction of targeted lncRNA function and its regulatory network. We validated the COIN approach by predicting the interaction networks of two well-known lncRNAs: HOTAIR and HOTTIP. Moreover, reliability of its predictions was also confirmed by a quantitative PCR in two independent DSCR9 over-expression cell lines.

Potential neurological pathway genes that relate DSCR9 to DS

DS is a neurobehavioral specificity disease [27]. Three pathways that are significantly enriched with DSCR9 co-expression genes associated with neurological functions, with top 10 qPCR validation genes in parentheses (see Fig. 4 and Table 1):

In the COIN analysis, we found 14 genes that functioned in more than one neuro-related pathway, including HTR4, P2RX2, and ADCY1 (Table 1). ADCY1 was found to encode a form of adenylate cyclase that is expressed in the brain and was involved in regulatory processes in the central nervous system. Previous studies indicated that it might play significant roles in memory acquisition and learning. Overall, combining the COIN analysis results and literature reports, we believe that the three pathways in which the DSCR9 co-expression genes were enriched are essential for the nervous system development. HTR4 is another gene induced by DSCR9 over-expression. HTR4 is a glycosylated transmembrane protein that functions in both the peripheral and central nervous system to modulate the release of various neurotransmitters. Previous studies indicated that HTR4 might be associated with neurological diseases such as Parkinson's disease 5 [28, 29]. The product of P2RX2 gene belongs to the family of purinoceptors for ATP.

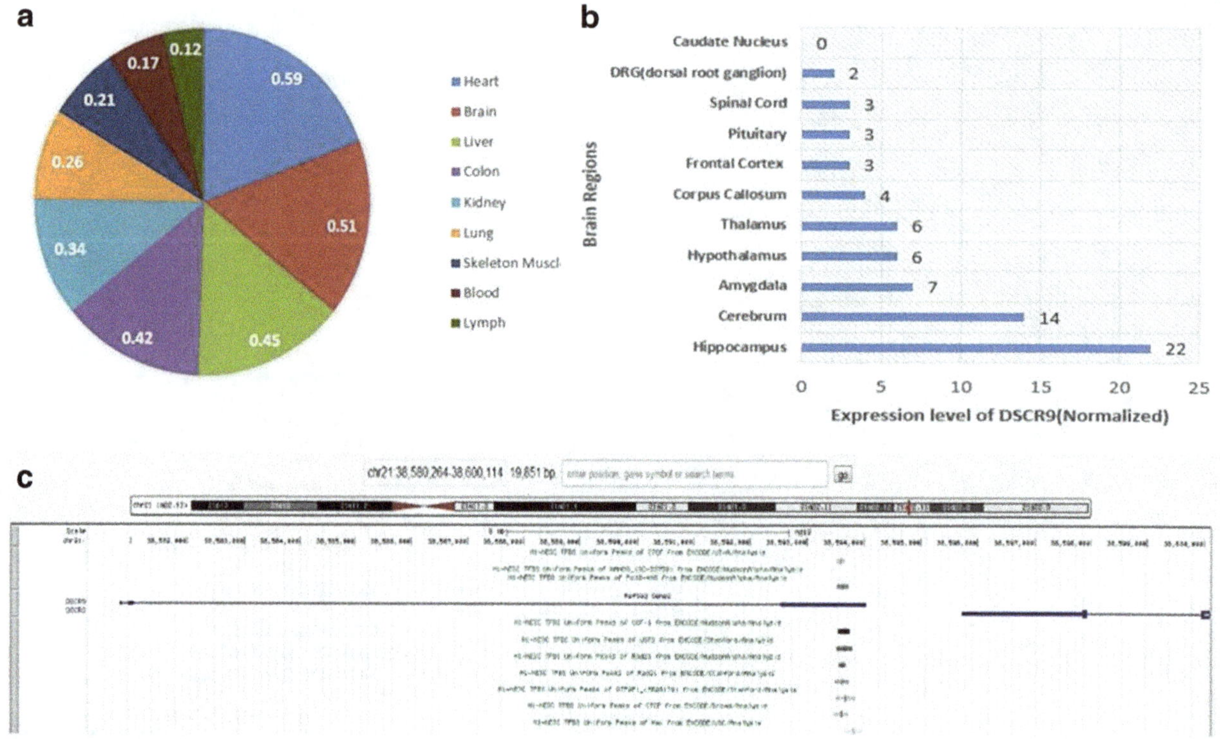

Fig. 6 Expression and regulation of DSCR9 in human tissues and brain regions. **a** DSCR9 expression levels (shown by FPKM levels, fragment per kilometer) in a series of human tissues (data obtained from NHPRTR project). **b** DSCR9 expression levels in different brain regions (data obtained from molecularbrain.org). **c** DSCR9 with four transcription-factor-binding-sites (TFBS) in the third DSCR9 exon displayed with UCSC genome browser

This receptor functions as a ligand-gated ion channel. Binding to ATP mediates synaptic transmission between neurons and from neurons to smooth muscle [28, 29].

Components of the neuro-active ligand-receptor interaction pathway showed high relevance to DS acute lymphoblastic leukemia [30]. Furthermore, it had been well studied that calcium ions generated versatile intracellular signals, supporting that calcium signaling pathway also played an important role in all types of neurons [31]. Impaired calcineurin activity was already linked to many human diseases including DS, Alzheimer's disease, brain ischemia, cardiac hypertrophy, and maybe more [32]. Interestingly, DSCR1 gene was upregulated in DS patients and encoded a protein that suppressed VEGF-mediated angiogenic signaling by the calcineurin pathway [25]. Previous report also suggested that presynaptic calcium channels might serve as the regulatory node in a dynamic, multilayered signaling network that exerted short-term control of neurotransmission in response to synaptic activity [33]. A transient rise of the calcium level in dendritic spines was essential for the induction of activity-dependent synaptic plasticity.

In addition, our experimental data supported the predicted DSCR9 regulation of its target genes in two different cancer cell lines, i.e., A549 and U251. Over 73% of our predicted DSCR9 target genes were upregulated when DSCR9 was over-expressed. Consistent with our gene ontology (GO) analysis, two of the unregulated genes, say, CACNG4 [34] and CACNA1F [35], encoded the subunits of calcium channels, which were involved in the calcium signaling pathway.

According to the co-expressed genes of DSCR9 in the resulting network, we predicted that DSCR9 might also contribute to the cardiovascular- and cerebral-related diseases. Consistently, data from NHPRTR project indicated that the highest expression levels of this lncRNA were detected in heart and brain (Fig. 6a). Moreover, compared to other regions of the brain, the abundance of DSCR9 seemed to accumulate in hippocampus (Fig. 6b). This supported our prediction that this lncRNA might be a factor leading to abnormal dendritic branching and spine number, and further reduction of brain volume as it had been previously reported that a reduction in brain volume in patients with DS was attributed to impaired dendritic and synaptic maturation [36]. Dendritic branching and spine number were dramatically reduced in pyramidal neurons in the hippocampus, visual cortex, and motor cortex after 4 months postnatal age in individuals with DS [28, 29].

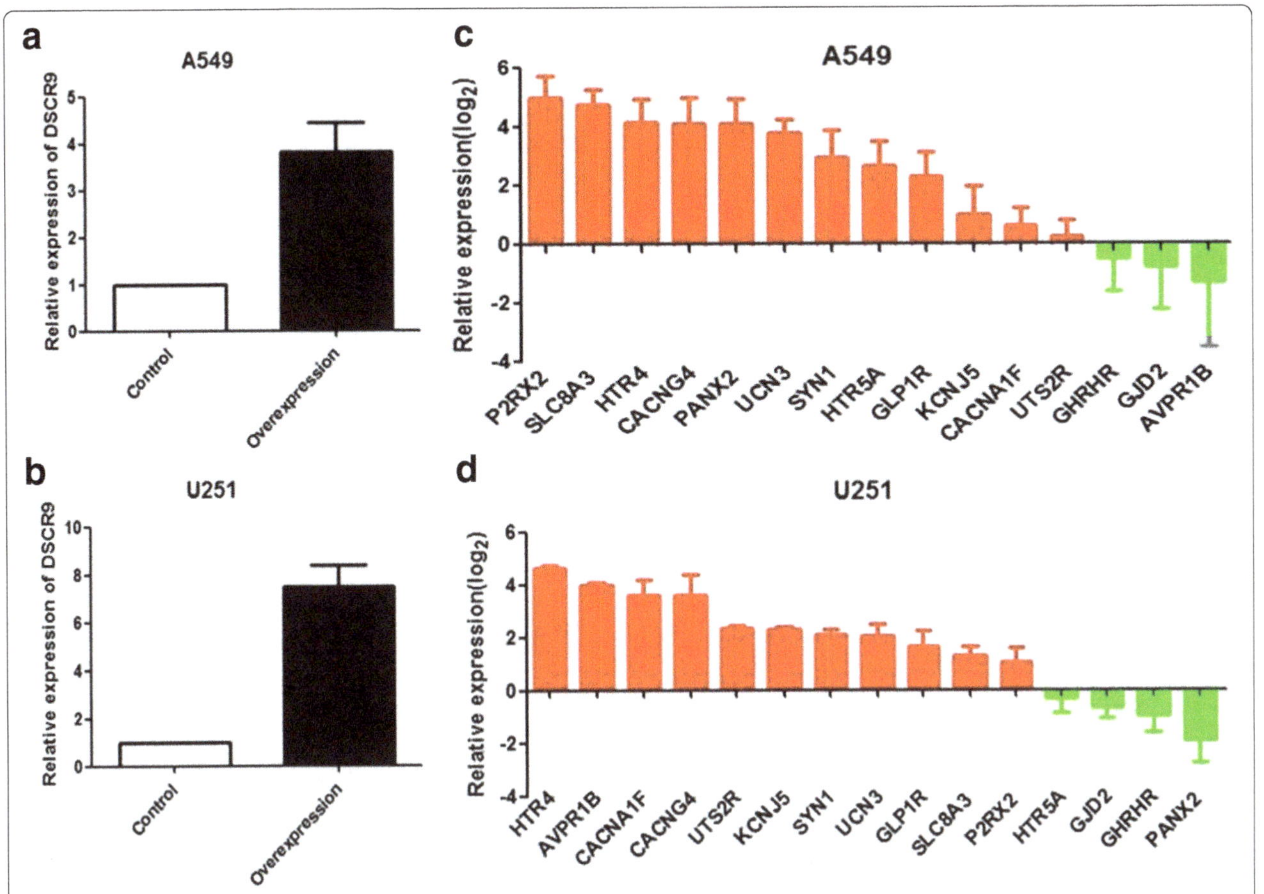

Fig. 7 Bioinformatics predictions were validated by QPCR. **a** The DSCR9 stable overexpression A549 and U251 cell lines were constructed. **b** QPCR analysis showing the predicted co-expressed genes were upregulated in both A549 and U251 upon DSCR9 overexpression

Potential transcription factors that relate DSCR9 to DS

In the COIN analysis of DSCR9, we identified three transcription factors including Rad21, USF1, and USF2 that were highly correlated with DSCR9 (Fig. 6c), and we further noticed that putative binding sites of these transcription factors were present in the third exon of DSCR9. Intriguingly, these transcription factors had been shown to be related to DS. Rad21 gene was found to be highly mutated in DS [37], and its expression level was elevated in DS patients [38]. Similarly, USF1 expression level was positively related with dCK gene [39], which was a well-documented DS leukemia-related gene [40]. Moreover, previous reports suggested that USF1 played a trans-activating role on the CBS-1b promoter [41], and CBS-1b gene had been accepted as a DS risk factor [42, 43]. The third transcription factor that is related to DSCR9 was USF2, whose expression was shown to be increasing during aging [44]. Interestingly, DS patients exhibited an increased risk of many chronic diseases, which were typically associated with aging. Previous reports suggested that trisomy 21 was linked to clinical manifestations of accelerated aging, and DSCR9 appeared to be a negative epigenetic clock controlling tissue aging in the brain [45]. Supporting its potential activities and expression regulation in DS, it was also found that DSCR9 displayed abnormal methylation pattern in DS patients [46], and the methylated sites were mainly located in the third exon [47]. All together, these results implicated a DS-related transcription regulation of DSCR9 lncRNA, which may involve altered DNA methylation patterns and/or chromatin structure as well as the transcription factors USF1/USF2/Rad21 binding to the exon 3 of DSCR9.

Conclusions

In conclusions, our studies established a solid bioinformatics pipeline for functional predictions of DS transcriptome associations. Our qPCR assay showed that the expression of these genes were induced by DSCR9, implying that these genes might be regulated by DSCR9. The results provided a valuable guidance for further investigations on the regulatory mechanism of DSCR9 as well as its relevance to DS and other neurological diseases. Dysregulation of DSCR9 and/or its target genes in these pathways might be responsible for the pathogenesis and progressing of DS. As extension of this work, generalized approach can be adapted for other disease-related transcriptome association studies.

Abbreviations

COIN: Correlation-interaction-network; DS: Down syndrome; DSCR: Down syndrome critical region; lncRNA: Long non-coding RNA; PPI: Protein–protein interaction

Acknowledgements

We thank Mr. Mark Deiparine for technical supports and writing assistance.

Funding

The study was supported by the Guangzhou Science and Technology Program (No. 2014234), the Guangdong Science and Technology Program (No. 2013B022000005), and the National Natural Science Foundation of China (No. 81671470). This work was partly supported by Youth Scientific Project Foundation of Guangzhou Medical University (2015A21) and Guangdong Natural Science Foundation (2015A030313474).

Author's contributions

MC, AC, and DC conceived and designed the experiments. KH, YJL, and AC analyzed the data. XS, YHL, and JL performed the molecular experiments. MC, YJL, XS, KH, and AC wrote the paper. JYW, ZFL, SYX, and HMG edited the paper and figures. All authors read and approved the final manuscript.

Competing interests

The authors declare that they have no competing interests.

Author details

[1]Department of Fetal Medicine and Prenatal Diagnosis, the Third Affiliated Hospital of Guangzhou Medical University, Guangzhou 510150, China. [2]Obstetrics and Gynecology Institute of Guangzhou, Guangzhou 510150, China. [3]The Medical Centre for Critical Pregnant Women in Guangzhou, Guangzhou 510150, China. [4]Key Laboratory for Major Obstetric Diseases of Guangdong Province, Guangzhou 510150, China. [5]Key Laboratory for Reproduction and Genetics of Guangdong Higher Education Institutes, Guangzhou, China. [6]Department of Mathematics, University of California, Berkeley, CA 94720, USA. [7]Mendel Genes Inc, Manhattan Beach, CA, Manhattan Beach, CA 90266, USA. [8]Department of Thoracic Surgery, Nanfang Hospital, Southern Medical University, Guangzhou, China. [9]State Key Laboratory of Respiratory Disease, The First Affiliated Hospital of Guangzhou Medical University, National Clinical Research Center for Respiratory Disease, Guangzhou, China.

References

1. Megarbane A, et al. The 50th anniversary of the discovery of trisomy 21: the past, present, and future of research and treatment of Down syndrome. Genet Med. 2009;11(9):611–6.
2. Luke S, Gandhi S, Verma RS. Conservation of the Down syndrome critical region in humans and great apes. Gene. 1995;161(2):283–5.
3. Wang ET, et al. Alternative isoform regulation in human tissue transcriptomes. Nature. 2008;456(7221):470–6.
4. Clark MB, Mattick JS. Long noncoding RNAs in cell biology. Semin Cell Dev Biol. 2011;22(4):366–76.
5. Gutschner T, Diederichs S. The hallmarks of cancer: a long non-coding RNA point of view. RNA Biol. 2012;9(6):703–19.
6. Fenoglio C, et al. An emerging role for long non-coding RNA dysregulation in neurological disorders. Int J Mol Sci. 2013;14(10):20427–42.
7. Willingham AT, et al. A strategy for probing the function of noncoding RNAs finds a repressor of NFAT. Science. 2005;309(5740):1570–3.
8. Pawel Foszner AG, Polanski A, Marczyk M, Jaksik R, Polanska J. Efficient algorithm for microarray probes re-annotation. Lect Notes Comput Sci. 2011;6923:281–9.
9. Rustici G, et al. ArrayExpress update—trends in database growth and links to data analysis tools. Nucleic Acids Res. 2013;41(Database issue):D987–90.
10. Gentleman RC, et al. Bioconductor: open software development for computational biology and bioinformatics. Genome Biol. 2004;5(10):R80.
11. Lindgren D, et al. Integrated genomic and gene expression profiling identifies two major genomic circuits in urothelial carcinoma. PLoS One. 2012;7(6):e38863.
12. Dabney A, Storey JD. W. G. R. qvalue: Q-value estimation for false discovery rate control. R Package Version, 2011. https://doi.org/10.18129/B9.bioc.qvalue.
13. Mathivanan S, et al. An evaluation of human protein-protein interaction data in the public domain. BMC Bioinformatics. 2006;7(Suppl 5):S19.
14. Shannon P, et al. Cytoscape: a software environment for integrated models of biomolecular interaction networks. Genome Res. 2003;13(11):2498–504.
15. Kanehisa M, et al. KEGG for integration and interpretation of large-scale molecular data sets. Nucleic Acids Res. 2012;40(Database issue):D109–14.
16. Croft D, et al. The reactome pathway knowledgebase. Nucleic Acids Res. 2014;42(Database issue):D472–7.
17. CE B. Teoria statistica delle classi e calcolo delle probabilit. Pubblicazioni del R Istituto Superiore di Scienze Economiche e Commerciali di Firenze. 1936;8:3–62.
18. Peng X, et al. Tissue-specific transcriptome sequencing analysis expands the non-human primate reference transcriptome resource (NHPRTR). Nucleic Acids Res. 2015;43(Database issue):D737–42.
19. Wang KC, et al. A long noncoding RNA maintains active chromatin to coordinate homeotic gene expression. Nature. 2011;472(7341):120–4.
20. Rinn JL, et al. Functional demarcation of active and silent chromatin domains in human HOX loci by noncoding RNAs. Cell. 2007;129(7):1311–23.
21. Nieto-Estevez V, et al. A global transcriptome analysis reveals molecular hallmarks of neural stem cell death, survival, and differentiation in response to partial FGF-2 and EGF deprivation. PLoS One. 2013;8(1):e53594.
22. Ling KH, et al. Functional transcriptome analysis of the postnatal brain of the Ts1Cje mouse model for Down syndrome reveals global disruption of interferon-related molecular networks. BMC Genomics. 2014;15:624.
23. Satge D, et al. A lack of neuroblastoma in Down syndrome: a study from 11 European countries. Cancer Res. 1998;58(3):448–52.
24. Patja K, et al. Cancer incidence of persons with Down syndrome in Finland: a population-based study. Int J Cancer. 2006;118(7):1769–72.
25. Baek KH, et al. Down's syndrome suppression of tumour growth and the role of the calcineurin inhibitor DSCR1. Nature. 2009;459(7250):1126–30.
26. Dehm SM, Bonham K. SRC gene expression in human cancer: the role of transcriptional activation. Biochem Cell Biol. 2004;82(2):263–74.
27. Jean-Adolphe Rondal P, Perera J. Down syndrome: neurobehavioural specificity. Down syndrome: Neurobehavioural specificity. 2006;
28. Marin-Padilla M. Pyramidal cell abnormalities in the motor cortex of a child with Down's syndrome. A Golgi study. J Comp Neurol. 1976;167(1):63–81.
29. Takashima S, et al. Abnormal neuronal development in the visual cortex of the human fetus and infant with Down's syndrome. A quantitative and qualitative Golgi study. Brain Res. 1981;225(1):1–21.
30. Loudin MG, Wang J, Leung HC, Gurusiddappa S, Meyer J, Condos G, et al. Genomic profiling in Down syndrome acute lymphoblastic leukemia identifies histone gene deletions associated with altered methylation profiles. Leukemia. 2011;25(10):1555–63.
31. Grienberger C, Konnerth A. Imaging calcium in neurons. Neuron. 2012;73(5): 862–85.
32. Bindu LH, Rani PU, Reddy PP. Calcineurin activity in children with mental handicap. Indian J Clin Biochem. 2007;22(2):32–5.
33. Catterall WA, Few AP. Calcium channel regulation and presynaptic plasticity. Neuron. 2008;59(6):882–901.
34. Burgess DL, et al. Identification of three novel ca(2+) channel gamma subunit genes reveals molecular diversification by tandem and chromosome duplication. Genome Res. 1999;9(12):1204–13.
35. Jalkanen R, et al. A novel CACNA1F gene mutation causes Aland Island eye disease. Invest Ophthalmol Vis Sci. 2007;48(6):2498–502.
36. Becker LE, Armstrong DL, Chan F. Dendritic atrophy in children with Down's syndrome. Ann Neurol. 1986;20(4):520–6.
37. Yoshida K, et al. The landscape of somatic mutations in Down syndrome-related myeloid disorders. Nat Genet. 2013;45(11):1293–9.
38. Engidawork E, et al. Aberrant protein expression in cerebral cortex of fetus with Down syndrome. Neuroscience. 2003;122(1):145–54.
39. Ge Y, et al. Role of USF1 in the differential expression of the human deoxycytidine kinase gene in acute myeloid leukemia. Leukemia. 2005;19(4):677–9.

40. Taub JW, et al. Cystathionine-beta-synthase cDNA transfection alters the sensitivity and metabolism of 1-beta-D-arabinofuranosylcytosine in CCRF-CEM leukemia cells in vitro and in vivo: a model of leukemia in Down syndrome. Cancer Res. 2000;60(22):6421–6.
41. Ge Y, Jensen TL, Matherly LH, Taub JW, et al. Synergistic regulation of human cystathionine-beta-synthase-1b promoter by transcription factors NF-YA isoforms and Sp1. Biochim Biophys Acta. 2002;1579(2–3):73–80.
42. Fintelman-Rodrigues N, et al. Investigation of CBS, MTR, RFC-1 and TC polymorphisms as maternal risk factors for Down syndrome. Dis Markers. 2009;26(4):155–61.
43. Wang SS, et al. Polymorphisms in genes RFC-1/CBS as maternal risk factors for Down syndrome in China. Arch Gynecol Obstet. 2013;288(2):273–7.
44. Singh K, Prasad S. Differential expression of Fmr-1 mRNA and FMRP in female mice brain during aging. Mol Biol Rep. 2008;35(4):677–84.
45. Horvath S, et al. Accelerated epigenetic aging in Down syndrome. Aging Cell. 2015;14(3):491–5.
46. Zhang Y, et al. DNA methylation analysis of chromosome 21 gene promoters at single base pair and single allele resolution. PLoS Genet. 2009; 5(3):e1000438.
47. Yegnasubramanian S, et al. Chromosome-wide mapping of DNA methylation patterns in normal and malignant prostate cells reveals pervasive methylation of gene-associated and conserved intergenic sequences. BMC Genomics. 2011;12:313.

14

Public attitudes in Japan toward participation in whole genome sequencing studies

Taketoshi Okita[1*†], Noriko Ohashi[2†], Daijiro Kabata[3], Ayumi Shintani[3] and Kazuto Kato[2]

Abstract

Background: Recent innovations in gene analysis technology have allowed for rapid and inexpensive sequencing of entire genomes. Thus, both conducting a study using whole genome sequencing (WGS) in a large population and the clinical application of research findings from such studies are currently feasible. However, to promote WGS studies, understanding and voluntary participation by the general public is needed. Therefore, it is essential to investigate the general public's attitude toward and understanding of WGS studies. The primary goal of our research is to investigate these issues and to discover how they relate to research participation in WGS studies.

Methods: A survey of awareness regarding WGS and studies using WGS was conducted with a sample of 2000 or more participants using a self-administered questionnaire posted on the Internet between February 20 and 21, 2015. Prior to the survey, we briefly explained WGS and WGS study-related issues to the respondents in order to provide them with the minimum knowledge required to answer the questionnaire. We then conducted an analysis, including cross-classification.

Results: For the question regarding interest in WGS, 46.6% of participants responded "Yes." 70.7% of all respondents said that they were interested in some kinds of findings that could be obtained from WGS studies. Regarding participation in WGS studies, 29.0% were interested in participating. The demographic factors significantly related to attitudes toward research participation were age, level of education, and employment status. The results also suggest that concerns about WGS have a positive effect on people's willingness to participate. Furthermore, it was shown that for people who were not interested in their gene-related information, concerns about WGS negatively impacted their willingness to participate. However, for people who were interested in their gene-related information, their concerns might not have impacted their willingness to participate.

Discussion and conclusions: This research has shown several key factors that affect the willingness of the general public for the participation to the WGS studies. One of the unexpected findings is that concerns toward WGS studies generally have positive effect on the peoples' attitude. It will be interesting to further investigate into the various types of concerns that people in different groups have about WGS.

Keywords: Whole genome sequencing studies, Public attitudes, Japan, Participation in WGS studies, Gene-related information

* Correspondence: okita@med.tohoku.ac.jp
†Equal contributors
[1]Graduate School of Medicine, Tohoku University, 2-1 Seiryo-machi, Aoba-ku, Sendai 980-8575, Japan
Full list of author information is available at the end of the article

Background

Significant advances in genome analysis have led to increased understanding of diseases related to genome mutation [1]. Such advances have also allowed researchers to obtain a variety of gene-related information, including information associated with drug responsiveness, physical constitution, and blood relationships [2, 3]. In particular, recent innovations in gene analysis technology have allowed for rapid and inexpensive sequencing of entire genomes [4]. Thanks to these developments, conducting studies using whole genome sequencing (WGS) in a large population and applying these research findings in clinical settings have become possible. However, to promote WGS studies and use their findings more widely in clinical settings, increased participation from the general public in WGS studies is needed. In addition, the need for domestic and international research collaboration and data sharing across projects and institutions has been rapidly growing [5].

In response to this trend, new issues associated with WGS studies have emerged, including the sharing of genome information obtained through WGS (data sharing) and returning personal genetic information to subjects (return of results).

In order to conduct future investigations using WGS in a large study population in Japan, and to make effective regulatory policies, we believe that knowledge regarding attitudes toward and understanding of WGS and studies using WGS needs to be obtained. Several reports on WGS awareness surveys conducted in Japan and in other countries have been published [6, 7], and public attitudes on WGS data sharing [8–15] and return of such results [16–19] have been investigated. However, public attitudes toward participation in WGS studies have never been reported in Japan.

In this study, after the respondents had received a summary explanation of WGS, we conducted a survey of people's interest and concerns regarding WGS, their willingness to participate in WGS studies, and background factors correlated with their attitudes, in order to aid in the design of future WGS studies.

Methods

Study design

A survey of awareness around WGS and studies using WGS was conducted using a self-administered questionnaire posted on the Internet. Prior to each section where respondents choose their answers, we briefly explained WGS and WGS study-related issues, including an outline of WGS studies, data sharing, and return of results, to the respondents to provide them with the minimum knowledge required to answer the questionnaire.

The survey was conducted with a target number of 2000 or more participants, in accordance with the demographic distribution shown in the Japanese national census between February 20 and 21, 2015. Video Research Ltd. was in charge of the visual layout of the questionnaire and the implementation of the survey.

Recruitment participants

Participants were recruited from the general public who were registered as part-time survey assistants with Video Research Ltd., to which we outsourced part of the survey.

The subjects comprised males and females aged 16 years or older with a wide variety of attributes, but who matched with the age distribution of the Japanese population based on the Japanese government's national population census in 2010. Only the responses from the subjects who completed the entire questionnaire were included, and enrollment was considered to have been completed when the target number of the respondents was reached.

We received all the results of the survey from Video Research Ltd. and conducted an analysis, including cross-classification.

Statistical analysis

All data were summarized using medians and interquartile ranges for continuous variables whereas counts and percentages for categorical variables.

To identify factors associated with the willingness to participate in WGS studies, a proportional-odds logistic regression model was used with inclusion of age, sex, income, family number, education, employment, and marriage as explanatory variables. Non-linear restricted cubic splines were used to assess non-linearity of all continuous variables. Missing data were imputed using the multiple imputation methods using areg.impute function in rms package of R software [20].

In order to examine the effect of each variable indicating subjects with or without interest in and concerns regarding WGS, as well as those who did or did not wish to get the gene-related information on willingness to participate in WGS studies, similar analyses were performed.

To discover whether the effect of concerns about WGS was influenced by whether or not a subject wanted to get gene-related information, the interaction between these two variables was assessed with inclusion of their cross-product terms in the multivariable regression along with the main effect variables.

All statistical inferences were made two-sided at the 5% significance level. All statistical analyses were performed with the R software (version 3.2.2, R Foundation for Statistical Computing, Vienna, Austria) using the rms package.

IRB approval

Institutional review board (IRB) approval was obtained at the Osaka University and Tohoku University prior to implementation of this study.

Results

Demographic characteristics (socio-demographic)

The demographic characteristics of the participants are presented in Table 1. A total of 2399 participants were included in the analysis.

The population comprises 1196 males (49.9%) and 1203 females (50.1%), with a mean age of 44.91 years. While it is difficult to make direct comparisons with the national census data and the results from our study, concerning educational background, we note the following differences: while those who studied until high school make up for 46.5% of people in the 2010 national census [21], in our study, only 29.6% of respondents ended their education with high school. Furthermore, while university and graduate school graduates only make up 19.9% of people in the census data, 44.5% of our respondents had graduated from either university or graduate school. It is clear from the above that the respondents in our survey had comparatively higher educational backgrounds than the general population. Concerning employment, the 2015 national census showed that approximately 16.7% (20 million) of people had part-time or temporary employment, while approximately 27.5% (33 million) had permanent employment [22]. In our survey, 17.3% of respondents were non-regularly employed and 36.0% had regular employment. Compared to the general population, more of the respondents had permanent employment. However, unemployment rates were higher among respondents in our survey (10.5%) than in the general population (3.4%), according to the 2015 national census.

Table 1 Participant characteristics

Characteristic	All participants ($N = 2399$)
Age	45.00 (33.00, 57.00)
Gender	
Male	1196 (49.9%)
Female	1203 (50.1%)
Education	
Junior high school	71 (3.0%)
High school	711 (29.6%)
Professional school	267 (11.1%)
Technical college	24 (1.0%)
Junior college	250 (10.4%)
University	952 (39.7%)
Graduate school	114 (4.8%)
Other	10 (0.4%)
Employment status	
Permanent employment	864 (36.0%)
Part-time or temporary employment	416 (17.3%)
Student	165 (6.9%)
Full-time homeworker	454 (18.9%)
Retired	80 (3.3%)
Unemployed	251 (10.5%)
Other	169 (7.0%)
Annual income, JPY	
< 2,000,000	277 (14.0%)
2,000,000–3,000,000	252 (12.8%)
3,000,000–4,000,000	278 (14.1%)
4,000,000–5,000,000	278 (14.1%)
5,000,000–6,000,000	213 (10.8%)
6,000,000–7,000,000	193 (9.8%)
7,000,000–8,000,000	127 (6.4%)
8,000,000–10,000,000	165 (8.4%)
≧ 10,000,000	192 (9.7%)
Number of people in household	3.00 (2.00, 4.00)
Marital status	
Unmarried	926 (38.6%)
Married	1309 (54.6%)
Divorced or widowed	164 (6.8%)

Interest in and concerns about WGS

After we provided a brief explanation of WGS, the participants answered the questions, including questions regarding their interest in and concerns about WGS, and what type of information they would have interest in if their own sample was to be analyzed using WGS (see Tables 2 and 3). However, these questions were not limited to participants' interest or concern about studies specifically (such as returning incidental findings), but rather attempted to clarify their attitude toward or opinion of WGS in general.

For the question about interest in WGS, the participants responded as follows: 46.6% ($n = 1118$) "Yes (Very much or Moderate)," 26.1% ($n = 625$) "Neutral," and 27.4% ($n = 656$) "No (A little or Not at all)." For the question of whether they have concerns regarding the establishment of WGS technology: 28.9% ($n = 692$) chose "Yes (Very much or Moderate)," 39.5% ($n = 947$) chose "Neutral," and 31.7% ($n = 760$) chose "No (A little or Not at all)."

For the question, "what type of information would you wish to know if your own sample was analyzed using

Table 2 Interest in and concerns about whole genome sequencing

	All participants ($N = 2399$)
Interest	
Very much	277 (11.5%)
Moderately	841 (35.1%)
Neutral	625 (26.1%)
A little	275 (11.5%)
Not at all	381 (15.9%)
Concern	
Very much	119 (5.0%)
Moderately	573 (23.9%)
Neutral	947 (39.5%)
A little	403 (16.8%)
Not at all	357 (14.9%)

WGS" (multiple answers), the most frequent answer was "disease-related information (55.9%, $n = 1342$)," the second was "information on physical constitution, such as types prone to obesity and inebriety (43.6%, $n = 1047$)," the third most frequent was "information on gene-related drug effectiveness and side effects (29.1%, $n = 698$)," the fourth was "information on aptitudes and abilities (26.3%, $n = 631$)," and the fifth was "information on family relations (7.1%, $n = 171$)." Overall, the top 3 were information related to physical conditions. Notably, the answer "there is no gene-related information I wish to know" accounted for a relatively high percentage of answers (29.3%, $n = 702$).

Willingness to participate in WGS studies

The participants were asked to answer questions regarding their willingness to participate in WGS studies after they were provided with a brief introduction to WGS-related issues, including data sharing and return of results. This survey revealed that 29.0% of the participants ($n = 696$) were willing ("Yes—Very much or Moderate)," 33.6% ($n = 806$) were "Neutral," and 37.4% ($n = 897$) were unwilling ("No—A little or Not at all)" (see Table 4). Regarding willingness to participate in WGS studies, more respondents were unwilling to participate than those willing to participate.

Table 3 Information participants wish to know

	All participants ($N = 2399$)
Disease-related information	1342 (55.9%)
Information on gene-related drug effectiveness and side effects	698 (29.1%)
Information on physical constitution, such as types prone to obesity or inebriety	1047 (43.6%)
Information on aptitudes and abilities	631 (26.3%)
Information on family relations	171 (7.1%)
There is no gene-related information I wish to know	702 (29.3%)

Table 4 Willingness to participate in the whole genome sequencing study

	All participants ($N = 2399$)
Very much/Moderately	696 (29.0%)
Neutral	806 (33.6%)
A little/Not at all	897 (37.4%)

Background factors correlated with willingness to participate in WGS studies: demographic factors

The demographic factors of age, level of education, and employment status were found to be significantly correlated with willingness to participate in WGS studies. However, we were unable to find any significant correlation with sex, income, number of family members, or marriage.

For age (see Fig. 1), willingness to participate in WGS studies decreased with increasing age ($p < 0.05$). However, as shown in Fig. 1, although the p value was marginal ($p = 0.062$), the curve based on logistic analysis revealed that the people around 40 years of age showed the most positive attitude toward participation in WGS studies, but this attitude became increasingly negative with age after this point.

Higher levels of education were associated with positive attitudes toward participation in WGS studies for some comparisons. More specifically, compared with university graduates, significant differences were observed for graduates from junior high schools ($p < 0.001$), high schools ($p < 0.001$), and professional schools ($p < 0.05$), but no difference was found for the graduates from

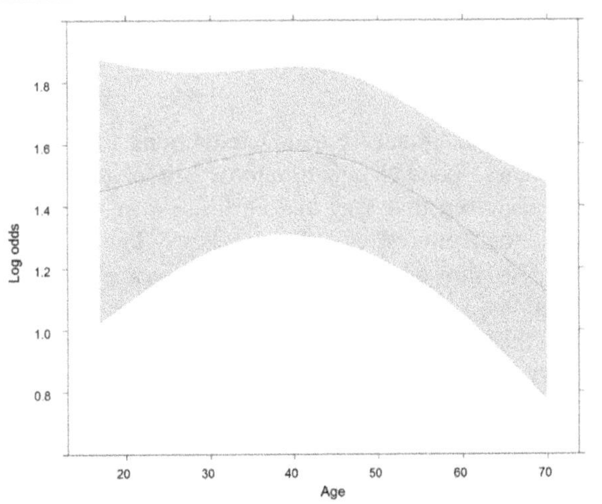

Fig. 1 The effects of age on willingness

technical colleges, junior colleges (which, in Japan, are considered as comparable to universities), and graduate schools. These results suggested that education levels, especially university graduate level or higher, could affect attitudes toward participation in WGS studies.

For employment status, compared with permanent employment as a reference, only the student group significantly showed a positive attitude ($p < 0.05$).

Background factors correlated with willingness to participate in WGS studies: attitudes toward WGS

For the relationship between the general interest in WGS and attitudes toward participation in WGS studies, the p value was significant ($p < 0.001$). The people who had concerns about WGS were also more willing to participate in WGS studies ($p < 0.001$).

For the relationship between interest in the information available from WGS and attitudes toward participation in WGS studies, distinct differences were observed between the people who wished to know gene-related information and the people who did not wish to know any gene-related information. The respondents who answered that they were interested in knowing their gene-related information (70.7%, $n = 1697$) showed more positive attitudes toward participation in WGS studies than those who did not have interest in such information (29.3%, $n = 702$) (odds rate = 10.3, $p < 0.001$).

Furthermore, analysis of the effect of concerns about WGS on the attitudes toward participation in WGS studies revealed that in the group of people who were interested in their gene-related information, any such concerns were unlikely to significantly influence willingness to participate in WGS studies ($p = 0.943$). In contrast, in the group of people not interested in any information from WGS, willingness to participate in WGS studies was significantly lower in the people with concerns about WGS than in those with no such concerns ($p = 0.021$).

Discussion

In this awareness survey conducted among members of the Japanese general public, almost half of the participants answered that they had an interest in WGS. The level of their interest was not necessarily low. However, many also showed negative attitudes toward participation in WGS studies. Our survey has showed that in Japan, there might exist a wide gap between people's interest in WGS and their willingness to participate in WGS studies.

However, since we can observe significantly positive attitude toward participation in WGS studies among students and younger people, if their attitudes toward WGS continue as they are, it is possible that the number of active participants in WGS research will increase in the future. Incidentally, the willingness to participate in WGS research was the strongest among people in their 40s. However, this trend showed a decrease past this point. One possible explanation for this is that people are getting married later and later in Japan, and thus, many people in their 40s are most concerned about the responsibility for their families and society. This fact may have some positive impact on their willingness to participate in WGS studies.

Another study performed in Japan (Ishiyama et al.) showed that men had significantly more positive attitudes toward genomic studies than women [6]. However, in our research, we were unable to observe this trend. Since both our method and subject differ from those of Ishiyama et al., it is difficult to make a simple comparison of our results. One possibility is that sex is becoming a weaker factor in terms of attitudes toward genomic and other kinds of scientific research.

The main factor correlated to a respondent's willingness to participate in WGS studies was whether or not they had interest in gene-related information. The group which expressed interest in gene-related information—compared to the group which did not—was much more likely to have a positive attitude toward participation in WGS research.

Additionally, the results suggest that for the group of people who are not interested in their gene-related information, concerns about WGS negatively impact their willingness to participate in WGS studies. However, if people are interested in knowing their gene-related information, their concerns might not impact their willingness to participate in WGS studies, or these concerns might even positively impact their interest in the information to be gained from WGS and their willingness to participate in WGS studies. Therefore, we can safely say that having an interest in gene-related information functions as a very strong and stable factor in whether or not a person is willing to participate in WGS research.

Our findings regarding the effect of people's concerns toward WGS were also of interest. Generally speaking, concerns about WGS would be expected to have a negative effect overall on people's willingness to participate. However, in our study, these concerns exert different effects depending on whether the person is interested in gene-related information or not. For those who are interested in gene-related information, these concerns do not affect their willingness to participate—in fact, they may positively influence the person's willingness. For those who are not interested in gene-related information, concerns about WGS seem to exert a negative influence on their willingness to participate. Considering the complexity of these results, it is necessary to understand and analyze the specific kinds of concerns that people have about WGS and whether these concerns are the same among those interested in gene-related information and those who are not.

Limitations

The level of education of the respondents in this study was higher than that shown in the national population census. This might be because the survey was conducted using a self-administered questionnaire posted on the web and because a scientific research topic, WGS, was the subject of the survey. In addition, the unemployment rate referenced in this survey (10.5%) seems much higher than Japan's actual rate of unemployment (3.4% at the time of writing). We suspect that this number may include some part-time workers. However, we are unable to investigate this statistic further within the scope of this paper.

Conclusion

The results suggest, based on the correlation analysis between the background factors and willingness to participate in WGS studies, that if a person has interest in knowing their gene-related information, this could be a major factor in improving their willingness to participate in WGS studies. Additionally, concerns about WGS did not always negatively impact the willingness to participate in WGS studies and thus could be a potential factor in increasing willingness. For these reasons, it is essential that we continue our investigation into the kinds of concerns that people have about WGS, as well as whether these concerns are shared between people who are interested in gene-related information and those who are not.

Acknowledgements

This work was supported by the Japanese Grant-in-Aid for Scientific Research on Innovative Areas from MEXT (Grant Number JP221S0002).
The authors would like to thank Joshua Wittig for the editorial assistance during the preparation of the manuscript.

Funding

This work was partly supported by the Grant-in-Aid for Scientific Research on Innovative Areas from MEXT, Japan (Grant Number JP221S0002).

Authors' contributions

TO and NO contributed equally as first authors. TO, NO, and KK conceived the study, participated in the research design, performed the research, analyzed the data, and wrote the manuscript. DK and AS analyzed the data and wrote the manuscript. All authors read and approved the final manuscript.

Competing interests

The authors declare that they have no competing interests.

[1]Graduate School of Medicine, Tohoku University, 2-1 Seiryo-machi, Aoba-ku, Sendai 980-8575, Japan. [2]Graduate School of Medicine, Osaka University, 2-2 Yamadaoka, Suita, Osaka 565-0871, Japan. [3]Graduate School of Medicine, Osaka City University, 1-4-3, Asahi, Abeno, Osaka, Osaka 545-8585, Japan.

References

1. Fortune MD, Guo H, Burren O, Schofield E, Walker NM, Ban M, Sawcer SJ, Bowes J, Worthington J, Barton A, Eyre S, Todd JA, Wallace C. Statistical colocalization of genetic risk variants for related autoimmune diseases in the context of common controls. Nat Genet. 2015;47(7):839–46. https://doi.org/10.1038/ng.3330.
2. Relling MV, Evans WE. Pharmacogenomics in the clinic. Nature. 2015; 526(7573):343–50. https://doi.org/10.1038/nature15817.
3. 23 and Me. Available at: https://www.23andme.com/dna-health-ancestry/. Accessed 2 Apr 2018.
4. Hayden EC. Technology: the $1,000 genome. Nature. 2014;507(7492):294–5. https://doi.org/10.1038/507294a.
5. Global Alliance for Genomics and Health. A federated ecosystem for sharing genomic, clinical data. Sci. 2016;352(6291):1278–80. https://doi.org/10.1126/science.aaf6162.
6. Ishiyama I, Nagai A, Muto K, Tamakoshi A, Kokado M, Mimura K, Tanzawa T, Yamagata Z. Relationship between public attitudes toward genomic studies related to medicine and their level of genomic literacy in Japan. Am J Med Genet A. 2008;146A(13):1696–706. https://doi.org/10.1002/ajmg.a.32322.
7. Dodson DS, Goldenberg AJ, Davis MM, Singer DC, Tarini BA. Parent and public interest in whole-genome sequencing. Public Health Genomics. 2015;18(3):151–9. https://doi.org/10.1159/000375115.
8. McGuire AL, Hamilton JA, Lunstroth R, McCullough LB, Goldman A. DNA data sharing: research participants' perspectives. Genet Med. 2008;10(1):46–53. https://doi.org/10.1097/GIM.0b013e31815f1e00.
9. Lemke AA, Wolf WA, Herbert-Beirne J, Smith ME. Public and biobank participant attitudes toward genetic research participation and data sharing. Public Health Genomics. 2010;13(6):368–77. https://doi.org/10.1159/000276767.
10. Trinidad SB, Fullerton SM, Bares JM, Jarvik GP, Larson EB, Burke W. Genomic research and wide data sharing: views of prospective participants. Genet Med. 2010;12(8):486–95. https://doi.org/10.1097/GIM.0b013e3181e38f9e.
11. Haga SB, O'Daniel J. Public perspectives regarding data-sharing practices in genomic research. Public Health Genomics. 2011;14(6):319–24. https://doi.org/10.1159/000324705.
12. McGuire AL, Oliver JM, Slashinski MJ, Graves JL, Wang T, Kelly PA, Fisher W, Lau CC, Goss J, Okcu M, Treadwell-Deering D, Goldman AM, Noebels JL, Hilsenbeck SG. To share or not to share: a randomized trial of consent for data sharing in genome research. Genet Med. 2011;13(11):948–55. https://doi.org/10.1097/GIM.0b013e3182227589.
13. Oliver JM, Slashinski MJ, Wang T, Kelly PA, Hilsenbeck SG, McGuire AL. Balancing the risks and benefits of genomic data sharing: genome research participants' perspectives. Public Health Genomics. 2012;15(2):106–14. https://doi.org/10.1159/000334718.
14. Jamal L, Sapp JC, Lewis K, Yanes T, Facio FM, Biesecker LG, Biesecker BB. Research participants' attitudes towards the confidentiality of genomic sequence information. Eur J Hum Genet. 2014;22(8):964–8. https://doi.org/10.1038/ejhg.2013.276.
15. Global Alliance for Genomics & Health. White Paper: Creating a global alliance to enable responsible sharing of genomic and clinical data, 2013. https://www.broadinstitute.org/files/news/pdfs/GAWhitePaperJune3.pdf. Accessed 2 Apr 2018.
16. Murphy J, Scott J, Kaufman D, Geller G, LeRoy L, Hudson K. Public expectations for return of results from large-cohort genetic research. Am J Bioeth. 2008;8(11):36–43. https://doi.org/10.1080/15265160802513093.
17. Yu JH, Crouch J, Jamal SM, Tabor HK, Bamshad MJ. Attitudes of African Americans toward return of results from exome and whole genome sequencing. Am J Med Genet A. 2013;161A(5):1064–72. https://doi.org/10.1002/ajmg.a.35914 .
18. Yu JH, Crouch J, Jamal SM, Bamshad MJ, Tabor HK. Attitudes of non-African American focus group participants toward return of results from exome and whole genome sequencing. Am J Med Genet A. 2014;164A(9):2153–60. https://doi.org/10.1002/ajmg.a.36610.
19. Middleton A, Morley KI, Bragin E, Firth HV, Hurles ME, Wright CF, Parker M, DDD study. Attitudes of nearly 7000 health professionals, genomic

researchers and publics toward the return of incidental results from sequencing research. Eur J Hum Genet. 2016;24(1):21–9. https://doi.org/10.1038/ejhg.2015.58.
20. Package 'rms'. Available at: https://cran.r-project.org/web/packages/rms/rms.pdf. Accessed 2 Apr 2018.
21. Statistics Bureau, Ministry of Internal Affairs and Communications. The 2010 population census (Japan). http://www.stat.go.jp/english/data/kokusei/2010/summary.html. Accessed 2 Apr 2018.
22. Statistics Bureau, Ministry of Internal Affairs and Communications. The 2015 population census (Japan). http://www.stat.go.jp/english/data/kokusei/2015/summary.html. Accessed 2 Apr 2018.

15

Oxidative stress-induced chromosome breaks within the *ABL* gene: a model for chromosome rearrangement in nasopharyngeal carcinoma

Sang-Nee Tan[1], Sai-Peng Sim[1*] and Alan Soo-Beng Khoo[2]

Abstract

Background: The mechanism underlying chromosome rearrangement in nasopharyngeal carcinoma (NPC) remains elusive. It is known that most of the aetiological factors of NPC trigger oxidative stress. Oxidative stress is a potent apoptotic inducer. During apoptosis, chromatin cleavage and DNA fragmentation occur. However, cells may undergo DNA repair and survive apoptosis. Non-homologous end joining (NHEJ) pathway has been known as the primary DNA repair system in human cells. The NHEJ process may repair DNA ends without any homology, although region of microhomology (a few nucleotides) is usually utilised by this DNA repair system. Cells that evade apoptosis via erroneous DNA repair may carry chromosomal aberration. Apoptotic nuclease was found to be associated with nuclear matrix during apoptosis. Matrix association region/scaffold attachment region (MAR/SAR) is the binding site of the chromosomal DNA loop structure to the nuclear matrix. When apoptotic nuclease is associated with nuclear matrix during apoptosis, it potentially cleaves at MAR/SAR. Cells that survive apoptosis via compromised DNA repair may carry chromosome rearrangement contributing to NPC tumourigenesis. The Abelson murine leukaemia (*ABL*) gene at 9q34 was targeted in this study as 9q34 is a common region of loss in NPC. This study aimed to identify the chromosome breakages and/or rearrangements in the *ABL* gene in cells undergoing oxidative stress-induced apoptosis.

Results: In the present study, in silico prediction of MAR/SAR was performed in the *ABL* gene. More than 80% of the predicted MAR/SAR sites are closely associated with previously reported patient breakpoint cluster regions (BCR). By using inverse polymerase chain reaction (IPCR), we demonstrated that hydrogen peroxide (H_2O_2)-induced apoptosis in normal nasopharyngeal epithelial and NPC cells led to chromosomal breakages within the *ABL* BCR that contains a MAR/SAR. Intriguingly, we detected two translocations in H_2O_2-treated cells. Region of microhomology was found at the translocation junctions. This observation is consistent with the operation of microhomology-mediated NHEJ.

Conclusions: Our findings suggested that oxidative stress-induced apoptosis may participate in chromosome rearrangements of NPC. A revised model for oxidative stress-induced apoptosis mediating chromosome rearrangement in NPC is proposed.

Keywords: NPC, Oxidative stress, H_2O_2, Apoptosis, *ABL*, MAR/SAR

* Correspondence: spsim@unimas.my
[1]Department of Paraclinical Sciences, Faculty of Medicine and Health Sciences, Universiti Malaysia Sarawak, Sarawak, Malaysia
Full list of author information is available at the end of the article

Background

Nasopharyngeal carcinoma (NPC) is a malignant neoplasm derived from mucosal epithelium of the nasopharynx. According to the World Health Organization (WHO), NPC can be classified into three subtypes according to the degree of epithelial differentiation, namely keratinising squamous cell carcinoma (Type I), non-keratinising squamous cell carcinoma (Type II) and undifferentiated or poorly differentiated carcinoma (Type III) [1].

NPC is a rare malignancy in most parts of the world; the incidence rates are below one per 100,000 persons per year [2, 3]. However, there are a few well-known notable exceptions [3]. The intermediate rates were reported in South-Eastern Asia, Northern Africa, the Middle East and Arctic Region [3, 4]. The highest incidence rate was observed among Southern Chinese living in central Guangdong province. The annual incidence rates for males and females in central Guangdong province are 23.3 per 100,000 and 8.9 per 100,000, respectively [4]. The NPC incidence rates are generally increasing from Northern China to Southern China [3, 4]. In addition, an exceptionally high incidence rate has been reported among the Bidayuh people, the second biggest ethnic group in Sarawak, Malaysia. The age-adjusted rate of Sarawak residents is 13.5 per 100,000 and 6.2 per 100,000 in males and females, respectively. Although the average rate in Sarawak is intermediate, the incidence rate for Bidayuh people is about 50% higher than that in Hong Kong (a part of the Cantonese region of Guangdong province) [5].

NPC is strongly associated with Epstein-Barr Virus (EBV) infection [6–8] as well as dietary [9–11], environmental [12] and genetic factors [13, 14]. Several genetic aberrations have been reported to be related to the development of NPC, suggesting that NPC tumourigenesis involves multiple genetic changes. These include chromosomal gains or losses [15–19], loss of heterozygosity (LOH) [20–23], homozygous deletions [24–27], promoter hypermethylation of tumour suppressor genes [28–31] and shortening of chromosome telomeres [32, 33].

Although the consistent chromosome rearrangements have long been identified in NPC, the molecular mechanism underlying the chromosome rearrangements of NPC remains poorly understood. In addition to EBV infection, long-term exposures to nitrosamines, formaldehyde, cigarette smoke and wood dust have all been found to be associated with NPC [12, 34–36]. More recently, much concern has been raised about the association between chronic inflammation of sinonasal tract and NPC [37, 38]. It is remarkable that all these aetiological factors may trigger oxidative stress [39–43]. Oxidative stress is an imbalance of pro-oxidants and antioxidants resulting in a disruption of redox signalling and control. Pro-oxidants induce oxidative stress either through excessive production of reactive oxygen species (ROS) or inhibition of antioxidant systems [44]. ROS are chemically-reactive molecules containing oxygen which include peroxyl RO·, hydroxyl radical OH·, superoxide $O^{2}·-$ and hydrogen peroxide H_2O_2 [45]. ROS cause several kinds of DNA damages, including strand cleavage, base modification and DNA-protein cross-linkage [45, 46]. Importantly, formaldehyde and acrolein, a component of cigarette smoke, are reactive aldehydes. In addition, reactive aldehydes may also be produced endogenously during oxidative stress. Aldehydes may cause adduct formation that impairs the function of DNA, RNA and proteins via electrophile-nucleophile interaction. Exposure to environmental aldehydes has been shown to be associated with the onset and development of human diseases that involve oxidative stress. It has been suggested that environmental and endogenous aldehydes may interact additively and exacerbate the cellular oxidative damage [47].

An evaluation of the levels of 8-hydroxy-2′-deoxyguanosine (8-OHdG), a biomarker of oxidative DNA damage, had been done among NPC patients. The tissue and serum levels of 8-OHdG in NPC patients have been found to be significantly higher than those in control patients [48]. Oxidative stress was suggested to play an important role in carcinogenesis [49]. Since there is a strong link between the aetiological factors of NPC and oxidative stress, it is intriguing to investigate the role of oxidative stress in the molecular mechanisms underlying chromosome rearrangements of NPC.

Oxidative stress may induce apoptosis [50, 51]. H_2O_2 has been well known as an apoptotic inducer for various human cell types, including osteoblasts [52, 53], sarcoma cells [54], osteosarcoma cells [55], hepatoma cells [56], astrocytoma cells [57], Jurkat T lymphocytes [58] and Fanconi's anaemia cells [59]. Apoptosis or programmed cell death was first described by Kerr et al. (1972). Apoptosis is a type of genetically controlled cell suicide which occurs naturally in multicellular organisms in order to eliminate poisonous cells. Apoptosis is morphologically characterised by condensation of chromatin, fragmentation of nuclei, compaction of cytoplasmic organelles, cell shrinkage and cytoplasmic membrane blebbing [60–62]. Apoptosis is related to several biochemical events, including externalisation of phosphatidylserine (PS) on cell membrane, alteration in mitochondrial membrane potential (MMP), release of cytochrome *c* (cyt *c*) from mitochondria, caspase activation and internucleosomal cleavage of DNA [63].

The alteration of nuclear chromatin during apoptosis is often associated with fragmentation of the genomic DNA into high-molecular-weight (HMW) DNA of 30 to 50 and 200 to 300 kbp [64, 65]. These sizes of fragments have been suggested to be derived from the release of loops (50 kbp) or rosettes (300 kbp) of chromatin, probably when they become detached from their binding sites on the nuclear scaffold [66]. Further degradation of the HMW DNA produces the internucleosomal DNA fragments of 180 to 200 bp [67, 68].

Cells undergoing apoptosis may recover from the execution phase of apoptosis upon DNA repair [69, 70]. There are two major double-strand breaks (DSBs) repair pathways, namely homologous recombination (HR) and non-homologous end joining (NHEJ) [71, 72]. Chromosomal DSB repair by HR is predominant during late S/G2 phases of the cell cycle. NHEJ is the more frequently used pathway that can repair a DSB at any time during the cell cycle. These two pathways have different degree of requirement for DNA homology. The HR pathway requires sufficient homology, usually more than 100 bp. Given that the HR DNA repair system ligates two DNA ends with homologous sequences, it gives rise to precise DNA repair. The NHEJ DNA repair system joins two DNA ends without intensive requirement of sequence homology. This pathway joins two DNA terminals with microhomology of a few base pairs [71, 73, 74]. NHEJ pathway has been shown to be prone to cause erroneous repair of DSBs. This may in turn lead to chromosomal aberrations [75]. It has been suggested that interaction of the NHEJ DNA repair system with apoptosis can act as a mechanism leading to translocation in leukaemia [70].

Chromosomal breakage takes place in the initial stage of chromosome rearrangement and apoptotic DNA fragmentation. It has been observed that chromosome breaks do not randomly occur throughout a gene. Rather, chromosome breaks normally fall within certain regions that contain specific chromatin structures, such as matrix association region/scaffold attachment region (MAR/SAR) [76, 77]. MAR/SAR are DNA sequences where DNA loop structure attaches to nuclear scaffold/matrix proteins [78]. There are two breakpoint cluster regions (BCR) identified in the *AF9* gene. These two BCRs are bordered by two experimentally isolated MAR/SARs [76]. The BCR of the mixed lineage leukaemia (*MLL*) gene has also been found to contain two MAR/SAR sequences [78]. In addition, the most crucial apoptotic nuclease CAD has been reported to associate with the nuclear matrix of apoptotic cells [79].

We previously demonstrated that in normal nasopharyngeal epithelial and NPC cells, oxidative stress-induced apoptosis resulted in chromosome breaks in the *AF9* gene located on chromosome 9p22. We further demonstrated that caspase-activated DNase (CAD) may be a major player in mediating the oxidative stress-induced chromosomal cleavages. A few chromosome breaks were identified within the *AF9* region that was previously reported to participate in translocation in an acute lymphoblastic leukaemia (ALL) patient. These findings suggested that oxidative stress-induced apoptosis may play an important role in mediating chromosome rearrangements in NPC [80]. In the present study, we further investigated the potential role of oxidative stress-induced apoptosis by targeting the Abelson murine leukaemia viral oncogene homologue 1 (*ABL*) gene located on chromosome 9q34. This study targeted the *ABL* gene because 9q34 is a common region of loss in NPC [23].

The *ABL* gene is a proto-oncogene which encodes a 150 kDa nonreceptor protein tyrosine kinase. It was first recognised as the cellular homologue of the *v-abl* oncogene product of the Abelson murine leukaemia virus [81, 82]. The ABL protein has a complex structure that contains many domains. These domains are found in proteins which are involved in the formation of complexes in signal transduction pathway. It has been demonstrated that overexpression of *ABL* in fibroblast resulted in growth arrest [83]. The product of *ABL-BCR* fusion appears to be an abnormal kinase that stimulates the proliferation of myeloid cells leading to chronic myelogenous leukaemia (CML) [84]. The *ABL* gene is 173,795 bp in length and it consists of 11 exons [Ensembl:ENSG00000097007]. The description of exons and introns in the *ABL* gene is shown in Additional file 1.

By using MAR/SAR recognition signature (MRS), we predicted 12 possible MAR/SAR sites in the *ABL* gene. We demonstrated that oxidative stress-induced apoptosis resulted in chromosome breaks in the *ABL* BCR which contains a MAR/SAR site. We detected shift translocations in H_2O_2-treated normal nasopharyngeal epithelial cells. Interestingly, we found region of microhomology at the breakpoint junctions. This observation suggests a role for NHEJ DNA repair system in mediating the translocation. At last, we illustrated the possible role of oxidative stress-induced apoptosis in mediating chromosome rearrangements in NPC via NHEJ DNA repair system.

Results

In silico prediction of MAR/SAR by using MAR/SAR recognition signature (MRS)

Potential MAR/SAR sites in the *ABL* gene were predicted by using MRS. MRS is a bipartite sequence that is strongly associated with MAR/SAR [85]. This bipartite sequence consists of 16 bp nucleotide motif (AWWRTAANNWWGNNNC) within a distance of 200 bp of the 8 bp nucleotide motif (AATAAYAA). However, for our preliminary results in the *ABL* gene, we only found one MRS (MAR/SAR 9 in Table 1) in the biochemically identified SAR1 [77]. The distance between the 8 bp sequence element and the 16 bp sequence element was 248 bp. Therefore, in this study, we set the maximal distance between the two sequence elements at 250 bp.

By using MRS, we predicted 12 potential MAR/SAR sites in the *ABL* gene. The nucleotide positions of the MRSs with their sequence composition, relative orientation, distance

Table 1 MAR/SAR predicted in the *ABL* gene

Predicted MAR/SAR	AWWRTAANNWWGNNNC (16 bp)	Nucleotide position	AATAAYAA (8 bp)	Nucleotide position	Distance (bp)	Location in exon/intron
1–1	ATTGTAACCATATCTC (C)	26,055–26,070	AATAATAA (C)	26,118–26,125	+ 47	Intron 1
1–2	ATCATAACTTAGCAAC (C)	26,613–26,628	AATAACAA (W)	26,565–26,572	−40	Intron 1
2	AAAAAAATTTTGTACC (C)	29,601–29,616	AATAATAA (C)	29,629–29,636	+ 12	Intron 1
3–1	ATAATAATTATACAAC (C)	80,698–80,713	AATAATAA (W)	80,726–80,733	+ 12	Intron 1
	AATATAAATAAAGTGC (W)	80,732–80,747			Overlap	
3–2	ATTGTAACTAAGGTTC (C)	81,458–81,473	AATAACAA (C)	81,422–81,429	−28	Intron 1
4	ATAATAATAAAGAGAT (W)	99,259–99,274	AATAATAA (W)	99,261–99,268	Overlap	Intron 1
	AATATAATCAACTGAC (W)	99,447–99,462			− 178	
5	ATAAAAAGGAAGATTC (W)	105,498–105,513	AATAATAA (C)	105,616–105,623	+ 102	Intron 1
6	AAAAAAAAAAAGACTC (C)	108,315–108,330	AATAATAA (C)	108,484–108,491	+ 153	Intron 1
			AATAATAA (C)	108,487–108,494	+ 156	
			AATAATAA (C)	108,490–108,497	+ 159	
7	AATGTAACAGAGAGCC (C)	117,122–117,137	AATAACAA (W)	117,343–117,350	+ 205	Intron 1
8	AAAATAAACATATACC (W)	119,757–119,772	AATAATAA (W)	119,739–119,746	− 10	Intron 1
9	AAAGTAAAATTGAAAG (C)	133,546–133,561	AATAACAA (W)	133,810–133,817	+ 248	Intron 1
10	ATTACAAGTTTGGTAC (C)	144,212–144,227	AATAATAA (C)	143,996–144,003	− 208	Intron 3
11	ATAAAAACAAAGAAGC (C)	163,018–163,033	AATAACAA (W)	163,048–163,055	+ 14	Intron 7
12	AAAATAATAATGGCCA (W)	167,856–167,871	AATAATAA (W)	167,858–167,865	Overlap	Intron 10

Nucleotide positions of the MRSs with their sequence composition, relative orientation (C, Crick strand and W, Watson strand), distance between the two sequence elements and location of the MRSs in the exon or intron of the *ABL* gene are shown. A negative distance indicates that 8 bp sequence element precedes the 16 bp sequence element

between the two sequence elements and location of the MRSs in the exon or intron of the *ABL* gene are shown in Table 1. Out of the 12 predicted MAR/SAR sites, 9 were identified in intron 1 which is the largest intron (approximately 140 kb in length) in the *ABL* gene (approximately 175 kb in length) (MAR/SAR 1–9 in Table 1). One potential MAR/SAR site was separately found in intron 3 (MAR/SAR 10 in Table 1) and intron 10 (MAR/SAR 11 in Table 1). The distribution of the predicted MAR/SAR sites in the *ABL* gene is shown in Fig. 1. One MAR/SAR site (MAR/SAR 9 in Table 1) was predicted within the biochemically defined SAR1 which is located in the second intron 1 [77].

Apoptosis detection

NP69 cells were either left untreated or treated with 100 μM of H_2O_2 for 16 and 24 h while HK1 cells were either left untreated or treated with 50 μM for 4 and 8 h. Cells treated with CPT was included as a positive control. The cells were then subjected to flow cytometric analyses of PS externalisation and MMP loss.

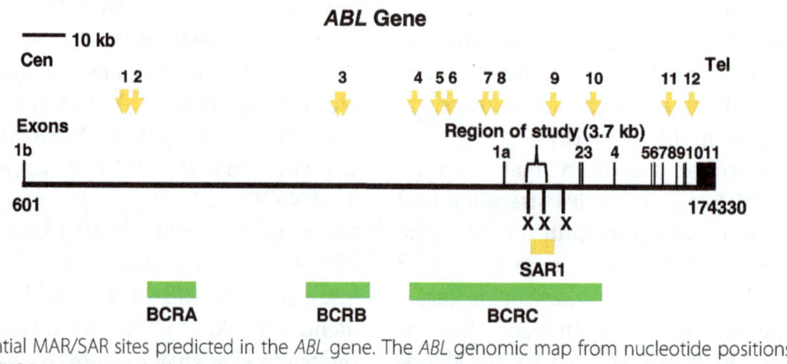

Fig. 1 Distribution of potential MAR/SAR sites predicted in the *ABL* gene. The *ABL* genomic map from nucleotide positions 601-174330 is illustrated above [Ensembl:ENSG00000097007]. The locations of exons 1 to 11 are shown. Green boxes represent the three previously reported patient breakpoints cluster regions which are designated as BCRA, BCRB and BCRC. Yellow box shows the previously biochemically extracted MAR/SAR which is designated as SAR1 [77]. Yellow arrows represent the potential MAR/SARs predicted by MRS. Clusters of more than one MRS within close proximity are regarded as a single potential MAR/SAR site. For instance, there were two MRSs predicted in BCRB, however, they were regarded as a single potential MAR/SAR site (MAR/SAR 3) because they were found in close proximity. There was one MAR/SAR site (MAR/SAR 9) predicted in the experimentally isolated SAR1

Phosphatidylserine (PS) externalisation

As shown in Fig. 2a i, the percentages of apoptotic cells detected in NP69 treated with 100 μM of H_2O_2 for 16 and 24 h were 2.82-fold ($p = 0.000170$) and 2.87-fold ($p = 3.4346E{-}8$) higher than that detected in the untreated control, respectively. The percentages of apoptotic cells detected in HK1 treated with 50 μM of H_2O_2 for 4 and 8 h were 1.48-fold ($p = 0.005735$) and 1.92-fold ($p = 0.000477$) higher than that detected in the untreated control, respectively (Fig. 2b i). Figure 2a i and b ii are the representative dot plot diagrams showing the apoptotic population of H_2O_2-treated NP69 and HK1 cells, respectively.

Mitochondrial membrane potential (MMP) loss

As shown in Fig. 3a i, the percentages of apoptotic cells detected in NP69 treated with 100 μM of H_2O_2 for 16 and 24 h were 2.45-fold ($p = 0.006$) and 2.25-fold ($p = 0.002$) higher than that detected in the untreated control, respectively. The percentages of apoptotic cells detected in HK1 treated with 50 μM of H_2O_2 for 4 and 8 h were 1.68-fold ($p = 0.009$) and 2.18-fold ($p = 0.007$) higher than that detected in the untreated control, respectively (Fig. 3b i). Figure 3a ii and b ii are the representative contour plot diagrams showing the apoptotic population of H_2O_2-treated NP69 and HK1 cells, respectively.

IPCR detection of chromosome breaks within the *ABL* gene mediated by stress-induced apoptosis

NP69 cells at confluency of 30–40% were treated with 10, 50 and 100 μM of H_2O_2 for 16 and 24 h while HK1 cells at optimal density were treated with 1, 10 or 50 μM of H_2O_2 for 2, 4, 6 and 8 h. For each cell line, an untreated sample was included to serve as a cell control. Nested IPCR was employed to identify chromosome breaks mediated by stress-induced apoptosis. The IPCR bands representing the *ABL* cleaved fragments detected were isolated, purified and sequenced.

Figures 4 and 5 show the IPCR results for H_2O_2-treated NP69 and HK1 cells, respectively. In the manipulation for

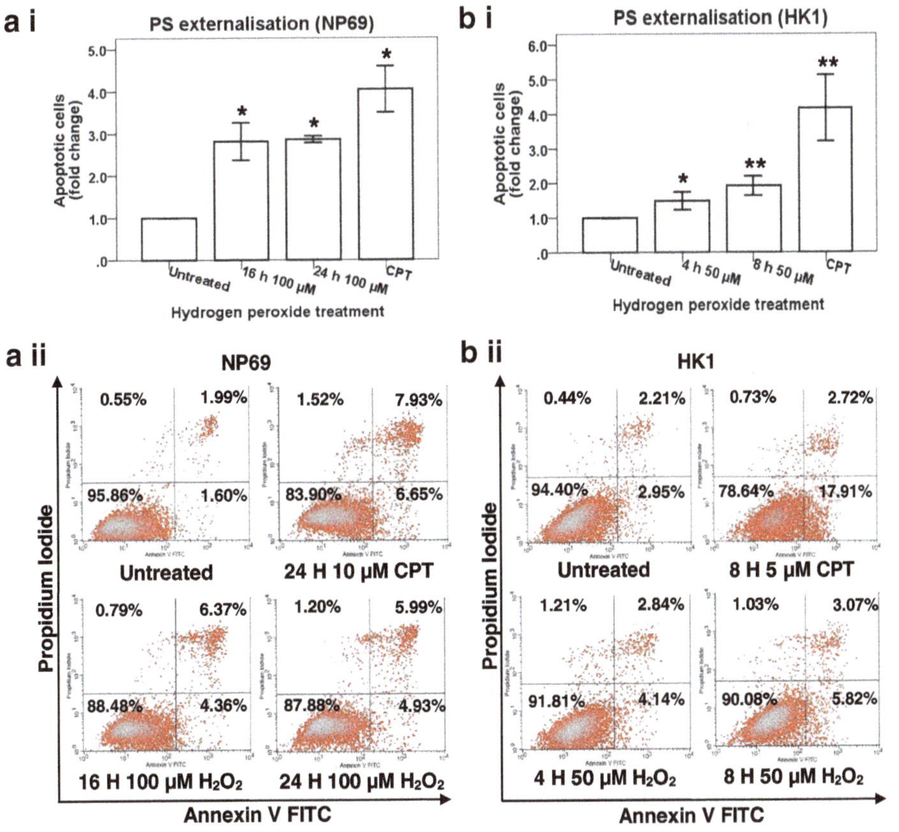

Fig. 2 Flow cytometric analysis of phosphatidylserine (PS) externalisation. NP69 cells were either left untreated or treated with 100 μM of H_2O_2 for 16 and 24 h while HK1 cells were either left untreated or treated with 50 μM for 4 and 8 h. Cells treated with CPT was included as a positive control. The percentage of cells showing PS externalisation was determined in H_2O_2-treated NP69 cells (**a i**) and HK1 cells (**b i**). Means and SD of three independent experiments performed in duplicate are shown. Data are expressed as fold change normalised to untreated control. $^{*}p < 0.01$, $^{**}p < 0.001$ (Student's *t* test). The representative dot plot diagrams indicating the apoptotic populations of (**a ii**) H_2O_2-treated NP69 cells and (**b ii**) H_2O_2-treated HK1 cells are shown. The lower left quadrants indicate healthy cells; the lower right quadrants indicate cells in early apoptosis; the upper right quadrants indicate cells in late apoptosis and necrosis

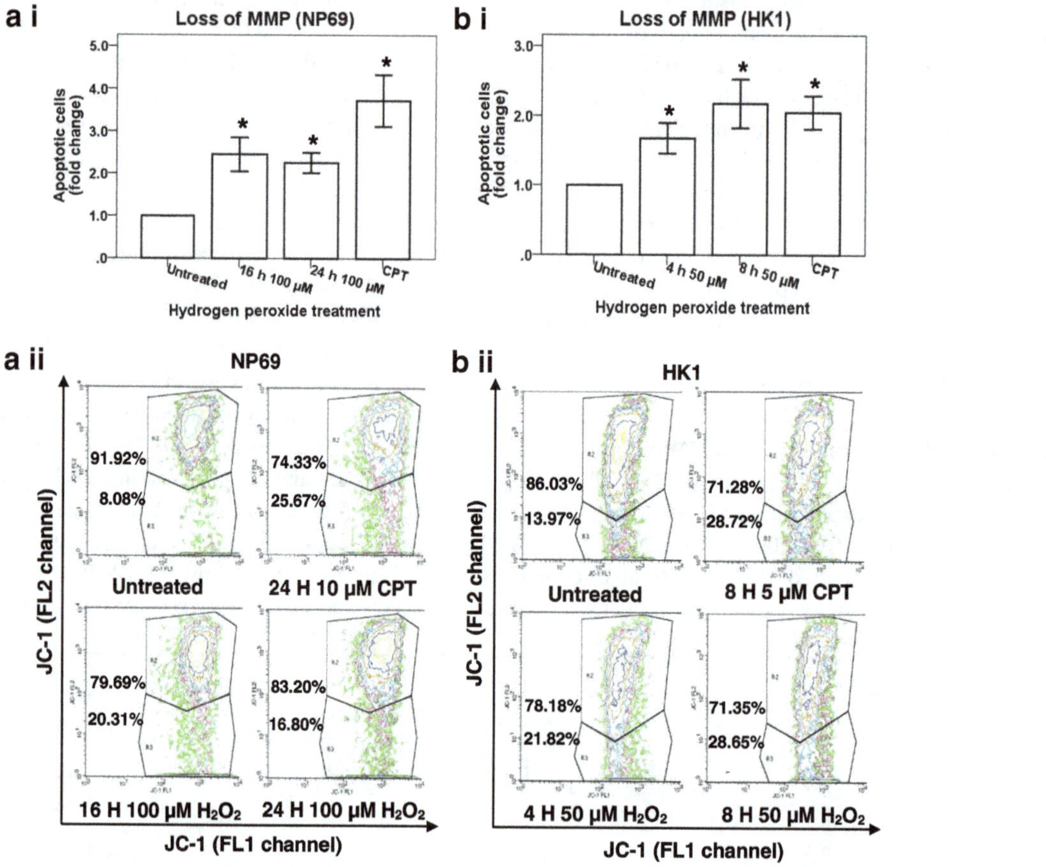

Fig. 3 Flow cytometric analysis of mitochondrial membrane potential (MMP) loss. NP69 cells were either left untreated or treated with 100 μM of H_2O_2 for 16 and 24 h while HK1 cells were either left untreated or treated with 50 μM for 4 and 8 h. Cells treated with CPT was included as a positive control. The percentage of cells showing MMP loss was determined in H_2O_2-treated NP69 cells (**a i**) and HK1 cells (**b i**). Means and SD of two independent experiments performed in duplicate are shown. Data are expressed as fold change normalised to untreated control. $^*p < 0.01$ (Student's *t* test). The representative contour plot diagrams indicating the apoptotic populations of (**a ii**) H_2O_2-treated NP69 cells and (**b ii**) H_2O_2-treated HK1 cells are shown. The upper quadrants indicate healthy cells whereas the lower quadrants indicate cells expressing MMP loss

nested IPCR, *Age* I (RE2 in Fig. 12) was used to linearise the cyclised DNA. If there is no breakage within the *ABL* gene, the IPCR product will be approximately 3 kb. On the contrary, if there is any breakage within the *ABL* gene, it should produce IPCR products which are smaller than 3 kb. As shown in Figures 4a and 5a, the *ABL* intact fragment of 3 kb is present in all of the samples. This amplification could serve as an internal control by proposing an optimal IPCR condition for the *ABL* gene. Besides, numerous IPCR bands of less than 3 kb were also obtained. However, these bands are less intense as compared with the intact fragment. This could be due to the competition between the intact fragments and the cleaved fragments for the amplification process. The intact fragments are usually more abundant as compared with the cleaved fragments. Consequently, the amplification of the cleaved fragments would be less efficient in the presence of the intact fragments. Therefore, double digestion with *Age* I and *Bsa*A I or *Age* I and *Eco*R I (RE3 in Fig. 12) was used to eliminate competition from the intact fragments for the nested IPCR reaction. These two different digestions gave rise to the detection of chromosome breaks within different regions. With double digestion of *Age* I and *Eco*R I, numerous distinct IPCR bands of less than 3 kb which represent the cleaved *ABL* fragment were detected in H_2O_2-treated NP69 (Fig. 4b, lanes 4–9) and H_2O_2-treated HK1 (Fig. 5b, lanes 4, 7, 9, 10, 11, 12 and 13) cells. A few cleavage bands were detected in the untreated NP69 cells (Fig. 4b, lane 1) which might be due to endogenous DNA breaks in the minority of untreated cells which were unhealthy. However, in general, there were more cleaved fragments detected in the treated samples compared with the untreated sample.

As shown in Fig. 4c, the intact fragment of 3 kb was still detected upon double digestion with *Age* I and *Bsa*A I. Most probably, this was due to incomplete digestion by *Age* I and *Bsa*A I. Regardless of the incomplete digestion, several cleavage bands of different sizes were detected in NP69 cells treated with various concentrations of H_2O_2 for different time points (Fig. 4c, lanes 4–9).

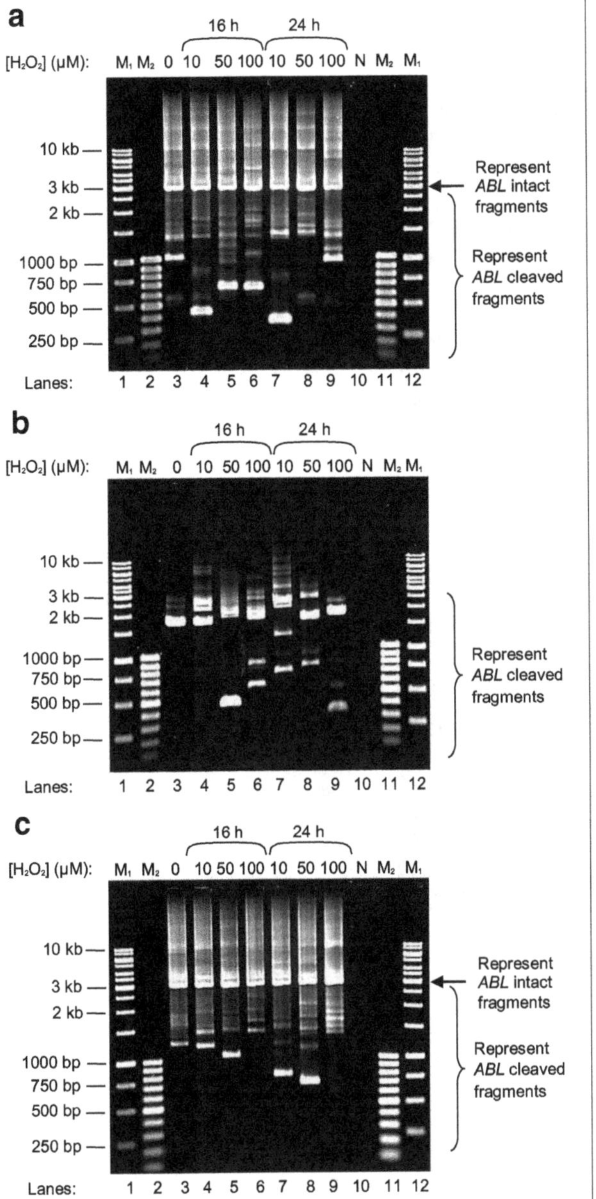

Fig. 4 Nested IPCR detection of DNA breakages within the *ABL* gene in H_2O_2-treated NP69. NP69 cells at 30–40% confluency were either untreated (lane 3) or treated with 10 μM (lanes 4 and 7), 50 μM (lanes 5 and 8) or 100 μM (lanes 6 and 9) of H_2O_2 for 16 h (lanes 4–6) and 24 h (lanes 7–9). Genomic DNA was isolated and manipulated for nested IPCR. In the manipulation for nested IPCR, the DNA samples were subjected to digestion with *Age* I (**a**), double digestion with *Age* I and *EcoR* I (**b**) or double digestion with *Age* I and *BsaA* I (**c**). The IPCR products were analysed on 1% agarose gel. Side arrows in panels **a** and **c** indicate the position of the 3 kb IPCR bands resulting from the amplification of the intact *ABL* gene. Side brackets in panels **a**, **b** and **c** indicate the possible IPCR bands from the *ABL* cleaved fragments. Negative control for PCR was included (lane 10). This IPCR result is representative of 2 repeats with similar results. M_1: 1 kb DNA ladder. M_2: 100 bp DNA ladder

DNA breakages were detected in cell samples treated with various concentrations of H_2O_2 at various time points. Based on the microscopic analysis and flow cytometric analyses, the optimal concentration and time point were determined. These optimal concentration and time point were used to repeat the experiments in NP69 and HK1 cells. For NP69 cells, we selected concentration of 100 μM with exposure time of 16 and 24 h. The microscopic analysis on H_2O_2-treated NP69 cells showed that cytoplasmic shrinkage was only observed in cells treated with 100 μM for 16 and 24 h (Additional file 2). In addition, we performed flow cytometric analyses of PS externalisation and MMP loss on NP69 cells treated with 100 μM for 16 and 24 h. In these two flow cytometric assays, both samples showed significantly higher percentage of apoptosis as compared with that of the untreated control (Figs. 2a i and 3a i).

For HK1 cells, we chose concentration of 50 μM with exposure time of 8 h. Cell blebbing, which is one of the major morphological changes, was relatively more prominent in HK1 cells treated with 50 μM of H_2O_2 for 8 h compared with those treated with lower concentration (1 and 10 μM) and shorter exposure time (2, 4 and 6 h) (Additional file 3). Moreover, the flow cytometric analysis of PS externalisation on sample treated with 50 μM of H_2O_2 for 8 h showed a more promising result compared with the sample treated with 50 μM of H_2O_2 for 4 h (Fig. 2b i).

H_2O_2 treatment was thus repeated in NP69 and HK1 cells with the selected concentration and time point. NP69 cells at confluency of 30–40% were either untreated or treated with 100 μM of H_2O_2 for 16 h while HK1 cells at confluency of 60–70% were either untreated or treated with 50 μM of H_2O_2 for 8 h. The cells were then harvested for gDNA extraction and nested IPCR. In the manipulation for nested IPCR, all the samples were subjected to double digestion with *Age* I and *Eco*R I (RE3 in Fig. 12).

Fig. 6a shows that numerous IPCR bands of less than 3 kb which represent the cleaved *ABL* gene detected in NP69 cells treated with H_2O_2 for 16 h (lanes 9, 10, 12 and 13) and 24 h (lanes 14–19). Three cleavage bands were identified in the untreated control (lanes 2 and 5). This might be due to spontaneous cell death of untreated cells as detected in our flow cytometric analyses. As shown in the bar chart in Fig. 6b, the cleavage frequencies of the *ABL* gene detected in NP69 cells treated with H_2O_2 for 16 and 24 h are 1.4-fold ($p = 0.004966$) and 1.8-fold ($p = 0.000009$) higher than that of the untreated control, respectively.

Similar findings were obtained from the H_2O_2 treatment of HK1 cells. Figure 7a is a representative gel picture showing the IPCR result obtained from this experiment. The untreated HK1 cells show a few cleavage bands (lanes

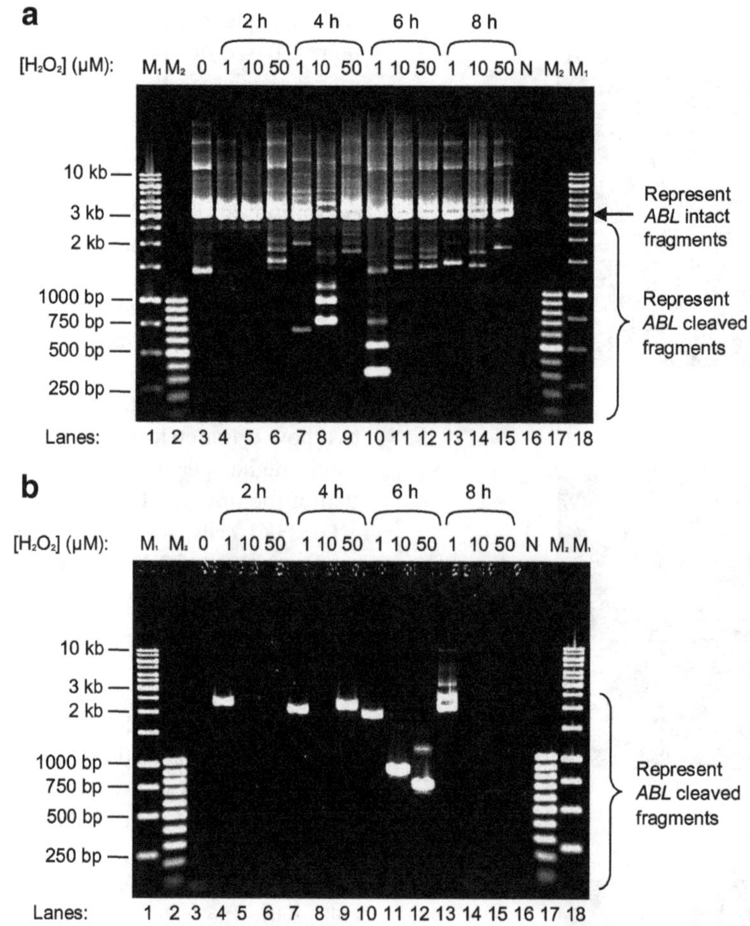

Fig. 5 Nested IPCR detection of DNA breakages within the *ABL* gene in H_2O_2-treated HK1. HK1 cells were seeded in 60-mm culture dishes and were grown to optimal density (60–70% confluency). The cells were then either untreated (lane 3) or treated with 1 µM (lanes 4, 7, 10 and 13), 10 µM (lanes 5, 8, 11 and 14) or 50 µM (lanes 6, 9, 12 and 15) of H_2O_2 for 2 h (lanes 4–6), 4 h (lanes 7–9), 6 h (lanes 10–12) and 8 h (lanes 13–15). Genomic DNA was isolated and manipulated for nested IPCR. In the modification for nested IPCR, the DNA samples were either subjected to digestion with *Age* I (**a**) or double digestion with *Age* I and *EcoR* I (**b**). The IPCR products were analysed on 1% agarose gel. Side arrow in panel **a** indicates the position of the 3 kb IPCR bands resulting from the amplification of the intact *ABL* gene. Side brackets in both panels **a** and **b** indicate the possible IPCR bands from the *ABL* cleaved fragments. Negative control for PCR was included (lane 16). This IPCR result is representative of 2 repeats with similar results. M_1: 1 kb DNA ladder. M_2: 100 bp DNA ladder

2–7) which most likely due to spontaneous cell death. In contrast, there were numerous cleavage bands identified in the H_2O_2-treated sample (lanes 8–13). The chart in Fig. 7b shows that the cleavage frequency of the *ABL* gene detected in H_2O_2-treated HK1 cells is 1.7-fold higher than that of untreated HK1 cells ($p = 0.000197$).

Sequencing results

In order to confirm that these fragments were derived from the cleaved *ABL* gene, some of the cleavage IPCR bands were extracted and sequenced. The sequencing results show that they were all derived from the cleaved *ABL* gene. Table 2 shows the breakpoints identified within the *ABL* gene in the H_2O_2-treated cells. A map illustrating the positions of chromosome breaks in HK1 and NP69 cells relative to the MAR/SAR sequences within the *ABL* gene is shown in Fig. 8.

Intriguingly, we detected two shift translocations in H_2O_2-treated NP69 cells. Translocation is rarely observed in NPC, compared with deletion and addition. The first shift translocation was identified in NP69 cells exposed to 100 µM of H_2O_2 for 16 h. As shown in Fig. 9a, the translocated segment was derived from the human lipoma *HMGIC* fusion partner-like 3 (*LHFPL3*) gene which is located on chromosome 7. The *LHFPL3* gene consisting of three exons is 578,576 bp in length. The description of exons and introns in the *LHFPL3* gene is shown in Additional file 4. The translocated segment (228 bp) of the *LHFPL3* gene is corresponding to coordinates 108006–108234 [Ensembl:ENSG00000187416]. The breakpoints (108,006 and 108,234) of the *LHFPL3*

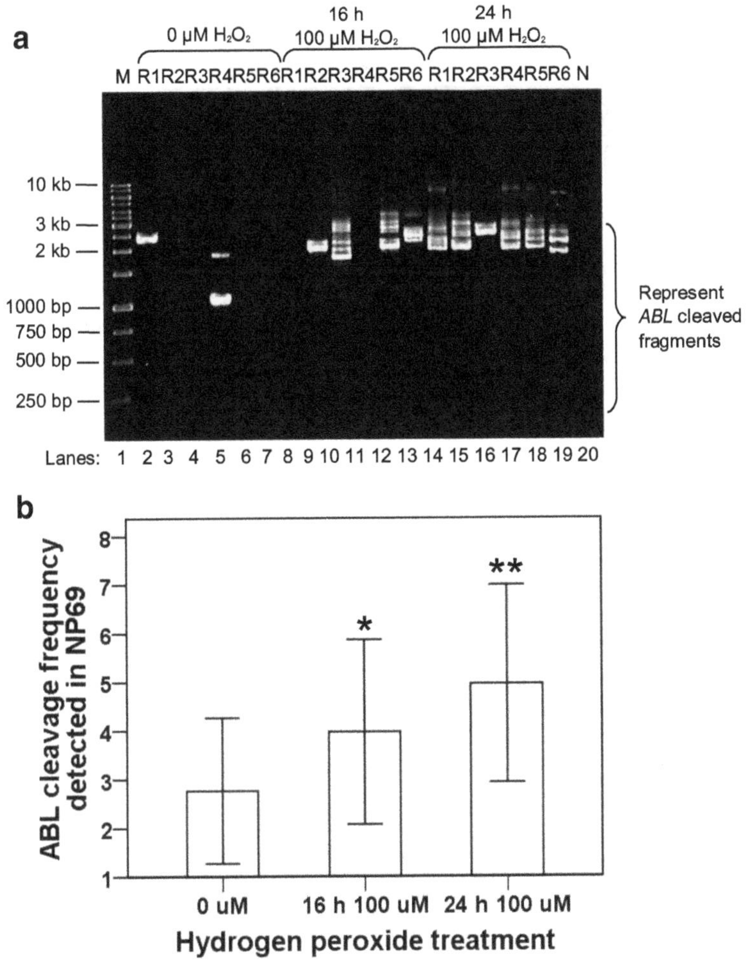

Fig. 6 IPCR analysis of H_2O_2-induced chromosome breaks within the *ABL* gene in NP69 cells. **a** IPCR result obtained from H_2O_2-treated NP69 cells. NP69 cells were either untreated (lanes 2–7) or treated with 100 μM of H_2O_2 for 16 h (lanes 8–13) and 24 h (lanes 14–19). Genomic DNA was isolated and manipulated for nested IPCR. Double digestion with *Age* I and *EcoR* I was employed to eliminate competition of the intact fragments in the amplification process. Each cell sample consisted of six replicates (R1–6) in the nested IPCR. The IPCR products were analysed on 1.0% agarose gel. Side bracket indicates the possible IPCR bands derived from the *ABL* cleaved chromosome. Negative control for PCR was included (Lane 20). M: 100 bp DNA ladder. **b** The average number of DNA cleavage detected within the *ABL* gene. The data was expressed as means and SD of three independent experiments. Each experiment consisted of 1–3 sets of IPCR. Each set of IPCR was performed in 4–7 IPCR replicates for each cell sample. *$p < 0.01$, **$p < 0.001$ (Student's *t* test)

gene were mapped within its first intron. Moreover, region of microhomology (TGCC) was found at the breakpoint junctions. The second shift translocation was identified in NP69 cells exposed to 10 μM of H_2O_2 for 24 h. The segment translocated to the *ABL* gene is derived from chromosome 5. The disabled homologue 2 (*DAB*) gene is 1,263,556 bp at the 5′ end of this segment while a gene encoding for hypothetical protein is 22,122 bp at the 3′ end (Fig. 9b).

Discussion

Oxidative stress increases genomic instability [86] which in turn contributes to carcinogenesis [87, 88]. Excessive production of ROS has been associated with mutation and alteration of gene expression [49]. Most of the aetiological factors of NPC were known to generate ROS. These aetiological factors include exposures to nitrosamines, cigarette smoke, formaldehyde and wood dust. EBV infection as well as chronic inflammation of sinonasal tract [39, 41–43].

In addition, formaldehyde and acrolein, a component of cigarette smoke, are reactive aldehydes which may impair the function of DNA, RNA and proteins through adduct formation. It has been suggested that the combined interactions of environmental aldehydes and endogenous aldehydes, which are produced during oxidative stress, may exacerbate the cellular oxidative damage [47].

Although the consistent chromosomal aberrations, such as deletion and addition, have long been identified in NPC, the underlying molecular mechanism requires further investigation. Apoptosis was suggested to participate

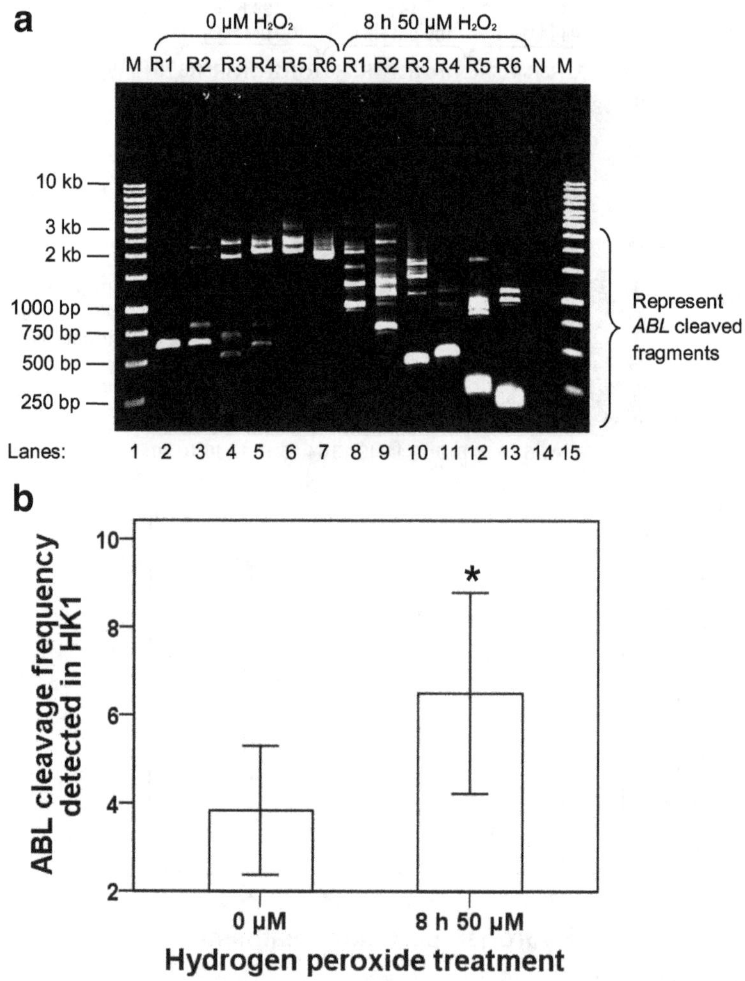

Fig. 7 IPCR analysis of H_2O_2-induced chromosome breaks within the *ABL* gene in HK1 cells. **a** IPCR result obtained from H_2O_2-treated HK1 cells. HK1 cells were either untreated (lanes 2–7) or treated with 50 μM of H_2O_2 for 8 h (lanes 8–13). Genomic DNA was isolated and manipulated for nested IPCR. In the manipulation for nested IPCR, samples were subjected to double digestion with *Age* I and *Eco*R I to eliminate the competition of the intact fragments for amplification process. Each cell sample consisted of six replicates in nested IPCR. The IPCR products were analysed on 1.0% agarose gel. Side bracket indicates the possible IPCR bands derived from the *ABL* cleaved chromosome. Negative control for PCR was included (lane 14). M: 100 bp DNA ladder. **b** The average number of DNA cleavage detected within the *ABL* gene. The data was expressed as means and SD of three independent experiments. Each experiment consisted of 1–3 sets of IPCR. Each set of IPCR was performed in 6 IPCR replicates for each cell sample. **Pp*< 0.001 (Student's *t* test)

in the chromosomal translocation process of leukaemia [70]. Given that there is a strong association between the aetiological factors of NPC and oxidative stress, we intended to investigate the role of oxidative stress-induced apoptosis in mediating the chromosome rearrangements of NPC.

We demonstrated that hydrogen peroxide (H_2O_2), a strong oxidising agent, was able to induce apoptosis in normal nasopharyngeal epithelial cells (NP69) and NPC cells (HK1). Both exposure of PS and disruption of MMP are key events of apoptosis [89, 90]. By using flow cytometric analyses of PS externalisation and MMP loss, we detected significantly higher percentages of apoptosis in H_2O_2-treated NP69 and HK1 cells as compared with the untreated controls. As compared with NPC cells (HK1), longer exposure time and higher dosage of H_2O_2 were needed to trigger apoptosis in normal nasopharyngeal epithelial cells (NP69). There may be several possibilities that lead to this variation. The intrachromosomal instability in cancer cells is usually higher than that in normal cells [91, 92]. Furthermore, cancer cells may have a defective DNA repair system which is unable to restore the genomic integrity [91]. These factors imply that cancer cells are more susceptible to DNA damage. When apoptosis is triggered by oxidative stress, DNA fragmentation occurs. Cells try to survive apoptosis through DNA repair. Therefore, as compared with normal cells, cancer cells which have a higher intrachromosomal instability or a

Table 2 The chromosome breaks identified within the *ABL* gene in cells treated with H_2O_2

Cell line treated with H_2O_2	Breakpoint
NP69	129,265
	129,287
	129,372
	129,408
	129,520
	129,534
	129,628
	129,823
	130,633
	130,634
	130,638
	130,687
	130,699
	130,719
	130,822
	130,864
	131,108
	131,232
HK1	129,152
	129,461
	129,739
	130,653
	130,696
	130,791
	130,854
	130,921
	131,042

The nucleotide positions of the chromosome breaks identified within the *ABL* gene were mapped according to the *ABL* sequence retrieved from Ensembl database [Ensembl:ENSG00000097007]

defective DNA repair system are usually more vulnerable to apoptosis.

In our previous report, we identified chromosomal breakages within the *AF9* gene in H_2O_2-treated NP69 and HK1 cells. In addition, inhibiting caspase-3 by caspase-3 inhibitor has abolished the *AF9* gene cleavages mediated by H_2O_2-induced apoptosis. Given that caspase-3 is the main activator of CAD-mediated DNA fragmentation in apoptosis, our findings suggested that CAD might be the major player which mediated the chromosomal breakages in H_2O_2-induced apoptosis [80]. It has been observed that CAD binds to the nuclear matrix during apoptosis [79]. Due to the fact that MAR/SAR sequences are the sites where DNA interacts with the nuclear matrix [93], it is likely that CAD cleaves the DNA at MAR/SAR sequences when it associates with nuclear matrix. Intriguingly, our previous report demonstrated that oxidative stress-induced apoptosis caused chromosomal breakages within the *AF9* BCR which is bordered by two MAR/SARs [80].

The present study targeted the *ABL* gene which is located on chromosome 9q34. This gene was targeted because 9q33-34 is one of the common deletion regions in NPC [23]. The *ABL* gene is the most common fusion partner gene with the breakpoint cluster region (*BCR*) gene which is located on chromosome 22q11 [94]. The reciprocal translocation t(9;22)(q34;q11) in CML was the first consistent chromosome rearrangement found in malignancy. The *ABL-BCR* fusion gene was named as the Philadelphia chromosome [95]. This reciprocal translocation was found in approximately 92% of CML patients. Thus, the *ABL-BCR* fusion gene is recognised as the cytogenetic hallmark of patients suffering from this disease [94, 96]. The presence of Philadelphia chromosome was also reported in 20 to 55% of adults and 2 to 10% of children with acute lymphoblastic leukaemia (ALL) [97] and rarely (1 to 2%) in acute non-lymphoblastic leukaemia (ANLL) [98]. There are three BCRs found within the *ABL* gene. The first BCR (BCRA) and the second BCR (BCRB) are located in intron 1b, whereas the third BCR (BCRC) spans through parts of introns 1b to 3. BCRC is the largest BCR of the *ABL* gene [77, 99].

One biochemically defined MAR/SAR has been previously identified within the BCRC of the *ABL* gene. This MAR/SAR was designated as SAR1. SAR1 was found within intron 1a [72]. In the present study, we predicted MAR/SAR sites within the *ABL* gene by using MRS which was proposed to be strongly associated with MAR/SAR [84]. It has been found that the two sequence elements of the MRS exist at a position near the dyad axis of the nucleosome. The wrapping of the DNA around the histone protein complex causes the two sequence elements of the MRS to be physically close together even if they are non-adjacent on the linear DNA. The close proximity between the two sequence elements of the MRS on the positioned nucleosome allows them to generate a protein-binding site in MAR/SAR [84].

The variation in the distance between the two sequence elements suggested a relation of the MRS to nucleosome organisation. In the *Drosophila* histone cluster, there was a MAR/SAR identified between the histone H1 and H3 genes. This MAR/SAR was found to contain a few nucleosomes and two MRSs. It was observed that the position of the two MRSs on their respective nucleosomes is similar. The first MRS, where the two sequence elements are overlapping, is found on the dyad axis of a nucleosome. The second MRS, where the two sequence elements are 145 bp apart, is located near the entry and exit sites of a nucleosome. Although the two sequence elements of the MRS are spatially distant, they are

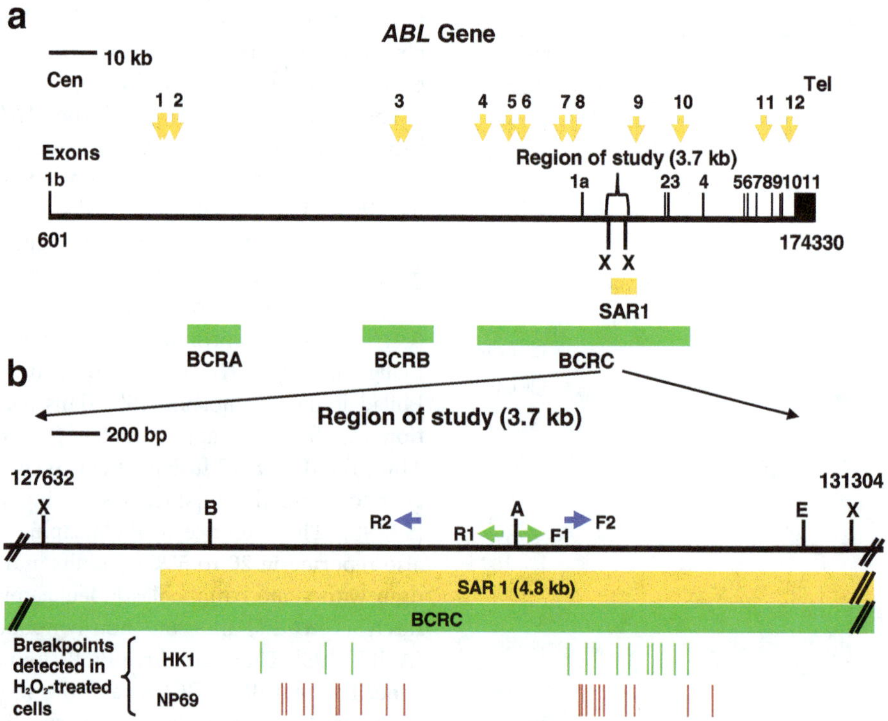

Fig. 8 A map representing the positions of H_2O_2-induced chromosome breaks within the *ABL* gene. **a** The *ABL* genomic map from nucleotide positions 601-174330 is illustrated above [Ensembl:ENSG00000097007]. The locations of exons 1–11 are shown. The green boxes indicate the three previously identified patient breakpoints cluster regions which are designated as BCRA, BCRB and BCRC. The yellow box shows the previously biochemically extracted MAR/SAR which is indicated as SAR1 [77]. The yellow arrows represent the potential MAR/SARs predicted by MRS in this study. **b** The region of study (3.7 kb). *Xba* I (X), *BsaA* I (B), *Age* I (A) and *EcoR* I (E) restriction sites are shown. The green and blue arrows represent the primers used in the first and second rounds of nested IPCR, respectively. The breakpoints identified in H_2O_2-treated HK1 and NP69 cells are indicated by the green and red vertical lines, respectively. All chromosome breaks were mapped within SAR1

brought close together when the DNA is turned around the histone core [85].

A nucleosome comprises a nucleosome core and a 'linker' DNA. The nucleosome core contains 145–147 bp of DNA wrapped around a core histone octamer. The histone octamer consists of two molecules each of the four core histones, namely, H2A, H2B, H3 and H4 [94]. It has been known that the length of 'linker' DNA ranges from 15 to 100 bp, depending on the cell types. The 'linker' DNA connects one nucleosome to the other (reviewed in [95]). The nucleosome repeat length (NRL) refers to the length of nucleosomal DNA (145–147 bp) plus the length of linker DNA (15–100 bp) [94, 95]. Using the micrococcal nuclease assay, the NRL has been reported to range from 160 to 240 bp [96, 97].

In the studies by van Drunen et al. (1999), the distance between the two sequence elements of the MRS has been suggested to be within 200 bp [85]. SAR prediction/SAR prediction presently performed in the *ABL* gene, there was only one MAR/SAR site (MAR/SAR 9 in Table 1) predicted in the experimentally isolated SAR1. The distance between the 8 bp sequence element and the 16 bp sequence element was found to be 248 bp. Given that the NRL may exceed 200 bp, for the mapping of MRS in the present study, the maximal distance between the 8 bp sequence element and the 16 bp sequence element was set at 250 bp. Besides, it is also possible that the two sequence elements, which are 248 bp apart, are located separately on two adjacent nucleosomes. In the positioned nucleosomes, interaction between two adjacent nucleosomes may happen. Thus, it seems possible that even if the two sequence elements are individually located on two adjacent nucleosomes, the wrapping of DNA around the histone protein complex may still cause them to be physically close together and enable them to generate a protein binding site.

We predicted 12 potential MAR/SAR sites within the *ABL* gene. One MAR/SAR site was predicted within the biochemically defined SAR1. Interestingly, 10 out of these 12 (> 80%) potential MAR/SAR sites are closely associated with the BCRs of the *ABL* gene (Fig. 1). MAR/SARs 1 and 2 were predicted next to BCRA. MAR/SAR 3 was found within BCRB. MAR/SARs 4 to 10 were identified within BCRC.

By using IPCR, we identified chromosome breaks in H_2O_2-treated NP69 and HK1 cells. The cleavage frequency

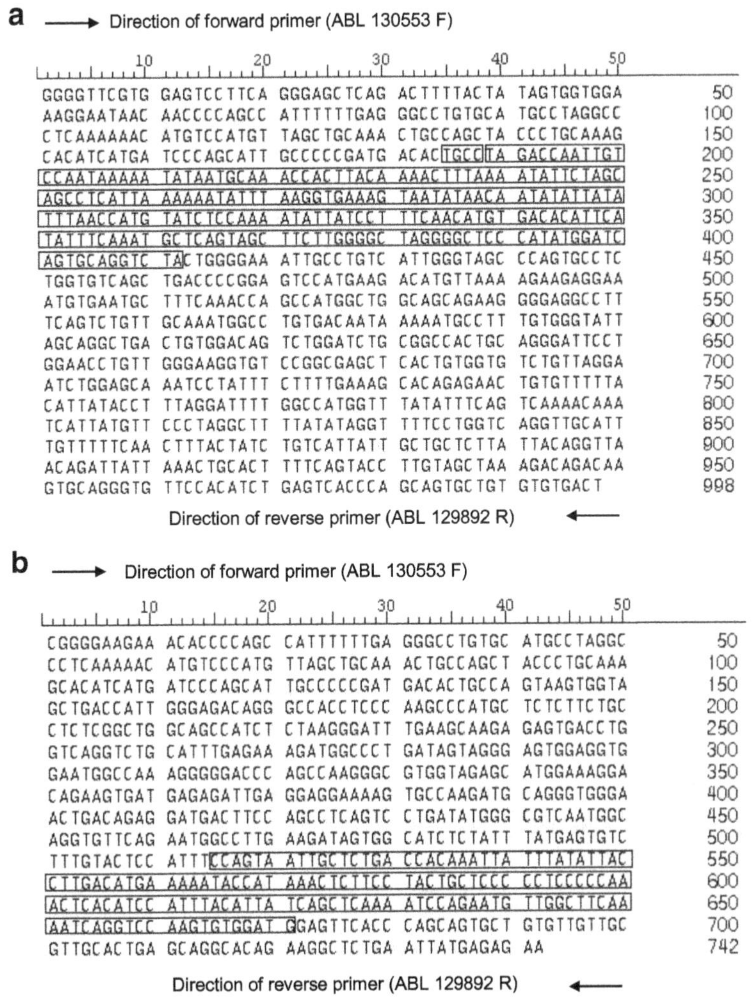

Fig. 9 Shift translocations detected in H_2O_2-treated NP69 cells. **a** Treatment of NP69 with 100 μM of H_2O_2 for 16 h resulted in shift translocation. The DNA sequences 1–184 and 413–998 (without the box) represent the sequence derived from the *ABL* gene. The DNA sequence 185–412 (within the box) represents the sequence derived from the *LHFPL3* gene which locates at chromosome 7. Region of microhomology (185–188, TGCC) was found at the breakpoint junctions. The translocated fragment (228 bp) of *LHFPL3* gene is corresponding to coordinates 108,006–108,234 [Ensembl:ENSG00000187416]. **b** Treatment of NP69 with 10 μM of H_2O_2 for 24 h resulted in shift translocation. The DNA sequences 1–524 and 672–742 (without the box) represent the sequence derived from the *ABL* gene. The DNA sequence 525–671 (within the box) represents the sequence of the fragment translocated to the *ABL* gene. This translocated fragment (147 bp) is derived from chromosome 5. The disabled homologue 2 (*DAB*) gene is 1,263,556 bp at the 5′ end of this translocated fragment while a gene encoding for a hypothetical protein is 22,122 bp at the 3′ end

of the *ABL* gene in H_2O_2-treated cells was significantly higher than that in the untreated control cells. This is true for both NP69 and HK1 cell lines. These results reaffirm our previous findings which demonstrated that oxidative stress-induced apoptosis resulted in chromosomal breakages in normal nasopharyngeal epithelial and NPC cells [80]. Taken together, our findings are consistent with other studies which discovered that H_2O_2 induced apoptotic DNA fragmentation. It has been demonstrated that H_2O_2 induced excision of chromosomal DNA loops mediated by topoisomerase II in U937 leukaemic cells [100]. The production of these HMW DNA fragments (50–100 kb loop-sized DNA fragments) is an initial event of apoptosis [65]. It has also shown that in caspase-3-expressing MCF-7 breast carcinoma cells, H_2O_2 activated DNA fragmentation with nucleosomal intervals [101]. The fragmentation of nuclear DNA into nucleosomal DNA ladders is another hallmark of apoptosis [102].

Our sequencing results have confirmed that the IPCR bands were derived from the cleaved *ABL* gene. All of the breakpoints were mapped within the biochemically defined SAR1 of the *ABL* gene. SAR1 is located in BCRC, the largest BCR of the *ABL* gene [77]. MAR/SAR is thought to be one of the common chromatin structures within BCRs. The BCRs of *AF9*, *MLL* and *AF4* genes have all been found to associate with MAR/SAR, suggesting a role for MAR/SAR in non-homologous recombination (NHR) [76–78]. MAR/SAR sequences were found to possess DNA unwinding properties [103, 104]. These properties allow them to facilitate the entry of protein factors that take part in chromosome condensation, apoptosis, transcription and replication [104, 105]. However, these unwinding properties also cause MAR/SAR sequences to be more prone to DNA breakage [103, 104].

In addition, two shift translocations were detected in H_2O_2-treated NP69 cells. One of the translocated segments was derived from the *LHFPL3* gene which locates at chromosome 7.

The other translocated segment was derived from chromosome 5. The disabled homologue 2 (*DAB*) gene is 1,263,556 bp at the 5′ end of this translocated segment while a gene encoding for hypothetical protein is 22,122 bp at the 3′ end.

The *LHFPL3* gene is one of the family members of LHFP-like genes. This gene family consists of six family members. All of the family members have been implicated in human diseases. Members of this family are transmembrane proteins which play important roles in extracellular matrix formation, differentiation and proliferation. Most of them have been associated with tumours [106]. The first member, *LHFP* on chromosome 13q12, was identified, for the first time, as a translocation partner of *HMGIC* gene on chromosome 12q15 in human lipoma with t(12;13)(q15;q12). Thus, it was annotated as lipoma *HMGIC* fusion partner (*LHFP*) gene [107]. The *LHFPL1* gene on chromosome Xq23 has been implicated in liver tumour [108]. The *LHFPL2* gene on chromosome 5q14.1 was found to be highly expressed in the novel subgroup of ALL [109, 110] and in patients who succumbed fatally to serous epithelial ovarian cancers (SEOC) [111]. The *LHFPL4* gene at 3p25.3 was identified as a novel methylation target specific for cervical cancer [112]. Mutation in the *LHFPL5* gene (on chromosome 6p21.31) which is also known as tetraspan membrane protein of hair cell stereocilia (*TMHS*) gene has been found to cause autosomal recessive nonsyndromic deafness [113].

The *LHFPL3* gene is located on chromosome 7q22.1. Deletions involving chromosome 7q22 are commonly observed in uterine leiomyoma (UL). Four distinct deletion intervals have been identified. One of the microdeletions contains the *LHFPL3* gene. The single deleted marker in the microdeletion was mapped within the first intron of the *LHFPL3* gene. These findings suggested that the *LHFPL*3 gene is a candidate tumour suppressor gene (TSG) for UL [106]. Deletion of 7q22 has also been associated with leukaemia. A commonly deleted segment of chromosome 7q22 has been identified in patients with a malignant myeloid disease. The *LHFPL3* gene is one of the candidate TSGs residing in this deletion interval [114]. More recently, the alteration of *LHFPL3* gene has been suggested to be a hallmark of primary glioblastoma [115].

Intriguingly, region of microhomology (four nucleotides) was found at the breakpoint junctions. This observation suggested that the shift translocation of the *LHFPL3* gene might be mediated by NHEJ DNA repair pathway. Based on the analysis of our sequencing data, we illustrated the potential model for the shift translocation of the *LHFPL3* gene (Fig. 10). As proposed by Betti and colleagues (2001), the interaction of the NHEJ DNA

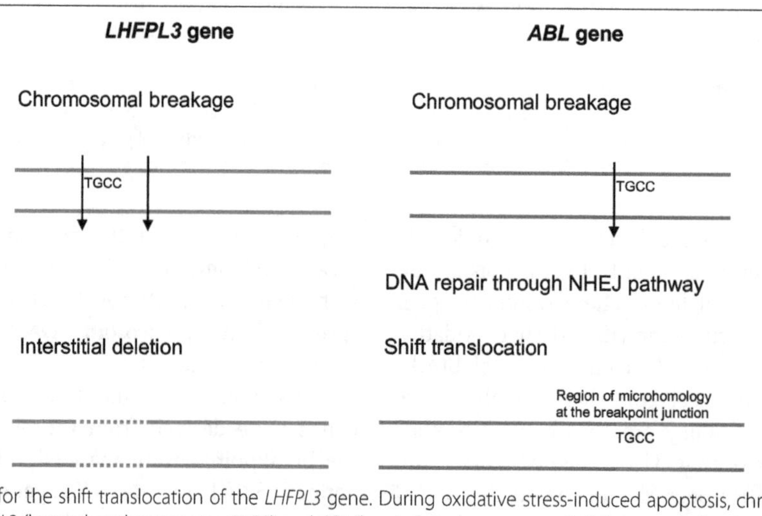

Fig. 10 A potential model for the shift translocation of the *LHFPL3* gene. During oxidative stress-induced apoptosis, chromosomal breakages occur within both the *LHFPL3* (located at chromosome 7q22) and *ABL* (located at chromosome 9q34) genes. Following that, interstitial deletion occurs within the *LHFPL3* gene. When the cells try to survive apoptosis, DNA repair takes place. By utilising the region of microhomology, TGCC, that was found at the breakpoint junctions of both the *LHFPL3* and *ABL* genes, the two DNA ends were joined. Subsequently, cells that survive apoptosis may carry the *ABL* gene with the shift translocation of a segment of the *LHFPL3* gene

repair pathway with apoptosis can act as a mechanism leading to translocation in leukaemia. They found that translocation junctions between the *MLL* gene and the partner DNA contain regions of microhomology consistent with the operation of NHEJ repair process [70]. In addition, it has been found that cells that survive apoptosis may contain rearranged chromosomes that contribute to leukaemogenesis [69]. Taken together, the findings of ours and of others support the notion that the interaction of the NHEJ DNA repair system with oxidative stress-induced apoptosis may be a possible mechanism leading to chromosome rearrangements in NPC.

We previously proposed a potential model for oxidative stress-induced chromosome rearrangements in NPC involving the *AF9* gene [80]. Based on the findings of the *ABL* gene in the present study and additional findings from the literature, we proposed a revised model (Fig. 11). The revision of this model enables us to further elucidate the potential role of oxidative stress-induced apoptosis in mediating chromosome rearrangements in NPC. We propose that oxidative stress plays an essential role in NPC aetiological factors. These include exposure to nitrosamine, wood dust, formaldehyde and cigarette smoke. EBV infection as well as chronic inflammation of sinonasal tract. Oxidative stress-induced apoptosis is initiated by apoptotic signalling. This includes PS externalisation and MMP loss. The apoptotic signalling may in turn result in the activation of the main effector caspase, caspase-3. Caspase-3 cleaves ICAD that contains two caspase-3 cleavage sites. Subsequently, CAD is being released from its chaperone, ICAD. Chromosomal DNA is cleaved by the activated CAD, presumably at MAR/SAR sites. Double strand breaks are primarily repaired through NHEJ pathway which is prone to cause erroneous DNA repair. Cells that evade apoptosis may harbour chromosome rearrangements such as translocation, deletion, addition and inversion. Repeated exposure to these aetiological factors that provoke oxidative stress may therefore contribute to tumourigenesis of NPC.

In the present study, we only focused on the SAR region of the *ABL* gene. It is difficult to draw a solid conclusion on the role of MAR/SAR in defining the positions of the chromosome breakages. Therefore, for the future work, comparison in the cleavage frequency between the *ABL* SAR region and non-SAR region may be carried out. This may allow a further elucidation of the potential role of MAR/SAR in mediating the chromosome breakages and rearrangements in oxidative stress-induced apoptosis.

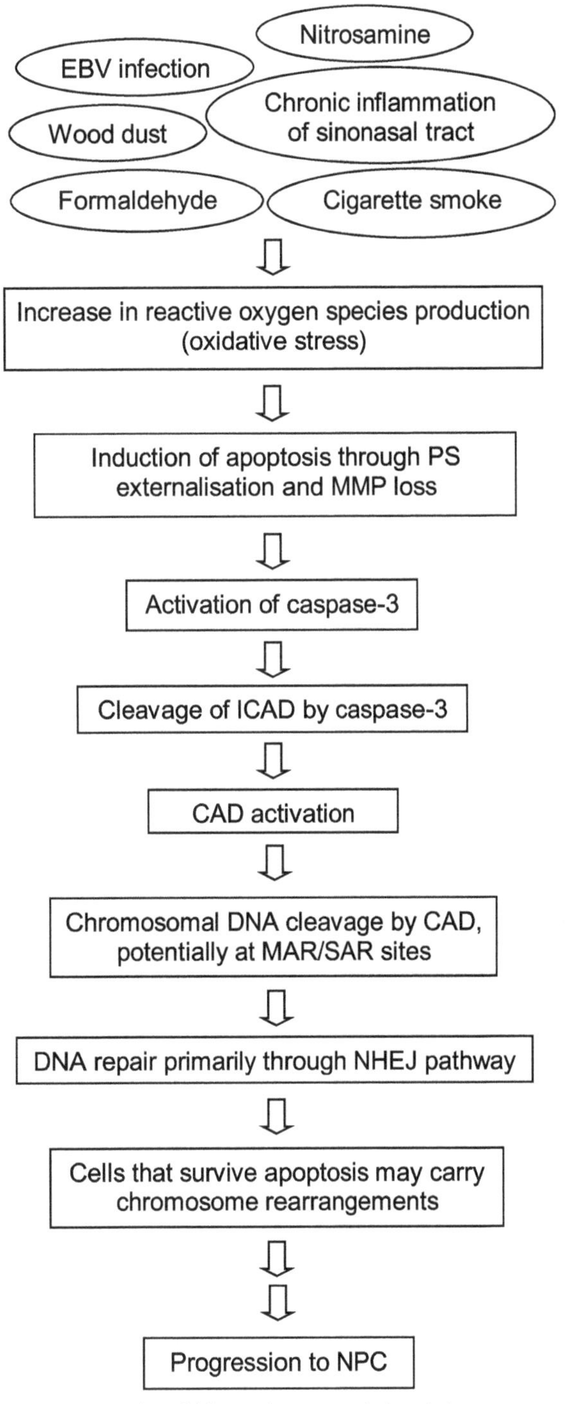

Fig. 11 A revised model for oxidative stress-induced chromosome rearrangement in NPC

Conclusions

Our findings demonstrated that oxidative stress-induced apoptosis may be a potential mechanism that leads to chromosome rearrangements in NPC. Our results also suggested that NHEJ system is potentially involved in DNA repair in cells undergoing oxidative stress-induced apoptosis. The interaction between NHEJ DNA repair system and oxidative stress-induced apoptosis may lead to chromosome rearrangements in surviving cells. A revised model for oxidative stress-induced apoptosis in mediating chromosome rearrangement in NPC is proposed.

Methods

Cell lines

NP69 normal nasopharyngeal epithelial cell line and HK1 NPC cell line were kindly provided by Prof. Tsao Sai Wah (The University of Hong Kong, Hong Kong, China) and Prof. Lo Kwok Wai (The Chinese University of Hong Kong, Hong Kong, China). NP69 is an immortalised nasopharyngeal epithelial cell line which was established by transfection with SV40 large T oncogene. It retains some characteristics of normal nasopharyngeal epithelial cells and is non-tumourigenic. This cell line may provide potential nasopharyngeal epithelial cell model for investigating mechanisms involved in NPC tumourigenesis [116]. HK1 was derived from a Chinese male patient with recurrent squamous NPC 17 ½ years after radiation therapy [117].

Chemicals

Hydrogen peroxide (H_2O_2) was bought from MP Biomedicals, USA. Keratinocyte-SFM medium, RPMI 1640 medium, penicillin, streptomycin, fetal bovine serum and L-glutamine were purchased from GIBCO, Invitrogen, USA. Annexin V-Fluorescein isothiocyanate (FITC) Apoptosis Detection Kit I (BD Pharmingen™) and Flow Cytometry Mitochondrial Membrane Potential Detection Kit were bought from BD™ MitoScreen, Becton–Dickinson Biosciences, USA. Camptothecin (CPT) was purchased from Santa Cruz Biotechnology, CA, USA. Ammonium acetate was bought from Merck, Germany. Chloroform was bought from R&M Chemicals, UK. Phenol and Sodium dodecyl sulfate (SDS) were procured from Amresco, USA. Isoamyl alchohol was purchased from Fluka, Switzerland. Phusion High-Fidelity DNA Polymerase was procured from Finnzymes, Finland. PCR primers were from First Base Laboratories. QIAquick Gel Extraction Kit and QIAquick Nucleotide Removal Kit were bought from QIAGEN, Germany. DNA Polymerase I Large (Klenow) Fragment, restriction enzymes and T4 DNA Ligase were obtained from New England Biolabs (NEB), USA. dNTP mix was purchased from Promega, USA.

Cell cultures

NP69 cells were grown in Keratinocyte-SFM medium supplemented with 100 μg/ml streptomycin, 100 U/ml penicillin, 40–50 μg/ml Bovine Pituitary Extract (BPE), 4–5 ng/ml recombinant Epidermal Growth Factor (rEGF) and 2% (*v*/*v*) heat-inactivated fetal bovine serum. HK1 cells were cultured in RPMI 1640 medium supplemented with 100 μg/ml streptomycin, 100 U/ml penicillin, 2 mM L-glutamine and 10% (*v*/*v*) heat-inactivated fetal bovine serum. Cells were cultured at 37 °C with 5% CO_2.

In silico prediction of MAR/SAR

The whole sequence of the *ABL* gene was retrieved from Ensembl (http://www.ensembl.org/index.html) database [Ensembl:ENSG00000097007]. The location of the experimentally defined MAR/SAR was determined from the previous report [77]. By using DNASTAR software (Lasergene, USA), we predicted the possible MAR/SAR sites within the *ABL* gene. The prediction of MAR/SAR site was performed by searching MRS which comprises two nucleotide motifs. The first nucleotide motif is an 8 bp degenerate sequence, AATAAYAA, where Y = C or T. The second nucleotide motif is a 16 bp degenerate sequence, AWWRTAANNWWGNNNC, where N = A, C, G or T; R = A or G; W = A or T. One mismatch is allowed in the 16 bp degenerate sequence. The 8 bp degenerate sequence has to be exactly matched. The two sequence elements of the MRS should be found within 200 bp apart. The two sequence elements can be present on either Watson or Crick strand and in either order. The two sequence elements may also be overlapping. When there are more than one motif of either 8 or 16 bp found within a distance of 200 bp, they are considered as a single MRS. In addition, when there is more than one MRS identified within close proximity, they are regarded as a single potential MAR/SAR site [85].

Apoptosis detection

Phosphatidylserine (PS) externalisation

NP69 cells (1.5×10^5) were plated in 150-mm culture dishes containing 15 ml of complete media. When NP69 cells reached confluency of 30–40% on the third day, NP69 cells were either left untreated or treated with 100 μM of H_2O_2 for 16 and 24 h. HK1 cells (5.5×10^5) were seeded in 150-mm culture dishes containing 15 ml of complete media. When HK1 cells reached confluency of 60–70% on the fourth day, HK1 cells were incubated with 50 μM of H_2O_2 for 4 and 8 h. NP69 and HK1 cells treated with camptothecin (CPT) were included as positive controls. After incubation, the cells were collected by using StemPro ACCUTASE Cell Dissociation Reagent. The percentage of apoptotic cells was determined by using Annexin V-FITC Apoptosis Detection Kit I as previously described [80].

Mitochondrial membrane potential (MMP) loss

NP69 and HK1 cells were treated and collected as described above. The percentage of MMP loss in the harvested cells was determined by using Flow Cytometry Mitochondrial Membrane Potential Detection Kit as previously described [80].

IPCR detection of chromosome breaks within the *ABL* gene

Induction of apoptosis in normal nasopharyngeal epithelial and NPC cells with H_2O_2

NP69 (2×10^4) and HK1 (8×10^4) cells were seeded in 60-mm culture plates containing 4 ml of complete media. When NP69 cells reached confluency of 30–40% on the third day, NP69 cells were either left untreated or treated with 10, 50 or 100 μM for 16 and 24 h. When HK1 cells reached confluency of 60–70% on the fourth day, HK1 cells were either left untreated or treated with 1, 10 or 50 μM of H_2O_2 for 2, 4, 6 and 8 h.

Genomic DNA extraction

At the end of the indicated exposure times, the used medium was discarded. The cells were washed once with cold 1× phosphate-buffered saline (PBS). Genomic DNA extraction was performed as previously described [80].

Manipulation of the extracted gDNA for nested IPCR

The extracted gDNA was manipulated as described previously [80] with minor modifications. Figure 12 shows the manipulation steps. Digestion of the gDNA was performed at 37 °C for 16 h with 100 U of *Xba* I (RE1 in Fig. 12). The staggered four base pairs (CTAG) 5′ overhang was produced by *Xba* I digestion. The blunt ends were generated by the apoptotic nuclease such as CAD [118]. After *Xba* I digestion, both ends of the intact targeted DNA fragment were *Xba* I sites with staggered overhangs. As for the cleaved targeted DNA fragment, one end was the blunt end produced by the apoptotic nuclease, and the other end was the staggered overhang generated by *Xba* I. To produce blunt-ended fragments, Klenow fill-in was performed with two μg of DNA template, two units of DNA Polymerase I Large (Klenow) Fragment and 33 μM of dNTP mix at 25 °C for 15 min. Cyclisation was then performed with 2000 U of T4 DNA ligase at 16 °C for 16 h. Ethanol precipitation was carried out with 3 M sodium acetate (NaAc) (one volume), pH 5.2 and ice cold absolute ethanol (2.5 volumes). Seventy percent ethanol was used to wash the DNA pellet. The DNA pellet was then air-dried and dissolved in TE, pH 8.0. The DNA sample was divided into three. The DNA samples of tubes 1, 2 and 3 were subjected to digestion with 10 U of *Age* I (RE2 in Fig. 12), double digestion with 10 U of each *Age* I and *Bsa*A I (RE3 in Fig. 12), and double digestion with 10 U of each *Age* I and *Eco*R I (RE3 in Fig. 12), respectively. These RE digestions were performed at 37 °C for 16 h. Digestion with *Age* I was used to linearise the cyclised DNA. Double digestion with *Age* I and *Bsa*A I or *Age* I and *Eco*R I was used to eliminate competition from the intact fragments during IPCR. The double digestion with

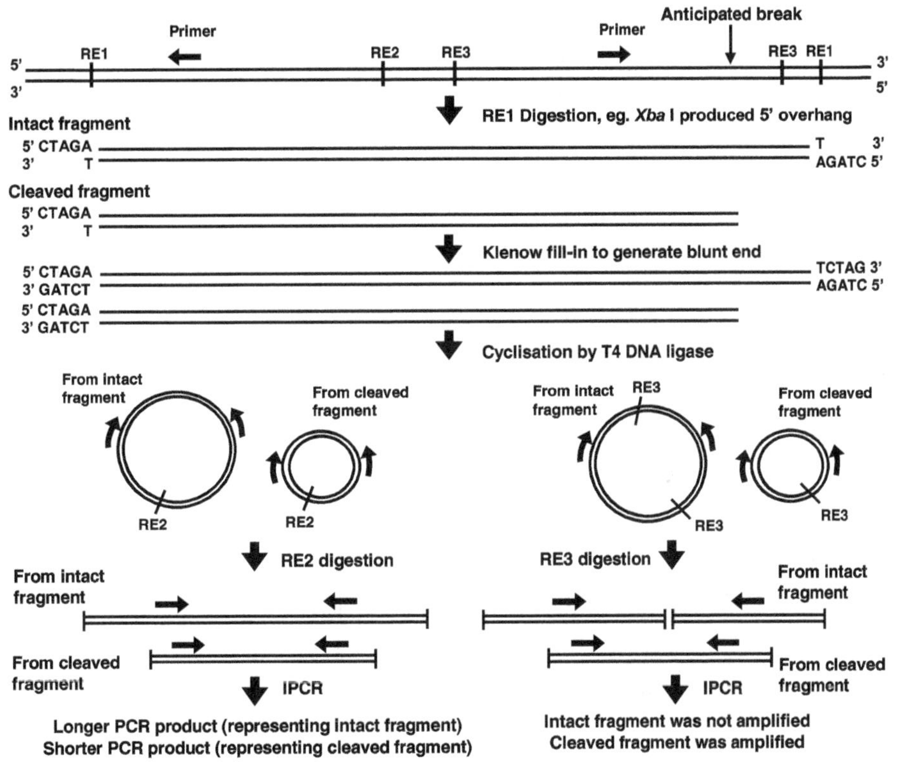

Fig. 12 A flowchart showing the manipulation steps in the preparation of genomic DNA for IPCR. The genomic DNA was subjected to RE digestions, Klenow fill-in and ligation prior to IPCR as reported before [80]

Age I and *BsaA* I enabled the detection of DNA cleavages occurred within the amplified region towards the 3′ end. The double digestion with *Age* I and *EcoR* I enabled the detection of DNA cleavages occurred within the amplified region towards the 5′ end. According to the manufacturer's protocol, QIAquick Nucleotide Removal Kit (QIAGEN) was used to purify the digested DNA.

Nested IPCR

The optical density (O.D.) of the purified DNA sample was measured by using an ultraviolet-visible micro-volume spectrophotometer (ND-1000, NanoDrop, USA). Nested IPCR was performed with 1× of HF buffer (containing 1.5 mM of $MgCl_2$), 0.5 μM of each reverse primer and forward primer, 200 μM of dNTP mix, 0.4 U of Phusion High-Fidelity DNA Polymerase and 200 ng of DNA template. To serve as a negative control, sterile ultrapure water was used to replace the DNA template. Cycle condition used in the first round was: 30 s of 98 °C for 1 cycle (initial denaturation), followed by 30 cycles of 98 °C for 10 s (denaturation), 64 °C for 30 s (annealing), 72 °C for 55 s (extension), followed by 1 cycle of 72 °C for 10 min (final extension). Similar cycle condition was used in the second round of IPCR, except that the extension time was 50 s. Two microlitres of 5-fold diluted IPCR product of the first round was used as DNA template. The primers used in the first round of IPCR were 5′-GGTACCTGGTGTCT GTCTCTATC-3′ (reverse) and 5′-AGAAGGTTTATGGG AGATGG-3′ (forward), whereas the primers used in the second round were 5′-TCTCTCATATCTCAGAGCC TTC-3′ (reverse) and 5′-CTTCAGGAGCTCAGACTTT TAC-3′ (forward). The IPCR assays were done by using a Veriti 96 Well Thermal Cycler (Applied Biosystems, USA).

Agarose gel electrophoresis and DNA sequencing

The PCR products were analysed on 1% agarose gel. The agarose gel electrophoresis was performed at 90 V for 1 h and 30 min. The agarose gel was briefly stained with ethidium bromide (0.5 μg/ml) and destained with distilled water. This was followed by visualisation of the gel on an ultraviolet (UV) transilluminator (Vilber Lourmat). The gel image was captured and analysed using a gel documentation (gel doc) and image analysis system (Syngene). The IPCR bands representing cleaved DNA fragments of the *ABL* gene were purified by using QIAquick Gel Extraction Kit (QIAGEN) according to the manufacturer's protocol and sequenced. By blasting the human genome database (Genomic BLAST, https://blast.ncbi.nlm.nih.gov/Blast.cgi), the sequencing data obtained was annotated. To identify the breakpoints of the cleaved fragments, the sequencing data was analysed and aligned with the published *ABL* gene sequence [Ensembl:ENSG00000097007] by using Seqman DNASTAR software (Lasergene, USA). The positions of DNA breaks identified were compared with the location of the MAR/SAR sequence isolated experimentally in the previous study [77] and the MRS identified in the present study. A genomic map was constructed to depict the positions of the detected DNA breaks relative to the location of the MAR/SAR.

Quantification of gene cleavage frequency

In each set of IPCR, four to seven IPCR replicates were prepared per cell sample. Each experiment consisted of one to three sets of IPCR. The number of IPCR bands representing the *ABL* cleaved fragments was counted. Gene cleavage frequency expresses the average number of *ABL* cleaved fragments detected in three independent experiments.

Statistical analysis

Experiments were repeated three to five times. The significance of differences in the gene cleavage frequency detected by nested IPCR was evaluated by Student's t test. Data for IPCR are expressed as mean and standard deviation (SD). Differences were considered statistically significant at p value < 0.05. All statistical tests are two sided.

Abbreviations

ALL: Acute lymphoblastic leukaemia; BCR: Breakpoint cluster region; CAD: Caspase-activated deoxyribonuclease; CML: Chronic myelogenous leukaemia; *DAB*: Disabled homologue 2; DSB: DNA double-strand breaks; EBV: Epstein-Barr virus; H_2O_2: Hydrogen peroxide; HMW: High-molecular-weight; HR: Homologous recombination; ICAD: Inhibitor of caspase-activated deoxyribonuclease; IPCR: Inverse polymerase chain reaction; *LHFPL3*: Human lipoma HMGIC fusion partner-like 3; MAR/SAR: Matrix association region/ scaffold attachment region; *MLL*: Mixed lineage leukaemia; MMP: Mitochondrial membrane potential; NHEJ: Non-homologous end joining; NPC: Nasopharyngeal carcinoma; PS: Phosphatidylserine; ROS: Reactive oxygen species; UL: Uterine leiomyoma

Acknowledgements

We would like to thank the Director-General of Health Malaysia for his permission to publish this article and the Director of Institute for Medical Research Malaysia for her support. We gratefully acknowledge Prof. Tsao Sai Wah (The University of Hong Kong, Hong Kong, China) and Prof. Lo Kwok Wai (The Chinese University of Hong Kong, Hong Kong, China) for providing the normal nasopharyngeal epithelial cell line (NP69) and NPC cell line (HK1). We are especially grateful to Dr. Cornelis M. van Drunen (Academic Medical Center, Amsterdam, The Netherlands) for his precious opinions. We express our gratitude to Universiti Malaysia Sarawak for all infrastructure support. This is also another project by the members of Malaysian Nasopharyngeal Carcinoma Group.

Funding
This project was funded by Ministry of Health Malaysia (grant number: 06-065), and the consent for publication has been granted.

Authors' contributions
SNT carried out all the experiments, performed in silico prediction of MAR/SAR, conducted statistical analyses, constructed genomic maps and wrote the manuscript. SPS conceptualised the main idea of the project, designed the study, interpreted the data, edited the manuscript and supervised the whole project. ASBK contributed to intellectual formulation of the project. All authors read and approved the final manuscript.

Consent for publication
Not applicable.

Competing interests
The authors declare that they have no competing interests.

Author details
[1]Department of Paraclinical Sciences, Faculty of Medicine and Health Sciences, Universiti Malaysia Sarawak, Sarawak, Malaysia. [2]Molecular Pathology Unit, Cancer Research Centre, Institute for Medical Research, Kuala Lumpur, Malaysia.

References
1. Shanmugaratnam K. Histological typing of nasopharyngeal carcinoma. IARC Sci Publ. 1978;20:3–12.
2. Hirayama T. Descriptive and analytical epidemiology of nasopharyngeal cancer. IARC Sci Publ. 1978;20:167–89.
3. Chang ET, Adami HO. The enigmatic epidemiology of nasopharyngeal carcinoma. Cancer Epidemiol Biomark Prev. 2006;15:1765–77.
4. Huang DP, Lo KW. Aetiological factors and pathogenesis. In: van Hasselt CA, Gibb AG, editors. Nasopharyngeal carcinoma. Hong Kong: The Chinese University Press; 1999. p. 31–60.
5. Devi BC, Pisani P, Tang TS, Parkin DM. High incidence of nasopharyngeal carcinoma in native people of Sarawak, Borneo Island. *Cancer Epidemiol Biomarkers Prev*. 2004;13:482–6.
6. Henle W, Henle G. Evidence for an etiologic relation of the Epstein-Barr virus to human malignancies. Laryngoscope. 1977;87:467–73.
7. Dickens P, Srivastava G, Loke SL, Chan CW, Liu YT. Epstein-Barr virus DNA in nasopharyngeal carcinomas from Chinese patients in Hong Kong. J Clin Pathol. 1992;45:396–7.
8. Feng FT, Cui Q, Liu WS, Guo YM, Feng QS, Chen LZ, et al. A single nucleotide polymorphism in the Epstein-Barr virus genome is strongly associated with a high risk of nasopharyngeal carcinoma. Chin J Cancer. 2015;34:563–72.
9. Fong LY, Ho JH, Huang DP. Preserved foods as possible cancer hazards: WA rats fed salted fish have mutagenic urine. Int J Cancer. 1979;23:542–6.
10. Armstrong RW, Armstrong MJ, Yu MC, Henderson BE. Salted fish and inhalants as risk factors for nasopharyngeal carcinoma in Malaysian Chinese. Cancer Res. 1983;43:2967–70.
11. Yong SK, Ha TC, Yeo MCR, Gaborieau V, McKay JD, Wee J. Associations of lifestyle and diet with the risk of nasopharyngeal carcinoma in Singapore: a case–control study. Chin J Cancer. 2017;36:3.
12. Armstrong RW, Imrey PB, Lye MS, Armstrong MJ, Yu MC, Sani S. Nasopharyngeal carcinoma in Malaysian Chinese: occupational exposures to particles, formaldehyde and heat. Int J Epidemiol. 2000;29:991–8.
13. Lo KW, Huang DP. Genetic and epigenetic changes in nasopharyngeal carcinoma. Semin Cancer Biol. 2002;12:451–62.
14. Guo XC, Scott K, Liu Y, Dean M, David V, Nelson GW, et al. Genetic factors leading to chronic Epstein-Barr virus infection and nasopharyngeal carcinoma in South East China: study design, methods and feasibility. Hum Genomics. 2006;2:365–75.
15. Chen YJ, Ko JY, Chen PJ, Shu CH, Hsu MT, Tsai SF, et al. Chromosomal aberrations in nasopharyngeal carcinoma analyzed by comparative genomic hybridization. Genes Chromosomes Cancer. 1999;25:169–75.
16. Hui AB, Lo KW, Leung SF, Teo P, Fung MK, To KF, et al. Detection of recurrent chromosomal gains and losses in primary nasopharyngeal carcinoma by comparative genomic hybridisation. Int J Cancer. 1999;82: 498–503.
17. Fang Y, Guan X, Guo Y, Sham J, Deng M, Liang Q, et al. Analysis of genetic alterations in primary nasopharyngeal carcinoma by comparative genomic hybridization. Genes Chromosomes Cancer. 2001;30:254–60.
18. Yan J, Fang Y, Liang Q, Huang Y, Zeng Y. Novel chromosomal alterations detected in primary nasopharyngeal carcinoma by comparative genomic hybridization. Chin Med J. 2001;114:418–21.
19. Li X, Wang E, Zhao YD, Ren JQ, Jin P, Yao KT, et al. Chromosomal imbalances in nasopharyngeal carcinoma: a meta-analysis of comparative genomic hybridization results. J Transl Med. 2006;4:4.
20. Hui AB, Lo KW, Leung SF, Choi PH, Fong Y, Lee JC, et al. Loss of heterozygosity on the long arm of chromosome 11 in nasopharyngeal carcinoma. Cancer Res. 1996;56:3225–9.
21. Mutirangura A, Pornthanakasem W, Sriuranpong V, Supiyaphun P, Voravud N. Loss of heterozygosity on chromosome 14 in nasopharyngeal carcinoma. Int J Cancer. 1998;78:153–6.
22. Lo KW, Teo PM, Hui AB, To KF, Tsang YS, Chan SY, et al. High resolution allelotype of microdissected primary nasopharyngeal carcinoma. Cancer Res. 2000;60:3348–53.
23. Shao JY, Wang HY, Huang XM, Feng QS, Huang P, Feng BJ, et al. Genome-wide allelotype analysis of sporadic primary nasopharyngeal carcinoma from southern China. Int J Oncol. 2000;17:1267–75.
24. Huang DP, Lo KW, van Hasselt CA, Woo JK, Choi PH, Leung SF, et al. A region of homozygous deletion on chromosome 9p21-22 in primary nasopharyngeal carcinoma. Cancer Res. 1994;54:4003–6.
25. Lo KW, Huang DP, Lau KM. p16 gene alterations in nasopharyngeal carcinoma. Cancer Res. 1995;55:2039–43.
26. Hu LF, Eiriksdottir G, Lebedeva T, Kholodniouk I, Alimov A, Chen F, et al. Loss of heterozygosity on chromosome arm 3p in nasopharyngeal carcinoma. Genes Chromosomes Cancer. 1996;17:118–26.
27. Hui AB, Or YY, Takano H, Tsang RK, To KF, Guan XY, et al. Array-based comparative genomic hybridization analysis identified cyclin D1 as a target oncogene at 11q13.3 in nasopharyngeal carcinoma. Cancer Res. 2005;65:8125–33.
28. Lo KW, Cheung ST, Leung SF, van Hasselt A, Tsang YS, Mak KF, et al. Hypermethylation of the p16 gene in nasopharyngeal carcinoma. Cancer Res. 1996;56:2721–5.
29. Kwong J, Lo KW, To KF, Teo PM, Johnson PJ, Huang DP. Promoter hypermethylation of multiple genes in nasopharyngeal carcinoma. Clin Cancer Res. 2002;8:131–7.
30. Tan SH, Ida H, Goh BC, Hsieh W, Loh M, Ito Y. Analyses of promoter hypermethylation for RUNX3 and other tumor suppressor genes in nasopharyngeal carcinoma. Anticancer Res. 2006;26:4287–92.
31. Yanatatsaneejit P, Chalermchai T, Kerekhanjanarong V, Shotelersuk K, Supiyaphun P, Mutirangura A, et al. Promoter hypermethylation of CCNA1, RARRES1, and HRASLS3 in nasopharyngeal carcinoma. Oral Oncol. 2008;44:400–6.
32. Xiao L, Zhou H, Yan J, Liu Y, Song Y. Changes of telomere length in human nasopharyngeal carcinoma. Hua Xi Yi Ke Da Xue Xue Bao. 1999;30:262–4.
33. Wen Z, Xiao JY, Guo MH. Telomere shortening in the pathogenesis of nasopharyngeal carcinoma. Di Yi Jun Yi Da Xue Xue Bao. 2002;22:329–30.
34. Poirier S, Ohshima H, de The G, Hubert A, Bourgade MC, Bartsch H. Volatile nitrosamine levels in common foods from Tunisia, South China and Greenland, high-risk areas for nasopharyngeal carcinoma (NPC). Int J Cancer. 1987;39:293–6.
35. Cheng YJ, Hildesheim A, Hsu MM, Chen IH, Brinton LA, Levine PH, et al. Cigarette smoking, alcohol consumption and risk of nasopharyngeal carcinoma in Taiwan. Cancer Causes Control. 1999;10:201–7.
36. Vaughan TL, Stewart PA, Teschke K, Lynch CF, Swanson GM, Lyon JL, et al. Occupational exposure to formaldehyde and wood dust and nasopharyngeal carcinoma. Occup Environ Med. 2000;57:376–84.

37. Hung SH, Chen PY, Lin HC, Ting J, Chung SD. Association of rhinosinusitis with nasopharyngeal carcinoma: a population-based study. Laryngoscope. 2014;124:1515–20.
38. Tsou YA, Lin CC, Tai CJ, Tsai MH, Tsai TC, Chen CM. Chronic rhinosinusitis and the risk of nasopharyngeal cancer in a Taiwanese health study. Am J Rhinol Allergy. 2014;28:168–72.
39. Carnevali S, Petruzzelli S, Longoni B, Vanacore R, Barale R, Cipollini M, et al. Cigarette smoke extract induces oxidative stress and apoptosis in human lung fibroblasts. Am J Physiol Lung Cell Mol Physiol. 2003;284:L955–63.
40. Uslu C, Taysi S, Bakan N. Lipid peroxidation and antioxidant enzyme activities in experimental maxillary sinusitis. Ann Clin Lab Sci. 2003;33:18–22.
41. van d V, Postma DS, Timens W, ten Hacken NH. Acute effects of cigarette smoke on inflammation and oxidative stress: a review. Thorax. 2004;59:713–21.
42. Kum C, Kiral F, Sekkin S, Seyrek K, Boyacioglu M. Effects of xylene and formaldehyde inhalations on oxidative stress in adult and developing rats livers. Exp Anim. 2007;56:35–42.
43. Pylkkanen L, Stockmann-Juvala H, Alenius H, Husgafvel-Pursiainen K, Savolainen K. Wood dusts induce the production of reactive oxygen species and caspase-3 activity in human bronchial epithelial cells. Toxicology. 2009;262:265–70.
44. Jones DP. Redefining oxidative stress. Antioxid Redox Signal. 2006;8:1865–79.
45. Halliwell B, Cross CE. Oxygen-derived species: their relation to human disease and environmental stress. Environ Health Perspect. 1994; 102(Suppl 10):5–12.
46. Wiseman H, Halliwell B. Damage to DNA by reactive oxygen and nitrogen species: role in inflammatory disease and progression to cancer. Biochem J. 1996;313(Pt 1):17–29.
47. LoPachin RM, Gavin T. Molecular mechanisms of aldehyde toxicity: a chemical perspective. Chem Res Toxicol. 2014;27:1081–91.
48. Huang YJ, Zhang BB, Ma N, Murata M, Tang AZ, Huang GW. Nitrative and oxidative DNA damage as potential survival biomarkers for nasopharyngeal carcinoma. Med Oncol. 2011;28:377–84.
49. Klaunig JE, Kamendulis LM. The role of oxidative stress in carcinogenesis. Annu Rev Pharmacol Toxicol. 2003;44:239–67.
50. Koul M, Kumar A, Deshidi R, Sharma V, Singh RD, Singh J, et al. Cladosporol A triggers apoptosis sensitivity by ROS-mediated autophagic flux in human breast cancer cells. BMC Cell Biol. 2017;18:26.
51. Sun YY, Wang C, Wang LG, Dai ZB, Yang KB. Arsenic trioxide induces apoptosis and the formation of reactive oxygen species in rat glioma cells. Cellular & Molecular Biology Letters. 2018;23
52. Park BG, Yoo CI, Kim HT, Kwon CH, Kim YK. Role of mitogen-activated protein kinases in hydrogen peroxide-induced cell death in osteoblastic cells. Toxicology. 2005;215:115–25.
53. Senkoylu A, Yilmaz A, Ergun MA, Ilhan MN, Simsek A, Altun N, et al. Effect of strontium ranelate on hydrogen peroxide-induced apoptosis of CRL-11372 cells. Biochem Genet. 2008;46:197–205.
54. Yoon SO, Kim MM, Park SJ, Kim D, Chung J, Chung AS. Selenite suppresses hydrogen peroxide-induced cell apoptosis through inhibition of ASK1/JNK and activation of PI3-K/Akt pathways. FASEB J. 2002;16:111–3.
55. Ogawa Y, Takahashi T, Kobayashi T, Kariya S, Nishioka A, Mizobuchi H, et al. Mechanism of hydrogen peroxide-induced apoptosis of the human osteosarcoma cell line HS-Os-1. Int J Mol Med. 2003;12:459–63.
56. Li J, Huang CY, Zheng RL, Cui KR, Li JF. Hydrogen peroxide induces apoptosis in human hepatoma cells and alters cell redox status. Cell Biol Int. 2000;24:9–23.
57. Kwon D, Choi C, Lee J, Kim KO, Kim JD, Kim SJ, et al. Hydrogen peroxide triggers the expression of Fas/FasL in astrocytoma cell lines and augments apoptosis. J Neuroimmunol. 2001;113:1–9.
58. Stridh H, Kimland M, Jones DP, Orrenius S, Hampton MB. Cytochrome c release and caspase activation in hydrogen peroxide- and tributyltin-induced apoptosis. FEBS Lett. 1998;429:351–5.
59. Zunino A, Degan P, Vigo T, Abbondandolo A. Hydrogen peroxide: effects on DNA, chromosomes, cell cycle and apoptosis induction in Fanconi's anemia cell lines. Mutagenesis. 2001;16:283–8.
60. Wyllie AH, Kerr JF, Currie AR. Cell death: the significance of apoptosis. Int Rev Cytol. 1980;68:251–306.
61. Coleman ML, Sahai EA, Yeo M, Bosch M, Dewar A, Olson MF. Membrane blebbing during apoptosis results from caspase-mediated activation of ROCK I. Nat Cell Biol. 2001;3:339–45.
62. Shimizu T, Maeno E, Okada Y. Prerequisite role of persistent cell shrinkage in apoptosis of human epithelial cells. Sheng Li Xue Bao. 2007;59:512–6.
63. Iguchi K, Hirano K, Hamatake M, Ishida R. Phosphatidylserine induces apoptosis in adherent cells. Apoptosis. 2001;6:263–8.
64. Brown DG, Sun XM, Cohen GM. Dexamethasone-induced apoptosis involves cleavage of DNA to large fragments prior to internucleosomal fragmentation. J Biol Chem. 1993;268:3037–9.
65. Oberhammer F, Wilson JW, Dive C, Morris ID, Hickman JA, Wakeling AE, et al. Apoptotic death in epithelial cells: cleavage of DNA to 300 and/or 50 kb fragments prior to or in the absence of internucleosomal fragmentation. EMBO J. 1993;12:3679–84.
66. Filipski J, Leblanc J, Youdale T, Sikorska M, Walker PR. Periodicity of DNA folding in higher order chromatin structures. EMBO J. 1990;9:1319–27.
67. Compton MM. A biochemical hallmark of apoptosis: internucleosomal degradation of the genome. Cancer Metastasis Rev. 1992;11:105–19.
68. Bortner CD, Oldenburg NB, Cidlowski JA. The role of DNA fragmentation in apoptosis. Trends Cell Biol. 1995;5:21–6.
69. Vaughan AT, Betti CJ, Villalobos MJ. Surviving apoptosis. Apoptosis. 2002;7:173–7.
70. Betti CJ, Villalobos MJ, Diaz MO, Vaughan AT. Apoptotic triggers initiate translocations within the MLL gene involving the nonhomologous end joining repair system. Cancer Res. 2001;61:4550–5.
71. Paques F, Haber JE. Multiple pathways of recombination induced by double-strand breaks in Saccharomyces cerevisiae. Microbiol Mol Biol Rev. 1999;63:349–404.
72. Lieber MR, Gu J, Lu H, Shimazaki N, Tsai AG. Nonhomologous DNA end joining (NHEJ) and chromosomal translocations in humans. Subcell Biochem. 2010;50:279–96.
73. Weaver DT. What to do at an end: DNA double-strand-break repair. Trends Genet. 1995;11:388–92.
74. Takata M, Sasaki MS, Sonoda E, Morrison C, Hashimoto M, Utsumi H, et al. Homologous recombination and non-homologous end-joining pathways of DNA double-strand break repair have overlapping roles in the maintenance of chromosomal integrity in vertebrate cells. EMBO J. 1998;17:5497–508.
75. Weinstock DM, Richardson CA, Elliott B, Jasin M. Modeling oncogenic translocations: distinct roles for double-strand break repair pathways in translocation formation in mammalian cells. DNA Repair (Amst). 2006;5:1065–74.
76. Strissel PL, Strick R, Tomek RJ, Roe BA, Rowley JD, Zeleznik L. DNA structural properties of AF9 are similar to MLL and could act as recombination hot spots resulting in MLL/AF9 translocations and leukemogenesis. Hum Mol Genet. 2000;9:1671–9.
77. Strick R, Zhang Y, Emmanuel N, Strissel PL. Common chromatin structures at breakpoint cluster regions may lead to chromosomal translocations found in chronic and acute leukemias. Hum Genet. 2006;119:479–95.
78. Broeker PL, Super HG, Thirman MJ, Pomykala H, Yonebayashi Y, Tanabe S, et al. Distribution of 11q23 breakpoints within the MLL breakpoint cluster region in de novo acute leukemia and in treatment-related acute myeloid leukemia: correlation with scaffold attachment regions and topoisomerase II consensus binding sites. *Blood.* 1996;87:1912–22.
79. Lechardeur D, Xu M, Lukacs GL. Contrasting nuclear dynamics of the caspase-activated DNase (CAD) in dividing and apoptotic cells. J Cell Biol. 2004;167:851–62.
80. Tan SN, Sim SP, Khoo AS. Potential role of oxidative stress-induced apoptosis in mediating chromosomal rearrangements in nasopharyngeal carcinoma. Cell Biosci. 2016;6:35.
81. Goff SP, Gilboa E, Witte ON, Baltimore D. Structure of the Abelson murine leukemia virus genome and the homologous cellular gene: studies with cloned viral DNA. Cell. 1980;22:777–85.
82. Wang JY, Ledley F, Goff S, Lee R, Groner Y, Baltimore D. The mouse c-abl locus: molecular cloning and characterization. Cell. 1984;36:349–56.
83. Sawyers CL, McLaughlin J, Goga A, Havlik M, Witte O. The nuclear tyrosine kinase c-Abl negatively regulates cell growth. Cell. 1994;77:121–31.
84. Lugo TG, Pendergast AM, Muller AJ, Witte ON. Tyrosine kinase activity and transformation potency of bcr-abl oncogene products. Science. 1990;247:1079–82.
85. van Drunen CM, Sewalt RG, Oosterling RW, Weisbeek PJ, Smeekens SC, van Driel R. A bipartite sequence element associated with matrix/scaffold attachment regions. Nucleic Acids Res. 1999;27:2924–30.
86. Sallmyr A, Fan J, Rassool FV. Genomic instability in myeloid malignancies: increased reactive oxygen species (ROS), DNA double strand breaks (DSBs) and error-prone repair. Cancer Lett. 2008;270:1–9.
87. Strunnikov AV. One-hit wonders of genomic instability. Cell Div. 2010;5
88. Giam M, Rancati G. Aneuploidy and chromosomal instability in cancer: a jackpot to chaos. Cell Div. 2015;10

89. Boichuk SV, Minnebaev MM, Mustafin IG. Key role of mitochondria in apoptosis of lymphocytes. Bull Exp Biol Med. 2001;132:1166–8.
90. Fadok VA, Voelker DR, Campbell PA, Cohen JJ, Bratton DL, Henson PM. Exposure of phosphatidylserine on the surface of apoptotic lymphocytes triggers specific recognition and removal by macrophages. J Immunol. 1992;148:2207–16.
91. Anderson GR. Genomic instability in cancer. Curr Sci. 2001;81:501–7.
92. Shen Z. Genomic instability and cancer: an introduction. J Mol Cell Biol. 2011;3:1–3.
93. Cockerill PN, Garrard WT. Chromosomal loop anchorage of the kappa immunoglobulin gene occurs next to the enhancer in a region containing topoisomerase II sites. Cell. 1986;44:273–82.
94. Bartram CR, de Klein A, Hagemeijer A, Grosveld G, Heisterkamp N, Groffen J. Localization of the human c-sis oncogene in Ph1-positive and Ph1-negative chronic myelocytic leukemia by in situ hybridization. Blood. 1984;63:223–5.
95. Nowell PC, Hungerford DA. Chromosome studies on normal and leukemic human leukocytes. J Natl Cancer Inst. 1960;25:85–109.
96. Shtivelman E, Gale RP, Dreazen O, Berrebi A, Zaizov R, Kubonishi I, et al. bcr-abl RNA in patients with chronic myelogenous leukemia. Blood. 1987;69:971–3.
97. Melo JV, Gordon DE, Tuszynski A, Dhut S, Young BD, Goldman JM. Expression of the ABL-BCR fusion gene in Philadelphia-positive acute lymphoblastic leukemia. Blood. 1993;81:2488–91.
98. Chen SJ, Chen Z, Font MP, d'Auriol L, Larsen CJ, Berger R. Structural alterations of the BCR and ABL genes in Ph1 positive acute leukemias with rearrangements in the BCR gene first intron: further evidence implicating Alu sequences in the chromosome translocation. Nucleic Acids Res. 1989;17:7631–42.
99. Jiang XY, Trujillo JM, Liang JC. Chromosomal breakpoints within the first intron of the ABL gene are nonrandom in patients with chronic myelogenous leukemia. Blood. 1990;76:597–601.
100. Li TK, Chen AY, Yu C, Mao Y, Wang H, Liu LF. Activation of topoisomerase II-mediated excision of chromosomal DNA loops during oxidative stress. Genes Dev. 1999;13:1553–60.
101. Kim DK, Cho ES, Um HD. Caspase-dependent and -independent events in apoptosis induced by hydrogen peroxide. Exp Cell Res. 2000;257:82–8.
102. Sakahira H, Enari M, Nagata S. Cleavage of CAD inhibitor in CAD activation and DNA degradation during apoptosis. Nature. 1998;391:96–9.
103. Legault J, Tremblay A, Mirault ME. Preferential localization of DNA damage induced by depurination and bleomycin in a plasmid containing a scaffold-associated region. Biochem Cell Biol. 1997;75:369–75.
104. Bode J, Benham C, Ernst E, Knopp A, Marschalek R, Strick R, et al. Fatal connections: when DNA ends meet on the nuclear matrix. J Cell Biochem Suppl. 2000;Suppl 35:3–22.
105. Gohring F, Schwab BL, Nicotera P, Leist M, Fackelmayer FO. The novel SAR-binding domain of scaffold attachment factor A (SAF-A) is a target in apoptotic nuclear breakdown. EMBO J. 1997;16:7361–71.
106. Ptacek T, Song C, Walker CL, Sell SM. Physical mapping of distinct 7q22 deletions in uterine leiomyoma and analysis of a recently annotated 7q22 candidate gene. Cancer Genet Cytogenet. 2007;174:116–20.
107. Petit MM, Schoenmakers EF, Huysmans C, Geurts JM, Mandahl N, Van de Ven WJ. LHFP, a novel translocation partner gene of HMGIC in a lipoma, is a member of a new family of LHFP-like genes. Genomics. 1999;57:438–41.
108. Huang C, Guo J, Liu S, Shan Y, Wu S, Cai Y, et al. Isolation, tissue distribution and prokaryotic expression of a novel human X-linked gene LHFPL1. DNA Seq. 2004;15:299–302.
109. Yeoh EJ, Ross ME, Shurtleff SA, Williams WK, Patel D, Mahfouz R, et al. Classification, subtype discovery, and prediction of outcome in pediatric acute lymphoblastic leukemia by gene expression profiling. Cancer Cell. 2002;1:133–43.
110. Schwaenen C, Wessendorf S, Kestler HA, Dohner H, Lichter P, Bentz M. DNA microarray analysis in malignant lymphomas. Ann Hematol. 2003;82:323–32.
111. Collins Y, Tan DF, Pejovic T, Mor G, Qian F, Rutherford T, et al. Identification of differentially expressed genes in clinically distinct groups of serous ovarian carcinomas using cDNA microarray. Int J Mol Med. 2004;14:43–53.
112. Wang SS, Smiraglia DJ, Wu YZ, Ghosh S, Rader JS, Cho KR, et al. Identification of novel methylation markers in cervical cancer using restriction landmark genomic scanning. Cancer Res. 2008;68:2489–97.
113. Kalay E, Li Y, Uzumcu A, Uyguner O, Collin RW, Caylan R, et al. Mutations in the lipoma HMGIC fusion partner-like 5 (LHFPL5) gene cause autosomal recessive nonsyndromic hearing loss. Hum Mutat. 2006;27:633–9.
114. Curtissa NP, Bonifasa JM, Lauchlea JO, Balkmana JD, Kratza CP, Emerlinga BM, et al. Isolation and analysis of candidate myeloid tumor suppressor genes from a commonly deleted segment of 7q22. Genomics. 2005;85:600–7.
115. Milinkovic V, Bankovic J, Rakic M, Stankovic T, Skender-Gazibara M, Ruzdijic S, et al. Identification of novel genetic alterations in samples of malignant glioma patients. PLoS One. 2013;8:e82108.
116. Tsao SW, Wang X, Liu Y, Cheung YC, Feng H, Zheng Z, et al. Establishment of two immortalized nasopharyngeal epithelial cell lines using SV40 large T and HPV16E6/E7 viral oncogenes. Biochim Biophys Acta. 2002;1590:150–8.
117. Huang DP, Ho JH, Poon YF, Chew EC, Saw D, Lui M, et al. Establishment of a cell line (NPC/HK1) from a differentiated squamous carcinoma of the nasopharynx. Int J Cancer. 1980;26:127–32.
118. Widlak P, Garrard WT. Discovery, regulation, and action of the major apoptotic nucleases DFF40/CAD and endonuclease G. J Cell Biochem. 2005;94:1078–87.

Remotely acting *SMCHD1* gene regulatory elements: in silico prediction and identification of potential regulatory variants in patients with FSHD

Mary B. Mayes[1], Taniesha Morgan[2], Jincy Winston[2], Daniel S. Buxton[1], Mihir Anant Kamat[1,5], Debbie Smith[3], Maggie Williams[3], Rebecca L. Martin[1], Dirk A. Kleinjan[4], David N. Cooper[2], Meena Upadhyaya[2] and Nadia Chuzhanova[1*]

Abstract

Background: Facioscapulohumeral dystrophy (FSHD) is commonly associated with contraction of the D4Z4 macro-satellite repeat on chromosome 4q35 (FSHD1) or mutations in the *SMCHD1* gene (FSHD2). Recent studies have shown that the clinical manifestation of FSHD1 can be modified by mutations in the *SMCHD1* gene within a given family. The absence of either D4Z4 contraction or *SMCHD1* mutations in a small cohort of patients suggests that the disease could also be due to disruption of gene regulation. In this study, we postulated that mutations responsible for exerting a modifier effect on FSHD might reside within remotely acting regulatory elements that have the potential to interact at a distance with their cognate gene promoter via chromatin looping. To explore this postulate, genome-wide Hi-C data were used to identify genomic fragments displaying the strongest interaction with the *SMCHD1* gene. These fragments were then narrowed down to shorter regions using ENCODE and FANTOM data on transcription factor binding sites and epigenetic marks characteristic of promoters, enhancers and silencers.

Results: We identified two regions, located respectively ~14 and ~85 kb upstream of the *SMCHD1* gene, which were then sequenced in 229 FSHD/FSHD-like patients (200 with D4Z4 repeat units <11). Three heterozygous sequence variants were found ~14 kb upstream of the *SMCHD1* gene. One of these variants was found to be of potential functional significance based on DNA methylation analysis. Further functional ascertainment will be required in order to establish the clinical/functional significance of the variants found.

Conclusions: In this study, we propose an improved approach to predict the possible locations of remotely acting regulatory elements that might influence the transcriptional regulation of their associated gene(s). It represents a new way to screen for disease-relevant mutations beyond the immediate vicinity of the specific disease gene. It promises to be useful for investigating disorders in which mutations could occur in remotely acting regulatory elements.

Keywords: Regulatory mutations, Regulatory elements, Hi-C data, *SMCHD1*, FSHD

* Correspondence: nadia.chuzhanova@ntu.ac.uk
[1]School of Science and Technology, Nottingham Trent University, Clifton Lane, Nottingham NG11 8NS, UK
Full list of author information is available at the end of the article

Background

Facioscapulohumeral muscular dystrophy (FSHD) is the most prevalent of the nine primary types of muscular dystrophy affecting adults and children; it is characterized by the weakness and atrophy of the facial and shoulder girdle muscle extending to the abdominal and lower limb muscle [1]. Two genetic loci are associated with the disease. The FSHD1 locus maps to 4q35 [2] and accounts for 95 % of clinical diseases in an FSHD context [1, 3]. A second FSHD locus, FSHD2, exists and is phenotypically indistinguishable from FSHD1 [2]. FSHD1 patients harbour a large deletion in the polymorphic *D4Z4* macro-satellite repeat array at 4q35 and invariably present with 1–10 repeats whereas non-affected individuals possess 11–150 repeats. Each 3.3-kb *D4Z4* unit contains a double homeobox 4 (*DUX4*) gene that, among others, is transcriptionally activated on contraction of the 4q35 repeat array as a consequence of the induction of chromatin remodelling of the 4qter region. A number of 4q subtelomeric sequence variants are now recognized, although FSHD1 only occurs in association with 'permissive' haplotypes, each of which is associated with a polyadenylation signal located immediately distal of the last *D4Z4* unit [4]. Approximately 5 % of FSHD patients lack a contraction of the *D4Z4* array, and the disease aetiology has been ascribed to a putative FSHD2 locus. Whole-exome sequencing identified the *SMCHD1* gene as the causative agent at the FSHD2 locus [5]. In FSHD2 families, the disease exhibits a more complex digenic inheritance because mutations in the chromosome-18-located *SMCHD1* gene segregate independently from the FSHD-permissive 4q haplotype [5]. *SMCHD1* is a member of a condensing/cohesion family of chromatin compact complexes that bind to the D4Z4 array [6]. However, not all FSHD2 patients can be explained by the lack of contraction of the *D4Z4* repeats or mutations in the *SMCHD1* gene, suggesting either that mutations may reside within the *SMCHD1* non-coding region or that the cause of FSHD in these families could be linked to yet another FSHD locus. Although FSHD1 and FSHD2 are characterized by different underlying genetic defects, they both appear to be caused by transcriptional de-repression of *DUX4* in the skeletal muscle [7].

DNA methylation changes the conformation of the chromatin and hence the accessibility of the encoded gene(s) in the vicinity. Hypermethylation generally leads to the compaction of chromatin, thereby reducing gene expression. Conversely, hypomethylation generally serves to relax the chromatin thereby upregulating gene expression. FSHD-affected individuals display hypomethylation at D4Z4 units, whereas unaffected subjects exhibit hypermethylation whilst FSHD non-manifesting carriers have an intermediate level of methylation [8]. The specific loss of histone H3 lysine 9 trimethylation and HP1 gamma/cohesion binding at D4Z4 repeats is reported to be associated with FSHD [6].

Recent studies have reported that the clinical manifestation of FSHD1 can be modified by mutations in the *SMCHD1* gene within a given family [9–11] although pathogenic FSHD mutations remained undetected in a majority of FSHD2 patients. The undetected mutations in such patients could, in principle, reside within regulatory elements, possibly at some distance from the *SMCHD1* gene, disrupting regulation of the gene.

Advances in techniques for capturing three-dimensional chromosome conformations (3C) have been prompted by the view that direct long-range interactions occur between gene promoters and distal genomic regions, bringing them into close spatial proximity through looping interactions [12], thereby explaining the impact of pathological mutations that are known to occur at some considerable distance from the genes whose function they influence [13]. Genome-wide mapping of long-range interactions using the Hi-C [14], the recently proposed in situ Hi-C [15] and Capture Hi-C [16] methods has made it possible to assess the propensity of genomic regions to form looping interactions by surveying their interaction frequencies; the latter two methods have yielded interaction maps of sufficiently high resolution to be practically useful in this regard.

Studies of the long-range interactions of ~22,000 promoters in two human blood cell lines, including lymphoblastoid cell line GM12878, employing the Capture Hi-C method [16] have revealed that interacting fragments are enriched in DNaseI hypersensitive sites (DHSs), a classical marker of regulatory regions [17]; the H3K4me3 histone mark, the tri-methylation of histone H3 at lysine 4, mainly associated with promoters that are active or poised to be activated ([18], reviewed in [19]); the H3K4me1 histone mark, the mono-methylation of histone H3 at lysine 4, a well-established feature of enhancers and promoters; and the H3K27ac mark, the acetylation of histone H3 at lysine 27, characteristic of active enhancers ([18], reviewed in [19]).

In this study, we postulated that mutations responsible for the clinical manifestation of FSHD or playing a disease-modifier role might reside within remotely acting regulatory elements that have the potential to interact with the DNA fragment containing the *SMCHD1* gene and that these elements might be located within regions enriched in epigenetic features characteristic of regulatory regions. We identified two remotely located putative regulatory elements residing within regions respectively strongly and weakly interacting with the fragment containing the *SMCHD1* gene. We screened these in silico-predicted regulatory regions for mutations in 229 FSHD patients (200 patients exhibiting <11 D4Z4 repeat units and 29 patients >11 D4Z4 repeat units). We report two novel sequence variants in a putative control region proximal to the *SMCHD1* gene (~14 kb upstream of the gene) in two FSHD families.

Results and discussion

In silico prediction of interacting fragments and potential remotely acting regulatory regions of *SMCHD1*

Hi-C data, available at www.ncbi.nlm.nih.gov/geo/ (accession number GSE18199), indicated that a 1-Mb fragment of chromosome 18, starting at position 2,000,000 and ending at position 2,999,999, which we have termed the *SMCHD1*-containing fragment since it contains the entire *SMCHD1* gene (positions 2,645,885–2,795,015; hg18), interacts most strongly with itself (3371 intra-chromosomal interactions with the *SMCHD1*-containing fragment were also recorded in Hi-C data) and with the adjacent 1-Mb downstream fragment on chromosome 18 (positions 3,000,000–3,999,999; 663 interactions). On a 100-kb scale, the *SMCHD1* gene was found to occupy two consecutive fragments (fragment 1: 2,600,000–2,699,999 and fragment 2: 2,700,000–2,799,999, Fig. 1); for each of these fragments, the number of intra-chromosomal interactions was also available from the Hi-C data. Fragment 1 and fragment 2 interacted most strongly with themselves (256 and 237 interactions, respectively) whereas the number of interactions with other 100-kb fragments on chromosome 18 did not exceed 30. The ENCODE data indicated the presence of two weak CTCF-binding sites (scores <445/1000) between positions 2,531,952 and 2,537,783 and a strong one (with score 719/1000) at positions 2,922,164 and 2,922,551. Among other functions, CTCF and its associated proteins that bind to the CTCF-binding sites are thought to play a role in forming barriers between regulatory regions but are also involved (together with other transcription factors) in controlling chromatin-looping interactions and structure (reviewed in [19]) although a recent study [16] indicated that insulator activities of CTCF could be tissue- or cell-line-specific. Because the first CTCF-binding site was found within the 100-kb region (2,500,000–2,599,999), which has only 23 interactions with the *SMCHD1*-containing region, we opted to extend the search for potential regulatory elements to positions 2,531,952–2,922,551.

The ENCODE data on histone modification marks, H3K4me1 and H3K27ac, available for eight cell lines, reveal the presence of these marks in the GM12878 cell line between positions 2,624,475 and 2,635,775. The ENCODE data also indicated the presence of an H3K4me3-enriched region, characteristic of an active promoter, somewhere between positions 2,644,486 and 2,648,785 on the interacting fragment (Fig. 1). There is a strong enhancer at positions 2,631,000–2,632,200 and a transcription factor binding site (TFBS)-enriched region between positions 2,631,527 and 2,632,188; the latter region, ~14 kb upstream of the *SMCHD1* gene, was selected for sequencing in FSHD patients.

A second region, distal (~84 kb) to the *SMCHD1* gene, was also selected for sequencing (positions 2,561,489–2,562,509) on the basis that it is enriched in H3K4me3 and TFBSs and corresponds to a peak of DNaseI sensitivity signal, characteristic of open chromatin and an active *cis*-regulatory region; it also shows a weak interaction with the *SMCHD1*-containing region. The

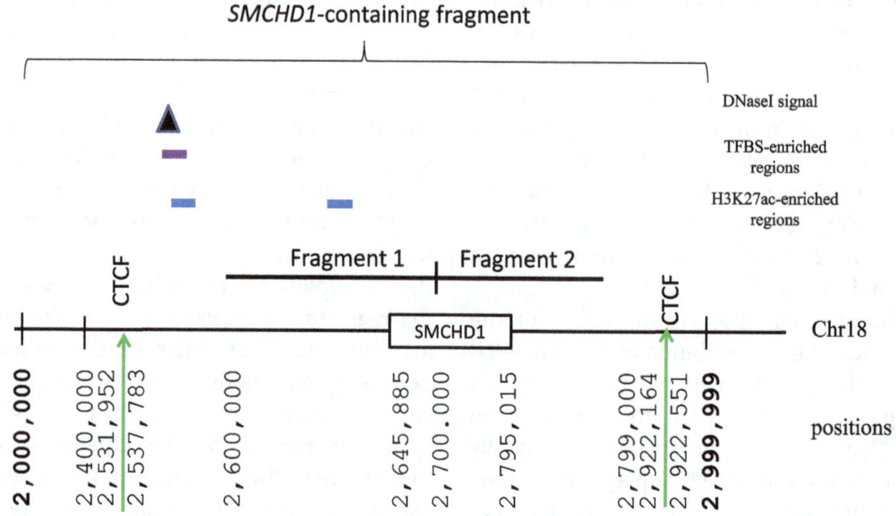

Fig. 1 Schematic representation of the chromosomal regions enriched in transcription factor binding sites (TFBS), histone mark H3K27ac and DNAseI hypersensitive signal within 1-Mb-long *SMCHD1*-containing DNA fragments (not to scale). Two consecutive 100-kb fragments harbouring the *SMCHD1* gene are denoted as *Fragments 1* and *2*. CTCF is a binding site for the insulator protein

ENCODE data on histone modification marks H3K4me1 and H3K27ac revealed the presence of these marks in eight cell lines, the exception being GM12878, between positions 2,557,575 and 2,566,775. ChromHMM track indicated the presence of an active promoter at positions 2,560,200–2,563,800, which corresponds precisely to the *METTL4* gene promoter.

Search for sequence variants within the potential remotely acting regulatory regions of *SMCHD1*

The two predicted regions, respectively ~14 kb and ~85 kb upstream of the *SMCHD1* gene, were sequenced in 229 FSHD patients of Central European descent (the DNA-sequencing procedure is described in the Additional file 1; the primers used are given in Table S1). DNA sequence data from these regions, derived from the 1000 Genomes Project, are available for 226 individuals from the CEU (Utah Residents with Northern and Western European Ancestry) population and were used as controls.

In addition, DNA from the parents of the FSHD probands, from families 1 and 2, was analysed. Family 1 was referred to us because the proband exhibited facial and shoulder girdle weakness; both parents were healthy. In family 2, the proband had early onset FSHD, the parents of the proband were healthy and non-consanguineous and no other member of the family was affected with FSHD. Neither parent showed any evidence of facial or other muscle weakness.

We identified variants in the region proximal (~14 kb upstream) to the *SMCHD1* gene in two probands from families 1 and 2. These variants were not present in dbSNP (http://www.ncbi.nlm.nih.gov/snp) or the 1000 Genomes Project data. The two variants, NCBI36/hg18: chr18:2,631,610 T > C and NCBI36/hg18:chr18:2,631,886 G > A, were found in probands from families 1 and 2, respectively (Figs. 2 and 3). They were also identified in the clinically unaffected mothers of these two patients.

Variant NCBI36/hg18:chr18:2,631,610 T > C was found in the proband who was originally tested due to facial and shoulder girdle muscle weakness. All three members of this family (family 1) were found to harbour >11 D4Z4 repeats, but only the proband was symptomatic. No mutations within the *SMCHD1*-coding region were detected. Both the proband and the unaffected mother harboured the heterozygous sequence change NCBI36/hg18:chr18:2,631,610 T > C. This nucleotide is conserved in primates (human, chimpanzee, gorilla, orang-utan, gibbon, rhesus monkey, crab-eating macaque, baboon, green monkey and squirrel monkey); the evolutionary conservation score, *phyloP* [20], calculated across 44 vertebrates, was 0.557, indicating a conserved nucleotide.

The second novel sequence change, NCBI36/hg18: 2,631,886 G > A, was found in the proband from family 2. Molecular testing based on *EcoRI/BlnI* analysis [10] identified two intact D4Z4 repeats in the proband and >11 in the mother. Interestingly, his unaffected mother was a mosaic for this change (Fig. 3). This nucleotide is conserved in primates (human, chimpanzee, orang-utan, rhesus monkey, baboon and marmoset). The evolutionary conservation score across 44 vertebrate species was 0.431. As the proband in this family has FSHD1, it is possible that this sequence variant could modify the clinical expression of FSHD1 as this patient exhibited weakness of facial and shoulder girdle muscle with early onset (8 years of age).

The third heterozygous sequence variant was found in the region proximal (~14 kb upstream) to the *SMCHD1* gene at position NCBI36/hg18:2,631,858; a surprisingly high proportion (182/229) of patients were found to have a C-allele at this position with 47/229 patients having a T-allele. A T/C polymorphism at this position is recorded in dbSNP (rs7229070) with C being the minor allele with MAF = 0.30 for a CEU (Utah Residents with Northern and Western European Ancestry) population (226 genomes) sequenced by the 1000 Genomes Project. According to dbSNP, the clinical significance of this SNP is unknown. This polymorphism occurs in a region which is evolutionarily conserved among primates; all the other primate species examined have nucleotide T in this position in their wild-type sequences. In addition, this nucleotide is also conserved in members of other mammalian orders including the artiodactyla, rodentia and carnivora (*phyloP* score across 44 vertebrate species is 0.691).

All three sequence changes, 2,631,610 T > C, 2,631,858 T > C and 2,631,886 G > A, occur in close proximity to, or within, the DNaseI hypersensitive cluster (positions 2,631,700–2,631,930) characterized by a high score (976 out of 1000) that corresponds to an open chromatin domain and rich in TFBSs (IRF1, STAT1, STAT2, RUNX3, BATF, SPI1) with IRF1 (positions 2,631,639–2,632,023), STAT1 (2,631,540–2,632,016), STAT2 (2,631,654–2,632,010) and RUNX3 (2,631,672–2,631,955) transcription factor binding sites having the highest score (1000/1000). All three sequence variants also occur in a region enriched in enhancer-associated marks, H3K4me1 and H3K27ac, observed in cell line GM12878.

Recently published data on looping interactions [15] reveal a strong looping interaction in the GM12878 cell line, between a 5-kb fragment corresponding to the promoter region of the *SMCHD1* gene (positions 2,640,000–2,645,000) and a 5-kb region (positions 2,630,000–2,635,000) harbouring the three variants found (note that all positions were 'lifted over' to the hg18 assembly). It was not possible to confirm the existence of a looping interaction between the *SMCHD1* gene promoter and the region ~14 kb upstream of the gene harbouring the variants using Capture Hi-C data [16] since only interactions between promoters and interacting fragments separated by >20 kb were reported.

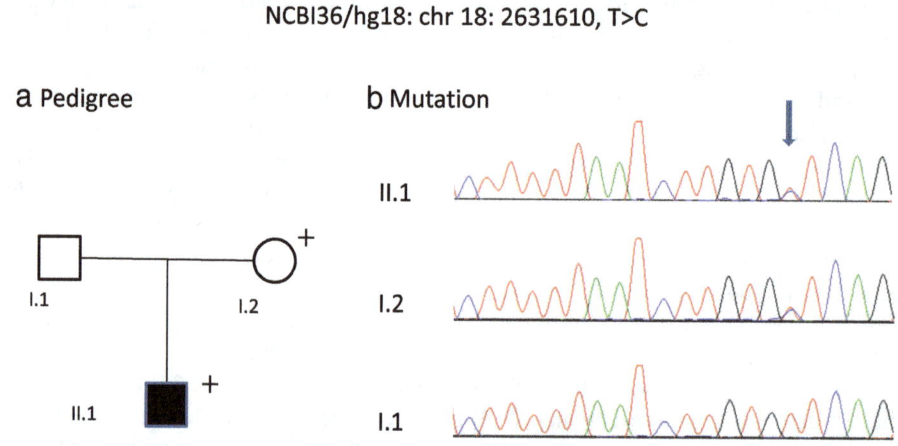

Fig. 2 Pedigree (**a**) and mutation (**b**) in the proximal promoter region of the *SMCHD1* gene, family 1

Methylation analysis of families 1 and 2

The SMCHD1 protein plays a role in the methylation of large chromosomal regions, including the X chromosome and the D4Z4 array. It is required for the maintenance of X inactivation in females and hypermethylation of CpG islands associated with the inactive X chromosome. Loss of *SMCHD1* activity in FSHD2 patients reduces methylation of the D4Z4 array, allowing the transcriptional machinery to gain access to the *DUX4* gene [5]. FSHD2 is caused by the co-inheritance of two independent events: an FSHD-permissive chromosome 4 haplotype (necessary for the polyadenylation of *DUX4* mRNA) and a variant in *SMCHD1* [5]. SMCHD1 regulates chromatin repression in a wide variety of different organisms. Given the wider role of *SMCHD1* in regulating methylation, it may be that *SMCHD1* serves as a modifier in human genetic disease [9–11].

Hartweck et al. [21] identified an extreme demethylation of region DR1 within the D4Z4 repeat and demonstrated that this region is hypomethylated in the majority of FSHD2 patients. On the basis of this finding, the methylation of this region was measured in blood samples from members of families 1 and 2 using bisulphite conversion followed by pyrosequencing; the ratios of C vs. T at each of the ten statistically validated sites within the DR1 region were averaged (see Additional file 1). The proband from family 2 has two D4Z4 repeat units and exhibits a methylation level of 24 %, compatible with the methylation level seen in FSHD2 patients [5]. The clinically unaffected mother of this patient was a mosaic for the variant, but her D4Z4 repeat was not hypomethylated. Based on a two D4Z4 unit array combined with the 4qA haplotype, hypomethylation and evolutionary conservation, we propose that the promoter

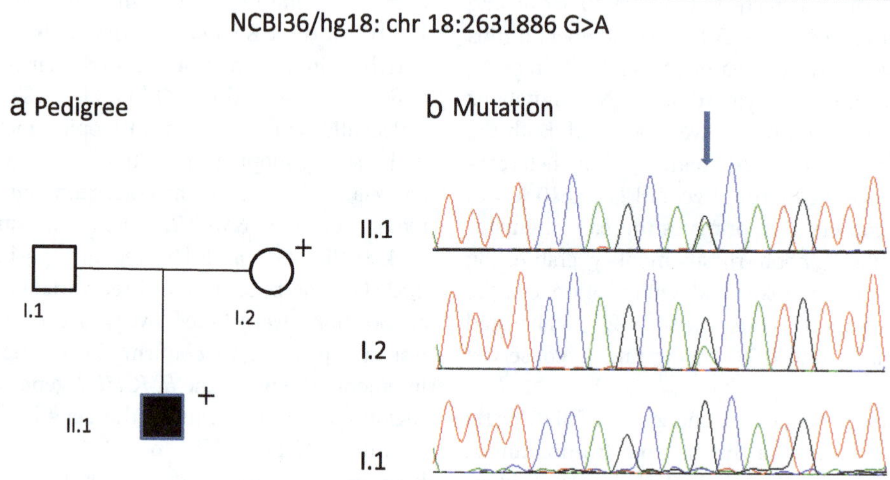

Fig. 3 Pedigree (**a**) and mutation (**b**) in the proximal promoter region of the *SMCHD1* gene, family 2

variant in *SMCHD1* (2,631,886 G > A) represents a plausible candidate for modifying the expression of FSHD1 in this patient.

The results of methylation analysis for the proband from family 1 were inconclusive; no hypomethylation of D4Z4 was observed in the DR1 region, but other regions (DR2 and DR3) were not tested. Based on an epigenetic signature, Jones et al. [22] have developed a new laboratory test that identifies symptomatic FSHD1 and 2 from non-penetrant FSHD carriers and other muscular dystrophies. In a recent study, Huichalaf et al. [23] have shown that DNA methylation does not correlate with the density of CpG dinucleotides within D4Z4. However, DNA methylation and histone de-acetylation are required to maintain the repression of the FSHD candidate gene. H3K27me3 was reported to be enriched at D4Z4 in healthy subjects and significantly decreased in FSHD patients. Further functional ascertainment will be required in order to establish the pathological significance of the variants identified in our study.

Methylation plays an important role in the marked clinical variability seen in FSHD patients. Recent studies indicate that a combination of genetic and epigenetic factors act on the D4Z4 repeat array to determine the probability of *DUX4* expression in the skeletal muscle and influence disease penetrance and progression [24]. In FSHD patients harbouring 1 to 6 D4Z4 units, the D4Z4 unit number appears to correlate inversely with the severity of the disease, but in patients with 7–10 D4Z4 units, severity depends upon the factors that regulate methylation [8]. In a family with a case of full-blown FSHD and a relative with a non-penetrant mutation bearing an identical number of D4Z4 units, the difference in clinical severity between these two family members appears to be due to methylation [8].

In FSHD2, the mutations that disrupt the open reading frame are believed to result in *SMCHD1* haploinsufficiency and to have less effect on D4Z4 methylation than mutations which maintain the *SMCHD1* open reading frame, possibly acting through a dominant negative mechanism [8]. These new developments are improving our understanding of the pathology of FSHD and emphasize the potential for combinatorial effects of genetic and epigenetic factors in influencing the onset and progression of the disease.

Conclusions

In this study, we propose an improved (as compared to the previously described [25]) approach for predicting the possible location of remotely acting regulatory elements that might influence the transcriptional regulation of their associated gene(s). This approach was successfully employed in the context of the in silico prediction of potential remotely acting regulatory elements for the *SMCHD1* gene. Subsequent sequencing of these predicted regions identified three sequence variants in FSHD patients which represent candidate variants of potential functional significance.

This approach demonstrates a novel means to screen for disease-relevant mutations that reside beyond the immediate vicinity of a given disease gene. It therefore promises not only to be useful in investigating disorders in which mutations may occur in remotely acting regulatory elements but also in identifying the causative noncoding mutations found by GWAS that are often distant from their target genes.

Methods

Long-range interaction data (Hi-C)

In December 2013, when this study commenced, the inter- and intra-chromosomal interaction data were available for, respectively, 1-Mb and 100-kb fragments from the human GM06990 lymphoblastoid cell line. These data were obtained by means of the chromosome conformation capture technique [14] and were available at www.ncbi.nlm.nih.gov/geo/ (accession number GSE18 199). Data, recently obtained using in situ Hi-C method [15] at a 1-kb resolution, together with Capture Hi-C data on looping interactions for ~22,000 promoters [16], were used to validate the initial results.

Epigenetic features, transcription factor binding sites and enhancers

Data on the occurrence and the strength of DNaseI hypersensitive sites, the occurrence and level of enrichment in the H3k4me1, H3k4me3 and H3k27ac methylation marks, assayed using various biochemical techniques, were downloaded from the ENCODE database ([26]; http://genome.ucsc.edu/index.html).

Genomic positions of 55 transcription factor binding sites (TFBSs), including an insulator CTCF-binding site, all assayed by ChIP-seq, were also downloaded from the ENCODE database. Genomic positions of active enhancers assayed using CAGE techniques were downloaded from the FANTOM (http://fantom.gsc.riken.jp/5/; [27]) and ENCODE (ChromHMM track) databases. It should be noted that biochemical assays used to map these functional elements usually capture extended DNA regions, often spanning several hundred base pairs, making it difficult to define the exact boundaries of functional elements.

In cases where data were available from multiple cell lines, the data for the karyotypically normal human lymphoblastoid cell line GM06990 were used. The choice of this cell line was predicated upon the availability of Hi-C data for this particular cell line. Data on H3k4me1 and H3K27ac methylation marks were not available for the GM06990 cell line; data for the GM12878 cell line were used instead.

Where necessary, all positions were 'lifted over' to the hg18 assembly using the Lift Genome Annotation program available at https://genome.ucsc.edu/cgi-bin/hgLiftOver.

In silico prediction of possible remotely acting regulatory regions and sites of mutation

To identify possible remotely acting regulatory regions and the sites of mutation within these regions, the following step-by-step procedure was adopted:

Step 1. Set the size of the fragment to 1 Mb or 100 kb.
Step 2. Use Hi-C data to find a fragment (bin) that harbours the gene/gene promoter in question, henceforth termed *a gene-containing fragment.*
Step 3. Use Hi-C data to find a fragment(s) with the highest number of interactions with the gene-containing fragment, henceforth termed *an interacting fragment(s).*
Step 4. Use ENCODE data to identify the location(s) of CTCF-binding sites within the interacting fragment. Refine the interacting fragment(s) to shorter regions bounded by the CTCF-binding site. If the interacting fragment is found upstream of the gene, a DNA fragment between the CTCF-binding site and the gene in question is considered as a region possibly harbouring remotely acting regulatory element(s). If the interacting fragment is found downstream of the gene, a DNA fragment between the gene and the CTCF-binding site is considered as a region possibly harbouring remotely acting regulatory element(s).
Step 5. Use the ENCODE and FANTOM data on the occurrence of epigenetic marks, enhancers, etc., to further refine regions possibly harbouring remotely acting regulatory element(s) by identifying shorter regions enriched in histone modification marks (H3K4me1, H3K27ac, H3K4me3) and transcription factor binding sites and corresponding to peaks of DNaseI hypersensitivity signal.
Step 6. Validate regions found using in situ and Capture Hi-C data.

The *SMCHD1* gene and clinical data

The *SMCHD1* gene, encoding the structural maintenance of chromosomes' flexible hinge domain-containing protein 1, occupies positions 2,645,885–2,795,015 on chromosome 18 (NCBI build 37, UCSC hg18 and Ensembl GRCh37).

Of the 229 FSHD patients analysed in this study, 29 had >11 D4Z4 repeat units whereas 2 of these patients harboured novel FSHD-causing (or disease-modifying) sequence variants in the *SMCHD1* gene identified in FSHD patients by Winston et al. [10] that could modulate the expression of FSHD1.

DNA sequencing and methylation analysis

The DNA from patients was digested with *Eco*RI and *Bln*I, size-fractionated by electrophoresis, Southern blotted and hybridized with DNA marker p13E11 [1]; sequencing was performed as described in Additional file 1. Methylation quantification was performed as described by Hartweck et al. [21]; see Additional file 1 for details.

Competing interests

The authors declare that they have no competing interests.

Authors' contributions

NC conceived of the approach. MBM, DSB, MAK and RLM performed the in silico analysis. MU, TM and JW performed the PCR and sequencing. DS and MW performed the methylation analysis. NC, MU, DAK and DNC interpreted the results. NC drafted the manuscript. All authors participated in the preparation of the manuscript, read and approved the final manuscript.

Acknowledgements

The work of RLM was supported by the Wellcome Trust summer vacation scholarship.

Author details

[1]School of Science and Technology, Nottingham Trent University, Clifton Lane, Nottingham NG11 8NS, UK. [2]Institute of Medical Genetics, School of Medicine, Cardiff University, Heath Park, Cardiff CF14 4XN, UK. [3]Blood Sciences Department and Bristol Genetics Laboratory, Southmead Hospital, Westbury-on-Trym, Bristol BS10 5NB, UK. [4]MRC Human Genetics Unit, Institute of Genetics and Molecular Medicine, University of Edinburgh, Edinburgh EH4 2XU, UK. [5]Present address: Cardiovascular Epidemiology Unit, Department of Public Health and Primary Care, University of Cambridge, Cambridge CB1 8RN, UK.

References

1. Richards M, Coppée F, Thomas N, Belayew A, Upadhyaya M. Facioscapulohumeral muscular dystrophy (FSHD): an enigma unravelled? Hum Genet. 2013;131:325–40.
2. Wijmenga C, Hewitt JE, Sandkuijl LA, Clark LN, Wright TJ, Dauwerse HG, et al. Chromosome 4q DNA rearrangements associated with facioscapulohumeral muscular dystrophy. Nat Genet. 1992;2:26–30.
3. Upadhyaya M, Cooper DN. Facioscapulohumeral muscular dystrophy. Clinical medicine and molecular cell biology. Oxford, UK: Garland Science/BIOS Scientific Publishers; 2004.
4. Lemmers RJ, van der Vliet PJ, Klooster R, Sacconi S, Camaño P, Dauwerse JG, et al. A unifying genetic model for facioscapulohumeral muscular dystrophy. Science. 2010;329:1650–3.
5. Lemmers RJ, Tawil R, Petek LM, Balog J, Block GJ, Santen GW, et al. Digenic inheritance of an *SMCHD1* mutation and an FSHD-permissive D4Z4 allele causes facioscapulohumeral muscular dystrophy type 2. Nat Genet. 2012;44:1370–4.
6. Zeng W, de Greef JC, Chen YY, Chien R, Kong X, Gregson HC, et al. Specific loss of histone H3 lysine 9 trimethylation and HP1gamma/cohesin binding at D4Z4 repeats is associated with facioscapulohumeral dystrophy (FSHD). PLoS Genet. 2009;5:e1000559.
7. Yao Z, Snider L, Balog J, Lemmers RJ, Van Der Maarel SM, Tawil R, et al. DUX4-induced gene expression is the major molecular signature in FSHD skeletal muscle. Hum Mol Genet. 2014;23:5342–52.
8. Lemmers RJ, Goeman JJ, van der Vliet PJ, van Nieuwenhuizen MP, Balog J, Vos-Versteeg M, et al. Inter-individual differences in CpG methylation at D4Z4 correlate with clinical variability in FSHD1 and FSHD2. Hum Mol Genet. 2015;24:659–69.

9. Sacconi S, Lemmers RJ, Balog J, van der Vliet PJ, Lahaut P, van Nieuwenhuizen MP, et al. The FSHD2 gene *SMCHD1* is a modifier of disease severity in families affected by FSHD1. Am J Hum Genet. 2013;93:744–51.
10. Winston J, Duerden L, Mort M, Frayling IM, Rogers MT, Upadhyaya M. Identification of two novel *SMCHD1* sequence variants in families with FSHD-like muscular dystrophy. Eur J Hum Genet. 2015;23:67–71.
11. Larsen M, Rost S, El Hajj N, Ferbert A, Deschauer M, Walter MC, et al. Diagnostic approach for FSHD revisited: SMCHD1 mutations cause FSHD2 and act as modifiers of disease severity in FSHD1. Eur J Hum Genet. 2015;23:808–16.
12. Dekker J. Gene regulation in the third dimension. Science. 2008;319:1793–4.
13. Bhatia S, Kleinjan DA. Disruption of long-range gene regulation in human genetic disease: a kaleidoscope of general principles, diverse mechanisms and unique phenotypic consequences. Hum Genet. 2014;133:815–45.
14. Lieberman-Aiden E, van Berkum NL, Williams L, Imakaev M, Ragoczy T, Telling A, et al. Comprehensive mapping of long-range interactions reveals folding principles of the human genome. Science. 2009;326:289–93.
15. Rao SS, Huntley MH, Durand NC, Stamenova EK, Bochkov ID, Robinson JT, et al. A 3D map of the human genome at kilobase resolution reveals principles of chromatin looping. Cell. 2014;59:1665–80.
16. Mifsud B, Tavares-Cadete F, Young AN, Sugar R, Schoenfelder S, Ferreira L, et al. Mapping long-range promoter contacts in human cells with high-resolution capture Hi-C. Nat Genet. 2015;47:598–606.
17. Thurman RE, Rynes E, Humbert R, Vierstra J, Maurano MT, Haugen E, et al. The accessible chromatin landscape of the human genome. Nature. 2012;489:75–82.
18. Creyghton MP, Cheng AW, Welstead GG, Kooistra T, Carey BW, Steine EJ, et al. Histone H3K27ac separates active from poised enhancers and predicts developmental state. Proc Natl Acad Sci U S A. 2010;107:21931–6.
19. Harmston N, Lenhard B. Chromatin and epigenetic features of long-range gene regulation. Nucleic Acids Res. 2013;41:7185–99.
20. Pollard KS, Hubisz MJ, Siepel A. Detection of non-neutral substitution rates on mammalian phylogenies. Genome Res. 2010;20:110–21.
21. Hartweck LM, Anderson LJ, Lemmers RJ, Dandapat A, Toso EA, Dalton JC, et al. A focal domain of extreme demethylation within D4Z4 in FSHD2. Neurology. 2013;80:392–9.
22. Jones TI, Yan C, Sapp PC, McKenna-Yasek D, Kang PB, Quinn C, et al. Identifying diagnostic DNA methylation profiles for facioscapulohumeral muscular dystrophy in blood and saliva using bisulfite sequencing. Clin Epigenetics. 2014;6:23.
23. Huichalaf C, Micheloni S, Ferri G, Caccia R, Gabellini D. DNA methylation analysis of the macrosatellite repeat associated with FSHD muscular dystrophy at single nucleotide level. PLoS One. 2014;9:e115278.
24. Daxinger L, Tapscott SJ, van der Maarel SM. Genetic and epigenetic contributors to FSHD. Curr Opin Genet Dev. 2015;33:56–61.
25. Hamby SE, Reviriego P, Cooper DN, Upadhyaya M, Chuzhanova N. Screening in silico predicted remotely acting *NF1* gene regulatory elements for mutations in patients with neurofibromatosis type 1. Hum Genomics. 2013;7:18.
26. Dunham I, Kundaje A, Aldred SF, Collins PJ, Davis CA, Doyle F, et al. ENCODE Project Consortium. An integrated encyclopedia of DNA elements in the human genome. Nature. 2012;489:57–74.
27. Andersson R, Gebhard C, Miguel-Escalada I, Hoof I, Bornholdt J, Boyd M, et al. An atlas of active enhancers across human cell types and tissues. Nature. 2014;507:455–61.

17

On the stability of the Bayenv method in assessing human SNP-environment associations

Lily M Blair[1*], Julie M Granka[1,2] and Marcus W Feldman[1]

Abstract

Background: Phenotypic variation along environmental gradients has been documented among and within many species, and in some cases, genetic variation has been shown to be associated with these gradients. Bayenv is a relatively new method developed to detect patterns of polymorphisms associated with environmental gradients. Using a Bayesian Markov Chain Monte Carlo (MCMC) approach, Bayenv evaluates whether a linear model relating population allele frequencies to environmental variables is more probable than a null model based on observed frequencies of neutral markers. Although this method has been used to detect environmental adaptation in a number of species, including humans, plants, fish, and mosquitoes, stability between independent runs of this MCMC algorithm has not been characterized. In this paper, we explore the variability of results between runs and the factors contributing to it.

Results: Independent runs of the Bayenv program were carried out using genome-wide single-nucleotide polymorphism (SNP) data from samples from 60 worldwide human populations following previous applications of the Bayenv method. To assess factors contributing to the method's stability, we used varying numbers of MCMC iterations and also analyzed a second modified data set that excluded two Siberian populations with extreme climate variables. Between any two runs, correlations between Bayes factors and the overlap of SNPs in the empirical *p* value tails were surprisingly low. Enrichments of genic versus non-genic SNPs in the empirical tails were more robust than the empirical *p* values; however, the significance of the enrichments for some environmental variables still varied among runs, contradicting previously published conclusions. Runs with a greater number of MCMC iterations slightly reduced run-to-run variability, and excluding the Siberian populations did not have a large effect on the stability of the runs.

Conclusions: Because of high run-to-run variability, we advise against making conclusions about genome-wide patterns of adaptation based on only one run of the Bayenv algorithm and recommend caution in interpreting previous studies that have used only one run. Moving forward, we suggest carrying out multiple independent runs of Bayenv and averaging Bayes factors between runs to produce more stable and reliable results. With these modifications, future discoveries of environmental adaptation within species using the Bayenv method will be more accurate, interpretable, and easily compared between studies.

Keywords: Environmental adaptation, Positive selection, Genome-wide scans, Human adaptation, Markov chain monte carlo, Natural selection

* Correspondence: lmblair@stanford.edu

[1]Department of Biology, Stanford University, Stanford, CA 94305, USA

Full list of author information is available at the end of the article

Background

Phenotypic variation along environmental gradients, such as limb length in animals increasing with colder climates [1], animal body size increasing in colder environments [2], and bird coloration becoming darker in more humid climates [3], was first documented in the 1800s. Since then, many associations between phenotypic variation and environmental variation, both among and between species, have been noted [4-20]. To define these patterns, Huxley [21] introduced the term 'cline', which refers to a continuous phenotypic or genetic gradient that is associated with a pattern of geographic and/or environmental heterogeneity.

Various methods have been developed to determine the contribution of environmental variation to allelic or genotypic frequency patterns. Early theory in population genetics was used to derive statistics that allowed variation in allele frequencies to distinguish selection from drift [22-24]. Additionally, it has been shown that correlations between features of the environment and allele frequencies can be detected [15,25-27]. Recently, the availability of genome-wide data on DNA polymorphisms has spurred interest in using haplotype frequencies or allele frequency spectra to infer the existence or extent of selection across genomes [28-37].

Coop et al. [38] developed a statistical procedure named Bayenv to detect genomic evidence of adaptation to specific environmental variation, and it is this method we address here. The method carries out a locus-by-locus Bayesian analysis to determine whether a pattern in which population allele frequencies vary linearly with environmental variables is more probable than a pattern in which allele frequencies are assumed to be neutral and unrelated to environmental variables.

Given a set of populations with associated environmental variables and genome-wide single-nucleotide polymorphism (SNP) allele frequencies, a null model is first estimated to describe how allele frequencies covary across populations. Sample allele frequencies are drawn from a set of underlying population frequencies, which are assumed to be distributed according to a multivariate normal distribution around a transformed global allele frequency with a variance-covariance matrix representing the genetic structure of the populations. These transformed frequencies are not constrained between 0 and 1. The prior on the variance-covariance matrix is an inverse Wishart distribution, and Markov chain Monte Carlo (MCMC) computation is used to explore the posterior distribution of the covariance matrix, given the ancestral allele frequency and population allele frequencies.

After the covariance of allele frequencies between populations is estimated, the posterior probability of an alternative model for an individual SNP is determined. The alternative model predicts that the transformed population allele frequencies at a SNP are normally distributed around the ancestral allele frequency according to a linear model:

$$P(\theta|\Omega, \alpha, \beta) \sim N(\alpha + \beta Y, \alpha(1-\alpha)\Omega),$$

where θ is the transformed allele frequency, α is the ancestral allele frequency, Y is the environmental variable, and Ω is the variance-covariance matrix that was estimated in the first step of this procedure.

For each SNP, the Bayes factor (BF) is the posterior probability of the data at that SNP under the alternative model, integrated over all parameters, then divided by the posterior probability of the data under the null model, integrated over all parameters of the null model. Using a single run of the MCMC under the null distribution and importance sampling, the Bayes factors for multiple environmental variables are estimated quickly for each of the SNPs assayed (see additional details in Coop et al. [38]). This method has been shown to have good power compared to a range of other approaches [39].

The Bayenv method has been applied to many species including humans [40-44], plants [45-49], fish [50,51], and mosquitoes [52]. In all cases, the objective was to determine the extent to which a species' pattern of genomic variation can be inferred to be a response to environmental variation - for example climate, geography, photoperiod, or soil type - and hence could be regarded as a signal of adaptation.

Hancock et al. [40] applied the Bayenv method to genome-wide data from 61 worldwide populations for evidence of adaptation to 11 local climatic variables. Analyzing the pattern of variation of genome-wide SNP allele frequencies across the populations, they found evidence for adaptation to several variables, the most significant being latitude, summer relative humidity, summer solar radiation, and winter relative humidity. They reported signs of adaptation at SNPs previously identified by genome-wide association studies (GWAS) to be significantly related to phenotypes associated with environmental gradients, as well as an enrichment of genic versus non-genic SNPs among the SNPs most strongly correlated with the climate variables.

Although the Bayenv method has been applied to different species and environmental variables and has produced some results agreeing with other analytical methods, the stability of the results across multiple MCMC runs is not known. It is also unknown how the results may be affected by the inclusion of specific populations from extreme environments, which could potentially introduce bias [53]. In this study, we carried out many independent runs of the Bayenv program using data from Hancock et al. [40] to determine the degree of variability between runs and the factors that may contribute to any variability.

During the writing of this manuscript, an updated version of this program, Bayenv 2.0 [53], was released. While the Bayes factor estimation procedure is the same, it adds features that allow for the input of pooled population sequencing data and the computation of non-parametric correlations of the allele frequencies with the environmental variables. Although using this new version is advisable, the analyses of the original Bayenv program presented here remain relevant both to further applications of the method as well as to studies using results of previous Bayenv analyses (i.e., Fraser [54]).

Results

All 60 populations; 100,000 MCMC iterations

First, we analyzed the full data set of 60 populations and assessed correlations with 11 climate variables using 100,000 MCMC iterations (as in Hancock et al. [40]). We compared results between pairs of five independent runs as well as between each of our runs and Hancock et al.'s [40] published results. We report comparisons of our runs with an updated set of results made available online by the authors (which we refer to as Hancock2) that used a different number of MCMC burn-ins from the original set of results made available (Hancock1) (see 'Methods' section for details). The 16 pairs analyzed here are shown by a's in Table 1.

We found that Bayenv produced slightly different results in independent runs of the program. For a given climate variable, pairwise correlations of the logarithms of Bayes factors (BF's) were computed for all SNPs from each run, and correlations were averaged across all pairs of runs mentioned above (shown in Table 1). The averaged correlation coefficients ranged from 0.66 to 0.89 (Table 2a), depending on the climate variable. Despite analyzing only 60 of the 61 populations analyzed in Hancock et al. [40], correlations among our runs were similar to correlations with results from Hancock et al. [40]. Correlations between the two versions of Hancock et al.'s [40] published results were also comparable.

Correlations of empirical p values for all SNPs between the pairs of runs listed in Table 1 were also calculated; these correlation coefficients ranged from 0.63 to 0.85 (Table 2e), depending on the climate variable. Again, all pairs of runs compared displayed similar correlation values. For both the BF and p value correlations, the longitude variable had the lowest correlation among runs while winter radiation flux had the highest.

To examine the consistency of SNPs present in the empirical tails of each run, the overlap of SNPs in the tails with p value cutoffs of 0.05, 0.01, 0.005, and 0.001 was calculated. Overlap was defined as the proportion of overlapping SNPs relative to the total number of SNPs in the tail; this proportion was averaged over all climate variables. All pairs of runs showed comparable proportions of overlaps, both when compared among the Blair runs and between Blair runs and Hancock2. These overlap proportions ranged from 0.24 in the 0.001 tails of both runs to 0.57 in the 0.05 tails of both runs when Blair1 was compared to Hancock2 (Table 3a), with similar values for overlap between all Blair runs. In general, greater SNP overlap occurred when the SNPs in one run's tail were compared to the less extreme tail in another run. (In other words, more SNPs in the 0.001 tail of run A overlapped with the SNPs in the 0.005 tail of run B than the 0.001 tail of run B, and even more overlapped with the SNPs in the 0.01 tail of run B).

To assess whether SNPs with more extreme p values had greater overlap among runs, we ordered the empirical p values from the lowest to the highest for each climate variable and binned the SNPs in groups of 1,000. Each bin was compared to the corresponding ordered bin in each of the other Blair runs. Bins of lower p values had the highest overlap and overlap decreased as the empirical p values increased. The mean overlap for the top 1,000 SNPs was 0.44, and overlap was much lower for less significant SNPs (Table 4).

There was also variation among the results at individual SNPs. For example, SNPs in the CORIN gene region (rs4558846, rs6447571, rs17601068) gave strong signals ($\log_{10}$BF = 21.9, 28.7, and 20.8 and empirical p values = 2.0×10^{-5}, 3.1×10^{-5}, and 2.1×10^{-5}) of association with minimum winter temperature as reported in Hancock et al. [40] (which we refer to as Hancock1), but show low signals in other runs: Blair1 ($\log_{10}$BF = −0.32, −0.50, and −0.48 and empirical p values = 0.40, 0.57, and 0.44) and Hancock2 ($\log_{10}$BF = −0.33, −0.07, and −0.16 and empirical p values = 0.36, 0.22, and 0.20). Some of this variability could be due to the limited burn-in in Hancock1 (see 'Methods' section). Other SNPs, such as rs2075756 in TRIP6, show more consistency. Blair1 obtained a $\log_{10}$BF of 4.0 and empirical p value of 7.9×10^{-4}, while Hancock2 reported a $\log_{10}$BF of 1.4 and empirical p value of 3.2×10^{-3} for this SNP. This SNP shows a strong signal with absolute latitude, the variable for which we saw consistently strong signals of genic/non-genic enrichment among SNPs with low empirical p values (see below).

We then assessed whether genic SNPs were enriched among SNPs with the strongest signals by comparing the fraction of genic SNPs to non-genic SNPs in the empirical tails. This enrichment was compared across all five of our runs as well as with Hancock1 and Hancock2. The absolute enrichment values (calculated as described in the 'Methods' section) varied across runs and differed from Hancock's [40] published results (results not shown). The significance values of the genic/non-genic enrichments, which represent the proportion of bootstrapped samples

Table 1 Pairs of Bayenv runs compared

	Hancock1	Hancock2	Blair1	Blair2	Blair3	Blair4	Blair5	LongBlair1	LongBlair2	LongBlair3	LongBlair4	LongBlair5	W/O_ Sib1	W/O_ Sib2	W/O_ Sib3	W/O_ Sib4	W/O_ Sib5
Hancock1																	
Hancock2	a																
Blair1		a															
Blair2		a	a														
Blair3		a	a	a													
Blair4		a	a	a	a												
Blair5		a	a	a	a	a											
LongBlair1																	
LongBlair2								b									
LongBlair3								b	b								
LongBlair4								b	b	b							
LongBlair5								b	b	b	b						
W/O_Sib1		d															
W/O_Sib2		d											c				
W/O_Sib3		d											c	c			
W/O_Sib4		d											c	c	c		
W/O_Sib5		d											c	c	c	c	

Letters indicate the parts of Tables 2 and 3 in which results for the indicated comparisons are averaged and presented. In Table 2, letters a, b, c, and d also correspond to parts e, f, g, and h, respectively. Hancock1 refers to the original Climate file posted on the dbCline website and published in Hancock et al. [40]; Hancock2 refers to file Climate.2 posted on dbCline website in autumn 2012 that uses the appropriate MCMC burn-in (see 'Methods' section); Blair1-5 refer to our five separate runs using 100,000 MCMC iterations; LongBlair1-5 refer to our five separate runs using 500,000 MCMC iterations; W/O_Sib1-5 refer to our five runs using a modified data set excluding the Siberian populations.

Table 2 Average correlations of the log of the Bayes factors (BF) and average correlations of the empirical *p* values between runs for each climate variable

	Latitude	Absolute value latitude	Longitude	Min temp win	Max temp sum	Prec. rate sum	Prec. rate win	Rad flux sum	Rad flux win	Rel humid sum	Rel humid Win
Log of BF											
(a) All populations (100,000 iterations) incl Hancock	0.87	0.88	0.66	0.83	0.84	0.73	0.75	0.74	0.89	0.73	0.79
(b) All populations (500,000 iterations)	0.93	0.92	0.82	0.88	0.90	0.85	0.85	0.83	0.93	0.85	0.88
(c) Without Siberia (150,000 iterations)	0.87	0.87	0.74	0.82	0.86	0.79	0.81	0.78	0.87	0.80	0.85
(d) Without Siberia (150,000 iterations) vs. Hancock2	0.79	0.82	0.70	0.81	0.81	0.70	0.72	0.75	0.84	0.78	0.76
Empirical *p* values											
(e) All populations (100,000 iterations) incl Hancock	0.81	0.84	0.63	0.79	0.81	0.68	0.70	0.70	0.85	0.70	0.75
(f) All populations (500,000 iterations)	0.91	0.91	0.81	0.88	0.90	0.82	0.83	0.82	0.92	0.84	0.86
(g) Without Siberia (150,000 iterations)	0.82	0.81	0.70	0.80	0.82	0.73	0.75	0.74	0.82	0.76	0.80
(h) Without Siberia (150,000 iterations) vs. Hancock2	0.61	0.65	0.67	0.74	0.66	0.62	0.68	0.70	0.67	0.72	0.75

'Win' refers to winter, 'sum' refers to summer, 'prec. rate' refers to precipitation rate, 'rad. flux' refers to radiation flux, and 'rel. humid' refers to relative humidity. Runs compared in each part are shown in Table 1. Incl Hancock means that the average is taken over pairs of Blair1-5 (100,000 iterations) and Hancock1-2, shown by a's in Table 1. Hancock1-2 were run with 150,000 iterations.

Table 3 Empirical *p* value tail comparisons between indicated runs

		.05	.01	.005	.001
(a) All populations (100,000 iterations) - Hancock2 vs. Blair1	**.05**	0.55	0.79	0.83	0.88
	.01	0.75	0.40	0.51	0.66
	.005	0.78	0.49	0.34	0.54
	.001	0.83	0.62	0.51	0.24
(b) All populations (500,000 iterations) - LongBlair1 vs. LongBlair3	**.05**	0.64	0.85	0.89	0.93
	.01	0.88	0.51	0.62	0.78
	.005	0.91	0.65	0.46	0.67
	.001	0.94	0.80	0.70	0.37
(c) Without Siberia (150,000 iterations) - W/O_Sib1 vs. W/O_Sib3	**.05**	0.60	0.83	0.86	0.92
	.01	0.81	0.45	0.56	0.73
	.005	0.85	0.56	0.40	0.61
	.001	0.90	0.71	0.59	0.30
(d) Without Siberia (150,000 iterations) vs. Hancock2 - Hancock2 vs. W/O_Sib1	**.05**	0.51	0.72	0.75	0.80
	.01	0.75	0.36	0.44	0.56
	.005	0.81	0.46	0.30	0.44
	.001	0.87	0.62	0.48	0.20

The table reports the fraction of SNPs from the smaller tail of one run that is present in the larger tail of the second run, averaged over climate variables. Two identical runs produce values of 1. Within each part (a to d), all pairs of runs (see Table 1) have similar fractions of overlap, and representative tables are shown here. For a given part, the first run's empirical tails are shown as rows (e.g., Hancock2 in part a), and the second run's are shown as columns (e.g., Blair1 in part a). Also, 623,318 SNPs are included in this analysis.

that are also enriched, also varied across runs (i.e., for summer solar radiation) (Table 5).

The absolute latitude and winter relative humidity variables were almost always significant, with the exception of the 0.005 tail in Blair5; the summer precipitation variable was never significant. All three of these findings agreed with Hancock et al.'s published results [40]. However, summer solar radiation and humidity were not consistently significant, though reported to be significant in Hancock et al. [40]. Among the remaining variables, results varied among runs, and our runs showed less significance on average than Hancock et al.'s published results [40].

Table 4 Overlap of ranked SNPs (ranked by empirical *p* value from lowest to highest and binned in groups of 1,000) between runs

Range of ranked SNPs	Overlap proportion
1–1,000	0.44 ± 0.020
100,001–101,000	0.0073 ± 0.0033
200,001–201,000	0.0051 ± 0.0023
300,001–301,000	0.0043 ± 0.0016
400,001–401,000	0.0042 ± 0.0014
500,001–501,000	0.0035 ± 0.0014
600,001–601,000	0.014 ± 0.0098

For each bin, we calculated the proportion of overlapping SNPs between each corresponding bin for all ten pairs of runs (Blair1-5). Table reports the mean and standard deviation of the overlap proportion across all pairs of runs. Only 1 bin per 100,000 SNPs is shown. Also, 623,318 total SNPs were used in the analysis.

All 60 populations; 500,000 MCMC iterations

In order to determine whether the observed variability between runs was due to the MCMC not converging, the number of MCMC iterations was increased fivefold to 500,000. Five independent runs were performed, each with different seeds, but with the same full data set of 60 populations. All pairs of the five runs with 500,000 iterations were compared (shown as b's in Table 1).

Between independent runs, both the Bayes factors and *p* values showed higher correlation coefficients compared to those between independent runs with 100,000 iterations. When averaged across all pairs of runs, the Bayes factor correlation coefficients ranged from 0.82 to 0.95 (Table 2b), and the correlation coefficients of the *p* values ranged from 0.81 to 0.92 (Table 2f). These correlations are significantly higher than among runs of 100,000 iterations for all environmental variables: paired Wilcoxon tests of these correlations, which for each climate variable are averaged across pairs of runs, give a *p* value of 0.0038 for the Bayes factors and a *p* value of 0.0038 for the empirical *p* values. The SNPs in the empirical *p* value tails were compared as before, and although the overlap was greater between long runs than between runs with 100,000 iterations, the proportion of overlapping SNPs in the tails was still not high, ranging from 0.64 in the 0.05 tails of two runs to 0.36 in the 0.001 tails of two runs (Table 3b).

Genic/non-genic enrichment significance values also stabilized somewhat in the runs with 500,000 iterations but still did not show the same significance values from run to run (Table 5). Genic/non-genic enrichments were highly significant for the absolute latitude and relative humidity winter variables and again were not significant for the summer precipitation variable across all five runs. Interestingly, none of the five runs showed significant genic/non-genic enrichment for winter precipitation rate, except in the 0.01 tail in Blair5. As before, varying levels of significance were seen for summer solar radiation and summer relative humidity. This differs from Hancock et al. [40], who reported high significance across all tails for these environmental variables.

In searching for the source of the differences described above, biases from allele frequency, SNP ascertainment panel, and method of calculating empirical *p* values were investigated, but none of these potential biases was shown to explain the variability among runs. First, the data were re-analyzed using empirical *p* values that were not binned by allele frequency. *P* value overlap did not

Table 5 Genic/non-genic enrichments for each climate variable in the indicated empirical tails

Climate variable	Data set	Empirical tail		
		0.05	0.01	0.005
Absolute latitude	Hancock1 (published results)	3	3	3
	Hancock1 (our analysis)	1	2	2
	Hancock2	3	3	3
	Blair1	0	2	3
	Blair2	0	3	3
	Blair3	1	1	2
	Blair4	1	3	3
	Blair5	1	1	3
	LongBlair1	3	3	3
	LongBlair2	2	2	1
	LongBlair3	1	2	2
	LongBlair4	3	2	1
	LongBlair5	2	2	3
	W/O_Sib1	1	0	0
	W/O_Sib2	0	1	2
	W/O_Sib3	1	0	1
	W/O_Sib4	0	0	0
	W/O_Sib5	0	0	0
Summer maximum temperature	Hancock1 (published results)	0	0	2
	Hancock1 (our analysis)	0	2	3
	Hancock2	2	3	3
	Blair1	0	0	0
	Blair2	0	0	0
	Blair3	0	1	0
	Blair4	0	0	0
	Blair5	1	0	0
	LongBlair1	0	0	1
	LongBlair2	0	1	0
	LongBlair3	0	0	1
	LongBlair4	0	2	2
	LongBlair5	0	0	1
	W/O_Sib1	3	2	2
	W/O_Sib2	0	0	0
	W/O_Sib3	0	0	0
	W/O_Sib4	0	2	3
	W/O_Sib5	3	0	0
Summer precipitation rate	Hancock1 (published results)	0	0	0
	Hancock1 (our analysis)	0	0	0
	Hancock2	0	0	0
	Blair1	0	0	0
	Blair2	0	0	0
	Blair3	0	0	0
	Blair4	0	0	0
	Blair5	0	0	0
	LongBlair1	0	0	0
	LongBlair2	0	0	0
	LongBlair3	0	0	0
	LongBlair4	0	0	0
	LongBlair5	0	0	0
	W/O_Sib1	0	0	0
	W/O_Sib2	0	0	0
	W/O_Sib3	0	0	0
	W/O_Sib4	0	0	0
	W/O_Sib5	0	0	0
Summer solar radiation	Hancock1 (published results)	3	3	3
	Hancock1 (our analysis)	3	3	3
	Hancock2	0	1	3
	Blair1	2	2	0
	Blair2	0	0	0
	Blair3	1	0	0
	Blair4	0	0	0
	Blair5	3	0	0
	LongBlair1	2	3	3
	LongBlair2	0	3	3
	LongBlair3	2	0	0
	LongBlair4	0	2	1
	LongBlair5	0	3	3
	W/O_Sib1	3	3	3
	W/O_Sib2	1	0	0
	W/O_Sib3	0	0	0
	W/O_Sib4	1	0	0
	W/O_Sib5	3	3	2
Summer relative humidity	Hancock1 (published results)	2	3	3
	Hancock1 (our analysis)	2	3	3
	Hancock2	0	3	3
	Blair1	0	2	1
	Blair2	2	0	0
	Blair3	0	0	0
	Blair4	0	2	0
	Blair5	1	0	0
	LongBlair1	0	0	0
	LongBlair2	1	2	3
	LongBlair3	0	2	1

Table 5 Genic/non-genic enrichments for each climate variable in the indicated empirical tails *(Continued)*

	LongBlair4	0	0	1
	LongBlair5	1	1	1
	W/O_Sib1	0	3	3
	W/O_Sib2	2	3	2
	W/O_Sib3	0	0	0
	W/O_Sib4	1	2	1
	W/O_Sib5	0	0	1
Winter minimum temperature	Hancock1 (published results)	1	0	0
	Hancock1 (our analysis)	1	0	0
	Hancock2	0	3	3
	Blair1	0	1	2
	Blair2	0	2	2
	Blair3	0	1	2
	Blair4	0	0	1
	Blair5	0	1	1
	LongBlair1	0	3	2
	LongBlair2	0	2	2
	LongBlair3	0	3	2
	LongBlair4	0	2	2
	LongBlair5	0	3	3
	W/O_Sib1	0	0	0
	W/O_Sib2	0	2	0
	W/O_Sib3	1	0	2
	W/O_Sib4	0	1	3
	W/O_Sib5	0	1	2
Winter precipitation rate	Hancock1 (published results)	0	3	3
	Hancock1 (our analysis)	1	3	3
	Hancock2	2	0	0
	Blair1	2	0	0
	Blair2	0	1	0
	Blair3	2	1	0
	Blair4	2	3	3
	Blair5	0	0	0
	LongBlair1	0	0	0
	LongBlair2	0	0	0
	LongBlair3	0	0	0
	LongBlair4	0	0	0
	LongBlair5	0	2	0
	W/O_Sib1	0	0	2
	W/O_Sib2	0	2	3
	W/O_Sib3	0	0	0
	W/O_Sib4	0	3	2
	W/O_Sib5	3	1	2

Table 5 Genic/non-genic enrichments for each climate variable in the indicated empirical tails *(Continued)*

Winter relative humidity	Hancock1 (published results)	3	3	3
	Hancock1 (our analysis)	3	2	2
	Hancock2	3	3	3
	Blair1	3	3	3
	Blair2	3	3	2
	Blair3	3	1	3
	Blair4	3	3	3
	Blair5	3	1	0
	LongBlair1	3	3	0
	LongBlair2	3	3	2
	LongBlair3	3	3	2
	LongBlair4	2	0	0
	LongBlair5	3	2	1
	W/O_Sib1	3	2	2
	W/O_Sib2	3	2	3
	W/O_Sib3	3	3	2
	W/O_Sib4	3	3	3
	W/O_Sib5	3	3	0
Winter solar radiation	Hancock1 (published results)	3	1	1
	Hancock1 (our analysis)	3	0	2
	Hancock2	3	2	0
	Blair1	3	1	0
	Blair2	3	2	1
	Blair3	3	0	0
	Blair4	3	0	0
	Blair5	2	1	0
	LongBlair1	3	0	0
	LongBlair2	3	2	0
	LongBlair3	3	2	1
	LongBlair4	3	3	2
	LongBlair5	3	1	0
	W/O_Sib1	3	0	0
	W/O_Sib2	3	1	2
	W/O_Sib3	3	0	0
	W/O_Sib4	3	3	2
	W/O_Sib5	2	0	1

The table shows the number of stars of significance for genic/non-genic enrichments in the 0.05, 0.01, and 0.005 tails of the empirical *p* value distribution for each indicated run (defined as in Table 1 and in the 'Methods' section). '1' denotes 95%–97.5% of bootstraps had an enrichment of genic compared to non-genic SNPs in that empirical tail, '2' denotes 97.5%–99%, and '3' denotes >99%. '0' denotes that a genic enrichment is not significant (i.e., that <95% of the bootstraps had an enrichment). Enrichments for the nine environmental variables that were shown in Hancock et al. [40] are shown here.

increase when binned vs. un-binned p values were compared. Second, the proportions of SNPs from each ascertainment panel in each tail were calculated, and these proportions were found to be similar to the overall proportions of SNPs in each ascertainment panel. Additionally, when each ascertainment panel was considered separately, the overlap in the tails of p values was similar to the overlap in the tails when all three ascertainment panels were considered together.

Without Siberian populations; 150,000 iterations

Because the Siberian populations, the Naukan Yup'ik and Maritime Chukchee, live in some of the most extreme climatic regions among the 60 sampled populations, it may be possible that removing these populations from the data set could affect the variability of Bayes factors, empirical p values, and genic/non-genic enrichments among independent runs. All pairs of runs using this modified data set were compared (shown as c's in Table 1). Pairwise correlations of the log of the Bayes factors (Table 2c) and p values (Table 2g) as well as fractions of overlap of SNPs in the 0.05, 0.01, 0.005, and 0.001 tails of the empirical p values (Table 3c) were comparable to the correlations between the runs of the full data set (60 populations with 100,000 iterations (Table 2a)). This suggests that the inclusion of populations with extreme environmental variables may not affect run-to-run variability.

To determine whether excluding the Siberian populations affected the results of the Bayenv method, each of the runs without the Siberian populations was compared to the updated Hancock results published on the dbCline website (Hancock2) (shown by d's in Table 1). Pairwise correlations between the log of the Bayes factors of these pairs (Table 2d), averaged across all pairs, were similar to correlations between pairs of non-Siberian runs (Table 2c). However, correlations of empirical p values between Hancock2 and non-Siberian runs (Table 2h) were lower than correlations between pairs of non-Siberian runs (Table 2g) and also lower than the correlations between pairs of runs using the full data set (Table 2e). Similarly, overlap in the empirical tails is lower when non-Siberian runs are compared to the results from Hancock2 (Table 3d), demonstrating a slight difference in results when these populations were excluded.

Although genic/non-genic enrichment significance values varied across runs of the non-Siberian data set (Table 5), some patterns differed from the results that included the full data set of 60 populations. The genic/non-genic enrichment for the absolute latitude variable was less significant than in runs analyzing all populations (Table 5), showing a small effect of exclusion of the Siberian populations. In addition, winter precipitation rate was sometimes significant in the non-Siberian runs; this variable was significant in Hancock et al.'s [40] published data but not in our runs with the full population set using 500,000 MCMC iterations.

Discussion

We have explored the computational replicability of the Bayenv program [38] in determining evidence for positive selection in response to climatic variation. We used genome-wide SNP data from 60 worldwide human populations and 11 climate variables as in Hancock et al. [40] and carried out independent Bayenv runs with 100,000 and 500,000 MCMC iterations, as well as with 150,000 iterations while omitting Siberian populations with extreme climatic variables. In comparing pairs of our runs as well as each of our runs with the results previously published in Hancock et al. [40], results from the Bayenv method appear to be variable across runs in their Bayes factors, empirical p values, and in the values and significances of enrichments of genic SNPs in the empirical tails. Although some run-to-run variability is expected in any MCMC algorithm, we show here that the differences between runs are sufficient to lead to varying biological conclusions regarding human adaptation.

In runs with 100,000 MCMC iterations, correlations of Bayes factors between replicate runs were low (0.66–0.88) between pairs of our runs, between our runs and the results of Hancock et al. [40], and between the different versions of results posted by Hancock et al. (Table 2a). To explore the possibility that the MCMC had failed to converge, we increased the number of MCMC iterations to 500,000; although these correlations increased as a result (0.82–0.93) (Table 2b), there was still substantial run-to-run variability in further analyses.

Because Bayes factors are dependent on the accuracy of the null model and, therefore, may be substantially inflated due to imperfections of this model [40], we focused further analyses on the empirical p values. Hancock et al. [40] write, 'Since we cannot expect the null model to account fully for the effects of population structure, we emphasize that we cannot take the BFs themselves at face value, nor can they be directly compared across climate variables. Therefore, we took a conservative approach and conducted subsequent analyses by comparing each SNP to the empirical distribution.' In our analysis, empirical p values were seen to be no more stable than the Bayes factors themselves, as correlation coefficients of the empirical p values were in the same range as the correlation coefficients of the Bayes factors (Table 2e). As before, the correlation of p values between runs improved when the MCMC iterations were increased to 500,000 (Table 2f). Although we recommend using a large number of MCMC iterations, this does not completely stabilize Bayenv results from run to run.

Amounts of SNP overlap in the tails of the empirical *p* value distributions showed that the relative ranking of the SNPs associated with environmental variables varied across runs, even with a larger number of MCMC iterations. While the overlap of SNPs between the extreme tails (i.e., top 0.01%) of two runs was found to be low, we recognize that this slightly overestimates run-to-run variability. For example, 24% of the 0.001 tail of Hancock2 overlapped with the 0.001 tail of Blair1; that proportion increased to 83% with the 0.005 tail of Blair1 (Table 3a). This suggests that the SNPs with the lowest empirical *p* values have generally low *p* values in all runs. Analysis of the overlap of SNPs with low empirical *p* values versus the overlap of SNPs with high or midrange empirical *p* values shows that strong signals are also more reproducible from run-to-run than weaker signals.

Despite this trend, a SNP's significance and relative ranking nearly always vary from one run of Bayenv to another. SNPs inferred to be significant using some pre-defined *p*-value threshold in one run may not be found to be significant in another run. Thus, the low fractions of empirical *p*-value tail overlap between runs suggest that caution is advisable in drawing inferences based on only one run, especially when making conclusions regarding individual SNPs or genes (see 'Results' section).

Finally, we assessed the stability of inferences regarding the enrichment of genic SNPs among SNPs strongly associated with a given climate variable. Evidence of such an enrichment has been argued as evidence of strong selection on genic regions in response to an associated environmental pressure [40,45]. Because the stability of these significances depends only on the ratio of genic to non-genic SNPs in the empirical tails, as opposed to the rank of each individual SNP, this measure could be consistent despite variability of Bayes factors and empirical *p* values. However, the significances of climate variables including summer maximum temperature, summer relative humidity, summer solar radiation, winter minimum temperature, winter precipitation rate, and winter solar radiation varied widely from run to run (Table 5).

Of particular interest are summer solar radiation and summer relative humidity, both claimed to be significant by Hancock et al. [40] but for which significance did not hold in our analyses. For these variables, as well as others, neither the enrichment values nor the significance of the enrichment values increased or decreased systematically when the tail cut-off was more extreme, as suggested in Hancock et al. [40]. The lack of significance for these variables in the Hancock2 analysis, which includes all 61 populations, shows that the exclusion of the data from Australian Aborigines from our analysis is not the cause for this difference in significance. Thus, replicate runs demonstrate that these two variables may have been less crucial in human adaptation than suggested previously [40].

Despite this variability between runs, certain climate variables show some consistency in genic/non-genic enrichments and agree with conclusions of Hancock et al. [40] (Table 5). Winter relative humidity is significant across all runs, including those in which the Siberian populations are excluded, suggesting that this may be a robust signal of adaptation in response to this climatic variable. Absolute latitude also shows a strong signal and is significant across all runs that include all populations. That we were able to replicate the significance of both of these variables, despite the fact that we did not include the Australian Aborigines (included in Hancock et al. [40]) underscores their potential importance in human adaptation.

We note that populations with extreme climate variables may indeed have an effect on some conclusions drawn from Bayenv analyses. For example, without the Siberian populations, an enrichment of genic SNPs was no longer significant for latitude (Table 5). Thus, while the inclusion of outlying populations may affect biological conclusions, it does not appear to affect run-to-run variability (see 'Results' section).

In summary, our analysis of the consistency of genic enrichments between runs demonstrates the importance of performing multiple runs of Bayenv. Two independent runs can potentially lead to very different biological conclusions. Significance of a particular environmental variable in multiple Bayenv runs constitutes much stronger evidence for its relevance and importance for human adaptation.

A new version of the Bayenv method, Bayenv 2.0 [53], has recently been released. Running the new version with 100,000 MCMC iterations produces slightly more stable results among runs than the old version; however, the variability is comparable to the runs of the old version with 500,000 MCMC iterations (results not shown). Gunther et al. note that their inclusion of non-parametric analyses in Bayenv 2.0 helps to account for potential false positives due to outlying populations [53]; however, our results suggest that results run on the previous version of the Bayenv program that do *not* include outlying populations may still lead to varying results across runs. Thus, while we suggest using the new version, the results described in this manuscript remain relevant to the application of Bayenv 2.0.

It can be difficult to disentangle stability and specificity, and we recognize that some of the run-to-run variability of this method may be due to less strong signals of environmental adaptation (i.e., winter minimum temperature as compared to latitude). However, as every run may be different, it is still advisable to carry out multiple runs when making biological conclusions, especially when the signal tested may not be strong to begin with. Averaging Bayes factors of two or more independent runs before calculating the empirical *p* values produces more stable empirical *p* values and genic/non-genic enrichments (results not shown). Additionally, individual SNPs that show large Bayes factors in at least one run could be rerun with more MCMC iterations to determine a Bayes factor value to which all iterations converge. Unfortunately, this process would not be as useful for a genome-wide analysis, as a large set of SNPs is necessary for the calculation of the empirical *p* values and the genic/non-genic enrichments.

Conclusions

We conclude that, in using the Bayenv method, relying on the relative rankings of SNPs calculated from only one run of the program may lead to biological conclusions that might not have been reached with another independent run of the program. Neither increasing the number of MCMC iterations nor removing outlying populations completely eliminated this variability. Thus, we suggest carrying out several independent runs of the Bayenv analysis to assess variability of the Bayes factors and empirical *p* values across runs. We also suggest averaging Bayes factors from independent runs to produce more stable results. With these modifications, future discoveries of environmental adaptation within species using the Bayenv method will be more accurate, interpretable, and easily compared between studies.

Methods

Populations

Allele frequency data on 623,318 SNPs from samples of 60 human populations from the freq_data file (61pops_climate_freqs.txt) on the dbCline website (http://genapps2.uchicago.edu:8081/dbcline/main.jsp) were analyzed. The data set included samples from 52 HGDP populations plus Vasekela !Kung, Amhara, Naukan Yup'ik, Maritime Chukchee, Luhya, Maasai, Tuscans, and Gujarati. These were the same populations used by Hancock et al. [40], with the exception of the Australian Aborigines, who were not included in the allele frequency file on dbCline. To translate allele frequencies into allele counts, the sample sizes from Li et al. [55] were used for the HGDP populations and from the supplementary information of Hancock et al. [40] for the additional eight populations.

Climate variables

The climate data consisted of the 11 variables in Hancock et al. [40] that are posted on the dbCline website: latitude, absolute latitude, longitude, winter minimum temperature, summer maximum temperature, summer precipitation rate, winter precipitation rate, summer radiation flux, winter radiation flux, summer relative humidity, and winter relative humidity. Climate data were obtained for each population directly from the dbCline website (61pops_climate_freqs.txt).

Covariance matrix estimation

For the set of 60 populations, three covariance matrices, each one representing a different ascertainment panel (Illumina 250 K, Illumina 300 K, or Illumina AFR), were generated using the Bayenv program (as in Hancock et al. [40]). The program was run for 100,000 MCMC iterations, and the entries of each matrix were averaged over the last three output matrices (output every 5,000 iterations). Visual comparison showed the matrices to be qualitatively similar to those in Coop et al. [38], and the matrices were stable from run to run. For our analysis, one covariance matrix for each ascertainment panel was estimated as described here and used for all runs.

Bayenv runs

Five independent runs of the Bayenv program were carried out using the full data set of 60 populations and 11 climate variables. As with the covariance matrices, SNPs were analyzed with their corresponding ascertainment panel. Each run used a different random seed but the same input data set and 100,000 MCMC iterations. Bayes factors produced from the three ascertainment panels were combined into one file for each run. These runs are referred to as Blair1 - Blair5 (see Table 1).

In a separate analysis, five independent runs of the Bayenv program were carried out as above, using the same full data set but increasing the MCMC iterations to 500,000 (referred to as LongBlair1 - LongBlair5) (see Table 1). For all runs, we use the default MCMC burn-in values.

Bayenv runs without Siberia

Five independent runs of the Bayenv program were carried out using a modified data set and 150,000 MCMC iterations. In this modified data set, all 11 climate variables were included, but the Siberian populations (Naukan Yup'ik and Maritime Chukchee) were excluded (referred to as W/O Siberia1 - W/O Siberia5; see Table 1). The aim here was to explore how the Siberian populations, who live in some of the most extreme climatic regions, might affect the stability and conclusions drawn from Bayes factors, empirical *p* values, and genic/non-genic enrichment values (see below).

Hancock runs

Bayes factors and empirical *p* values from Hancock et al. [40] were obtained from the dbCline website. Following the Hancock et al. [40] publication, the authors posted a set of results, which we refer to as Hancock1, which used fewer burn-ins for the MCMC procedure and thus failed to converge (Anna Di Rienzo, personal correspondence). After further correspondence with the authors, a second set of results that used a longer burn-in was posted in autumn 2013 (which we refer to as Hancock2). We examined BF correlations and empirical p value correlations and overlap between these two data sets (described below). For calculations of genic/non-genic enrichment (described below), we performed the analysis on both Hancock1 and Hancock2. In general, our analysis of the Hancock1 data set gave similar results to those published by Hancock et al. [40].

SNP p values

As in Hancock et al. [40], we calculated empirical p values by binning SNPs based on their global allele frequency (across 60 populations). For each SNP, a transformed rank statistic was computed by comparing the computed Bayes factor for the SNP with the Bayes factors from other SNPs in the same global allele frequency bin. For this process, all SNPs were analyzed together, and no distinction was made for SNPs belonging to different ascertainment panels.

Genic/non-genic enrichments

To calculate genic/non-genic enrichments, a SNP was defined as genic if it was within 10 kb of a gene and non-genic if it was greater than 50 kb from a gene (as in Hancock et al. [40], personal correspondence). Enrichment in an empirical tail was calculated with the following equation:

$$\text{Enrichment} = \frac{\frac{n_g}{n_{ng}}}{\frac{N_g}{N_{ng}}},$$

where n_g and n_{ng} are the number of genic and non-genic SNPs, respectively, in the empirical tail, and N_g and N_{ng} are the number of genic and non-genic SNPs, respectively, among all tested SNPs. To determine the significance of the genic/non-genic enrichments, 500-kb regions from the genome were re-sampled with replacement (also as in Hancock et al. [40]). Then, given the p values of the SNPs in the re-sampled regions, we calculated enrichment values in each empirical tail. This bootstrap re-sampling procedure was repeated 10,000 times. Significance values were determined by counting the fraction of bootstrapped samples with enrichment values above 1. In Table 5, one star denotes that 95%–97.5% of bootstrapped samples were enriched, two stars denote 97.5%–99%, and three stars denote >99%; zero stars denotes that <95% of bootstrap samples were enriched, and thus that the genic enrichment is not significant.

Comparison of runs

Each run produces a Bayes factor for each climate variable for each SNP. From those, the empirical p values were calculated. For each climate variable, the Pearson correlations of the log of the Bayes factors and the Pearson correlations of the empirical p values were computed for each pair of runs compared (see Table 1). For each set of comparisons (i.e., Blair1-5, or W/O_Sib1-5 vs. Hancock2) the correlation coefficients were then averaged across all relevant pairs of runs (see letters of Table 1). For all pairs shown in Table 1, empirical p values were also compared by calculating the fraction of SNPs in the tail of one run that was present in the tail of the other run. In another analysis, to assess differences in run stability by empirical p value, SNPs were ordered by empirical p value from lowest to highest and binned into groups of 1,000. Overlaps between the corresponding bins of different runs (Blair1-5) were calculated. (This assessed what proportion of the 1,000 most-significant SNPs in Blair1 were present in the 1,000 most-significant SNPs in Blair2, what proportion of SNPs ranked 1,001–2,000 in Blair1 were present in SNPs ranked 1,001–2,000 of Blair2, and so forth.) Finally, significances of genic/non-genic enrichment values were calculated for each run, as described above, and compared.

Abbreviations

BF: Bayes factor; MCMC: Markov Chain Monte Carlo; SNP: Single-nucleotide polymorphism.

Competing interests

The authors declare that they have no competing interests.

Authors' contributions

LB participated in the design of the study and performed the statistical analysis. JG and MF participated in its design and coordination. All authors read and approved the final manuscript.

Acknowledgements

We would like to thank Jonathan Pritchard, Graham Coop, Brenna Henn, and Anna Di Rienzo for helpful discussion and comments, as well as two anonymous reviewers for their input. This work was supported in part by the Center for Computational, Evolutionary and Human Genomics at Stanford University.

Author details

[1]Department of Biology, Stanford University, Stanford, CA 94305, USA.
[2]AncestryDNA, 153 Townsend St. Ste. 800, San Francisco, CA 94107, USA.

References

1. Allen JA: **The influence of physical conditions in the genesis of species.** *Radical Rev* 1877, **1**:108–140.
2. Bergmann C: **Über die Verhältnisse der wärmeökonomie der Thiere zu ihrer Grösse.** *Göttinger Studien* 1847, **3**:595–708.
3. Gloger CL: *Das Abändern der Vögel durch Einfluss des Klimas.* August Schulz: Breslau; 1833.
4. Ashton KG, Tracy MC, de Queiroz A: **Is Bergmann's rule valid for mammals?** *Am Nat* 2000, **156**:390–415.
5. Brown JH, Lee AK: **Bergmann's rule and climatic adaptation in woodrats (*Neotoma*).** *Evolution* 1969, **23**:329–338.
6. Coyne JA, Beecham E: **Heritability of two morphological characters within and among natural populations of *Drosophila melanogaster*.** *Genetics* 1987, **117**:727–737.
7. Freckleton RP, Harvey PH, Pagel M: **Bergmann's rule and body size in mammals.** *Am Nat* 2003, **161**:821–825.
8. Harcourt AH, Schreier BM: **Diversity, body mass, and latitudinal gradients in primates.** *Int J Primatol* 2009, **30**:283–300.

9. Huey RB, Gilchrist GW, Carlson ML, Berrigan D, Serra L: **Rapid evolution of a geographic cline in size in an introduced fly.** *Science* 2000, **287**:308–309.
10. Jablonski NG, Chaplin G: **The evolution of human skin coloration.** *J Hum Evol* 2000, **39**:57–106.
11. Jain S, Bradshaw A: **Evolutionary divergence among adjacent plant populations. I. The evidence and its theoretical analysis.** *Heredity* 1966, **21**:407–441.
12. Johnston RE, Selander RK: **Evolution in the house sparrow II. Adaptive differentiation in North American populations.** *Evolution* 1971, **25**:1–28.
13. Katzmarzyk PT, Leonard WR: **Climatic influences on human body size and proportions: ecological adaptations and secular trends.** *Am J Phys Anthropol* 1998, **106**:483–503.
14. Leonard WR, Sorensen MV, Galloway VA, Spencer GJ, Mosher MJ, Osipova L, Spitsyn VA: **Climatic influences on basal metabolic rates among circumpolar populations.** *Am J Hum Biol* 2002, **14**:609–620.
15. Mullen LM, Hoekstra HE: **Natural selection along an environmental gradient: a classic cline in mouse pigmentation.** *Evolution* 2008, **62**:1555–1570.
16. Pool JE, Aquadro CF: **The genetic basis of adaptive pigmentation variation in *Drosophila melanogaster*.** *Mol Ecol* 2007, **16**:2844–2851.
17. Storz JF, Balasingh J, Bhat HR, Nathan PT, Doss DPS, Prakash AA, Kunz TH: **Clinal variation in body size and sexual dimorphism in an Indian fruit bat, *Cynopterus sphinx* (Chiroptera: Pteropodidae).** *Biol J Linnean Soc* 2001, **72**:17–31.
18. Relethford JH: **Apportionment of global human genetic diversity based on craniometrics and skin color.** *Am J Phys Anthropol* 2002, **118**:393–398.
19. Sumner F: **The analysis of a concrete case of intergradation between two subspecies.** *Proc Natl Acad Sci* 1929, **15**:110–120.
20. Roberts DF: **Body weight, race and climate.** *Am J Phys Anthropol* 1953, **11**:533–558.
21. Huxley J: **Clines: an auxiliary taxonomic principle.** *Nature* 1839, **142**:219–220.
22. Cavalli-Sforza L: **Population structure and human evolution.** *Proc R Soc Lond B Biol Sci* 1966, **164**:362–379.
23. Endler JA: *Natural Selection in the Wild.* Princeton University Press: Princeton; 1986.
24. Lewontin R, Krakauer J: **Distribution of gene frequency as a test of the theory of the selective neutrality of polymorphisms.** *Genetics* 1973, **74**:175–195.
25. Haldane JB: **The theory of a cline.** *J Genet* 1948, **48**:277–284.
26. Slatkin M: **Gene flow and selection in a cline.** *Genetics* 1973, **75**:733–756.
27. Lenormand T: **Gene flow and the limits to natural selection.** *Trends Ecol Evol* 2002, **17**:183–189.
28. Akey JM, Zhang G, Zhang K, Jin L, Shriver MD: **Interrogating a high-density SNP map for signatures of natural selection.** *Genome Res* 2002, **12**:1805–1814.
29. Barreiro L, Laval G, Quach H, Patin E, Quintana-Murci L: **Natural selection has driven population differentiation in modern humans.** *Nat Genet* 2008, **40**:340–345.
30. Bonin A: **Explorative genome scan to detect candidate loci for adaptation along a gradient of altitude in the common frog (*Rana temporaria*).** *Mol Ecol* 2006, **23**:773–783.
31. Coop G, Pickrell JK, Novembre J, Kudaravalli S, Li J, Absher D, Myers R, Cavalli-Sforza LL, Feldman MW, Pritchard JK: **The role of geography in human adaptation.** *PLoS Genet* 2009, **5**:e1000500.
32. Fournier-Level A, Korte A, Cooper MD, Nordborg M, Schmitt J, Wilczek AM: **A map of local adaptation in *Arabidopsis thaliana*.** *Science* 2011, **334**:86–89.
33. Kohn MH, Pelz HJ, Wayne RK: **Locus-specific genetic differentiation at Rw among warfarin resistant rat (*Rattus norvegicus*) populations.** *Genetics* 2003, **164**:1055–1070.
34. Pickrell JK, Coop G, Novembre J, Kudaravalli S, Li JZ, Absher D, Srinivasan BS, Barsh GS, Myers RM, Feldman MW, Pritchard JK: **Signals of recent positive selection in a worldwide sample of human populations.** *Genome Res* 2009, **19**:826–837.
35. Sabeti PC, Schaffner SF, Fry B, Lohmueller J, Varilly P, Shamovsky O, Palma A, Mikkelsen TS, Altshuler D, Lander ES: **Positive natural selection in the human lineage.** *Science* 2006, **312**:1614–1620.
36. Tang K, Thornton KR, Stoneking M: **A new approach for using genome scans to detect recent positive selection in the human genome.** *PLoS Biol* 2007, **5**:e171.
37. Voight BF, Kudaravalli S, Wen X, Pritchard JK: **A map of recent positive selection in the human genome.** *PLoS Biol* 2006, **4**(3):e72.
38. Coop G, Witonsky D, Di Rienzo A, Pritchard JK: **Using environmental correlations to identify loci underlying local adaptation.** *Genetics* 2010, **185**:1411–1423.
39. De Mita S, Thuillet AC, Gay L, Ahmadi N, Manel S, Ronfort J, Vigouroux Y: **Detecting selection along environmental gradients: analysis of eight methods and their effectiveness for outbreeding and selfing populations.** *Mol Ecol* 2013, **22**:1383–1399.
40. Hancock AM, Witonsky DB, Alkorta-Aranburu G, Beall CM, Gebremedhin A, Sukernik R, Utermann G, Pritchard JK, Coop G, Di Rienzo A: **Adaptations to climate-mediated selective pressures in humans.** *PLoS Genet* 2011, **7**(4):e1001375. doi:10.1371/journal.pgen.1001375.
41. Hancock AM, Witonsky DB, Ehler E, Alkorta-Aranburu G, Beall C, Gebremedhin A, Sukernik R, Utermann G, Pritchard J, Coop G, Di Reinzo A: **Human adaptations to diet, subsistence, and ecoregion are due to subtle shifts in allele frequency.** *Proc Natl Acad Sci* 2010, **107**(Suppl 2):8924–8930.
42. Hancock AM, Clark VJ, Qian Y, Di Rienzo A: **Population genetic analysis of the uncoupling proteins supports a role for UCP3 in human cold resistance.** *Mol Biol Evol* 2011, **28**:601–14.
43. Hancock A, Witonsky D, Gordon AS, Eshel G, Pritchard JK, Coop G, Di Rienzo A: **Adaptations to climate in candidate genes for common metabolic disorders.** *PLoS Genet* 2008, **4**:e32.
44. Fumagalli M, Sironi M, Pozzoli U, Ferrer-Admettla A, Pattini L, Nielsen R: **Signatures of environmental genetic adaptation pinpoint pathogens as the main selective pressure through human evolution.** *PLoS Genet* 2011, **7**(11):e1002355.
45. Hancock AM, Brachi B, Faure N, Horton MW, Jarymowycz LB, Sperone FG, Toomajian C, Roux F, Berelson J: **Adaptation to climate across the *Arabidopsis thaliana* genome.** *Science* 2011, **334**:83–86.
46. Eckert AJ, Bower AD, Gonzalez-Martinez SC, Wegrzyn JL, Coop G, Neale DB: **Back to nature: ecological genomics of loblolly pine (*Pinus taeda*, Pinaceae).** *Mol Ecol* 2010, **2010**(19):3789–805.
47. Fang Z, Pyhäjärvi T, Weber AL, Dawe RK, Glaubitz JC, Gonzalez JJ, Ross-Ibarra C, Doebley J, Morrell PL, Ross-Ibarra J: **Megabase-scale inversion polymorphism in the wild ancestor of maize.** *Genetics* 2012, **191**:883–894.
48. Keller SR, Levsen N, Olson MS, Tiffin P: **Local adaptation in the flowering- time gene network of balsam poplar, *Populus balsamifera L*.** *Mol Biol Evol* 2012, **29**:3143–52.
49. Pyhajarvi T, Hufford MB, Mezmouk S, Ross-Ibarra J: **Complex patterns of local adaptation in teosinte.** *Genome Biol Evol* 2013, **5**(9):1594–1609.
50. Jones FC, Chan YF, Schmutz J, Grimwood J, Brady SD, Southwick AM, Absher DM, Myers RM, Riemchen TE, Deagle BE, Schluter D, Kingsley DM: **A genome-wide SNP genotyping array reveals patterns of global and repeated species-pair divergence in sticklebacks.** *Curr Biol* 2011, **2011** (22):83–90.
51. Limborg MT, Blankenship SM, Young SF, Utter FM, Seeb LW, Hansen MHH, Seeb JE: **Signatures of natural selection among lineages and habitats in *Oncorhynchus mykiss*.** *Ecol Evol* 2012, **2**:1–18.
52. Cheng C, White BJ, Kamdem C, Mockaitis K, Costantini C, Hahn MW, Besansky NJ: **Ecological genomics of *Anopheles gambiae* along a latitudinal cline in cameroon: a population resequencing approach.** *Genetics* 2012, **190**:1417–1432.
53. Gunther T, Coop G: **Robust identification of local adaptation from allele frequencies.** *Genetics* 2013, **195**:205–220.
54. Fraser HB: **Gene expression drives local adaptation in humans.** *Genome Res* 2013, **23**:1089–1096.
55. Li JZ, Absher DM, Tang H, Southwick AM, Casto AM, Ramachandran S, Cann HM, Barsh GS, Feldman M, Cavalli-Sforza LL, Myers RM: **Worldwide human relationships inferred from genome-wide patterns of variation.** *Science* 2008, **319**:1100–1104.

Forced swim test induces divergent global transcriptomic alterations in the hippocampus of high versus low novelty-seeker rats

Pothitos M Pitychoutis[1*†], Despina Sanoudou[2†], Margarita Papandreou[2], Dimitris Nasias[2], Marianna Kouskou[2], Craig R Tomlinson[3,4], Panagiotis A Tsonis[1†] and Zeta Papadopoulou-Daifoti[2†]

Abstract

Background: Many neuropsychiatric disorders, including stress-related mood disorders, are complex multi-parametric syndromes. Susceptibility to stress and depression is individually different. The best animal model of individual differences that can be used to study the neurobiology of affect regards spontaneous reactions to novelty. Experimentally, when naive rats are exposed to the stress of a novel environment, they display a highly variable exploratory activity and are classified as high or low responders (HR or LR, respectively). Importantly, HR and LR rats do not seem to exhibit a substantial differentiation in relation to their 'depressive-like' *status* in the forced swim test (FST), a widely used animal model of 'behavioral despair'. In the present study, we investigated whether FST exposure would be accompanied by phenotype-dependent differences in hippocampal gene expression in HR and LR rats.

Results: HR and LR rats present a distinct behavioral pattern in the pre-test session but develop comparable depressive-like status in the second FST session. At 24 h following the second FST session, HR and LR rats (stressed and unstressed controls) were sacrificed and hippocampal samples were independently analyzed on whole rat genome Illumina arrays. Functional analysis into pathways and networks was performed using Ingenuity Pathway Analysis (IPA) software. Notably, hippocampal gene expression signatures between HR and LR rats were markedly divergent, despite their comparable depressive-like status in the FST. These molecular differences are reflected in both the extent of transcriptional remodeling (number of significantly changed genes) and the types of molecular pathways affected following FST exposure. A markedly higher number of genes (i.e., 2.28-fold) were statistically significantly changed following FST in LR rats, as compared to their HR counterparts. Notably, genes associated with neurogenesis and synaptic plasticity were induced in the hippocampus of LR rats in response to FST, whereas in HR rats, FST induced pathways directly or indirectly associated with induction of apoptotic mechanisms.

Conclusions: The markedly divergent gene expression signatures exposed herein support the notion that the hippocampus of HR and LR rats undergoes distinct transcriptional remodeling in response to the same stress regimen, thus yielding a different FST-related 'endophenotype', despite the seemingly similar depressive-like phenotype.

Keywords: Stress, Gene expression, Rats, Depression, Individual differences, Hippocampus, FST, Neurogenesis, Neuroplasticity, Apoptosis

* Correspondence: ppitychoutis1@udayton.edu
†Equal contributors
[1]Department of Biology & Center for Tissue Regeneration and Engineering (TREND), University of Dayton, 300 College Park, Dayton, OH 45469-2320, USA
Full list of author information is available at the end of the article

Background

Many neuropsychiatric disorders, including stress-related mood disorders, are complex multi-parametric syndromes. Strikingly, susceptibility to depression and stress differs between individuals. Accurate diagnosis is hard to establish and current pharmacotherapeutic strategies suffer from significant variability in effectiveness, making the understanding of inter-individual variations crucial to unveiling effective new treatments [1].

The best animal model of individual differences that can be used to study the neurobiology of affect regards spontaneous reactions to novelty [2]. Experimentally, when naive rats are exposed to the stress of a novel environment, they display a highly variable exploratory activity; some rats are characterized by high rates of locomotor reactivity (high responders (HR)) whereas others by low rates (low responders (LR)) (for a detailed review on the HR/LR model, the reader is referred to [2,3]). Novelty-seeking is an affect-related trait that has been associated with anxiety and emotional reactivity [4] as well as with depressive symptomatology [5,6] in humans and in rodents.

The forced swim test (FST) [7] is a widely implemented animal model of behavioral despair with high predictive validity for agents with antidepressive potential [8,9]. Increased passive behavioral responses in FST, such as immobility, and decreased active behaviors like swimming or climbing are thought to be a clear indication of 'depressive-like' symptomatology [10,11]. Interestingly, male HR and LR rats do not exhibit a substantial differentiation in relation to their depressive-like *status* in the FST [4,12,13]. However, previous results from our lab indicate that HR and LR rats may exhibit a different behavioral pattern characterized by distinct active behavioral reactions expressed during either the first or the second exposure to water [4]. Moreover, we recently reported that chronic antidepressant treatment with clomipramine may attenuate depressive-like symptomatology in FST in male HR but not in their LR counterparts [12]. This evidence combined suggests that idiosyncratic responses to novelty may lead to a phenotypically similar depressive-like outcome, through different, yet unknown, molecular mechanisms.

In search of the brain structures likely to be involved in these processes, the hippocampus stands out due to its central role in the detection of novelty and the guidance of behavioral responses on the basis of familiarity [14]. Furthermore, the hippocampus has been the focus of multiple studies on the pathophysiology and treatment of stress-related disorders [15-17]. Indeed, stress often has detrimental effects on the integrity and function of the hippocampus. These effects include alterations in the activity of monoaminergic systems, in dendritic and synaptic remodeling, as well as in levels of adult hippocampal neurogenesis [18].

In the present study, we investigated whether FST exposure would be accompanied by phenotype-dependent differences in hippocampal gene expression in HR and LR rats. The markedly divergent gene expression signatures exposed herein support the notion that the hippocampus of HR and LR rats undergoes distinct transcriptional remodeling in response to the same stress regimen, thus yielding a different FST-related 'endophenotype', despite the seemingly similar depressive-like phenotype.

Results

HR and LR rats present with distinct behavioral pattern in the pre-test session but comparable depressive-like status in the second FST session

A one-way analysis of variance (ANOVA) showed that HR rats are characterized by higher vertical activity as compared to their LR counterparts ($F_{(1,23)} = 36.234$; $p < 0.001$; Figure 1a). This level of significance was maintained when these initial groups were dichotomized in order for the control and FST groups to be formed (data not shown). Most importantly, during the pre-test FST session, a one-way ANOVA showed that LR rats spent less time swimming around the cylinder, as compared to their HR counterparts ($F_{(1,13)} = 4.957$; $p = 0.046$; Figure 1b). In LR rats, this decrease in swimming was attributed to a non-significant increase of immobility. The same analysis showed that the structure of HR and LR rats' behavior was similar during the second FST session (Figure 1c). Specifically, both HR and LR rats exhibited increased floating behavior and decreased active behaviors (i.e., climbing and swimming) during the second session, which reflect the establishment of depressive-like behavior in this test.

Hippocampal gene expression signature changes in HR and LR rats following exposure to FST

The global gene expression signatures in the hippocampus of HR and LR rats were assessed in rats exposed to FST versus their respective unstressed counterparts. Bioinformatic analysis revealed significant changes in 258 transcripts (243 upregulated; 15 downregulated) in HR rats upon FST exposure (Additional files 1 and 2). On the other hand, in LR rats, 589 transcripts were statistically significantly changed, all of which were upregulated. Despite the large number of significant changes, only 20 transcripts were affected in both novelty-seeking phenotypes (Figure 2; Additional file 3).

Gene Ontology (GO) classification

To determine the biological processes and molecular functions likely to be affected by the significant gene expression changes, a comparative analysis tool provided by the 'GeneCodis3' software was used for the Gene Ontology

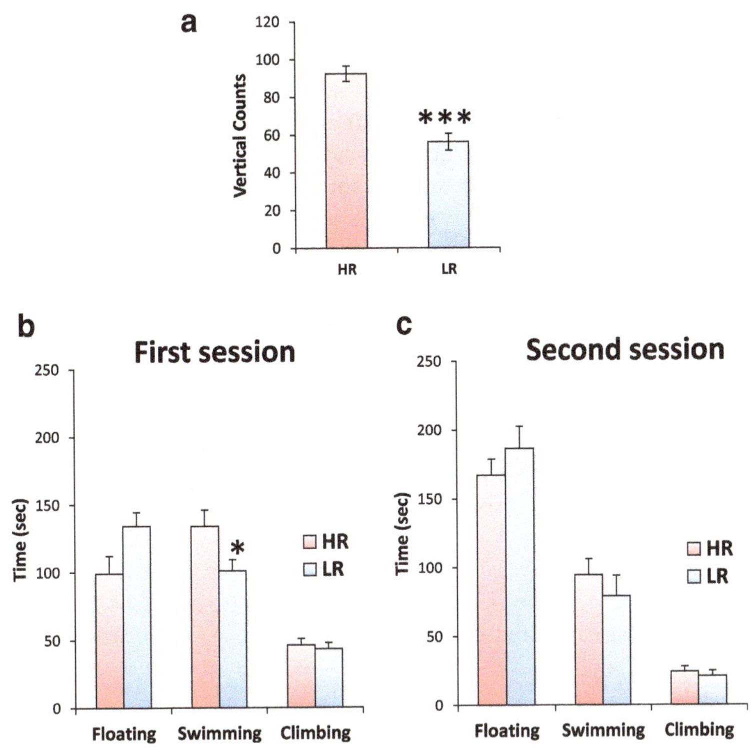

Figure 1 Behavioral data. (a) HR/LR classification: the initial classification of rats into HR or LR phenotype on the basis of their vertical locomotor activity (i.e., vertical counts) in an open field ($N = 12$ per phenotype) was assessed with one-way ANOVAs. The structure of HR and LR rats' behavior (i.e., immobility, swimming, climbing) ($N = 7$ per group) in the **(b)** first (pre-test) and **(c)** second (test) FST sessions was assessed with separate one-way ANOVAs. *Bars* represent means ± SEM; *$p < 0.05$; ***$p < 0.001$.

classification of the differentially expressed genes in both groups. GO level 5 annotations were obtained for 117 unique genes in HR rats and 354 unique genes in LR rats (only 9 genes were common in both groups). Focusing on the 'biological process' (BP) of the unique genes, 50 processes (annotations) in HR and 189 processes in LR rats were found to be statistically significantly changed. In HR rats, the BPs most highly affected were the 'cellular macromolecule metabolic process' (56 genes) and the 'cellular biosynthetic process' (42 genes) whereas in LR rats, the 'regulation of cellular process' (154 genes) and 'cellular macromolecule metabolic process' (125 genes). As far as the 9 common genes are concerned, no annotations were significantly enriched.

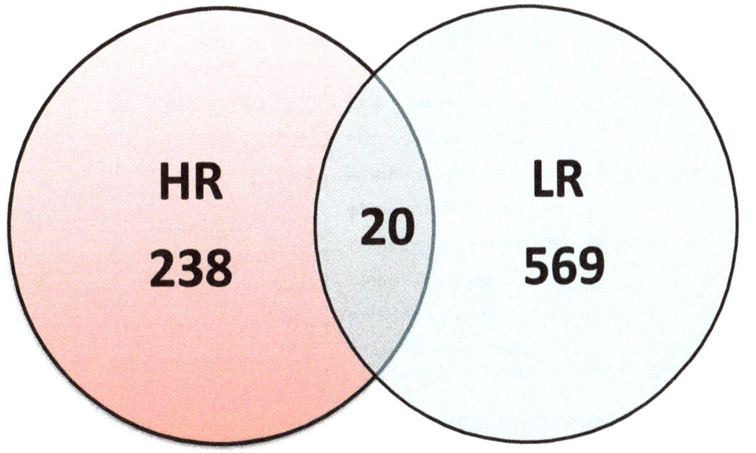

Figure 2 Exposure to FST affected a different number of transcripts in HR versus LR rats (238 versus 569). Notably, only 20 common transcripts were affected in both novelty-seeking phenotypes.

In order to depict those processes orchestrating possible hippocampal structural and/or functional alterations in response to FST in the two novelty-seeking phenotypes, an additional set of selection criteria was applied, depicting statistically significant GO categories selected by one or more of the following keywords: neuro-, synaptic, stress, plasticity, cell death, cytoskeleton, migration, and adhesion (Figure 3). This approach unveiled even greater differences between the molecular response of HR and LR rats to FST exposure. Notably, 'synaptic transmission' (21 genes), 'cell migration' (15 genes), 'cell-cell adhesion' (10 genes), 'neurotransmitter transport' (8 genes), and 'neuroblast proliferation' (4 genes) are among the GO BPs found to be represented only in the LR group, while 'regulation of cell death' (11 HR versus 21 LR genes), 'cellular response to stress' (11 HR versus 20 LR genes), and 'actin cytoskeleton organization' (5 HR versus 11 LR genes) were found to be affected in both groups.

Ingenuity Pathway Analysis (IPA)

The GO findings were confirmed and further enriched through IPA. In HR rats, IPA analysis indicated that the observed gene expression changes were associated with 73 statistically significant functional categories including 'cellular growth and proliferation' (45 genes), 'cellular development' (24 genes), 'cellular movement' (9 genes), 'cell morphology' (19 genes), and 'nervous system development and function' (11 genes). Similarly, 76 IPA functions were affected in LR rats with 'cellular assembly and organization' (73 genes), 'cellular function and maintenance' (77 genes), 'cell-to-cell signaling and interaction' (53 genes), 'nervous system development and function' (83 genes), 'cell morphology' (74 genes), and 'cellular development' (30 genes) being among the functions overrepresented. It is noteworthy that both the GO and IPA analyses confirmed that FST exposure affected the transcription of genes related to hippocampal cellular proliferation in both novelty-seeking phenotypes.

Zooming in the specific molecular pathways associated with the observed gene expression changes, data mining through IPA exposed markedly divergent mechanisms in the hippocampus of the two novelty-seeking phenotypes in response to FST stress. Overall, 49 canonical IPA

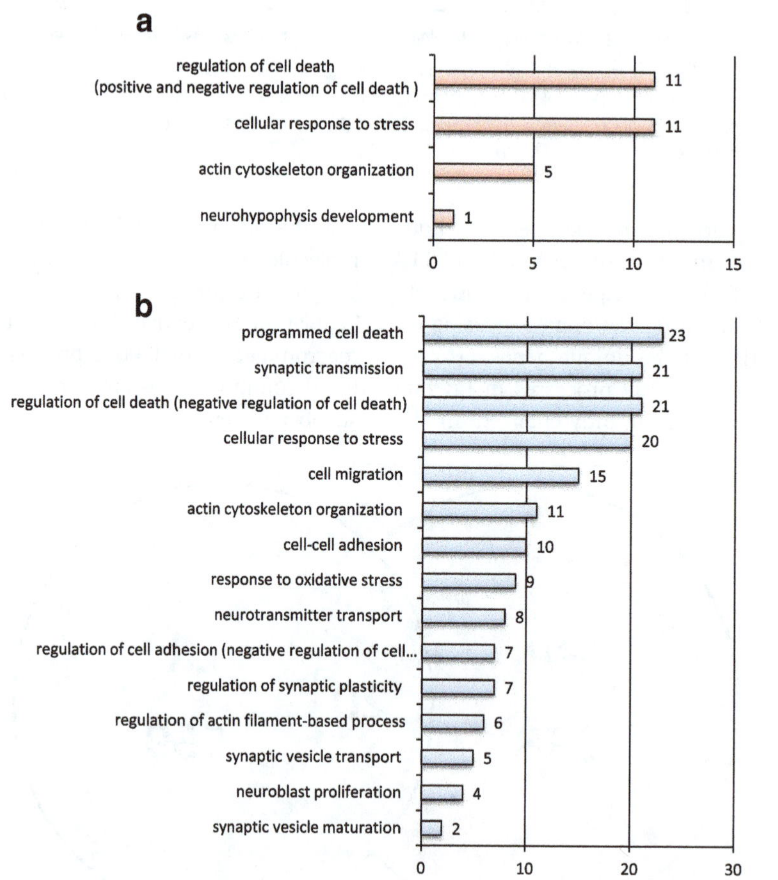

Figure 3 Distribution of differentially expressed genes according to GO BP in (a) HR and (b) LR rats. GeneCodis functional GO BP level 5 categories that meet conservative statistical criteria (hypergeometric distribution and false discovery rate (FDR) test, $p < 0.05$) and include any one of the selected relevant keywords (i.e., neuro-, synaptic, stress, plasticity, death, cytoskeleton, migration, and adhesion). *Numbers* in the chart denote the number of genes grouped in each annotation.

pathways comprising 1 to 11 genes each were depicted as significantly changed following FST in HR rats, versus 62 pathways comprising 1 to 16 genes each in LR rats. Only 15 pathways were common in both HR and LR rats, and these were largely represented by different genes due to the low number of common genes affected in both novelty-seeking phenotypes. Notably, in HR rats, 'EIF2 Signaling' and 'Axonal Guidance Signaling' were among the top five pathways affected upon FST exposure, whereas in LR rats, these were 'Ephrin Receptor Signaling', 'Clathrin-mediated Endocytosis Signaling', and 'ERK/MAPK Signaling' (Additional file 4).

Mammalian Adult Neurogenesis Gene Ontology (MANGO) analysis

In order to decipher the impact of FST exposure on the regulation of neurogenesis-related mechanisms, the MANGO was implemented [19]. In particular, the MANGO database was screened for genes that were statistically significantly changed in our data. According to this form of analysis, FST exposure induced a phenotype-dependent expression of neurogenesis-related transcripts in the hippocampus of HR and LR rats. In particular, seven neurogenesis-related genes were statistically significantly upregulated in LR (i.e., *Ephb2, Nog, Ntf3, Tgfb1, Smad7, Sox2,* and *Srr*), as compared to only three genes in HR rats (i.e., *Bmpr1a, Gsk3b,* and *Jag1*). This analysis revealed that exposure to the same FST regimen induced a phenotype-dependent regulation of unique neurogenesis-related genes in the hippocampus of the two novelty-seeking phenotypes.

IPA network analysis

Finally, network analysis was performed to unravel the intricate relationships between the different significantly changed pathways. In HR rats, 13 networks were generated by IPA versus 25 in LR rats. Consistently with all other levels of analysis, the molecular networks that emerged for the HR and LR responses to FST exposure were markedly different. This indicates that not only the individual genes and pathways are characteristic of each of the two behavioral phenotypes, but also the cascades of molecular events (i.e., the networks of genes from different pathways) giving rise to the depressive-like phenotype are distinct. The top networks of genes whose expression was significantly altered in the hippocampus of the two novelty-seeking phenotypes are presented in Figure 4.

Discussion

Herein, we report for the first time the marked divergence of the FST-induced hippocampal gene expression signatures between HR and LR rats, despite their comparable depressive-like status during the FST. These molecular differences are reflected in both the extent of transcriptional remodeling (number of significantly changed genes) and the types of molecular pathways affected following FST exposure. A markedly higher number of genes (i.e., 2.28-fold) were statistically significantly changed following FST in LR as compared to their HR counterparts (i.e., 589 versus 258), while only a strikingly low number of genes (i.e., 20 genes) were commonly altered in both novelty-seeking phenotypes. These distinct gene expression signatures support the notion that the hippocampus of HR and LR rats undergoes distinct transcriptional remodeling in response to the same environmentally induced stress, thus yielding a different FST-related endophenotype, despite the seemingly similar depressive-like phenotype in this test.

Structure of HR and LR rats' behavior in the FST

HR and LR rats present a distinct behavioral pattern in the pre-test session but develop comparable depressive-like status in the second FST session. Notably, LR rats spent less time swimming around the cylinder during the pre-test session, as compared to their HR counterparts. Present behavioral findings essentially replicate previous results from our lab; Antoniou et al. [4] reported that HR and LR rats develop a different behavioral FST pattern characterized by a common behavioral state of 'despair', but distinct active behavioral reactions are expressed during the first exposure to water [4]. The structure of HR and LR rats' behavior during the second FST session is similar between the two novelty-seeking phenotypes, as shown herein and as previously reported [4,12,13]. The majority of studies in this field typically quantify only the second FST session, as it is relevant to the screening of antidepressant drugs [12]. However, it is noteworthy that the pre-test session reflects rats' responsiveness to the presentation of a novel swim stressor that may ultimately lead to the modulation of behavior during subsequent exposure to the stressful stimulus (e.g., induction of learned helplessness) or even to the differential regulation of relevant endophenotypes (i.e., neurobiological indices) without apparent effect in the behaviors assessed [12].

FST induces differential regulation of neurogenesis-related transcripts in HR compared to LR

In the present study, both the GO and IPA analyses demonstrate that FST exposure affected the transcription of genes related to hippocampal cellular proliferation and homeostasis in both novelty-seeking phenotypes. According to the MANGO analysis, FST exposure induced a phenotype-dependent expression of neurogenesis-related genes in the hippocampus of HR and LR rats. In particular, only three neurogenesis-related genes were statistically significantly upregulated in HR (i.e., *Bmpr1a, Gsk3b,* and

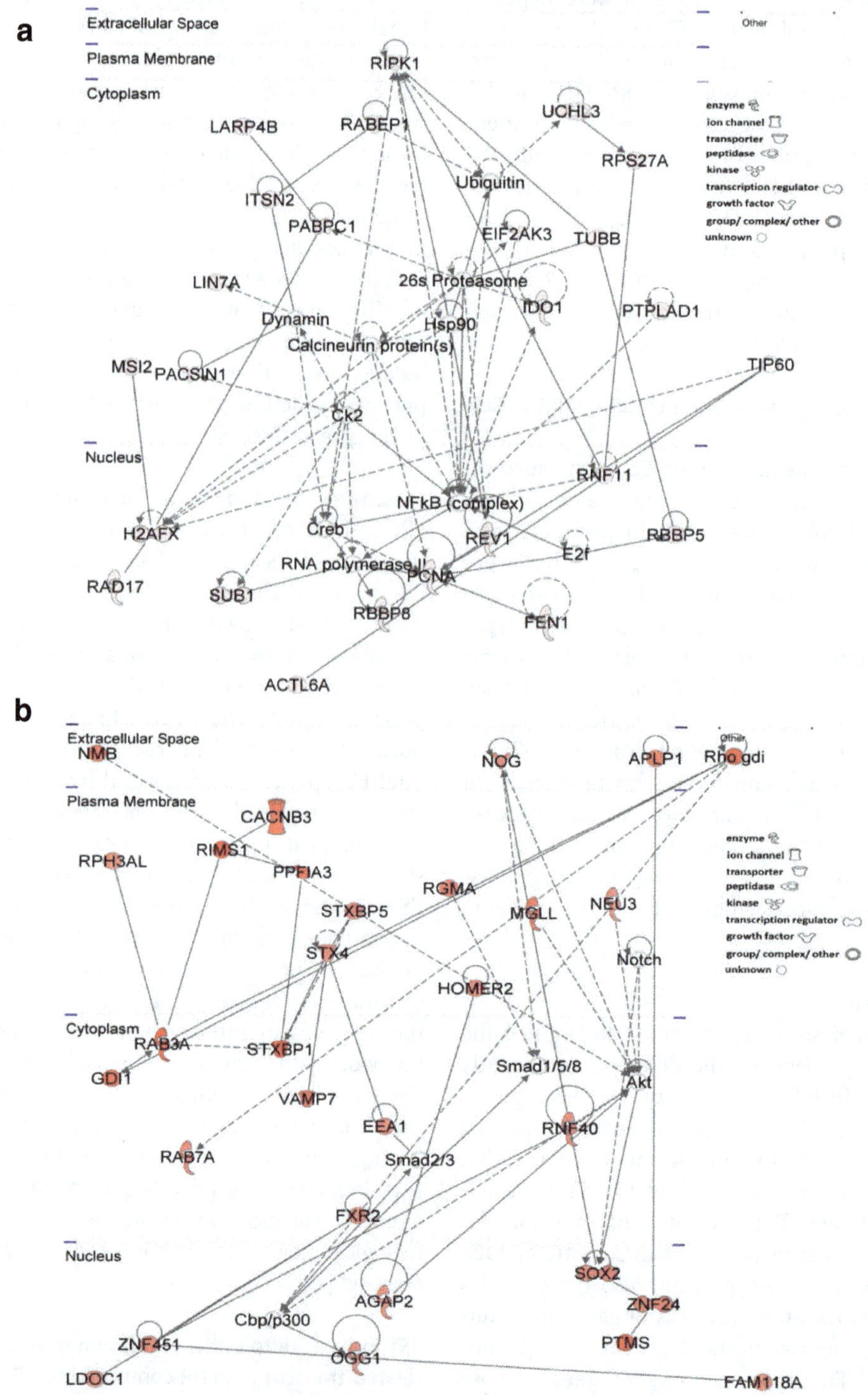

Figure 4 Top Ingenuity networks of genes whose expression was significantly altered in the hippocampus upon FST exposure. (a) HR (24 upregulated transcripts) and **(b)** LR (30 upregulated transcripts) rats. Upregulated molecules are shown in *red* and increasing color intensity reflects higher fold changes. Genes with baseline expression are shown in *white*. *Arrows* point to the activated molecule in a signaling pathway or the product of a metabolic pathway. *Solid lines* indicate direct and *dashed lines* indirect interactions. *Geometrical shapes* (nodes) reflect the key functions of the specific protein.

Jag1), as compared to seven genes in LR rats (i.e., *Ephb2, Nog, Ntf3, Tgfb1, Smad7, Sox2,* and *Srr*). Importantly, exposure of HR rats to the FST paradigm induced hippocampal *Gsk3b* (glycogen synthase kinase 3 beta; +2.089) mRNA levels. Gsk3b is a negative regulator of the b-catenin and the canonical Wnt signaling pathway, which modulate hippocampal neurogenesis [20,21], but has also been reported to modulate Notch signaling by enhancing its stability [22]. Notably, overexpression of *Gsk3b* in the murine hippocampus causes dramatic alterations in both the dendritic tree morphology and the postsynaptic densities of newborn neurons and has been proposed as a target for the treatment of Alzheimer's disease [23]. Moreover, FST exposure in HR rats upregulated *Bmpr1a* (bone morphogenetic protein receptor, type IA; +2.275) and *Jag1* (jagged; +2.163; the ligand of Notch1), which have been reported to affect neurogenesis in the adult hippocampus either directly or indirectly [24,25].

In LR rats, the FST induced the transcriptional activation of genes that positively regulate neurogenesis in the adult hippocampus. Notably, *Ntf3* (neurotrophin 3; +2.469) facilitates hippocampal plasticity by promoting neurogenesis in the dentate gyrus [26], *Nog* (noggin; +2.177) is a BMP antagonist that diverts stem cells from a glial to a neuronal fate [27], and *Sox2* (SRY (sex-determining region Y)-box 2; +2.201) maintains neural stem cell properties, including proliferation/survival, self-renewal, and neurogenesis [28]. Moreover, FST exposure led to overexpression of *Tgfb1* (transforming growth factor, beta 1; +2.219) and *Smad7* (SMAD family member 7; +2.443) mRNA in the hippocampus of LR rats. Interestingly, TGFb1 has been recognized as a negative regulator of adult neurogenesis since infusion of TGFb1 into the ventricles of the adult rat brain reduced the number of proliferating cells in the hippocampus [29], while SMAD7 may either directly antagonize TGF receptor signaling [30] or act in a TGF receptor-independent manner and further suppress neurogenesis [31]. Although not registered in the MANGO database, several fibroblast growth factor (FGF)-related genes were upregulated in the hippocampus of LR rats following FST, namely *Fgf12* (fibroblast growth factor 12; +3.116), *Fgf13* (fibroblast growth factor 13; +2.465), *Fgfr2* (fibroblast growth factor receptor 2; +2.088), and *Frs3* (fibroblast growth factor receptor substrate 3; +2.113). Intriguingly, upregulation of *Fgfr2* upon FST exposure may mediate the neuroprotective effects of FGF2 on hippocampal formation [32]. Thus, it could be hypothesized that FST has a beneficial effect in the hippocampus of LR rats by promoting neuroprotective processes.

Intriguingly, FST induced phenotype-dependent effects on the transcription of prominent serotonin (5-HT)-related enzymes. In the hippocampus of LR rats, FST upregulated *Tph2* (tryptophan hydroxylase 2; +2.145), the rate-limiting enzyme of 5-HT biosynthesis, that is possibly indicative of serotonergic activation in this brain region. On the other hand, in HR rats, *Ido1* (indoleamine 2,3-dioxygenase; +2.302) levels were induced. It is noteworthy that IDO1 is a proinflammatory enzyme that converts L-tryptophan (the precursor of serotonin; 5-HT) to kynurenine and its neurotoxic metabolites that are able to produce oxidative stress by increasing the production of reactive oxygen species (ROS) or to overstimulate hippocampal *N*-methyl-D-aspartate (NMDA) receptors and lead to apoptosis and hippocampal atrophy [33].

Overall, FST exposure induced a phenotype-dependent modulation of neurogenesis-related genes in the hippocampus of the two novelty-seeking phenotypes. Notably, the hippocampus is a stress- and glucocorticoid (GC)-sensitive brain region. In particular, within the dentate gyrus (DG), there exists a high density of glucocorticoid receptors (GRs) that respond to enhanced circulating GCs [34]. According to previous reports from our group and others, serum corticosterone concentrations are enhanced by FST exposure irrespective of the HR/LR phenotype [12,13] and this could negatively affect hippocampal cell proliferation [35]. Moreover, it has been shown that hippocampal DNA damage occurs immediately after exposure of male Wistar rats to FST [36] and that exposure of female rats to the same regimen may reduce cell survival in the hippocampus, probably by increasing corticosterone levels [37]. Indeed, this profile nicely fits in with the transcriptomic remodeling observed in HR rats in response to FST in the current study. On the other hand, recent data also support a beneficial role for acute stress on the hippocampus. In particular, exposure of rats to restraint stress increased cell proliferation and astrocytic fibroblast growth factor 2 (*fgf2*) expression in the dorsal hippocampus [38]. Remarkably, in the present study, exposure of LR rats to FST resulted in enhanced expression of hippocampal *Ffgr2* mRNA levels. Importantly, FST appears to promote neurogenic processes in LR rats. These findings are in accordance with the enhanced neurogenesis observed in the DG of LR rats at baseline. As shown before, cell proliferation in the DG of LR Wistar rats, evaluated by the incorporation of 5-bromo-2′-deoxyuridine (Brdu) in progenitors, was twice that observed in their HR counterparts [39]. Similarly, selectively bred HR and LR (bHR/bLR) lines of Sprague-Dawley rats revealed, through Ki67 immunohistochemistry, enhanced cell proliferation in the DG of developing bLR versus bHR pups [40]. In conclusion, the increased rates of hippocampal neurogenesis in LR rats at baseline appear to be maintained in response to FST-induced stress, with the activation of numerous relevant processes. These molecular changes contribute to the markedly different molecular *milieu* of LR compared to HR rats that may predispose to distinct long-term behavioral or pharmacological responses.

FST induced upregulation of neuroplasticity-related transcripts in LR rats

Neuroplasticity refers to the brain's ability to change its structure and function during maturation, learning, environmental challenges, pathology, and stress. Synaptic pruning and remodeling of the postsynaptic cytoskeleton comprise structural changes that are possible in the adult brain. Interestingly, exposure to FST upregulated genes directly or indirectly involved in neuroplasticity in a phenotype-specific manner in LR rats. Ephrins and their receptors govern the topographic guidance of axons during central nervous system (CNS) development but are also implicated in neuronal plasticity in the adult brain [41]. Herein, the "Ephrin Receptor Signaling" pathway was the second among the top five canonical IPA pathways affected in the hippocampus of LR rats upon FST exposure (Figure 5). Specifically, FST resulted in the upregulation of both *Ephb2* (+2.167) and *Epha5* (+2.879) transcripts. Ephrin receptors are the largest known family of receptor tyrosine kinases in the mammalian genome and are divided into A and B subclasses based on affinity for their membrane-associated ligands, ephrin-As and ephrin-Bs. Notably, *Ephα5* is expressed at both the protein and mRNA levels in the adult murine hippocampus where it has been hypothesized to promote synaptic plasticity [42]. Most importantly, *Ephb2* is expressed in the adult hippocampus and regulates, among others, synaptic structure, NMDA receptor clustering and function, and LTP [43].

In line with this evidence, the IPA functions predicted to be statistically significantly enhanced in LR rats upon FST exposure involved 'growth of neurites' (*z*-score 2.825), 'LTP of synapse' (*z*-score 2.396), 'microtubule dynamics' (*z*-score 2.453), 'organization of cytoplasm' (*z*-score 2.276), and 'organization of cytoskeleton' (*z*-score 2.269), all of which are directly related with neuroplasticity. Of particular interest was the upregulation of a number of neuroplasticity-related synaptic players in the hippocampus of LR rats, such as *Ddn* (dendrin; +2.225), a postsynaptic regulator of the structure of synaptic cytoskeleton, and *Stx4* (syntaxin 4; +2.178), an essential postsynaptic component for synaptic plasticity in hippocampal neurons [44]. Moreover, the upregulation of both *Pfn* (profilin) isoforms (i.e., 1 and 2; +2.709 and 2.396, respectively) that are prominent regulators of actin dynamics in the CNS is indicative of the cytoskeletal reorganization that possibly takes place upon FST exposure in the

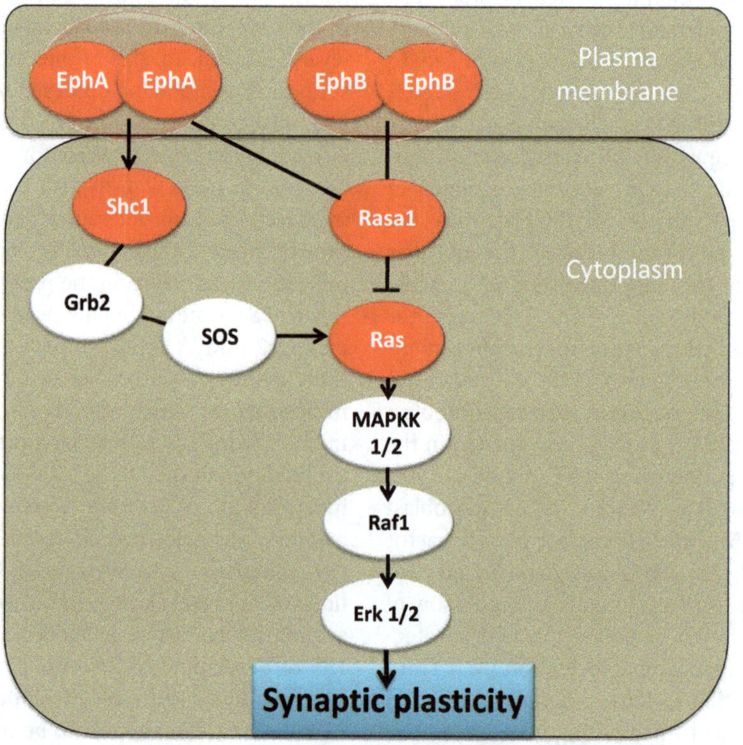

Figure 5 'Ephrin Receptor Signaling' was the second canonical IPA pathway induced in LR rats upon FST exposure. Upregulated molecules are shown in *red*. *Blind lines* denote binding, *arrows* signify activation, whereas *T-like lines* indicate inhibition. *EphA5* Ephrin A5 receptor, *EphB2* Ephrin B2 receptor, *Erk1/2* mitogen-activated protein kinase 1, *GRB2* growth factor receptor-bound protein 2, *MAPKK1/2* mitogen-activated protein kinase 2, *RASA1* RAS p21 protein activator (GTPase-activating protein) 1, *RAF1* v-raf-1 murine leukemia viral oncogene homolog 1, *Shc1* Src homology 2 domain containing) transforming protein 1, *SOS* son of sevenless homolog 2.

hippocampus of LR rats [45]. Moreover, FST appears to stimulate neurotransmitters' release in LR rats, as reflected by the upregulation of *Dnm1* (dynamin 1; +2.388) which is involved in exo-endocytosis of synaptic vesicles [46] and *Syt12* (synaptotagmin 12; +2.352) which is a synaptic vesicle phosphoprotein that modulates neurotransmitter release [47]. Hippocampal synaptotagmin mRNA levels were found to be enhanced upon immobilization stress in rats, in view of the effects of stress on synaptic plasticity [48]. Remarkably, exposure of LR rats to FST resulted in the upregulation of genes implicated in core neuroadaptive processes in the adult hippocampus.

FST induced apoptosis-related mechanisms in the hippocampus of HR rats

The top canonical IPA pathway affected in HR rats upon FST exposure was that of eIF2 (eukaryotic translation initiation factor 2) signaling. The eIF2 complex is essential in all eukaryotes for protein synthesis, since it recruits the initiator methionyl tRNA (Met-tRNA) to ribosomes to begin translation. In addition, multiple endoplasmic reticulum (ER) stress-related genes were overexpressed in the hippocampus of HR rats following FST exposure, including *Perk* (*eif2ak3*; eIF2a kinase 3; +2.053), *Sgpp2* (sphingosine-1-phosphate (S1P) phosphatase 2; +3.121), and *Mbtps2* (membrane-bound transcription factor peptidase 2; +2.593). As depicted in Figure 6, *Perk* upregulation could lead to phosphorylation-induced inhibition of eIF2α that could in turn result in global reduction of protein translation. In addition, upregulation of *Gsk3b* (+2.089) may inhibit the activity of eIF2b (eukaryotic initiation factor 2B) that could also lead to inhibition of protein translation in neuronal cells [49]. Interestingly, it has been proposed that inhibition of protein translation is a mode of inducing neuronal apoptosis and neurodegeneration in Alzheimer's disease [50,51]. Notably, induction of *Mbtps2* that encodes an intramembrane zinc metalloprotease involved in ER stress response may enhance the activation of target genes related to the ER stress pathway through an *Atf-6* (activating transcription factor 6)-dependent mechanism. In addition, *Sgpp2* is an enzyme localized in cell and ER membranes that forms sphingosine from S1P that in turn is N-acylated to ceramide; both ceramide and sphingosine have been associated with induction of ER stress, growth arrest, and apoptosis in mammalian cells [52-54]. FST also upregulated *Rip1* (TNF receptor (TNFRSF)-interacting serine-threonine kinase 1; +2.249), a kinase originally discovered via its interaction with the death domain of TNFR1 [55]. Importantly, *Rip1* is a central initiator of cell death and may also play a role in the ceramide death pathway [51]. Overall, it appears that in HR rats FST induced the overexpression of multiple genes directly and

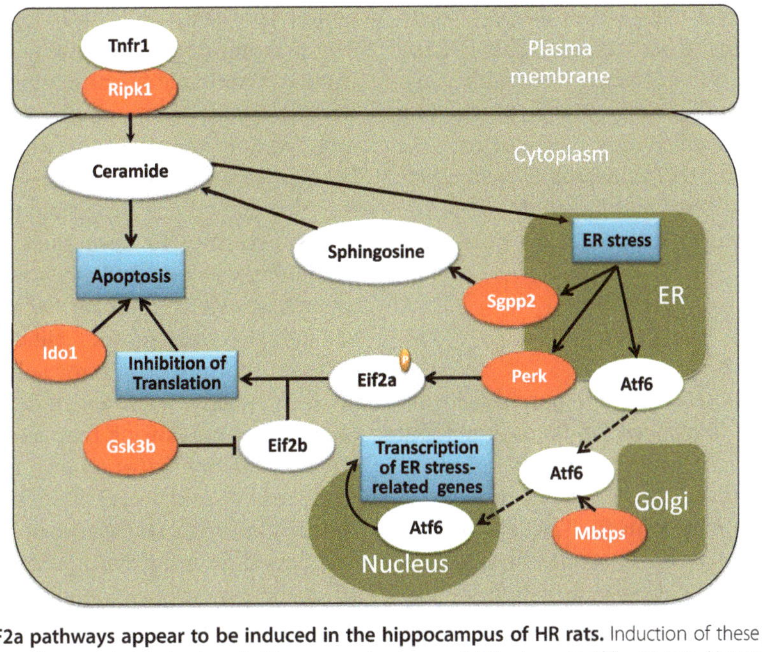

Figure 6 ER stress and eIF2a pathways appear to be induced in the hippocampus of HR rats. Induction of these pathways could be implicated in the mobilization of apoptotic mechanisms in hippocampal neurons of HR rats upon FST exposure. Upregulated genes are shown in *red*. *Blind lines* denote binding, *arrows* signify activation, whereas *T-like lines* indicate inhibition. *ATF-6* activation transcription factor 6, *eIF2a* eukaryotic initiation factor 2a, *eIF2b* eukaryotic initiation factor 2b, *Gsk3b* glycogen synthase kinase-3 beta, *Ido1* indoleamino-2,3-dioxygenase, *MBTPS* membrane-bound transcription factor peptidase, *Perk* eIF2a kinase 3, *Sgpp2* sphingosine-1-phosphate (S1P) phosphatase 2, *Ripk1* TNF receptor (TNFRSF)-interacting serine-threonine kinase 1, *TNFR1* tumor necrosis factor receptor superfamily.

indirectly associated with apoptotic mechanisms in the adult hippocampus.

Conclusions

Herein, we report that FST-induced hippocampal gene expression signatures between HR and LR rats are markedly divergent. The two novelty-seeking phenotypes present a distinct behavioral pattern in the pre-test session but develop comparable depressive-like status in the second session of the FST. At 24 h post-FST exposure, hippocampal gene expression signatures between HR and LR rats were markedly divergent, despite their similar FST performance. These molecular differences are reflected in both the extent of transcriptional remodeling (i.e., number of significantly changed genes) and the types of molecular pathways affected following FST exposure. A markedly higher number of genes (i.e., 2.28-fold) were statistically significantly changed following FST in LR rats, as compared to their HR counterparts. Specifically, global transcriptomic alterations suggest that following FST exposure, neurogenesis and neuroplasticity-related processes are enhanced in LR rats, while pathways that are directly or indirectly implicated in apoptotic cascades are induced in their HR counterparts. Importantly, the present data extend previous findings on the differential modulation of neurogenesis between the two novelty-seeking phenotypes at baseline.

Overall, the present data suggest a complex pattern of molecular modifications that likely reflect distinct neurobiological processes occurring between the adult HR and LR hippocampus upon FST exposure. Strikingly, susceptibility to stress is individually different. Gaining an insight into individual differences may elucidate the neurobiological substrate for this vulnerability, help clarify pathophysiological mechanisms, and hopefully aid in developing strategies to discover more effective pharmacotherapies [2].

Methods

Animals

Male *Sprague-Dawley* rats ($N = 24$) were purchased from Athens' Pasteur Institute. Rats aged 80–90 days old and weighed 250–300 g at the beginning of the experiments. Rats were group housed in plastic cages measuring 570 mm × 380 mm × 200 mm, with food pellets and tap water available *ad libitum*, under controlled laboratory conditions (i.e., 12-h light/dark with lights on at 0700 hours and a constant temperature of 21°C ± 1°C), as previously described [12]. Approval to conduct the experiments described has been obtained from the animal subjects review board of our institution. All animal experiments have been carried out in accordance with the European Communities Council Directive of 24 November 1986 (86/609/EEC). Efforts were made to minimize the numbers of animals used and to reduce their suffering.

Classification into HR/LR groups

All rats were classified as HR and LR according to their frequency of rearing (counts registered when the rat's body inclined vertically with hind paws on the floor and forepaws on the wall of the cage) in an activity chamber, as in previous studies [4,12,56,57]. Vertical activity has been used, apart from general locomotion, as a sole criterion for assignment of rats into groups during their exposure to novelty [58,59]. The rats were ranked using the frequency of vertical counts (rearing). Animals above the median were designated as HR, while the rest were classified as LR. Testing was conducted in two clear Plexiglas activity chambers measuring 430 × 430 × 300 mm (Med Associates Inc., St Albans, VT, USA), and behavior was recorded for a 15-min observation period, as previously described [12,56].

Forced swim test (FST)

All rats were acclimatized to the test room for 1 h before the beginning of the experiment. HR and LR rats ($N = 7$ per phenotype) were individually placed in a cylindrical tank measuring 50 cm in height and 20 cm in width or were left undisturbed in their home cage and served as controls ($N = 5$ per phenotype). The tank was filled with water (24°C ± 1°C), and water was changed after each session. The animals were forced to swim for a 15-min period (pre-test) and 24 h later were subjected to a 5-min swimming session (test) [60]. The total duration of floating (immobility), swimming, and climbing periods were scored by the same observer for the first 5 min of each session (pre-test and test), as mentioned before [12,61]. Rats were considered to show immobility when they floated without struggling, making only those movements necessary to keep their heads above the water. Swimming was recorded when they actively swam around in circles, while climbing was scored when the rats climbed the walls of the cylinder. Following swimming sessions, the rats were removed from the tank, carefully dried in heated cages, and then returned to their home cages. At 24 h following the second session of the FST, all HR and LR rats (stressed and respective controls) were sacrificed by rapid decapitation with the aid of a guillotine. Notably, in earlier studies, we and others have chosen this wait period in order to examine prolonged stress-induced effects on gene expression alterations within a short time period following the stress session [17,61-63].

RNA isolation and qualitative and quantitative analyses

Following decapitation, the brain was removed from the skull and the region of the hippocampus was rapidly isolated on ice. Sample preparation, hybridization, washing, and staining were performed using the direct hybridization gene expression assay by Illumina, involving a first- and second-strand reverse transcription step, followed by a single *in vitro* transcription (IVT) amplification that incorporates biotin-labeled nucleotides, array hybridization, washing, blocking, and streptavadin-Cy3 staining. Hippocampal samples ($N = 5$ per group per phenotype) were analyzed independently on whole rat genome Illumina arrays (RatRef-12 Expression BeadChip), containing 22,523 probes selected primarily from the NCBI RefSeq database (Release 16), representing approximately 22,260 coding transcripts. The arrays were scanned using the BeadArray Reader (Illumina, San Diego, CA, USA).

Statistics, bioinformatic analysis, and data mining

Statistical analysis of behavioral data was performed with SPSS version 20 statistical software (SPSS Inc., Chicago, IL, USA). Results were analyzed using one-way analysis of variance (ANOVA) for the factor of phenotype (HR versus LR). Specifically, separate one-way ANOVAs were performed for each FST session. All data were first tested against ANOVA data assumptions.

The resulting gene expression data sets from the 20 hippocampal samples were scanned using the Illumina GenomeStudio Gene Expression Module (version 2010.2) at default specifications and thresholds. The scanned microarray image files were pre-processed using quantile normalization without background correction [64]. Bioinformatic analysis was performed using the BRB-Array Tools Version 4.2.1 [65]. All intensity values were transformed to the log2 basis. Differentially expressed genes were then identified using a random-variance *t* test, with a p value threshold of ≤ 0.01. To increase stringency, only genes with ≥ 2-fold change were further considered for the purposes of this study. Significantly changed genes were annotated according to the Gene Ontology classification system into 'biological process' (BP), 'molecular function' (MF), and 'cellular component' (CC) by using the GeneCodis3 software, a web tool for functional interpretation of experimental data. The application is freely available at http://genecodis.cnb.csic.es [66-68]. In particular, unique gene symbols (ENTREZ) corresponding to HR and LR data sets, respectively, were processed in the application 'Comparative analysis' of GeneCodis3 for *Rattus norvegicus*, the GO level (level 5), and the statistical parameters of the analysis. The submission of two different lists at the same time results in simultaneous modular and singular enrichment analyses for each one. Hypergeometric distribution was used for the calculation of p values and the FDR method to correct p values for multiple hypothesis testing. Statistical significance for both tests was determined at the $p < 0.05$ level. Annotations were then retrieved for each data set in order for unique and common genes to be further analyzed.

Literature mining and biological interpretation of significantly changed genes was performed with the Ingenuity Pathway Analysis software (IPA, Ingenuity Systems, www.ingenuity.com). Data of analysis are experimentally observed and come from Ingenuity Knowledge Base. Fisher's exact test is used for the grouping of analyzed genes into biofunctions and canonical pathways by calculating the p value of a given category. Statistical significance association of genes to the functional or pathway categories is considered for p values less than 0.05 and is expressed as the negative log of the p value. Additionally, the activation state of a functional category is predicted by calculating the z-score. A predicted increase in activation of a function is considered when the z-score is ≥ 2, and accordingly, a predicted decrease is considered when the z-score is ≤ 2.

In order to decipher the impact of FST exposure on the regulation of neurogenesis-related mechanisms, the Mammalian Adult Neurogenesis Gene Ontology (MANGO) was further implemented [19]. The MANGO accounts for a database of genes mapped to cell types and processes that have been curated from the literature concerning adult neurogenesis [19]. Herein, the MANGO database was screened for genes that were statistically significantly changed in our data sets.

Competing interests

The authors declare that they have no competing interests.

Authors' contributions

PMP, DS, PAT, and ZPD conceived the study and participated in its design and coordination, the interpretation of results, and the drafting of the manuscript. MP, DN, and MK performed the bioinfomatic analysis of the microarray data. CRT participated in the generation of the microarray data. All authors read and approved the final draft.

Acknowledgements

This work was supported by grants from the Special Account for Research Grants of the University of Athens (Greece) and from the National Institutes of Health grant EY16707. The input of Dr. R. Cowper-Sal-lari in the initial stages of the microarray analysis is greatly appreciated. The funders had no role in the study design, data collection and analysis, decision to publish, or preparation of the manuscript.

Author details

[1]Department of Biology & Center for Tissue Regeneration and Engineering (TREND), University of Dayton, 300 College Park, Dayton, OH 45469-2320, USA. [2]Department of Pharmacology, Medical School, National & Kapodistrian University of Athens, 75 Mikras Asias StreetGoudi, Athens 11527, Greece. [3]Department of Medicine, Dartmouth Hitchcock Medical Center, Dartmouth College, Lebanon, NH 03756, USA. [4]Department of Pharmacology and Toxicology, Dartmouth Hitchcock Medical Center, Dartmouth College, Lebanon, NH 03756, USA.

References

1. Duclot F, Hollis F, Darcy MJ, Kabbaj M: **Individual differences in novelty-seeking behavior in rats as a model for psychosocial stress-related mood disorders.** *Physiol Behav* 2011, **104:**296–305.
2. Harro J: **Inter-individual differences in neurobiology as vulnerability factors for affective disorders: implications for psychopharmacology.** *Pharmacol Ther* 2010, **125:**402–422.
3. Blanchard MM, Mendelsohn D, Stamp JA: **The HR/LR model: further evidence as an animal model of sensation seeking.** *Neurosci Biobehav Rev* 2009, **33:**1145–1154.
4. Antoniou K, Papathanasiou G, Papalexi E, Hyphantis T, Nomikos GG, Spyraki C, Papadopoulou-Daifoti Z: **Individual responses to novelty are associated with differences in behavioral and neurochemical profiles.** *Behav Brain Res* 2008, **187:**462–472.
5. Hansenne M, Reggers J, Pinto E, Kjiri K, Ajamier A, Ansseau M: **Temperament and character inventory (TCI) and depression.** *J Psychiatr Res* 1999, **33:**31–36.
6. Hur JW, Kim YK: **Comparison of clinical features and personality dimensions between patients with major depressive disorder and normal control.** *Psychiatry Investig* 2009, **6:**150–155.
7. Porsolt RD, Anton G, Blavet N, Jalfre M: **Behavioural despair in rats: a new model sensitive to antidepressant treatments.** *Eur J Pharmacol* 1978, **47:**379–391.
8. Borsini F: **Role of the serotonergic system in the forced swimming test.** *Neurosci Biobehav Rev* 1995, **19:**377–395.
9. Cryan JF, Markou A, Lucki I: **Assessing antidepressant activity in rodents: recent developments and future needs.** *Trends Pharmacol Sci* 2002, **23:**238–245.
10. Detke MJ, Rickels M, Lucki I: **Active behaviors in the rat forced swimming test differentially produced by serotonergic and noradrenergic antidepressants.** *Psychopharmacology (Berl)* 1995, **121:**66–72.
11. Lopez-Rubalcava C, Lucki I: **Strain differences in the behavioral effects of antidepressant drugs in the rat forced swimming test.** *Neuropsychopharmacology* 2000, **22:**191–199.
12. Pitychoutis PM, Pallis EG, Mikail HG, Papadopoulou-Daifoti Z: **Individual differences in novelty-seeking predict differential responses to chronic antidepressant treatment through sex- and phenotype-dependent neurochemical signatures.** *Behav Brain Res* 2011, **223:**154–168.
13. Jama A, Cecchi M, Calvo N, Watson SJ, Akil H: **Inter-individual differences in novelty-seeking behavior in rats predict differential responses to desipramine in the forced swim test.** *Psychopharmacology* 2008, **198:**333–340.
14. Knight RT, Nakada T: **Cortico-limbic circuits and novelty: a review of EEG and blood flow data.** *Rev Neurosci* 1998, **9:**57–70.
15. Nestler EJ, Barrot M, DiLeone RJ, Eisch AJ, Gold SJ, Monteggia LM: **Neurobiology of depression.** *Neuron* 2002, **34:**13–25.
16. Pittenger C, Duman RS: **Stress, depression, and neuroplasticity: a convergence of mechanisms.** *Neuropsychopharmacology* 2008, **33:**88–109.
17. Pitychoutis PM, Dalla C, Sideris AC, Tsonis PA, Papadopoulou-Daifoti Z: **5-HT (1A), 5-HT(2A), and 5-HT(2C) receptor mRNA modulation by antidepressant treatment in the chronic mild stress model of depression: sex differences exposed.** *Neuroscience* 2012, **210:**152–167.
18. Dalla C, Pitychoutis PM, Kokras N, Papadopoulou-Daifoti Z: **Sex differences in response to stress and expression of depressive-like behaviours in the rat.** *Curr Top Behav Neurosci* 2011, **8:**97–118.
19. Overall RW, Paszkowski-Rogacz M, Kempermann G: **The mammalian adult neurogenesis gene ontology (MANGO) provides a structural framework for published information on genes regulating adult hippocampal neurogenesis.** *PloS One* 2012, **7:**e48527.
20. Hur EM, Zhou FQ: **GSK3 signalling in neural development.** *Nat Rev Neurosci* 2010, **11:**539–551.
21. Guo W, Murthy AC, Zhang L, Johnson EB, Schaller EG, Allan AM, Zhao X: **Inhibition of GSK3beta improves hippocampus-dependent learning and rescues neurogenesis in a mouse model of fragile X syndrome.** *Hum Mol Genet* 2012, **21:**681–691.
22. Foltz DR, Santiago MC, Berechid BE, Nye JS: **Glycogen synthase kinase-3beta modulates notch signaling and stability.** *Curr Biol* 2002, **12:**1006–1011.
23. Llorens-Martin M, Fuster-Matanzo A, Teixeira CM, Jurado-Arjona J, Ulloa F, Defelipe J, Rabano A, Hernandez F, Soriano E, Avila J: **GSK-3beta overexpression causes reversible alterations on postsynaptic densities and dendritic morphology of hippocampal granule neurons in vivo.** *Mol Psychiatry* 2013, **18:**451–460.
24. Mira H, Andreu Z, Suh H, Lie DC, Jessberger S, Consiglio A, San Emeterio J, Hortiguela R, Marques-Torrejon MA, Nakashima K, Colak D, Götz M, Fariñas I, Gage FH: **Signaling through BMPR-IA regulates quiescence and long-term activity of neural stem cells in the adult hippocampus.** *Cell Stem Cell* 2010, **7:**78–89.
25. Ables JL, Breunig JJ, Eisch AJ, Rakic P: **Not(ch) just development: Notch signalling in the adult brain.** *Nat Rev Neurosci* 2011, **12:**269–283.
26. Shimazu K, Zhao M, Sakata K, Akbarian S, Bates B, Jaenisch R, Lu B: **NT-3 facilitates hippocampal plasticity and learning and memory by regulating neurogenesis.** *Learn Mem* 2006, **13:**307–315.
27. Bonaguidi MA, Peng CY, McGuire T, Falciglia G, Gobeske KT, Czeisler C, Kessler JA: **Noggin expands neural stem cells in the adult hippocampus.** *J Neurosci* 2008, **28:**9194–9204.
28. Favaro R, Valotta M, Ferri AL, Latorre E, Mariani J, Giachino C, Lancini C, Tosetti V, Ottolenghi S, Taylor V, Nicolis SK: **Hippocampal development and neural stem cell maintenance require Sox2-dependent regulation of Shh.** *Nat Neurosci* 2009, **12:**1248–1256.
29. Wachs FP, Winner B, Couillard-Despres S, Schiller T, Aigner R, Winkler J, Bogdahn U, Aigner L: **Transforming growth factor-beta1 is a negative modulator of adult neurogenesis.** *J Neuropathol Exp Neurol* 2006, **65:**358–370.
30. Kandasamy M, Reilmann R, Winkler J, Bogdahn U, Aigner L: **Transforming growth factor-beta signaling in the neural stem cell niche: a therapeutic target for Huntington's disease.** *Neurol Res Int* 2011, **2011:**124256.
31. Krampert M, Chirasani SR, Wachs FP, Aigner R, Bogdahn U, Yingling JM, Heldin CH, Aigner L, Heuchel R: **Smad7 regulates the adult neural stem/progenitor cell pool in a transforming growth factor beta- and bone morphogenetic protein-independent manner.** *Mol Cell Biol* 2010, **30:**3685–3694.
32. Chadashvili T, Peterson DA: **Cytoarchitecture of fibroblast growth factor receptor 2 (FGFR-2) immunoreactivity in astrocytes of neurogenic and non-neurogenic regions of the young adult and aged rat brain.** *J Comp Neurol* 2006, **498:**1–15.
33. Wichers MC, Maes M: **The role of indoleamine 2,3-dioxygenase (IDO) in the pathophysiology of interferon-alpha-induced depression.** *J Psychiatry Neurosci* 2004, **29:**11–17.
34. Sousa N, Almeida OF: **Corticosteroids: sculptors of the hippocampal formation.** *Rev Neurosci* 2002, **13:**59–84.
35. Cameron HA, Tanapat P, Gould E: **Adrenal steroids and N-methyl-D-aspartate receptor activation regulate neurogenesis in the dentate gyrus of adult rats through a common pathway.** *Neurosci* 1998, **82:**349–354.
36. Consiglio AR, Ramos AL, Henriques JA, Picada JN: **DNA brain damage after stress in rats.** *Prog Neuropsychopharmacol Biol Psychiatry* 2010, **34:**652–656.
37. Vega-Rivera NM, Fernandez-Guasti A, Ramirez-Rodriguez G, Estrada-Camarena E: **Acute stress further decreases the effect of ovariectomy on immobility behavior and hippocampal cell survival in rats.** *Psychoneuroendocrinology* 2013, **38:**1407–1417.
38. Kirby ED, Muroy SE, Sun WG, Covarrubias D, Leong MJ, Barchas LA, Covarrubias D, Leong MJ, Barchas LA, Kaufer D: **Acute stress enhances adult rat hippocampal neurogenesis and activation of newborn neurons via secreted astrocytic FGF2.** *eLife* 2013, **2:**e00362.
39. Lemaire V, Aurousseau C, Le Moal M, Abrous DN: **Behavioural trait of reactivity to novelty is related to hippocampal neurogenesis.** *Eur J Neurosci* 1999, **11:**4006–4014.
40. Clinton SM, Stead JD, Miller S, Watson SJ, Akil H: **Developmental underpinnings of differences in rodent novelty-seeking and emotional reactivity.** *Eur J Neurosci* 2011, **34:**994–1005.
41. Xiao D, Miller GM, Jassen A, Westmoreland SV, Pauley D, Madras BK: **Ephrin/Eph receptor expression in brain of adult nonhuman primates: implications for neuroadaptation.** *Brain Res* 2006, **1067:**67–77.
42. Cooper MA, Crockett DP, Nowakowski RS, Gale NW, Zhou R: **Distribution of EphA5 receptor protein in the developing and adult mouse nervous system.** *J Comp Neurol* 2009, **514:**310–328.
43. Henderson JT, Georgiou J, Jia Z, Robertson J, Elowe S, Roder JC, Pawson T: **The receptor tyrosine kinase EphB2 regulates NMDA-dependent synaptic function.** *Neuron* 2001, **32:**1041–1056.
44. Mohanasundaram P, Shanmugam MM: **Role of syntaxin 4 in activity-dependent exocytosis and synaptic plasticity in hippocampal neurons.** *Sci Signal* 2010, **3:**jc7.

45. Murk K, Wittenmayer N, Michaelsen-Preusse K, Dresbach T, Schoenenberger CA, Korte M, Jockusch BM, Rothkegel M: **Neuronal profilin isoforms are addressed by different signalling pathways.** *PLoS One* 2012, **7**:e34167.
46. Mallozzi C, D'Amore C, Camerini S, Macchia G, Crescenzi M, Petrucci TC, Di Stasi AM: **Phosphorylation and nitration of tyrosine residues affect functional properties of Synaptophysin and Dynamin I, two proteins involved in exo-endocytosis of synaptic vesicles.** *Biochimica Et Biophysica Acta* 2013, **1833**:110–121.
47. Maximov A, Shin OH, Liu X, Sudhof TC: **Synaptotagmin-12, a synaptic vesicle phosphoprotein that modulates spontaneous neurotransmitter release.** *J Cell Biol* 2007, **176**:113–124.
48. Thome J, Pesold B, Baader M, Hu M, Gewirtz JC, Duman RS, Henn FA: **Stress differentially regulates synaptophysin and synaptotagmin expression in hippocampus.** *Biol Psychiatry* 2001, **50**:809–812.
49. Webb BL, Proud CG: **Eukaryotic initiation factor 2B (eIF2B).** *Int J Biochem Cell Biol* 1997, **29**:1127–1131.
50. Chang RC, Wong AK, Ng HK, Hugon J: **Phosphorylation of eukaryotic initiation factor-2alpha (eIF2alpha) is associated with neuronal degeneration in Alzheimer's disease.** *Neuroreport* 2002, **13**:2429–2432.
51. Thon L, Mohlig H, Mathieu S, Lange A, Bulanova E, Winoto-Morbach S, Schutze S, Bulfone-Paus S, Adam D: **Ceramide mediates caspase-independent programmed cell death.** *FASEB J* 1945–1956, **2005**:19.
52. Spiegel S, Milstien S: **Sphingosine-1-phosphate: an enigmatic signalling lipid.** *Nat Rev Mol Cell Biol* 2003, **4**:397–407.
53. Mandala SM: **Sphingosine-1-phosphate phosphatases.** *Prostaglandins Other Lipid Mediat* 2001, **64**:143–156.
54. Chen CL, Lin CF, Chang WT, Huang WC, Teng CF, Lin YS: **Ceramide induces p38 MAPK and JNK activation through a mechanism involving a thioredoxin-interacting protein-mediated pathway.** *Blood* 2008, **111**:4365–4374.
55. Stanger BZ, Leder P, Lee TH, Kim E, Seed B: **RIP: a novel protein containing a death domain that interacts with Fas/APO-1 (CD95) in yeast and causes cell death.** *Cell* 1995, **81**:513–523.
56. Polissidis A, Chouliara O, Galanopoulos A, Rentesi G, Dosi M, Hyphantis T, Marselos M, Papadopoulou-Daifoti Z, Nomikos GG, Spyraki C, Tzavara ET, Antoniou K: **Individual differences in the effects of cannabinoids on motor activity, dopaminergic activity and DARPP-32 phosphorylation in distinct regions of the brain.** *Int J Neuropsychopharmacol* 2010, **13**:1175–1191.
57. Topic B, Dere E, Schulz D, de Souza Silva MA, Jocham G, Kart E, Huston JP: **Aged and adult rats compared in acquisition and extinction of escape from the water maze: focus on individual differences.** *Behav Neurosci* 2005, **119**:127–144.
58. Antoniou K, Papathanasiou G, Panagis G, Nomikos GG, Hyphantis T, Papadopoulou-Daifoti Z: **Individual responses to novelty predict qualitative differences in d-amphetamine-induced open field but not reward-related behaviors in rats.** *Neurosci* 2004, **123**:613–623.
59. Thiel CM, Muller CP, Huston JP, Schwarting RK: **High versus low reactivity to a novel environment: behavioural, pharmacological and neurochemical assessments.** *Neurosci* 1999, **93**:243–251.
60. Porsolt RD, Le Pichon M, Jalfre M: **Depression: a new animal model sensitive to antidepressant treatments.** *Nature* 1977, **266**:730–732.
61. Drossopoulou G, Antoniou K, Kitraki E, Papathanasiou G, Papalexi E, Dalla C, Papadopoulou-Daifoti Z: **Sex differences in behavioral, neurochemical and neuroendocrine effects induced by the forced swim test in rats.** *Neuroscience* 2004, **126**:849–857.
62. Lopez JF, Liberzon I, Vazquez DM, Young EA, Watson SJ: **Serotonin 1A receptor messenger RNA regulation in the hippocampus after acute stress.** *Biol Psychiatry* 1999, **45**:934–937.
63. Shishkina GT, Kalinina TS, Berezova IV, Bulygina VV, Dygalo NN: **Resistance to the development of stress-induced behavioral despair in the forced swim test associated with elevated hippocampal Bcl-xl expression.** *Behav Brain Res* 2010, **213**:218–224.
64. Wright GW, Simon RM: **A random variance model for detection of differential gene expression in small microarray experiments.** *Bioinformatics* 2003, **19**:2448–2455.
65. Smyth GK, Speed T: **Normalization of cDNA microarray data.** *Methods* 2003, **31**:265–273.
66. Tabas-Madrid D, Nogales-Cadenas R, Pascual-Montano A: **GeneCodis3: a non-redundant and modular enrichment analysis tool for functional genomics.** *Nucleic Acids Res* 2012, **40**:W478–W483.
67. Nogales-Cadenas R, Carmona-Saez P, Vazquez M, Vicente C, Yang X, Tirado F, Carazo JM, Pascual-Montano A: **GeneCodis: interpreting gene lists through enrichment analysis and integration of diverse biological information.** *Nucleic Acids Res* 2009, **37**:W317–W322.
68. Carmona-Saez P, Chagoyen M, Tirado F, Carazo JM, Pascual-Montano A: **GENECODIS: a web-based tool for finding significant concurrent annotations in gene lists.** *Genome Biol* 2007, **8**:R3.

New susceptibility locus for obesity and dyslipidaemia on chromosome 3q22.3

Maie Alshahid[1], Salma M Wakil[2], Mohammed Al-Najai[2], Nzioka P Muiya[2], Samar Elhawari[2], Daisy Gueco[2], Editha Andres[2], Samia Hagos[2], Nejat Mazhar[2], Brian F Meyer[1] and Nduna Dzimiri[2*]

Abstract

Background: The muscle Ras (*MRAS*) gene resides on chromosome 3q22.3 and encodes a member of the membrane-associated Ras small GTPase proteins, which function as signal transducers in multiple processes including cell growth and differentiation. Its role in cardiovascular disease is not fully understood yet. In a preliminary study in heterozygous familial hypercholesterolaemia, we identified a locus linking the early onset of coronary artery disease (CAD) to chromosome 3q.22 and elected to sequence the *MRAS* gene using the MegaBACE DNA analysis system. In the present study, we investigated the association of seven single-nucleotide polymorphisms (SNPs) at this locus with CAD and its dyslipidaemia-related risk traits in 4,650 Saudi angiographed individuals using TaqMan assays by the Applied Biosystems real-time Prism 7900HT Sequence Detection System.

Results: Among the studied SNPs, rs6782181 ($p = 0.017$) and rs9818870T ($p = 0.009$) were associated with CAD following adjustment for sex, age and other confounding risk factors. The rs6782181_GG also conferred risk for obesity (1,764 cases vs. 2,586 controls) [1.16(1.03–1.30); $p = 0.017$], hypercholesterolaemia (1,686 vs. 2,744) [1.23(1.02–1.47); $p = 0.019$], hypertriglyceridaemia (1,155 vs. 3,496) [1.29(1.01–1.45); $p = 0.043$] and low high-density lipoprotein-cholesterol (lHDL-chol) levels (1,935 vs. 2,401) [1.15(1.02–1.30); $p = 0.023$] after adjustment. Additionally, rs253662_(CT+TT) [1.16(1.01–1.32); $p = 0.030$] was associated with lHDL-chol levels. Interestingly, rs253662 ($p = 0.014$) and rs6782181 ($p = 0.019$) were protective against acquiring high low-density lipoprotein-cholesterol (hLDL-chol) levels ($p = 0.014$), while rs1720819 showed similar effects against CAD ($p < 0.0001$). More importantly, a 7-mer haplotype, ACCTGAC ($\chi^2 = 7.66$; $p = 0.0056$), constructed from the studied SNPs, its 6-mer derivative CCTGAC ($\chi^2 = 6.90$; $p = 0.0086$) and several other shorter derivatives conferred risk for obesity. hLDL-chol was weakly linked to CTAA ($\chi^2 = 3.79$; $p = 0.052$) and CCT ($\chi^2 = 4.32$; $p = 0.038$), while several other haplotypes were protective against both obesity and hLDL-chol level.

Conclusion: Our results demonstrate that the genomic locus for the *MRAS* gene confers risk for CAD, obesity and dyslipidaemia and point to the possible involvement of other genes or regulatory elements at this locus, rather than changes in the M-Ras protein function, in these events.

Keywords: *MRAS* gene polymorphism, Coronary artery disease, Obesity, Hypercholesterolaemia, Dyslipidaemia, Hypertriglyceridaemia, Linkage, Haplotype

* Correspondence: dzimiri@kfshrc.edu.sa
[2]Genetics Department, MBC-03, King Faisal Specialist Hospital and Research Centre, Riyadh 11211, Saudi Arabia
Full list of author information is available at the end of the article

Background

The *MRAS* gene encodes a member of the membrane-associated family of Ras small GTPase proteins [1], which function as signal transducers in multiple processes including cell growth and differentiation [2-6]. With the growing number of effectors being continually described for this GTP-binding protein superfamily [7,8], it is becoming increasingly evident that their signal transduction leads to the generation of a multitude of cellular signals. Among others, these proteins play a role in the tumour necrosis factor-alpha and MAP kinase signalling pathways [9-12]. Hence, the dysregulation of the Ras signalling has been associated with many types of cancers [13-17]. Notably, the M-Ras protein is ubiquitously expressed but very highly so in the cardiovascular system, particularly in the heart itself [18], intuitively pointing to a possibly pivotal role in cardiovascular function. However, its role in cardiovascular disease is not fully understood yet. A recent study by Erdmann et al. [19] revealed a region on 3q22.3, which encompasses the *MRAS* gene, as a risk factor for coronary artery disease (CAD), while another study by Ellis et al. [20] proposed that the rs9818870 at this locus is a predictor of cardiovascular risk in individuals free of overt heart disease. Besides, a previous study has shown that M-Ras is engaged in tumour necrosis factor-alpha-stimulated lymphocyte function-associated antigen 1 activation in splenocytes [10], thus suggesting a role in adhesion signalling, which is an important aspect of atherosclerotic pathways [21].

Currently, there is hardly any tangible data available in the literature pertaining to the possible role of the *MRAS* gene polymorphism in cardiovascular risk traits for CAD, such as obesity, hypertension (HTN), type 2 diabetes mellitus (T2DM) or dyslipidaemic disorders. In a linkage study on the early onset of CAD in heterozygous familial hypercholesterolaemia (HFH), we identified a locus on chromosome (chr) 3q.22 that was linked to both disorders (Dzimiri et al., unpublished data). Considering the fact that this locus has also been implicated in CAD, our present study addressed the question as to whether it might signify a link between dyslipidaemia-related disorders and atherosclerosis onset in general, in a homogenous cohort of 4,650 Saudi individuals harbouring CAD and its lipid-metabolic risk traits.

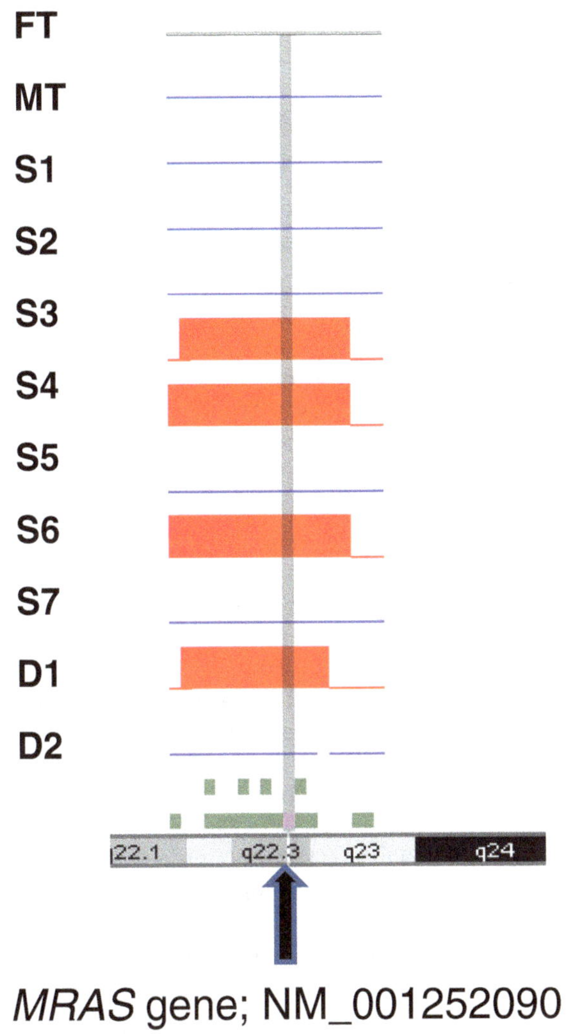

Figure 1 Homozygosity mapping for early onset of coronary artery disease in heterozygous familial hypercholesterolaemia. Affymetrix GT console mapping indicating the position of homozygosity for the four affected offsprings: S3, S4, S6 and D1. FT, father; MT, mother; S1–7, sons 1–7; D1 and 2, daughters 1 and 2.

Results

The initial linkage study using the Affymetrix Gene Chip 250 styl mapping array revealed several genomic loci, including one on chr 3q.22, as a potential risk locus for both the early onset of CAD and familial hypercholesterolaemia in a family of 11 members harbouring HFH (Figure 1). We then selected the *MRAS*, which resides at this locus, as the target gene for further studies on its role as a risk for both CAD and dyslipidaemic disorders. This was followed by sequencing the gene in the family members and 200 individuals from the general population in order to identify informative single-nucleotide polymorphisms (SNPs) of interest. Our results indicated that the *MRAS* gene is extensively polymorphic in the Saudi population, yielding several SNPs in both the coding and non-coding areas of the gene. Particularly noteworthy was the finding of a large number of intronic SNPs as well as those residing in the 3 prime untranslated region (3′-UTR) of the gene. This stimulated an interest for us to examine the likelihood that these regions may constitute significant risk loci for complex diseases, such as CAD. Accordingly, we then selected seven SNPs showing minor allele frequencies of >0.1 in these individuals, as potentially informative for their possible association with disease. Apart from their frequencies in the above groups, the selection of these SNPs was also partly based on available information on their impact on disease to allow comparison with other data in the literature. Selected SNPs were rs1199338_AC (1), rs166195_CT (2), rs253662_CT (3), rs1720819_TG (4), rs6782181_AG (5), rs3732837_TA (6) and rs9818870_CT (7), numbered sequentially according to their chromosomal positional arrangement (Figure 2). The linkage disequilibrium for the SNPs is given in Figure 3.

The association analysis was performed in a total of 4,650 angiographed individuals, of which 2,429 had CAD and 2,221 served as CAD-free controls. Among the studied SNPs, rs6782181 (p = 0.017) and rs9818870T (p = 0.009) were associated with CAD following adjustment for sex, age and the confounding effects of other risk factors (Table 1; see also Additional file 1). Notably, rs6782181GG also conferred risk for obesity (1,764 cases vs. 2,586 controls) [1.16(1.03–1.30); p = 0.017], hypercholesterolaemia (hChol; 1,686 vs. 2,744) [1.23(1.02–1.47); p = 0.019], hypertriglyceridaemia (hTG; 1,166 vs. 3,168) [1.29(1.01–1.45); p = 0.043] and low high-density lipoprotein-cholesterol (lHDL-chol) levels (1,935 vs. 2,401) [1.15(1.02–1.30); p = 0.023] following adjustment for the possible confounding effects of other risk factors (Table 2; Figure 4). Additionally, rs253662 (CT+TT) [1.16(1.01–1.32); p = 0.030] was also implicated in the harbouring of lHDL-chol levels. One other SNP, rs166195, lost its association with lHDL-chol following the adjustment for confounders. Interestingly, two of the SNPs, rs253662 (p = 0.014) and rs6782181 (p = 0.019), were protective against acquiring hLDL-chol, whereby the former showed similar but weaker properties towards hypercholesterolaemia. One other variant, rs1720819, was protective against CAD (p < 0.0001). Furthermore, none of the variants showed any delineable relationship with hypertension or type 2 diabetes mellitus.

We were also interested in evaluating whether any haplotype was identifiable in relationship to these risk traits. We employed the most common 7-mer haplotype, ACCTAAC (frequency = 0.558), constructed from the studied SNPs as a baseline for comparing their relationships with the disease traits. Interestingly, a 7-mer haplotype, ACCTGAC (χ^2 = 7.66; p = 0.0056), its 6-mer derivative CCTGAC (χ^2 = 6.90; p = 0.0086) and other shorter derivatives conferred risk for obesity. Another 7-mer haplotype, ACCTAAC (χ^2 = 4.63; p = 0.031), and its derivatives were protective against obesity (Table 3; Additional file 2). Besides, no notable causative association was observed between the haplotypes and changes in total cholesterol level. Notably, apart from a 3-mer CCT (χ^2 = 4.32; p = 0.038) that was positively associated with hLDL-chol levels, the only other haplotypes linked, but weakly so, to this trait were the 4-mer TAAC (p = 0.057) and its 3-mer derivative AAC (p = 0.052), both of which were actually protective towards obesity (χ^2 = 4.40; p = 0.036). Instead, two 7-mer haplotypes, ATTTGAC (χ^2 = 4.39; p = 0.036) and CCCGGTT (χ^2 = 4.31; p = 0.038), showed protective properties towards acquiring hLDL-chol levels.

Discussion

The present study evaluated the role of gene variants residing at the genomic locus of the *MRAS* gene in CAD and its metabolic risk factors. Our results indicate that at least two of the studied SNPs were associated with CAD. The finding of an association with CAD in the present study is in agreement with the association of a cluster of four SNPs consisting of rs1199338, rs2347252,

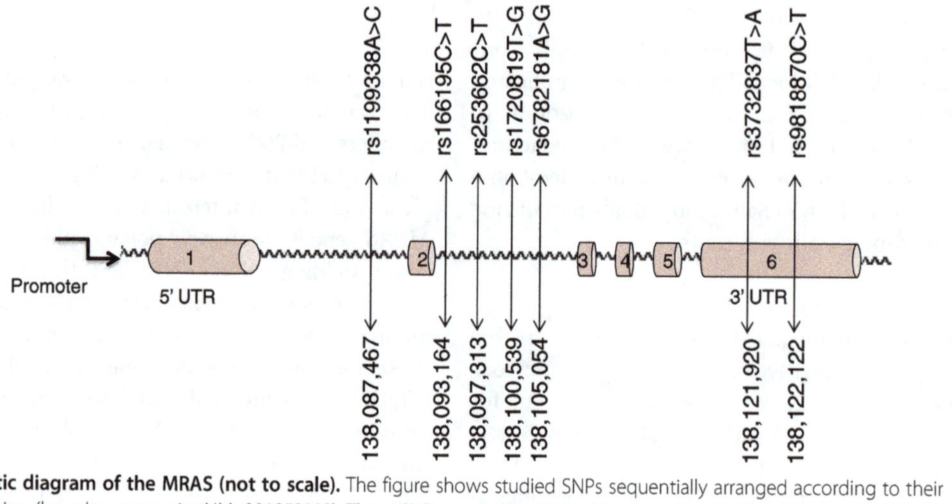

Figure 2 Schematic diagram of the MRAS (not to scale). The figure shows studied SNPs sequentially arranged according to their chromosomal position (based on transcript NM_001252090). These SNPs were detected with a prevalence of >0.1 in the general population.

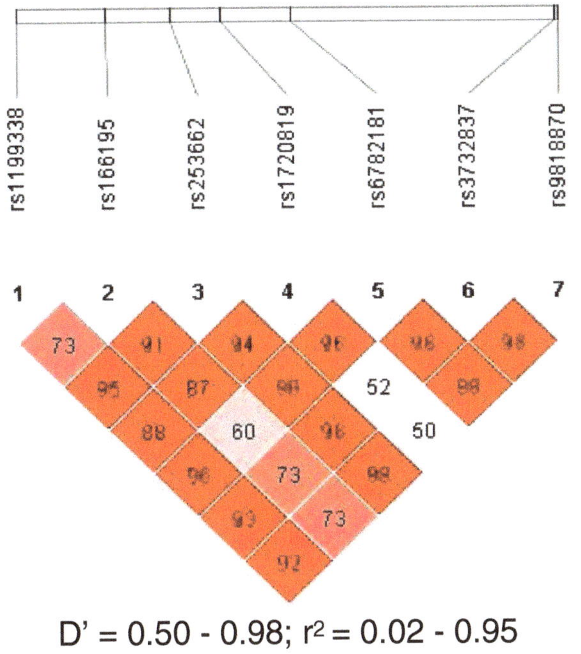

Figure 3 Linkage disequilibrium of the seven studied SNPs. The figure displays the coefficient of the linkage disequilibrium (*D'*) and the regression coefficient of the linkage disequilibrium (r^2) for the studied SNPs.

rs3732837 and rs9818870 in this chromosomal region with CAD in the study of Erdmann et al. [19]. More importantly, our study also revealed that, apart from CAD, this locus is linked to its important risk traits, obesity, and the lipidaemic disorders, hyperlipidaemia, hypertriglyceridaemia and the harbouring of low HDL-cholesterol levels. Notably, one predisposing SNP, rs6782181, was common to CAD, obesity and the dyslipidaemia traits, both independently and following correction for confounding factors. Thus, our data not only describe a relationship for this SNP with three important dyslipidaemia-related risk traits, but also suggest a possible genomic link for these cardiovascular disease traits at this locus. Besides, lHDL-chol levels were further associated with the dominant inheritance mode for rs253662 and weakly linked to that for the rs166195. Therefore, put together, our results point to the genomic locus for the *MRAS* gene as conferring risk for CAD and its important metabolic risk traits.

Interestingly, the association of rs253662 with the harbouring of low HDL-cholesterol levels is in direct contrast to its protective effects on the acquisition of high LDL-cholesterol and similar but weaker properties towards elevated total cholesterol levels. Currently, apart from a study implicating the rs9818870 as a predictor of

Table 1 Association of *MRAS* variants with coronary artery disease

Model	Unstandardized coefficient		Standardized coefficient	*t*	*p* value	95.0% C.I. for beta	
	Beta	S.E.	Beta			Lower bound	Upper bound
Sex	−0.107	0.009	−0.102	−12.05	0.000***	−0.125	−0.09
Age	0.003	0.000	0.081	8.53	0.000***	0.002	0.003
FH	−0.004	0.01	−0.003	−0.401	0.689	−0.024	0.016
T2DM	0.082	0.009	0.082	9.034	0.000***	0.064	0.1
MI	0.586	0.01	0.56	59.874	0.000***	0.567	0.605
HTN	0.03	0.01	0.026	2.907	0.004***	0.01	0.05
OBS	−0.018	0.008	−0.018	−2.177	0.03*	−0.035	−0.002
rs119338_C	0.002	0.023	0.002	0.106	0.915	−0.043	0.048
rs166195_C	0.053	0.049	0.045	1.076	0.282	−0.044	0.15
rs166195_T	0.038	0.048	0.033	0.785	0.433	−0.057	0.133
rs253662_T	−0.012	0.019	−0.009	−0.622	0.534	−0.049	0.026
rs1720819_T	−0.182	0.052	−0.119	−3.505	0.000***	−0.284	−0.08
rs1720819_G	−0.164	0.055	−0.104	−2.996	0.003**	−0.272	−0.057
rs6782181_AA	0.073	0.031	0.071	2.379	0.017*	0.013	0.133
rs6782181_T	0.057	0.031	0.055	1.816	0.069	−0.005	0.118
rs3732837_AA	0.013	0.022	0.009	0.58	0.562	−0.03	0.056
rs3732837_T	−0.027	0.043	−0.017	−0.643	0.52	−0.111	0.056
rs9818870_C	0.097	0.04	0.063	2.416	0.016*	0.018	0.176
rs9818870_T	0.14	0.054	0.087	2.612	0.009**	0.035	0.246

The table shows the significant association of the *MRAS* gene variants with CAD following Bonferroni correction. Beta, odds ratio coefficient; *FH* family history of CAD, *T2DM* type 2 diabetes mellitus, *MI* myocardial infarction, *HTN* hypertension, *OBS* obesity, *C.I* confidence interval, *S.E.* standard error; *t*, standard *t* test. $^*p < 0.05$; $^{**}p < 0.01$; $^{***}p < 0.005$.

Table 2 Association of *MRAS* gene with obesity and dyslipidaemia

Variant	Controls	Cases	Univariate analysis		Multivariate analysis	
			Exp(*B*) (95% C.I.)	*p* value	Exp(*B*) (95% C.I.)	*p* value
Obesity						
rs6782181_GG	0.153	0.397	1.15(1.02–1.29)	0.023*	1.16(1.03–1.30)	0.017*
Hypercholesterolaemia						
rs253662_TT	0.051	0.038	0.73(0.59–0.90)	0.041*	0.74(0.53–1.01)	0.054
rs6782181_GG	0.015	0.022	1.47(1.07–2.03)	0.018*	1.23(1.02–1.47)	0.019*
Low HDL levels						
rs166195_CT+TT	0.356	0.385	1.14(1.00–1.29)	0.047*	1.00(0.99–1.00)	0.961
rs253662_CT+TT	0.318	0.344	1.13(0.99–1.29)	0.064	1.16(1.01–1.32)	0.030*
rs6782181_GG	0.151	0.167	1.13(1.00–1.27)	0.045*	1.15(1.02–1.30)	0.023*
High LDL-cholesterol						
rs253662_T	0.192	0.162	0.82(0.70–0.96)	0.013*	0.82(0.69–0.96)	0.014*
rs253662_CT+TT	0.336	0.291	0.82(0.67–0.98)	0.028*	0.81(0.68–0.98)	0.030*
rs6782181_G	0.387	0.356	0.88(0.77–0.99)	0.038*	0.88(0.77–0.99)	0.040*
rs6782181_AG+GG	0.614	0.564	0.82(0.68–0.97	0.019*	0.81(0.68–0.97)	0.019*
Hypertriglyceridaemia						
rs6782181_G	0.152	0.175	1.19(1.00–1.42)	0.067	1.29(1.01–1.45)	0.043*

The table shows the *MRAS* variants associated with the disease traits. B, odds ratio; C.I., confidence interval. *$p < 0.05$.

cardiovascular risk in individuals free of overt heart disease, there is hardly any literature describing the relationship between the *MRAS* gene or this genomic locus and lipid metabolic disorders. Thus, to our knowledge, this is the first report implicating this region in obesity and dyslipidaemia.

We further evaluated the possibility that haplotypes might be more informative than the individual SNPs in discerning the impact of the gene on disease. It was remarkable that a haplotype constructed from these SNPs and several of its shorter derivatives conferred risk for obesity, displaying a much higher level of significance for the relationships than the individual variants. Even more unique was the observation that the only trait unequivocally associated with these haplotypes was obesity. If anything, for the other disease traits, the large majority of the associations were actually protective against acquiring high LDL-chol, in line with the observed protective actions of the individual SNPs. These observations can be conceived as confirming the importance of this genomic region as a specific risk for obesity. The weak association of haplotypes with the dyslipidaemic traits may indicate that the influence of the gene on dyslipidaemia-related events may be driven primarily through the interaction of the gene with obesity.

The interesting question is whether or not this locus might provide a mechanistic link between dyslipidaemia

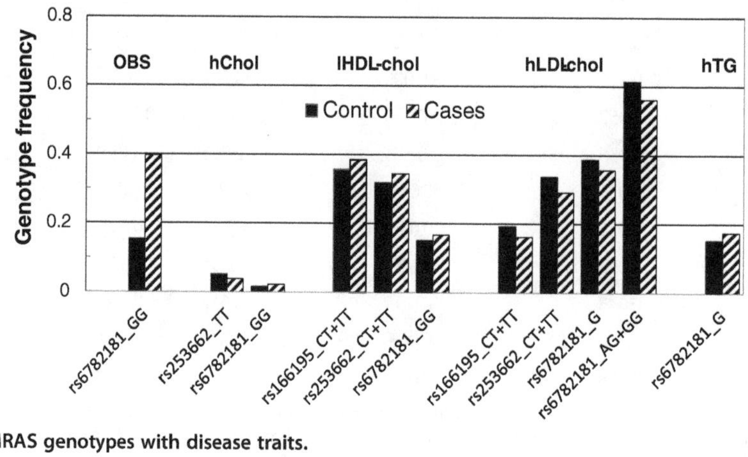

Figure 4 Association of MRAS genotypes with disease traits.

Table 3 Association of *MRAS* haplotypes with obesity and high low-density lipoprotein levels

Block	Haplotype	Pooled	Cases	Control	χ^2	*p* value
Obesity						
1–7	ACCTAAC	0.558	0.541	0.565	4.63	0.031
	ACCTGAC	0.035	0.042	0.031	7.66	0.0056*
1–6	ACCTAA	0.558	0.541	0.564	4.60	0.032
	ACCTGA	0.036	0.044	0.032	8.28	0.004**
2–7	CCTAAC	0.56	0.543	0.567	4.61	0.032
	CCTGAC	0.035	0.042	0.032	6.90	0.0086*
3–7	CTAAC	0.614	0.597	0.618	3.74	0.053
	CTGAC	0.04	0.047	0.036	7.15	0.0075*
1–5	ACCTA	0.558	0.541	0.564	4.36	0.039
	ACCTG	0.037	0.045	0.033	9.22	0.0024**
4–7	TAAC	0.618	0.601	0.623	4.40	0.036
High LDL-cholesterol						
1–7	ATTTGAC	0.172	0.153	0.177	4.39	0.036
	CCCGGTT	0.058	0.045	0.060	4.31	0.038
2–7	TTTGAC	0.173	0.152	0.179	5.11	0.024
	CCGGTT	0.062	0.046	0.064	6.27	0.012
1–6	ATTTGA	0.172	0.152	0.177	4.54	0.033
1–5	ATTTG	0.172	0.152	0.176	4.49	0.034
2–6	TTTGA	0.173	0.151	0.178	5.45	0.019
4–7	TAAC	0.618	0.640	0.612	3.61	0.057
	GGTT	0.061	0.048	0.064	4.81	0.028
1–4	ATTT	0.174	0.153	0.179	4.92	0.027
3–6	CTAA	0.613	0.635	0.606	3.79	0.052
	TTGA	0.181	0.158	0.187	6.02	0.014
2–5	TTTG	0.174	0.152	0.179	5.23	0.022
1–3	ATT	0.175	0.154	0.180	5.01	0.025
5–7	AAC	0.620	0.642	0.613	3.78	0.052
2–4	CCT	0.638	0.663	0.632	4.32	0.038

The table shows selected haplotypes constructed from combinations of the studied variants. The most frequent 7-mer haplotype, ACCTAAC (0.558), was employed as the baseline haplotype to determine the relative effects of the associated haplotypes. The minor allele frequency and control and case haplotype frequencies are proportions. Number 1 denotes rs1199338_AC, 2 is rs166195_CT, 3 is rs253662_CT, 4 is rs1720819_TG, 5 is rs6782181_AG, 6 is rs3732837_AT and 7 is rs9818870_CT, arranged sequentially by their chromosomal positions, and blocks represent the variant constituting the respective haplotypes. *$p < 0.01$; **$p < 0.005$ by the χ^2 test.

and the manifestation of CAD. A number of observations seem to suggest a direct link of the mechanisms involved in these events at the genomic level to some function of this locus. Particularly noteworthy is the fact that the studied variants are either intronic or reside in the 3′-UTR relative to the *MRAS* gene sequence. This implies that the effects of the studied variants are not likely to be directly related to functional changes of M-Ras protein *per se*. Thus, while no speculations can be made of the mechanisms involved in these interactions, based on the current findings alone, it can nonetheless be inferred that other yet unidentified genes/entities at the chromosome 3q.22 locus may ultimately be responsible for the observed manifestations and might offer a link between dyslipidaemia and pathways leading to atherosclerosis. Such mechanisms, particularly those related to SNPs in the 3′-UTR may involve some gene regulatory entities embedded in the function of this genomic region. Genetic hitchhiking might also explain the association of particularly the intragenic non-coding SNPs, as a result of the variants being linked with a causative mutation on the M-Ras through positive selection at this locus [22-24] or some spatial selection pressures [25-29]. However, further speculation on this possibility is beyond the scope of the present work. Hence, further studies are needed to look into these possibilities.

Conclusions

In conclusion, our study suggests that the genomic locus for the *MRAS* gene confers risk for CAD, obesity and dyslipidaemia. However, our findings point to the involvement of other genes or factors, rather than changes in the M-Ras protein function, as the possible underlying cause for the observed events.

Methods

Study population

The initial linkage study was performed in a family of 11 individuals with HFH, in which the primary proband underwent triple bypass surgery at the age of 14 years. Following the identification of chromosome 3q.22 as a potential locus for this disorder as well as the early onset of CAD, we elected to sequence the *MRAS* gene in the family and 200 other individuals from the general population. We then embarked on a case-control study in a total of 4,650 individuals consisting of 2,429 CAD patients (1,860 males and 569 females, mean age 55.6 ± 0.4 years) with angiographically determined narrowing of the coronary vessels by at least 50% and 2,221 angiographed controls (1,189 males and 1,032 females, mean age 53.8 ± 0.5 years). The controls for CAD were a group of individuals undergoing surgery for heart valvular diseases, and those who may have reported with chest pain, but were established to have no significant coronary stenosis by angiography. Among these, 3,060 individuals had established MI. The primary patient subset of interest in this population comprised 1,764 obese individuals with a body mass index of ≥30.0 kg/m^2 (Table 4). Furthermore, 3,541 individuals had primary (essential) HTN, defined as ≥140 mmHg systolic blood pressure or ≥90 mmHg diastolic pressure based on The Sixth Report of the Joint National Committee on Prevention, Detection, Evaluation, and Treatment of High Blood Pressure (JNC VI) criteria [30]. Accordingly, essential, primary or idiopathic hypertension is defined as high blood pressure in which secondary causes such as renovascular disease, renal failure, pheochromocytoma,

Table 4 Demographics and clinical data of the studied patients

	Controls			Cases		
	All	Male	Female	All	Male	Female
Demographics						
OBS	2,586	1,886(0.73)	700(0.27)	1,764	967(0.55)	797(0.45)
Age[a]	53.8 ± 0.5	56.4 ± 0.3	51.1 ± 0.7	55.6 ± 0.4	55.5 ± 0.4	55.7 ± 0.4
BMI[a]	25.3 ± 0.1	25.3 ± 0.1	25.2 ± 0.1	34.9 ± 0.2	33.8 ± 0.1	35.9 ± 0.2
CAD	2,221	1,189(0.54)	1,032(0.46)	2,429	1,860(0.66)	569(0.34)
MI	1,590	787(0.49)	803(0.51)	3,060	2,262(0.74)	798(0.26)
FH	3,730	2,449(0.66)	1,283(0.34)	920	602(0.65)	318(0.35)
T2DM	2,090	1,336(0.64)	754(0.36)	2,560	1,713(0.55)	847(0.45)
HTN	1,109	735(0.66)	374(0.34)	3,541	2,314(0.65)	1,227(0.35)
lHDLC	2,516	1,487(0.59)	1,029(0.41)	1,935	1,487(0.77)	448(0.23)
hLDLC	3,696	2,449(0.75)	1,029(0.25)	860	634(0.74)	226(0.26)
hChol	2,744	1,801(0.66)	943(0.34)	1,686	1,117(0.66)	569(0.34)
hTG	3,168	2,028(0.64)	1,140(0.36)	1,166	829(0.71)	337(0.29)
Smokers	2,791	1,314(0.47)	1,477(0.53)	1,771	1,694(0.96)	77(0.04)
Clinical characteristics (relative to coronary artery disease)						
Age	50.6 ± 0.4	51.2 ± 0.5	49.8 ± 0.5	60.3 ± 0.2	59.8 ± 0.3	61.8 ± 0.54
BMI	29.0 ± 0.2	27.97 ± 0.2	30.3 ± 0.3	28.9 ± 0.1	28.3 ± 0.1	31.0 ± 0.3
Total chol	4.51 ± 0.02	4.42 ± 0.03	4.62 ± 0.03	4.48 ± 0.02	4.43 ± 0.03	4.66 ± 0.05
HDL-chol	1.26 ± 0.01	1.18 ± 0.03	1.33 ± 0.01	1.15 ± 0.01	1.15 ± 0.01	1.25 ± 0.02
LDL-chol	2.76 ± 0.02	2.73 ± 0.03	2.80 ± 0.03	2.71 ± 0.02	2.68 ± 0.02	2.84 ± 0.06
TG	1.52 ± 0.02	1.60 ± 0.03	1.44 ± 0.03	1.78 ± 0.02	1.78 ± 0.03	1.78 ± 0.05
FG	6.87 ± 0.16	6.80 ± 0.23	6.92 ± 0.22	9.45 ± 0.31	9.27 ± 0.37	9.88 ± 0.56
BP	120/83	119/81	121/82	128/84	130/85	127/83

The numbers in brackets give the fraction of the total (all) values of the group. *BMI* body mass index, *CAD* coronary artery disease, *FH* family history of CAD, *MI* myocardial infarction, *FG* fasting glucose, T2DM type 2 diabetes mellitus, lHDLC low high-density lipoprotein-cholesterol, hLDLC high low-density lipoprotein-cholesterol, hTG hypertriglyceridaemia, hChol hypercholesterolaemia, HTN hypertension. [a]Values relative to obesity as the case group.

aldosteronism or other causes of secondary hypertension or Mendelian forms (monogenic) are not present [30]. The third subset of interest was composed of 2,560 individuals with T2DM (formerly known as non-insulin-dependent diabetes mellitus or adult-onset diabetes). The National Diabetes Data Group of the USA and the Second World Health Organization Expert Committee on Diabetes Mellitus [31] defined type 2 diabetes mellitus as a metabolic disorder that is characterized by high blood glucose (generally defined as a fasting glucose level >126 mg/dl) in the context of insulin resistance and relative insulin deficiency. Among these subsets of patients, some individuals harboured a combination of two or possibly three of the cardiovascular risk traits. The overall exclusion criteria for the disease cases were major cardiac rhythm disturbances, incapacitating or life-threatening illness, major psychiatric illness or substance abuse, history of cerebral vascular disease, neurological disorder and the administration of psychotropic medication. Exclusion criteria for the controls were, among others, diseases such as cancer, autoimmune disease or any other disorders likely to interfere with the variables under investigation. All participants signed an informed consent, and the study was performed in accordance with the regulations laid down by the King Faisal Specialist Hospital and Research Centre Ethics Committee in compliance with the Helsinki Declaration [32].

Linkage analysis and screening for mutations

Five millilitres of peripheral blood was sampled in EDTA tubes from each of the study individuals after obtaining their written consent, and genomic DNA was extracted from leukocytes by the standard salt method using Gentra PUREGENE DNA isolation kit (Qiagen Sciences, Germantown, MD, USA). For the genome-wide scanning with the Affymetrix Gene Chip 250 styl mapping array (Affymetrix Inc., Santa Clara, CA, USA), 250 ng of genomic DNA was digested with the restriction endonuclease *StyI*, mixed with Styl adaptors and ligated with T4 DNA ligase. The mixture was added to four separate polymerase chain reactions (PCRs), amplified, pooled and purified to remove the unincorporated dNTPs. The PCR product was then fragmented, biotinylated, hybridized to the 250 styl array for 18 h, washed, stained and scanned as recommended by the manufacturer. SNP genotypes, linear chromosomal locations and marker ordering were accomplished using the Affymetrix Genotyping Console (GC) Software version 3.02 (Affymetrix, Inc., Santa Clara, CA, USA). Extended blocks or regions of homozygosity (ROHs) were identified using the GC software. Conventionally, ROHs are defined as stretches of consecutive alleles in affected individuals and heterozygous or homozygous for the other allele in unaffected members of the same family. For the linkage analysis, multipoint parametric linkage analysis was performed using the GeneHunter Easy Linkage Analysis Software 4.0 module for estimating the LOD scores. The disease was assumed to be an autosomal dominant trait with 100% penetrance. Copy Number Analyzer for GeneChip® Ver. 3.0 (Affymetrix, Inc., Santa Clara, CA, USA) was employed to check the shared chromosomal regions of homozygosity.

Sequencing of the *MRAS* gene

The sequencing of the *MRAS* gene was accomplished using the MegaBACE DNA analysis system (Amersham Biosciences, Sunnyvale, CA, USA). Briefly, the DNA was subjected to PCR by standard methods described elsewhere. Five microlitres of PCR product was then treated with 2 μl of ExoSAP-IT (USB Corporation, Cleveland, OH, USA) at 37°C for 30 min to allow the hydrolytic removal of excess primers by exonuclease 1 and shrimp alkaline phosphatase. The enzymes were inactivated at 80°C for 15 min, and the sequencing reaction was initiated by mixing 2 μl of DNA, 1 μl of 5 μmol primer, 8 μl of DYEnamic ET Dye Terminator (Amersham Biosciences, Amersham, Buckinghamshire, UK) and 9 μl of distilled water. The mixture was thermally cycled 40× at 95°C for 20 s, 50°C for 15 s and 60°C for 1 min. Unincorporated dye-labelled terminators were removed by gel filtration through the DyeEx 96 plate (Qiagen, GmbH, Hilden, Germany). The eluent was vacuum-dried and dissolved in 10 μl of loading solution (GE Healthcare UK Ltd, Little Chalfont, Buckinghamshire, UK) for sequencing. Data were analyzed for SNPs using the Lasergene software (DNASTAR, Inc., Madison, WI, USA).

Association experiments

Once the SNPs of interest were identified, genotyping was achieved by TaqMan chemistry using the Applied Biosystems real-time Prism 7900HT Sequence Detection System (ABI Inc., Foster City, CA, USA). Primers and the TaqMan fluorogenic probes bearing a suitable reporter dye on the 5′-end and a quencher dye on the 3′-end were designed using the Primer Express Software V2.0 (ABI Inc., Foster City, CA, USA) and procured from Applied Biosystems (ABI, Warrington, UK). One probe (for allele 1) was labelled with VIC dye and the other (for allele 2) with FAM dye at the 5′-end, and serial dilutions were run to determine the optimal working concentration. For each reaction, a 25-μl reaction was prepared by mixing 5 μl of 50 ng DNA, 12.5 μl of 2× Universal mix (Eurogentec, Liege Science Park, Seraing, Belgium), 1.25 μl of 20× probe assay mix and 6.25 μl of DNase-free distilled water. Three no-template controls were included in each plate for the normalization of emission signal. The thermal profile for amplification for the first cycle occurred at 50°C for 2 min and 95°C for 10 min, followed by 40 cycles of 94°C for 15 s and 60°C for 30 s. The plates were then scanned for FRET signal using the 7900HT Sequence Detection System and data analyzed using the SDS 2.0 software (ABI Inc., Foster City, CA, USA).

Statistical analysis

Comparison of genotypes and alleles between different groups for continuous dependent variables was achieved by analysis of variance or Student's t test as appropriate. Categorical variables were analyzed by chi-square test. Univariate and multiple logistic regression analyses were used to compute odds ratios and their 95% confidence intervals as well as to determine the confounding effects of the different cardiovascular risk traits on the individual respective relationships. The haplo.stats package [33] in the R Statistical Computing software [34] was used to perform haplotype-based association analysis. Odds ratios for haplotypes were calculated using as reference the most frequent 7-mer haplotype, ACCTAAC (frequency = 0.558), and the Haplotype Score statistic for the association of a haplotype with the binary trait was calculated as in Schaid et al. [35] and Lake et al. [36]. The significance of association was determined between haplotypes and the case-control status - a binomial trait denoting whether or not a patient had the disease. All other statistical analyses were performed using the SPSS version 20 software (SPSS Inc., Chicago, IL, USA), and data are expressed as mean ± SEM. Associations with a two-tailed p value <0.05 were considered statistically significant.

Abbreviations

3'-UTR: 3 prime untranslated region; BP: Blood pressure; CAD: Coronary artery disease; FG: Fasting glucose; FH: Family history of CAD; hChol: Hypercholesterolaemia; HDLC: High-density lipoprotein-cholesterol; hTG: Hypertriglyceridaemia; HTN: Hypertension; LDLC: Low-density lipoprotein-cholesterol; MI: Myocardial infarction; *MRAS*: Membrane-associated muscle Ras small GTPase gene; OBS: Obesity; SNP: Single-nucleotide polymorphism; T2DM: Type 2 diabetes mellitus.

Competing interests

The authors declare that they have no competing interests.

Authors' contributions

MA supervised the recruitment of the patients and compliance with institutional ethical procedures. SMW performed the gene mapping analysis. MA-N was responsible for overall running the TaqMan assays. NPM was involved in running the Affymetrix assays, designing the probes, screening for gene mutations as well as participating in the write up of the manuscript. SE performed part of the sequencing experiments. SH ran the Affymetrix assays. DG contributed to running the TaqMan assays. EA performed the DNA isolation and maintained the patient database. NM was responsible for clinical patient data and material acquisition. BFM contributed to the write up of the manuscript. ND is the principal investigator, with the overall responsibility of the project and the preparation of the manuscript. All authors read and approved the final manuscript.

Acknowledgements

This study was supported through the Royal Cardiovascular Research Grant (RAC2030012) under the King Faisal Specialist Hospital and Research Centre for which the authors wish to express their gratitude.

Author details

[1]King Faisal Heart Institute, MBC-16, King Faisal Specialist Hospital and Research Centre, Riyadh 11211, Saudi Arabia. [2]Genetics Department, MBC-03, King Faisal Specialist Hospital and Research Centre, Riyadh 11211, Saudi Arabia.

References

1. Kimmelman A, Tolkacheva T, Lorenzi MV, Osada M, Chan AM: **Identification and characterization of R-ras3: a novel member of the RAS gene family with a non-ubiquitous pattern of tissue distribution.** *Oncogene* 1997, **15**(22):2675–2685.
2. Quilliam LA, Castro AF, Rogers-Graham KS, Martin CB, Der CJ, Bi C: **M-Ras/R-Ras3, a transforming ras protein regulated by Sos1, GRF1, and p120 Ras GTPase-activating protein, interacts with the putative Ras effector AF6.** *J Biol Chem* 1999, **274**(34):23850–23857.
3. Watanabe-Takano H, Takano K, Keduka E, Endo T: **M-Ras is activated by bone morphogenetic protein-2 and participates in osteoblastic determination, differentiation, and transdifferentiation.** *Exp Cell Res* 2010, **316**(3):477–490.
4. Harrisingh MC, Lloyd AC: **Ras/Raf/ERK signalling and NF1.** *Cell Cycle* 2004, **3**(10):1255–1258.
5. Liebmann C: **Regulation of MAP kinase activity by peptide receptor signalling pathway: paradigms of multiplicity.** *Cell Signal* 2001, **13**(11):777–785.
6. Kerkhoff E, Rapp UR: **Cell cycle targets of Ras/Raf signalling.** *Oncogene* 1998, **17**(11 Reviews):1457–1462.
7. Chakraborti S, Chakraborti T: **Oxidant-mediated activation of mitogen-activated protein kinases and nuclear transcription factors in the cardiovascular system: a brief overview.** *Cell Signal* 1998, **10**(10):675–683.
8. Chong H, Vikis HG, Guan KL: **Mechanisms of regulating the Raf kinase family.** *Cell Signal* 2003, **15**(5):463–469.
9. Ehrhardt GR, Leslie KB, Lee F, Wieler JS, Schrader JW: **M-Ras, a widely expressed 29-kD homologue of p21 Ras: expression of a constitutively active mutant results in factor-independent growth of an interleukin-3-dependent cell line.** *Blood* 1999, **94**(7):2433–2444.
10. Yoshikawa Y, Satoh T, Tamura T, Wei P, Bilasy SE, Edamatsu H, Aiba A, Katagiri K, Kinashi T, Nakao K, Kataoka T: **The M-Ras-RA-GEF-2-Rap1 pathway mediates tumor necrosis factor-alpha dependent regulation of integrin activation in splenocytes.** *Mol Biol Cell* 2007, **18**(8):2949–2959.
11. Kimmelman AC, Osada M, Chan AM: **R-Ras3, a brain-specific Ras-related protein, activates Akt and promotes cell survival in PC12 cells.** *Oncogene* 2000, **19**(16):2014–2022.
12. Louahed J, Grasso L, De Smet C, Van Roost E, Wildmann C, Nicolaides NC, Levitt RC, Renauld JC: **Interleukin-9-induced expression of M-Ras/R-Ras3 oncogene in T-helper clones.** *Blood* 1999, **94**(5):1701–1710.
13. Muto S, Katsuki M, Horie S: **Rapid induction of skin tumors in human but not mouse c-Ha-ras proto-oncogene transgenic mice by chemical carcinogenesis.** *Cancer Sci* 2006, **97**(9):842–847.
14. Abuli A, Fernandez-Rozadilla C, Giraldez MD, Munoz J, Gonzalo V, Bessa X, Bujanda L, Rene JM, Lanas A, Garcia AM, Salo J, Arguello L, Vilella A, Carreno R, Jover R, Xicola RM, Llor X, Carvajal-Carmona L, Tomlinson IP, Kerr DJ, Houlston RS, Pique JM, Carracedo A, Castells A, Andreu M, Ruiz-Ponte C, Castellvi-Bel S: **A two-phase case–control study for colorectal cancer genetic susceptibility: candidate genes from chromosomal regions 9q22 and 3q22.** *Br J Cancer* 2011, **105**(6):870–875.
15. Guo X, Schrader KA, Xu Y, Schrader JW: **Expression of a constitutively active mutant of M-Ras in normal bone marrow is sufficient for induction of a malignant mastocytosis/mast cell leukemia, distinct from the histiocytosis/monocytic leukemia induced by expression of activated H-Ras.** *Oncogene* 2005, **24**(14):2330–2342.
16. Zhang KX, Ward KR, Schrader JW: **Multiple aspects of the phenotype of mammary epithelial cells transformed by expression of activated M-Ras depend on an autocrine mechanism mediated by hepatocyte growth factor/scatter factor.** *Mol Cancer Res* 2004, **2**(4):242–255.

17. Wang D, Yang W, Du J, Devalaraja MN, Liang P, Matsumoto K, Tsubakimoto K, Endo T, Richmond A: **MGSA/GRO-mediated melanocyte transformation involves induction of Ras expression.** *Oncogene* 2000, **19**(40):4647–4659.
18. *MRAS (muscle RAS oncogene homolog).* http://biogps.org/#goto=genereport &id=22808.
19. Erdmann J, Grosshennig A, Braund PS, Konig IR, Hengstenberg C, Hall AS, Linsel-Nitschke P, Kathiresan S, Wright B, Tregouet DA, Cambien F, Bruse P, Aherrahrou Z, Wagner AK, Stark K, Schwartz SM, Salomaa V, Elosua R, Melander O, Voight BF, O'Donnell CJ, Peltonen L, Siscovick DS, Altshuler D, Merlini PA, Peyvandi F, Bernardinelli L, Ardissino D, Schillert A, Blankenberg S, *et al*: **New susceptibility locus for coronary artery disease on chromosome 3q22.3.** *Nat Genet* 2009, **41**(3):280–282.
20. Ellis KL, Frampton CM, Pilbrow AP, Troughton RW, Doughty RN, Whalley GA, Ellis CJ, Skelton L, Thomson J, Yandle TG, Richards AM, Cameron VA: **Genomic risk variants at 1p13.3, 1q41, and 3q22.3 are associated with subsequent cardiovascular outcomes in healthy controls and in established coronary artery disease.** *Circ Cardiovasc Genet* 2001, **4**(6):636–646.
21. Galkina E, Ley K: **Vascular adhesion molecules in atherosclerosis.** *Arterioscler Thromb Vasc Biol* 2007, **27**(11):2292–2301.
22. Mäkinen HS, Shikano T, Cano JM, Merilä J: **Hitchhiking mapping reveals a candidate genomic region for natural selection in three-spined stickleback chromosome VIII.** *Genetics* 2008, **178**(1):453–465.
23. Barton NH: **Genetic hitchhiking.** *Philos Trans R Soc Lond B Biol Sci* 2000, **355**(1403):1553–1562.
24. Yeaman S: **Genomic rearrangements and the evolution of clusters of locally adaptive loci.** *Proc Natl Acad Sci USA* 2013, **110**(19):E1743–E1751.
25. Schneider KA, Kim Y: **Genetic hitchhiking under heterogeneous spatial selection pressures.** *PLoS One* 2013, **8**(4):e61742.
26. Barton NH, Etheridge AM, Kelleher J, Veber A: **Genetic hitchhiking in spatially extended populations.** *Theor Popul Biol.* in press.
27. Desai MM, Walczak AM, Fisher DS: **Genetic diversity and the structure of genealogies in rapidly adapting populations.** *Genetics* 2013, **193**(2):565–585.
28. Flaxman SM, Feder JL, Nosil P: **Spatially explicit models of divergence and genome hitchhiking.** *J Evol Biol* 2012, **25**(12):2633–2650.
29. Kim Y: **Stochastic patterns of polymorphism after a selective sweep over a subdivided population.** *Genet Res* 2013. doi:10.1017/S0016672313000062.
30. Dzimiri N, Al-Najai M, Elhawari S, Gueco D, Vigilla MG, Andres E, Muiya P, Mazher N, Alshahid M, Meyer BF: **The interaction of Agt gene polymorphism with cardiovascular risk traits in atherosclerosis.** *Arterioscler Thromb Vasc Biol* 2012, **32**(5_MeetingAbstracts):A524.
31. American Heart Association: **Arteriosclerosis, Thrombosis, and Vascular Biology Annual Conference 2007.** *Arterioscler Thromb Vasc Biol* 2007, **27**(6):e35–e137.
32. *WMA Declaration of Helsinki - ethical principles for medical research involving human subjects.* http://www.wma.net/en/30publications/10policies/b3/index.html.
33. Schaid DJ: *Statistical genetics and genetic epidemiology.* http://mayoresearch.mayo.edu/mayo/research/schaid_lab/software.cfm.
34. *The R Project for Statistical Computing.* http://www.r-project.org/.
35. Schaid DJ, Rowland CM, Tines DE, Jacobson RM, Poland GA: **Score tests for association between traits and haplotypes when linkage phase is ambiguous.** *Am J Hum Genet* 2002, **70**(2):425–434.
36. Lake SL, Lyon H, Tantisira K, Silverman EK, Weiss ST, Laird NM, Schaid DJ: **Estimation and tests of haplotype-environment interaction when linkage phase is ambiguous.** *Hum Hered* 2003, **55**(1):56–65.

Permissions

All chapters in this book were first published in JNI, by BioMed Central; hereby published with permission under the Creative Commons Attribution License or equivalent. Every chapter published in this book has been scrutinized by our experts. Their significance has been extensively debated. The topics covered herein carry significant findings which will fuel the growth of the discipline. They may even be implemented as practical applications or may be referred to as a beginning point for another development.

The contributors of this book come from diverse backgrounds, making this book a truly international effort. This book will bring forth new frontiers with its revolutionizing research information and detailed analysis of the nascent developments around the world.

We would like to thank all the contributing authors for lending their expertise to make the book truly unique. They have played a crucial role in the development of this book. Without their invaluable contributions this book wouldn't have been possible. They have made vital efforts to compile up to date information on the varied aspects of this subject to make this book a valuable addition to the collection of many professionals and students.

This book was conceptualized with the vision of imparting up-to-date information and advanced data in this field. To ensure the same, a matchless editorial board was set up. Every individual on the board went through rigorous rounds of assessment to prove their worth. After which they invested a large part of their time researching and compiling the most relevant data for our readers.

The editorial board has been involved in producing this book since its inception. They have spent rigorous hours researching and exploring the diverse topics which have resulted in the successful publishing of this book. They have passed on their knowledge of decades through this book. To expedite this challenging task, the publisher supported the team at every step. A small team of assistant editors was also appointed to further simplify the editing procedure and attain best results for the readers.

Apart from the editorial board, the designing team has also invested a significant amount of their time in understanding the subject and creating the most relevant covers. They scrutinized every image to scout for the most suitable representation of the subject and create an appropriate cover for the book.

The publishing team has been an ardent support to the editorial, designing and production team. Their endless efforts to recruit the best for this project, has resulted in the accomplishment of this book. They are a veteran in the field of academics and their pool of knowledge is as vast as their experience in printing. Their expertise and guidance has proved useful at every step. Their uncompromising quality standards have made this book an exceptional effort. Their encouragement from time to time has been an inspiration for everyone.

The publisher and the editorial board hope that this book will prove to be a valuable piece of knowledge for researchers, students, practitioners and scholars across the globe.

List of Contributors

C George Priya Doss and N Nagasundaram
Medical Biotechnology Division, Centre for Nanobiotechnology, School of Biosciences and Technology, VIT University, Vellore 632014, Tamil Nadu 632014, India

Hailong Zhu
Department of Computer Sciences, Hong Kong Baptist University, Kowloon Tong, Hong Kong

Chiranjib Chakraborty
Department of Computer Sciences, Hong Kong Baptist University, Kowloon Tong, Hong Kong
Department of Bio-informatics, School of Computer and Information Sciences, Galgotias University, Uttar Pradesh, India

Luonan Chen
Medical Biotechnology Division, Centre for Nanobiotechnology, School of Biosciences and Technology, VIT University, Vellore 632014, Tamil Nadu 632014, India
Key Laboratory of Systems Biology, SIBS-Novo Nordisk Translational Research Centre for PreDiabetes, Shanghai Institutes of Biological Sciences, Chinese Academy of Sciences, Shanghai, China
Collaborative Research Center for Innovative Mathematical Modelling, Institute of Industrial Science, The University of Tokyo, Tokyo 153-8505, Japan

Shannon Bruse, Maria Picchi, Yong Lin, Raymond J. Langley, Steven A. Belinsky and Shuguang Leng
Lovelace Respiratory Research Institute, 2425 Ridgecrest Drive SE, Albuquerque, NM 87108, USA

Michael Moreau, Nan Wang, Hongseok Ha and Jinchuan Xing
Department of Genetics, Rutgers, The State University of New Jersey, 145 Bevier Road, Piscataway, NJ 08854, USA

Yana Bromberg
Department of Biochemistry and Microbiology, Rutgers, the State University of New Jersey, Piscataway, NJ, USA

Jun-Ho Jang and Toru Nyunoya
Lovelace Respiratory Research Institute, 2425 Ridgecrest Drive SE, Albuquerque, NM 87108, USA
Department of Internal Medicine, University of New Mexico and New Mexico VA Health Care System, Albuquerque, NM, USA

Jenny Mao
Department of Internal Medicine, University of New Mexico and New Mexico VA Health Care System, Albuquerque, NM, USA

Clifford Qualls
Biomedical Research Institute of New Mexico, Albuquerque, NM, USA

Julia Klensney-Tait and Joseph Zabner
Department of Medicine, University of Iowa, Iowa City, IA, USA

Hátylas Azevedo and Carlos Alberto Moreira-Filho
Department of Pediatrics, Faculdade de Medicina da Universidade de São Paulo (FMUSP), São Paulo, Brazil

Niels Olsen Saraiva Câmara
Laboratory of Transplantation Immunobiology, Department of Immunology, Institute of Biomedical Sciences, Universidade de São Paulo (USP), São Paulo, Brazil
Laboratory of Clinical and Experimental Immunology, Nephrology Division, Universidade Federal de São Paulo (UNIFESP), São Paulo, Brazil

Paulo Guilherme Renesto, Erika Naka and Marcos Antônio Cenedeze
Laboratory of Clinical and Experimental Immunology, Nephrology Division, Universidade Federal de São Paulo (UNIFESP), São Paulo, Brazil

Rogério Chinen, Ana Cristina Carvalho de Matos and Alvaro Pacheco-Silva
Laboratory of Clinical and Experimental Immunology, Nephrology Division, Universidade Federal de São Paulo (UNIFESP), São Paulo, Brazil
Instituto Israelita de Ensino e Pesquisa Albert Einstein, Hospital Albert Einstein, São Paulo, Brazil

Sara E. Patterson, Rangjiao Liu, Cara M. Statz, Daniel Durkin, Anuradha Lakshminarayana and Susan M. Mockus
The Jackson Laboratory for Genomic Medicine, 10 Discovery Dr., Farmington, CT 06032, USA

Dorota Monies, Mohamed Abouelhoda, Anas M. Alazami, Ewa Goljan, Banan Alyounes, Dyala Jaroudi, Abdulelah AlIssa, Khalid Alabdulrahman, Shazia Subhani, Mohamed El-Kalioby, Tariq Faquih, Salma M. Wakil, Nada A. Altassan and Brian F. Meyer
Department of Genetics, Research Centre, King Faisal Specialist Hospital and Research Centre, Riyadh 11211, Kingdom of Saudi Arabia
Saudi Human Genome Program, King Abdulaziz City for Science and Technology, Riyadh, Saudi Arabia

Saeed Bohlega
Saudi Human Genome Program, King Abdulaziz City for Science and Technology, Riyadh, Saudi Arabia
Department of Neuroscience, King Faisal Specialist Hospital and Research Centre, Riyadh, Saudi Arabia

Mohamed A. Almuhaizea
Department of Neuroscience, King Faisal Specialist Hospital and Research Centre, Riyadh, Saudi Arabia

Hindi N. Alhindi
Department of Pathology and Laboratory Medicine, King Faisal Specialist Hospital and Research Centre, Riyadh, Saudi Arabia

Lucia F. Jorge-Nebert, Susan M. Pinney and Daniel W. Nebert
Department of Environmental Health and Center for Environmental Genetics, University of Cincinnati Medical Center, Cincinnati, OH 45267-0056, USA

Ge Zhang
Department of Environmental Health and Center for Environmental Genetics, University of Cincinnati Medical Center, Cincinnati, OH 45267-0056, USA
Division of Human Genetics, Department of Pediatrics & Molecular Developmental Biology, Cincinnati Children's Hospital, Cincinnati, Ohio 45229-2899, USA

Zhengwen Jiang
Department of Environmental Health and Center for Environmental Genetics, University of Cincinnati Medical Center, Cincinnati, OH 45267-0056, USA
Genesky Diagnostics, Suzhou, China

Keith M. Wilson and Jack L. Gluckman
Department of Otolaryngology-Head and Neck Surgery, University of Cincinnati College of Medicine, Cincinnati, OH 45267-0528, USA

Randall Butler
Department of Pathology and Laboratory Medicine, University of Cincinnati College of Medicine, Cincinnati, OH 45267-0533, USA

Rayabarapu Pranavchand and Battini Mohan Reddy
Molecular Anthropology Group, Biological Anthropology Unit, Indian Statistical Institute, Hyderabad, India

Arramraju Sreenivas Kumar
Department of Cardiology, Care Hospitals, Banjara Hills, Hyderabad, India

Kyung-Chang Kim, Sunyoung Lee, Junseock Son, Younghyun Shin, Cheol-Hee Yoon, Chun Kang and Byeong-Sun Choi
Division of AIDS, Center for Immunology and Pathology, Korea National Institute of Health, Chung-buk 28160, Republic of Korea

Benjamin M. Helm
Division of Medical Genetics and Metabolism, Children's Hospital of The King's Daughters, Norfolk, VA 23507, USA
Department of Medical & Molecular Genetics, Riley Hospital for Children at IU Health, Indiana University School of Medicine, Indianapolis, IN 46202, USA

Jason R. Willer, Azita Sadeghpour, Nicholas Katsanis and Erica E. Davis
Center for Human Disease Modeling, Duke University Medical Center, Durham, NC 27701, USA

Christelle Golzio
Center for Human Disease Modeling, Duke University Medical Center, Durham, NC 27701, USA
Institute of Genetics and Molecular and Cellular Biology, 67404 Illkirch, France

Eric Crouch
Department of Ophthalmology, Children's Hospital of the King's Daughters, Norfolk, VA 23507, USA

Samantha Schrier Vergano
Division of Medical Genetics and Metabolism, Children's Hospital of The King's Daughters, Norfolk, VA 23507, USA
Department of Pediatrics, Eastern Virginia Medical School, Norfolk, VA 23507, USA

Sanaz Firouzi and Kaoru Uchimaru
Department of Computational Biology and Medical Sciences, Graduate School of Frontier Sciences, The University of Tokyo, Kashiwanoha, Kashiwa-shi, Chiba, Japan

Amir Farmanbar and Kenta Nakai
Department of Computational Biology and Medical Sciences, Graduate School of Frontier Sciences, The University of Tokyo, Kashiwanoha, Kashiwa-shi, Chiba, Japan
Laboratory of Functional Analysis in silico, Human Genome Center, The Institute of Medical Science, The University of Tokyo, Tokyo, Japan

Wojciech Makałowski
Institute of Bioinformatics, Faculty of Medicine, University of Muenster, Muenster, Germany

Masako Iwanaga
Department of Frontier Life Science, Graduate School of Biomedical Sciences, Nagasaki University, Nagasaki, Japan

Atae Utsunomiya
Department of Hematology, Imamura General Hospital, Kagoshima, Japan

Toshiki Watanabe
Department of Computational Biology and Medical Sciences, Graduate School of Frontier Sciences, The University of Tokyo, Kashiwanoha, Kashiwa-shi, Chiba, Japan
Department of Advanced Medical Innovation, St. Marianna University Graduate School of Medicine, Kanagawa, Japan

Junji Ohnishi
Department of Food and Nutrition, Tokyo Kasei University, Itabashi, Tokyo, Japan
Foundation for Advancement of International Science, Kasuga, Tsukuba, Japan

Shigeko Sakamoto and Miyo Hori
Foundation for Advancement of International Science, Kasuga, Tsukuba, Japan

Satoshi Ayuzawa
Division of Health Sciences, Tsukuba University of Technology, Kasuga, Tsukuba, Japan
Departmemt of Neurosugery, Center for Integrative Medicine, Kasuga, Tsukuba, Japan

Tomoko Sasaoka
Departmemt of Neurosugery, Center for Integrative Medicine, Kasuga, Tsukuba, Japan

Seiji Nakamura
DNA Chip Research Inc., Minato, Tokyo, Japan

Masakazu Tanatsugu
Graduate School of Medicine, Kyoto Prefectural University of Medicine, Kyoto, Japan

Eriko Takimoto-Ohnishi
Foundation for Advancement of International Science, Kasuga, Tsukuba, Japan
National Center for Child Health and Development, Tokyo, Japan

Kazuo Murakami
Foundation for Advancement of International Science, Kasuga, Tsukuba, Japan
The Institute of Esoteric Culture, Koyasan University, Ina, Wakayama, Japan

Konstantinos Mitropoulos
National and Kapodistrian University of Athens School of Medicine, Athens, Greece

Eleni Merkouri Papadima, Angeliki Balasopoulou, Kyriaki Charalampidou, Vasiliki Galani, Krystallia-Vassiliki Zafeiri, Marina Bartsakoulia, Evangelia-Eirini Tsermpini, Clint Mizzi and Theodora Katsila
Department of Pharmacy, University of Patras School of Health Sciences, Campus, Rion, GR-26504 Patras, Greece

George P. Patrinos
Department of Pharmacy, University of Patras School of Health Sciences, Campus, Rion, GR-26504 Patras, Greece
Department of Pathology, College of Medicine and Health Sciences, United Arab Emirates University, Al-Ain, UAE

Georgia Xiromerisiou, Efthymios Dardiotis, Styliani Ralli and Georgios M. Hadjigeorgiou
School of Medicine, University of Thessaly, Larisa, Greece

Georgia Deretzi
Papageorgiou hospital, Thessaloniki, Greece

Anne John and Bassam R. Ali
Department of Pathology, College of Medicine and Health Sciences, United Arab Emirates University, Al-Ain, UAE

Kyriaki Kydonopoulou, Elpida Papadopoulou and Spyridon Gerou
ANALYSI Diagnostic Laboratories S.A, Thessaloniki, Greece

Alba di Pardo and Evangelos Kiskinis
Departments of Neurology and Physiology, Northwestern University Feinberg School of Medicine, Chicago, IL, USA

Fulya Akcimen and Nazli Basak
Suna and Inan Kirac Foundation, NDAL, Bogazici University, Istanbul, Turkey
Institute of Medical Genetics, Cardiff University, Cardiff, UK

Sandro Orrù
Department of Medical Sciences and Public Health, University of Cagliari, Cagliari, Italy
CNR IRGB, Cagliari, Italy

Annalisa Loizedda
Department of Medical Sciences and Public Health, University of Cagliari, Cagliari, Italy
CNR IRGB, Cagliari, Italy

Valerija Dobričić and Vladimir S. Kostić
Institute of Neurology CCS, School of Medicine, University of Belgrade, Belgrade, Serbia

Ivana Novaković
Institute of Neurology CCS, School of Medicine, University of Belgrade, Belgrade, Serbia
Faculty of Medicine, Institute of Human Genetics, University of Belgrade, Belgrade, Serbia

Radoje Drmanac
Complete Genomics Inc., Mountain View, CA, USA

Brock A. Peters
Complete Genomics Inc., Mountain View, CA, USA
BGI Shenzhen, Shenzhen, People's Republic of China

David N. Cooper
Institute of Medical Genetics, Cardiff University, Cardiff, UK

Zhengfei Lai
Department of Fetal Medicine and Prenatal Diagnosis, the Third Affiliated Hospital of Guangzhou Medical University, Guangzhou 510150, China

Min Chen, Shuya Xue, Haimei Gao and Dunjin Chen
Department of Fetal Medicine and Prenatal Diagnosis, the Third Affiliated Hospital of Guangzhou Medical University, Guangzhou 510150, China
Obstetrics and Gynecology Institute of Guangzhou, Guangzhou 510150, China
The Medical Centre for Critical Pregnant Women in Guangzhou, Guangzhou 510150, China
Key Laboratory for Major Obstetric Diseases of Guangdong Province, Guangzhou 510150, China
Key Laboratory for Reproduction and Genetics of Guangdong Higher Education Institutes, Guangzhou, China

Jiayan Wang
Department of Fetal Medicine and Prenatal Diagnosis, the Third Affiliated Hospital of Guangzhou Medical University, Guangzhou 510150, China
Obstetrics and Gynecology Institute of Guangzhou, Guangzhou 510150, China
The Medical Centre for Critical Pregnant Women in Guangzhou, Guangzhou 510150, China
Key Laboratory for Major Obstetric Diseases of Guangdong Province, Guangzhou 510150, China

Allen Chen
Department of Mathematics, University of California, Berkeley, CA 94720, USA
Mendel Genes Inc, Manhattan Beach, CA, Manhattan Beach, CA 90266, USA

Yingjun Luo, Kailing Huang and Yanhui Liu
Mendel Genes Inc, Manhattan Beach, CA, Manhattan Beach, CA 90266, USA

Xiaoshun Shi
Department of Thoracic Surgery, Nanfang Hospital, Southern Medical University, Guangzhou, China

Jin Li
State Key Laboratory of Respiratory Disease, The First Affiliated Hospital of Guangzhou Medical University, National Clinical Research Center for Respiratory Disease, Guangzhou, China

Taketoshi Okita
Graduate School of Medicine, Tohoku University, 2-1 Seiryo-machi, Aoba-ku, Sendai 980-8575, Japan

Noriko Ohashi and Kazuto Kato
Graduate School of Medicine, Osaka University, 2-2 Yamadaoka, Suita, Osaka 565-0871, Japan

Daijiro Kabata and Ayumi Shintani
Graduate School of Medicine, Osaka City University, 1-4-3, Asahi, Abeno, Osaka, Osaka 545-8585, Japan

Sang-Nee Tan and Sai-Peng Sim
Department of Paraclinical Sciences, Faculty of Medicine and Health Sciences, Universiti Malaysia Sarawak, Sarawak, Malaysia

Alan Soo-Beng Khoo
Molecular Pathology Unit, Cancer Research Centre, Institute for Medical Research, Kuala Lumpur, Malaysia

Mary B. Mayes, Daniel S. Buxton, Rebecca L. Martin and Nadia Chuzhanova
School of Science and Technology, Nottingham Trent University, Clifton Lane, Nottingham NG11 8NS, UK

Taniesha Morgan, Jincy Winston, David N. Cooper and Meena Upadhyaya
Institute of Medical Genetics, School of Medicine, Cardiff University, Heath Park, Cardiff CF14 4XN, UK

Debbie Smith and Maggie Williams
Blood Sciences Department and Bristol Genetics Laboratory, Southmead Hospital, Westbury-on-Trym, Bristol BS10 5NB, UK

Dirk A. Kleinjan
MRC Human Genetics Unit, Institute of Genetics and Molecular Medicine, University of Edinburgh, Edinburgh EH4 2XU, UK

Mihir Anant Kamat
School of Science and Technology, Nottingham Trent University, Clifton Lane, Nottingham NG11 8NS, UK
Cardiovascular Epidemiology Unit, Department of Public Health and Primary Care, University of Cambridge, Cambridge CB1 8RN, UK

Lily M Blair and Marcus W Feldman
Department of Biology, Stanford University, Stanford, CA 94305, USA

Julie M Granka
Department of Biology, Stanford University, Stanford, CA 94305, USA
AncestryDNA, 153 Townsend St. Ste. 800, San Francisco, CA 94107, USA

Pothitos M Pitychoutis and Panagiotis A Tsonis
Department of Biology & Center for Tissue Regeneration and Engineering (TREND), University of Dayton, 300 College Park, Dayton, OH 45469-2320, USA

Despina Sanoudou, Margarita Papandreou, Dimitris Nasias, Marianna Kouskou and Zeta Papadopoulou-Daifoti
Department of Pharmacology, Medical School, National & Kapodistrian, University of Athens, 75 Mikras Asias StreetGoudi, Athens 11527, Greece

Craig R Tomlinson
Department of Medicine, Dartmouth Hitchcock Medical Center, Dartmouth College, Lebanon, NH 03756, USA
Department of Pharmacology and Toxicology, Dartmouth Hitchcock Medical Center, Dartmouth College, Lebanon, NH 03756, USA

Maie Alshahid and Brian F Meyer
King Faisal Heart Institute, MBC-16, King Faisal Specialist Hospital and Research Centre, Riyadh 11211, Saudi Arabia

Salma M Wakil, Mohammed Al-Najai, Nzioka P Muiya, Samar Elhawari, Daisy Gueco, Editha Andres, Samia Hagos, Nejat Mazhar and Nduna Dzimiri
Genetics Department, MBC-03, King Faisal Specialist Hospital and Research Centre, Riyadh 11211, Saudi Arabia

Index

www.ingramcontent.com/pod-product-compliance
Lightning Source LLC
LaVergne TN
LVHW080914210926
848485LV00004B/650

* 9 7 8 1 6 3 9 2 7 2 5 3 2 *